International Series in Operations Research & Management Science

Volume 299

More information about this series at http://www.springer.com/series/6161

Louis Anthony Cox Jr.

Quantitative Risk Analysis of Air Pollution Health Effects

Springer

Louis Anthony Cox Jr.
Cox Associates and University of Colorado
Denver, CO, USA

ISSN 0884-8289 ISSN 2214-7934 (electronic)
International Series in Operations Research & Management Science
ISBN 978-3-030-57357-7 ISBN 978-3-030-57358-4 (eBook)
https://doi.org/10.1007/978-3-030-57358-4

This Springer imprint is published by the registered company Springer Nature Switzerland AG
The registered company address is: Gewerbestrasse 11, 6330 Cham, Switzerland

To
Christine and Emeline, for many discussions of causality and regulatory science

and

Jim Boylan, Mark Frampton, Sabine Lange, Corey Masuca, Steve Packham, and Aaron Yeow for their courage, diligence, and integrity in upholding sound science, honest inquiry, constructive debate, and transparent process even under pressure

Preface

In most areas of applied research, sound science entails the use of clear definitions, explicit derivations of conclusions from data, independently reproducible tests of predictions against observations, and careful qualification of causal interpretations and conclusions to acknowledge remaining ambiguities or conflicts in the evidence. This book proposes that these principles of sound science can and should be applied to the study of air pollution health effects associated with exposures to air pollution. It describes and illustrates technical methods for doing so.

This view contrasts with an alternative weight-of-evidence (WoE) approach that does not require these restrictions, but that instead draws on authoritative judgments by selected experts to reach policy-relevant conclusions. These conclusions include causal determination judgments about effects caused by exposure. The WoE approach is currently very popular. It has dominated much regulatory risk assessment in the past two decades and has influenced numerous technical articles and policy recommendations. The authors of ostensibly scientific articles about air pollution health effects now routinely offer policy-relevant conclusions and recommendations based on their own judgments that associations, regression coefficients, and untested modeling assumptions in selected statistical models should be interpreted as if they provided reliable causal information about how changing exposures would change risks of health effects in exposed people. Yet, this approach does not provide a scientifically or statistically sound method for learning about how the world works from observations and empirical tests of falsifiable predictions. Correlation truly is not causality; human judgment is notoriously fallible and prone to heuristics and biases; and causally effective regulation requires understanding how changes in exposure change health risks in enough detail to make demonstrably accurate predictions. Characterizing expert judgments about associations between health risks and exposure levels is not an adequate substitute for scientific understanding of the changes in health effects caused by changes in exposures.

This book discusses the WoE and scientific approaches. It explains why the scientific approach, while perhaps less popular and more exacting than the WoE approach, can contribute to more useful regulatory risk analysis by using data to help reach valid conclusions about how public policy interventions affect real-world outcomes and by identifying causally effective policies to better protect public health. It explains and illustrates applications of machine learning, artificial intelligence, statistics, epidemiology, and biologically based simulation modeling to answer questions about how changing exposures affects health risks. The book is intended primarily for health risk analysts, public health researchers and policymakers, and air pollution health effects researchers who are not content to let air pollution health effects estimates and policy recommendations rest on the authoritative judgments, assumptions, and policy preferences of selected experts, but who want to understand for themselves how to analyze health effects data, build and test predictive risk models, and ascertain and communicate accurately what is known, what is only assumed or conjectured, and what is still uncertain about human health effects of air pollution. Researchers interested in applied artificial intelligence and machine learning, simulation of biological responses, and statistical and epidemiological modeling of health effects may also find value in some of the modern methods of data analysis and modeling covered in this book. Finally, the first two chapters and the last two chapters of the book are intended to be useful to a broader audience of regulatory scientists and analysts interested in avoiding common pitfalls in the WoE approach and in applying principles of sound science and risk analysts to develop more credible, accurate, and trustworthy assessments of human health risks caused by air pollution and more effective interventions and policies for reducing these risks.

Denver, CO, USA Louis Anthony Cox Jr.

Acknowledgments

This book has benefitted from many stimulating conversations with colleagues and students at the University of Colorado about how best to teach and practice statistical inference, epidemiology, risk analysis, artificial intelligence, and causal analysis while preserving a clear conceptual and mathematical distinction between statistically significant associations and causal effects with practical significance. I especially thank Marlene Smith, Gary Kochenberger, and Fred Glover from the University of Colorado. Julie Smith and Sarah McMillan at the United States Department of Agriculture provided stimulating challenges and support for developing and applying causal analysis and inference methods and software for evaluating the effects of interventions. Dale Drysdale and the National Stone, Sand, and Gravel Association (NSSGA) posed the challenging research question of how to better understand and model quantitative dose–response functions for respirable crystalline silica particles and asbestos fibers, leading to the material in Chaps. 2–6. I thank Dale and members of the NSSGA Science Advisory Board for extremely stimulating discussions of scientific and modeling issues and for their enthusiastic support of this research. Chapter 3 would not exist without the key suggestion from Julie Goodman that complex mathematical models could be explained simply via the key diagrams used to communicate results; it was a pleasure collaborating on the expository paper on which Chap. 3 is based. I am also grateful to members of the National Institute for Occupational Safety and Health (NIOSH) Board of Scientific Councilors for discussions of practical needs for occupational safety and to NIOSH for developing and documenting the regulatory risk analysis framework discussed in Chap. 3. I have benefitted from discussions of air pollution health effects literature with Roger McClellan and am very grateful for the extraordinarily fine reviewers and thoughtful comments that he obtained for the technical papers on which Chaps. 4–6, 8, 9, and 15 are based.

My understanding of the molecular and cell biology of lung diseases was greatly improved in the 2000s by years of research with my late friend Ted Sanders of Philip Morris Inc. on biomathematical modeling of biological mechanisms of lung diseases caused by cigarette smoking. I think Ted, had he lived, would have relished the exciting new discoveries related to the NLRP3 inflammasome discussed in

Chaps. 3–6. I also thank Susan Dudley for her enthusiasm and support for this work. Warner North and Michael Greenberg provided excellent advice and comments that improved the material in Chaps. 9 and 5, respectively; to Warner, I also owe an intellectual debt for prompting me to think more constructively about how causal analysis should be carried out in practice by practitioners with realistically imperfect data and limited causal knowledge. Douglas Popken assembled data sets and co-authored the journal articles on which Chaps. 8 and 17 are based; I appreciate his outstanding work in creating powerful data sets and analyses for understanding public health risks of air pollution.

The approaches to causal analysis and modeling in Chaps. 7–17 had a long genesis, including work at the American Institutes for Research on quasi-experiments and causal inference in the 1970s; work on epidemiology and attribution of risk in the presence of joint causes at Arthur D. Little and MIT in the 1980s; research in artificial intelligence for the telecommunications company US WEST in the 1980s and 1990s; and discussions with many graduate students in courses in epidemiology, risk analysis, and causal analysis that I have taught at the University of Colorado in more recent decades. Preparation of these chapters was supported by Cox Associates, LLC, which has received funding from the National Stone, Sand, and Gravel Association for research on causal modeling and review of approaches to regulatory risk assessment; from the United States Department of Agriculture for research on causal inference and program evaluation; and from the George Washington University Regulatory Studies Center, the American Petroleum Institute, and the American Chemistry Council for research to develop and apply software supporting objective (model-free) assessments of causality in regulatory risk assessment. Much of the research reported in this book was inspired by work undertaken by the author in serving as chair of the U.S. Environmental Protection Agency (EPA) Clean Air Scientific Advisory Committee (CASAC). None of these organizations reviewed this book prior to publication; the views expressed, research questions asked, and conclusions reached are solely those of the author.

Material from the following articles has been used with the kind permission of their publishers:

- Cox LA Jr, Popken DA. Should air pollution health effects assumptions be tested? Fine particulate matter and COVID-19 mortality as an example. Global Epidemiolgy 2020;100033. https://doi.org/10.1016/j.gloepi.2020.100033. (Chapter 1)
- Cox LA Jr. Implications of nonlinearity, confounding, and interactions for estimating exposure concentration-response functions in quantitative risk analysis. Environ Res. 2020;187:109638. (Chapter 2)
- Cox LA Jr, Goodman JE, Engel AM. Chronic inflammation, adverse outcome pathways, and risk assessment: a diagrammatic exposition. Regul Toxicol Pharmacol. 2020;114:104663. (Chapter 3)
- Cox LA Jr. Risk analysis implications of dose-response thresholds for NLRP3 inflammasome-mediated diseases: respirable crystalline silica and lung cancer as an example. Dose Response. 2019. (Chapter 4)

- Cox LA, Jr. Dose-response modeling of NLRP3 inflammasome-mediated diseases: asbestos, lung cancer, and malignant mesothelioma as examples. Crit Rev Toxicol. 2020:1–22. (Chapter 5)
- Cox LA, Jr. Nonlinear dose-time-response functions and health-protective exposure limits for inflammation-mediated diseases. Environ Res. 2020;182. (Chapter 6)
- Cox LA Jr. Why reduced-form regression models of health effects versus exposures should not replace QRA: livestock production and infant mortality as an example. Risk Anal. 2009;29(12):1664–71. (Chapter 7)
- Cox LA Jr., Popken DA, Berman W. Causal vs. spurious spatial exposure-response associations in health risk analysis. Crit Rev Toxicol. 2013;43(S1):26–38. (Chapter 8)
- Cox LA Jr. Modernizing the Bradford Hill criteria for assessing causal relationships in observational data. Crit Rev Toxicol. 2018;1–31. (Chapter 9)
- Cox LA Jr. Using Bayesian Networks to clarify interpretation of exposure-response regression coefficients: blood lead-mortality associations an example. Crit Rev Toxicol. Forthcoming 2020. (Chapter 10)
- Cox LAT Jr. Socioeconomic and particulate air pollution correlates of heart disease risk. Environ Res. 2018;167:386–92. https://doi.org/10.1016/j.envres.2018.07.023. (Chapter 13)
- Cox LA Jr. Reassessing the human health benefits from cleaner air. Risk Anal. 2012;32(5):816–29. (Chapter 14)
- Cox LA Jr. Do causal concentration-response functions exist? A critical review of associational and causal relations between fine particulate matter and mortality. Crit Rev Toxicol. 2017;47(7):603–31. (Chapter 15)
- Cox LA. Effects of exposure estimation errors on estimated exposure-response relations for PM2.5. Environ Res. 2018;164:636–46. https://doi.org/10.1016/j.envres.2018.03.038. (Chapter 16)
- Cox, LA Jr, Popken DA. Has reducing PM2.5 and ozone caused reduced mortality rates in the United States? Ann Epidemiol. 2015;25(3):162–73. (Chapter 17)
- Cox LA Jr. Improving causal determination. Global Epidemiol. 2019. https://www.globalepidemiology.com/article/S2590-1133(19)30004-5/pdf. (Chapter 18)
- Cox LA Jr. Communicating more clearly about deaths caused by air pollution. Global Epidemiol. 2019. (Chapter 19)

I thank the publishers and coauthors of these works.

Contents

Part I
Estimating and Simulating Dynamic Health Risks

Chapter 1
Scientific Method for Health Risk Analysis: The Example of Fine Particulate Matter Air Pollution and COVID-19 Mortality Risk

Introduction: Scientific Method for Quantitative Risk Assessment

Applied science is largely about how to use observations to learn, express, and verify predictive generalizations—causal laws stating that *if* certain antecedent conditions hold, *then* certain consequences will follow. Non-deterministic or incompletely known causal laws may only determine conditional probabilities or occurrence rates for consequences from known conditions (Spirtes 2010). For example, different exposure concentrations of air pollution might cause different mortality incidence rates or age-specific hazard rates for people with different values of causally relevant covariates. A defining characteristic of sound science is that causal laws and their predictions are formulated and expressed unambiguously, using clear operational definitions, so that they can be independently tested and verified by others and empirically confirmed, refuted, or refined as needed using new data as it becomes available. Comparing unambiguous predictions to observations (using statistics if the predictions are probabilistic) determines the extent to which they are empirically supported. The authority of valid scientific conclusions rests on their testability, potential falsifiability, and empirically demonstrated predictive validity when tested. Using new data to constantly question, test, verify, and if necessary correct and refine previous predictive generalizations, and wider theories and networks of assumptions into which they may fit, is a hallmark of sound science. Its practical benefit in risk analysis is better understanding of what truly protects people, and what does not—for example, the unexpected discovery that administering retinol and beta carotene to subjects at risk of lung cancer increased risk instead of decreasing it (Omenn et al. 1996; Goodman et al. 2004).

Recent proposals to apply this concept of science at the United States Environmental Protection Agency (EPA) focused on the following proposed normative principles for better understanding and communicating human health

L. A. Cox Jr., *Quantitative Risk Analysis of Air Pollution Health Effects*, International Series in Operations Research & Management Science 299, https://doi.org/10.1007/978-3-030-57358-4_1

effects associated with air pollution (Clean Air Scientific Advisory Committee (CASAC) 2019).

1. *Consider all of the relevant, high-quality data before forming conclusions.* For example, if some studies report that reducing current ambient levels of particulate air pollution substantially reduces all-cause mortality rates (Clancy et al. 2002), but other, equally good (or better) studies report that large reductions in pollution levels have produced no detectable effect on all-cause mortality rates (Dockery et al. 2013), then consider both sets of studies before drawing conclusions about how changing air pollution levels affects public health. Acknowledge apparent discrepancies in available data (Burns et al. 2020). Seek to understand and resolve them (Zigler and Dominici 2014). Such openness to observations, even when—or especially when—they conflict with what has previously been believed, helps to overcome confirmation bias and to learn how the world works (Kahneman 2011).
2. *Provide explicit, independently verifiable derivations of conclusions from data.* Sound scientific conclusions are derived from observations by explicit reasoning that can be checked by others. This is often done by testing null hypotheses implied by proposed causal laws (or implied by theories comprised of networks of such laws). For example, if exposure to a substance is hypothesized to be a cause of increased mortality risk, then testing the null hypothesis that mortality risk is conditionally independent of exposure, given the values of other variables (such as income, daily high and low temperatures in the weeks prior to mortality, or co-exposures and co-morbidities) and showing that the data allow confident rejection of this null hypothesis would support the causal hypothesis. Failing to reject the null hypothesis would not support it. Explicitly stating the null hypotheses tested, the data and procedures (and, where necessary, any assumptions) used to test them, and the results of the tests provides a dispassionate basis for stating, verifying, and defending conclusions. There is neither room nor need in this procedure for authoritative holistic judgments, which are often not as readily independently testable or falsifiable, as a basis for conclusions. Rather, the validity and credibility of conclusions rests on the data and methods used to test them, independent of the credentials, expertise, authority, beliefs, or preferences of those making them.
3. *Use terms with clear conceptual and operational definitions in stating conclusions.* For example does a determination that an observed statistical association between exposure to fine particulate matter (PM2.5) and mortality risk is "causal" imply that reducing PM2.5 alone would be sufficient to reduce mortality rates, other conditions remaining fixed ("manipulative causation"); or does it simply mean that the association satisfies some or all of the Bradford Hill considerations (see Chap. 9) for associations ("attributive causation")? (Neither of these implies the other, as discussed in more detail in Chaps. 9 and 15.) Must mortality risk be reduced by at least a certain minimal amount if exposure is removed for an association to be described as causal, or is no such minimum effect size intended by the description? More generally, those offering scientific

advice to policy makers and the public should define exactly what they mean by key terms, such as "causal" or "likely to be causal," used to communicate conclusions to policy makers. They should specify how observations can be used to determine whether these terms hold.

4. Most importantly, *show results from empirical tests of causal conclusions (predictive generalizations)* implied by causal theories or models against observational data. That is, do not simply report causal conclusions at the end of a study, but test them against data and report the results. Causal determinations, conclusions, and interpretations of observations should be tested systematically for both external validity (can the specific causal conclusions from individual studies be synthesized and generalized to yield valid general causal conclusions that apply across studies and to new conditions or interventions?) and internal validity (are stated causal conclusions implied by, or at least consistent with, the data from the specific studies that generated the data? Are rival (non-causal) conclusions refuted by data?) (Campbell and Stanley 1963). The results of these tests should be reported wherever they are available. Technical methods for systematically testing internal validity have been well developed since the early 1960s for quasi-experiments and other observational studies, using relevant comparison groups and adjustment sets (Campbell and Stanley 1963; Textor et al. 2016). Methods for assessing external validity and identifying externally valid ("transportable") causal conclusions from data have been developed more recently, based largely on conditional probabilities and empirical constraints that are invariant across settings and interventions (Schwartz et al. 2011; Bareinboim and Pearl 2013; Triantafillou and Tsamardinos 2015; Peters et al. 2016).
5. *Correctly qualify causal conclusions and interpretations.* Clearly identify any untested or unverified assumptions used in deriving stated causal conclusions, and state conclusions as being contingent on the validity of these assumptions. If a body of evidence is consistent with several alternative causal interpretations, perhaps under alternative assumptions, then present all of them; do not select just one. Where possible, use observations from multiple studies to rule out rival potential explanations for observations, narrowing the set of potential valid causal conclusions and interpretations that are consistent with the entire body of observations from multiple studies and populations (Campbell and Stanley 1963; Triantafillou and Tsamardinos 2015; Cox Jr. 2018). Do not characterize observations as supporting one interpretation, such as that exposure to air pollution causes increased mortality, if they support alternative interpretations at least as well, such as that both air pollution and mortality rates are correlated over time due to coincident trends, or that confounders (such as poverty, or daily temperature extremes in the week or two before death) explain the observed association between them.

This book describes and illustrates methods for putting these principles into practice. Chapters 2, 3, 4, 5, and 6 discuss methods for estimating and simulating exposure-related health risks, focusing primarily on events that take place within the human body because of exposure to air pollutants or chemicals in air—pharmacokinetics, pharmacodynamics, and induction of disease processes. In addition to such "micro" analyses of changes within an individual, Chap. 2 also introduces methods for "macro" analyses that estimate changes in the health risks of exposed populations using epidemiological data. These methods are discussed and applied further in Parts 2 and 3 of the book. Chapter 2 compares statistical regression models, which quantify statistical associations between exposure and response in the presence of other variables, with simulation and causal Bayesian network methods for elucidating why and how much changing some variables—especially, exposure concentrations and durations—will change others, including health risks. Chapter 3 provides a relatively accessible introduction to the key concepts of simulation-based modeling of exposure-related health effects, seeking to build intuition for how and why different time patterns of exposure affect risk differently, even if they deliver the same cumulative exposure per unit time (e.g., per week or per year). Chapters 4 and 5 develop and illustrate the simulation-based approach in more detail and apply it to respirable crystalline silica and asbestos, respectively—two occupational hazards of great current interest. Chapter 6 emphasizes the importance of nonlinear responses in exposure-related diseases, due largely to the fact that induction of chronic inflammation mediates many such diseases. Chronic inflammation involves biological mechanisms with clear thresholds and nonlinearities, in which doubling concentration more than doubles risk. Chapter 6 closes Part 1 of the book by considering how regulations to protect occupational health might be made more effective by paying more attention to variability in exposure concentrations over time, arguing that regulating average exposure concentrations and frequencies of excursions above permitted exposure concentration limits is often inadequate for limiting risks that are primarily driven by repeated high exposure concentrations spaced closely together in time.

Part 2 critically examines the use of statistical regression models in public health risk assessment, building on the insights and methods discussed in Chap. 2. Chapters 7, 8, 9, 10, 11, and 12 argue that regression modeling is not a suitable replacement for quantitative risk assessment (QRA) methods that model causal processes. These chapters apply the association-causation distinction to past studies of health effects associated with—but not necessarily caused by—emissions from factory farms (Chap. 7), spatially distributed sources of natural asbestos (Chap. 8), low levels of the heavy metals lead and molybdenum (Chaps. 10 and 11), and low-level occupational exposures to benzene (Chap. 12). These applications are powered by advances in theory and techniques for causal modeling. Chapter 9 reviews the traditional Bradford-Hill considerations for trying to make judgments about whether observed associations are best explained by causation. It suggests that quantitative methods of causal analysis and machine learning developed over the past century can add considerable value to the Bradford-Hill considerations, helping to shift the scientific basis for causal inferences away from the psychology of judgment emphasized by Hill, and toward more objective and reliable data-driven considerations, such as whether health effects are found to depend on exposures when causal analysis methods are used to hold the values of other variables fixed. The principles and methods of causal analysis reviewed in Part 2 emphasizes robust data-driven inferences, based largely on information theory and non-parametric tests, to minimize dependence on modeling assumptions. Other causal modeling techniques (such as regression-based and propensity score model-based methods) that rely on judgments about the plausibility of untested parametric modeling assumptions are not emphasized.

Part 3 turns to health effects associated with fine particulate matter (PM2.5) air pollution, consisting of atmospheric particulate matter (PM) with particles having a diameter of less than 2.5 μm. Decades of regression modeling, funded largely by the United States Environmental Protection Agency (US EPA), have produced a vast and still growing literature documenting statistical associations between PM2.5 levels and a wide variety of adverse effects, from autism to violent crime, in a wide variety of regression models (Chaps. 14, 15, and 16). Yet, the central paradox of this area of research is that there is very little clear evidence that reducing ambient levels of particulate air pollution actually causes reductions in the various adverse effects attributed to it (Chaps. 17, 18, and 19). Nonetheless, a vocal community of air pollution health effects investigators, activists, advocates, and reporters take it as axiomatic that PM2.5 air pollution kills, and that many deaths per year could be prevented by further reducing ambient concentrations of PM2.5 (Goldman and Dominici 2019). This firmly held and widely shared belief, constantly assumed in calculations of regulatory benefits (Chap. 14) and asserted for years as a well-established fact in countless news stories, op-eds, and scientific journals (Chap. 19), is not easily shaken by empirical observations and findings to the contrary (Henneman et al. 2017; Burns et al. 2020). Instead, expert "weight-of-evidence" (WoE) judgments and "causal determination" judgments (Chap. 18) are frequently advocated as a more dependable basis than causal analysis of data for choosing what should be believed or assumed and communicated to policy makers and the public (e.g., Goldman and Dominici 2019). Part 3 takes issue with this judgment-based approach. It notes that often-cited associations between air pollution and adverse health effects do not fully control for important confounders (Chaps. 13 and 15); that association-based studies of air pollution health effects have typically assumed causality without rigorously testing other explanations (Chaps. 14 and 15); that realistic errors in estimates of exposure concentrations distort the shapes of estimated concentration-response functions (Chap. 16); and that, in practice, reducing air pollution in many cities and areas has not caused the improvements in public health that were projected based on statistical associations (Chap. 17). Part 3 closes by arguing that both the essentially political practice of using consensus expert judgments to make "causal determinations" about health effects of exposures (Chap. 18) and the communication of such assumptions and judgments to the public and policy makers as if they were scientific facts can be substantially improved (Chap. 19). The key to better informing policy makers and the public is to better distinguish between trustworthy, empirically demonstrable and reproducible effects, and hypothesized effects based on unverified modeling assumptions and judgments. Headlines reporting model-based projections of deaths attributed to pollution do not imply that reducing pollution would reduce deaths, although the language used often obscures this fact. Chapter 19 concludes with recommendations for communicating more accurately about what is and is not implied by studies of air pollution-health effects associations that might be useful in crafting effective, well-informed regulations and policies.

This book makes repeated use of current data science and machine learning techniques, such as partial dependence plots estimated from random forest model ensembles, and Bayesian networks. For readers unfamiliar with these techniques, Chap. 2 provides a brief nontechnical introduction, and Chap. 9 and its appendixes and references provide more technical detail. In an effort to make the chapters relatively self-contained, relevant techniques are briefly reviewed in less detail in other chapters as needed. Similarly, the early chapters introduce examples and analyses that are expanded on in later chapters. Readers primarily interested in specific substances (e.g., crystalline silica, lead, asbestos, benzene, or fine particulate matter) should find sufficient explanation of the technical methods used in each chapter to follow the analyses without reading all of the preceding chapters. Conversely, readers more interested in general principles may wish to focus on Chaps. 2, 9, 18 and 19, as the remaining chapters mainly deal with applications of the principles and methods discussed in these methodological chapters.

Scientific Method vs. Weight-of-Evidence Consensus Judgments as Paradigms for Regulatory Risk Analysis

Efforts to apply the foregoing principles 1–5 to reviews of air pollution health effects have provoked considerable backlash, enthusiastic *ad hominem* attacks, and much reporting describing them as dangerous attacks on the established science of air pollution health effects research and an attempt to subvert years of progress in environmental regulation (Tollefson 2019; Drugmond 2020). For example, a 29 April, 2019 Editorial in *Nature* stated the following.

- "Research linking fine particulate pollution and premature deaths is under attack." A less dramatic account, and the point of view adopted in this book, is that decades of research *associating* exposure to fine particulate matter with increased mortality rates is well accepted, but has recently come under scrutiny to see what valid *causal* conclusions it can yield (Burns et al. 2019; Cox 2017; Henneman et al. 2017). In less contentious times, this reexamination might be viewed not as an "attack" on such research, but as an attempt to understand and use it correctly. Likewise, demands for clear definitions (principle 3 above)—especially whether the designation "causal" is intended to refer to manipulative causation or to Hill causation (or to something else)—are viewed in this book not as an attack on science, or an effort to create an impossible (or any) burden of proof (Goldman and Dominici 2019; Tollefson 2019), but rather as an effort to bring much-needed clarity to an area with a long history of using vague and ambiguous terms (such as pollution being "linked" to health effects) that have obscured the distinction between correlation and causality. Part 3 argues that such clarity is crucial for well-informed and causally effective policy-making and communication (Chap. 19); pursuing it advances rather than undermines sound science.

- "… Sceptics often argue that the epidemiological evidence cannot prove that air pollution causes premature deaths. But that is deliberately ignoring the weight of evidence from an array of rigorous epidemiological studies, aligned with other sources of evidence." This misrepresents crucial points. As explained in Part III of this book, the key scientific issue is not that epidemiological evidence *cannot* prove that reducing current ambient air pollution levels causes reduced death rates, but rather that, to date, multiple studies in multiple countries *do not* do so (Burns et al. 2020; Cox Jr. 2017; Henneman et al. 2017). More precisely, existing epidemiological data sets enable powerful statistical tests of specific null hypotheses, including whether mortality rate is conditionally independent of air pollution, given levels of known potential confounders (e.g., income and daily high and low temperatures in the weeks before death (Zhang et al. 2017)); and whether mortality rates have declined no more quickly where air pollution has decreased substantially than where it has increased or remained unchanged (Chap. 17). The view espoused in this book is that such null hypotheses can and should be directly tested using available data (e.g., via conditional independence tests and rate of change comparisons) (Chaps. 13, 15, and 17), and that the evidence base considered by EPA should be expanded to include results of more such studies (Clean Air Scientific Advisory Committee (CASAC) 2019). In multiple studies in multiple countries, the null hypothesis of *no causal effect* of changes in exposure on changes in mortality rate is not clearly rejected when rigorous tests are applied to epidemiological data, although the different null hypothesis of *no association* between exposure and mortality is often decisively rejected (e.g., Burns et al. 2019; Henneman et al. 2017; Stenlund et al. 2009; Zigler and Dominici 2014, Chap. 17). Those who assert the contrary—that data from famous studies of association, such as the Harvard Six Cities study, have already amply shown that ambient levels of air pollution cause increased mortality—conflate association and causation, often using ambiguous words such as "links" and "linking" that needlessly obscure this crucial distinction (e.g., Tollefson 2019). They refer to "rigorous epidemiological studies" without noting that most of these studies only address associations and do not fully control for obvious potential confounders, such as poverty or humidity and daily high and low temperatures lagged by more than a day or two. We follow other investigators who rely on data rather than authoritative judgment for their conclusions in remaining open to the possibility that reducing air pollution does in fact reduce mortality rates (e.g., Burns et al. 2019), but that the causal effect is small enough to have escaped detection (Chap. 17).
- "But questioning the evidence won't make it go away. …Now is not the time to undermine efforts to clean air—it is time to strengthen them." Much of the backlash against the principles of sound science proposed above seems to start with policy convictions, such as that now is the time to strengthen efforts to clean air; to then assert that it is well known and long established that current ambient levels of fine particulate matter kill, but without presenting causal analyses or

empirical data demonstrating that this is likely to be true; and to refer to ambiguous and conflicting evidence of association in the presence of incompletely controlled confounders as if it supported these conclusions (Goldman and Dominici 2019). But sounder science and better informed policy-making both require a different procedure that first seeks to understand what valid causal conclusions are justified by observations, as a prelude to understanding what policies are needed to fully and effectively protect public health, including the health of sensitive subpopulations, with an adequate margin of safety. This approach, developed in the chapters that follow, might support policy recommendations in either direction.

This book argues that in air pollution health effects research, as in other areas of applied science, investigators can best pursue objective scientific truth by using dispassionate, sound analysis of data to inform policy, rather than letting policy preferences and judgments inform or constrain selection and interpretation of evidence. This reversal has been vociferously opposed by advocates of a "weight of evidence" (WoE) framework that makes judgments of selected experts—and especially, "causal determination" judgments (Chap. 18)—central to interpretation of evidence and conclusions communicated to policy makers (Goldman and Dominici 2019; Krimsky 2005; Linkov et al. 2015). The WoE framework does not require use of the foregoing principles of sound science. However, as argued in the following chapters, following these principles better serves the discovery and communication of valid causal conclusions, the needs of policy makers to be informed about the public health consequences caused by alternative policy choices, and hence the public interest.

A Recent Example: PM2.5 and COVID-19 Mortality

To illustrate why statistical modeling of exposure-response associations accompanied by judgments about causal interpretations of statistical associations and regression coefficients—the current weight-of-evidence (WoE) approach favored in much current regulatory risk analysis for air pollutants—is not a valid basis for determining whether or to what extent risk of harm to human health would be reduced by reducing exposure, and why the traditional scientific method based on testing predictive generalizations against data remains a more reliable paradigm for risk analysis and risk management, we consider the question of whether past exposure to fine particulate matter (PM2.5) increases risk of COVID-19 mortality. As COVID-19 mortalities mounted worldwide in the first two quarters of 2020, headlines and scientific articles warned that fine particulate matter air pollution (PM2.5) increases risk of COVID-19-related illness and death. For example, Jiang et al. (2020) used a Poisson regression model to conclude that PM2.5 and humidity increased the risk of daily COVID-19 incidence in three Chinese cities, while coarse particulate air pollution (PM10) and temperature decreased this risk. Bashir et al. (2020) calculated significant ordinal correlations between PM2.5 and other air pollutants (PM10, SO2, NO2, and CO) and COVID-19 cases in California, and concluded that such correlations should encourage regulators to more tightly control pollution sources to prevent harm. Most famously, Wu et al. (2020) fit a negative binomial regression model to county-level data in the United States, and interpreted their finding of a significant positive regression coefficient for PM2.5 as implying that "A small increase in long-term exposure to PM2.5 leads to a large increase in the COVID-19 death rate." This interpretation attracted national headlines and widespread political concern (Friedman 2020).

These examples follow a common technical approach, also widely applied in many other areas of current regulatory risk assessment and public health, with the following steps:

1. Collect data on estimated air pollution levels, one or more adverse health outcomes of interest (such as COVID-19 mortality), and any covariates of interest (e.g., humidity, temperature, population density, etc.)
2. Fit one or more regression model to the data, treating air pollution levels as predictors and adverse health outcomes as dependent variables. Include other variables as predictors at the modeler's discretion.
3. If the regression coefficient for a pollutant as a predictor of an adverse health outcome is significantly positive in the one or more regression models, use judgment to interpret this as evidence that reducing levels of the pollutant would reduce risk of the adverse health outcome.
4. Communicate the results to policy makers and the press using the policy-relevant language of *causation* and *change*—that is, claim that a given reduction in pollution would create a corresponding reduction in adverse health outcomes—rather than in the (technically accurate) language of *association* and *difference*: that a given difference in estimated exposures is associated with a corresponding difference in the conditional expected value of a dependent variable predicted by the selected regression model.

Step 3 is based on a judgment that a positive regression coefficient in a modeler-selected regression model is evidence of a causal relationship: that it implies or suggests that reducing exposure would reduce risk, even if the experiment has not actually been made. In this respect, it incorporates the central principle of the WoE framework: that a well-informed expert scientist can make a useful judgment about whether the association indicated by a statistically significant positive regression coefficient is likely to be causal. This assumption is scrutinized next.

Do Positive Regression Coefficients Provide Evidence of Causation?

As noted by Dominici et al. (2014), either significant positive coefficients or significant negative regression coefficients (or no significant regression coefficient at all) for air pollution as a predictor of mortality risk can often be produced from the same data, depending on the modeling choices made; thus "There is a growing consensus in economics, political science, statistics, and other fields that the associational or regression approach to inferring causal relations—on the basis of adjustment with observable confounders—is unreliable in many settings." In the field of air pollution health effects research, however, investigators continue to rely on regression modeling in step 2 of the above approach. A skilled regression modeler can usually produce a model with a significant positive regression coefficient for exposure in step 2, allowing steps 3 and 4 to proceed. We illustrate how this can be done, using a data set on PM2.5 and COVID-19 mortality in the United States as an example. The data set, described and provided via a web link in Appendix, compiles county-level data on historical ambient PM2.5 concentration estimates, COVID-19 mortality rates and case rates (per 100,000 people) through April of 2020, along with numerous other county-level variables.

A key step in regression modeling is to select variables to include in the model. Figure 1.1 shows a random forest (nonparametric model ensemble) importance plot for county-level variables as predictors of COVID-19 mortality rates, where the "importance" of each variable is measured by the estimated percentage increase in mean squared prediction error if that variable is dropped as a predictor. The few most important predictors of COVID-19 mortality (*DeathRate100k*) are *PCT_BLACK*, the percentage of a county population that is Black; *PopDensity*, the average density of the population in the county (number of people per square mile) or its logarithm, *PopDensityLog* (the log transform makes little difference to nonparametric methods such as random forest, but can be important for parametric regression models); *Longitude*, time since first case in the county (*FirstCaseDays*), average

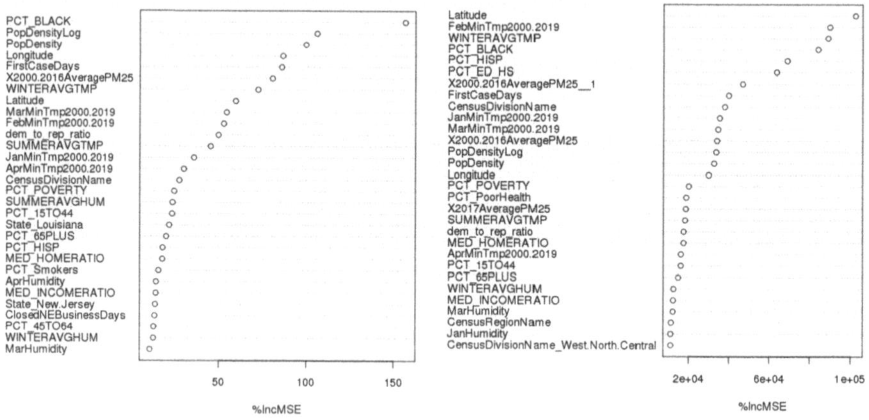

Fig 1.1 Importance plots for several variables as predictors of COVID-19 mortality (left) and case rates (right) per 100,000 people. The plots are generated by random forest nonparametric model ensembles that explain about 48% of the variance in mortality rates and 40% of the variability in case rates among counties in the United States as of April, 2020. Appendix provides details and data. "Importance" is measured as the percentage increase in mean squared prediction error if a variable is dropped as a predictor

estimated PM2.5 concentration between 2000 and 2016 (*X2000.2016AveragePM25*), average temperature during the winter months between 2000 and 2016 (*WINTERAVGTMP*), and *Latitude*. For the case rate (COVID-19 cases reported per 100,000 people), the most important predictors also include the average minimum temperature in February over the past two decades (*FebMinTmp2000.2019*), percent Hispanic (*PCT_HISP*), and percent of population with at least high school educations (*PCT_ED_HS*). These ten predictors alone explain about the same percentages of the variances in COVID-19 mortality and case rates across counties (48% and 40%, respectively) as the full set of over 60 variables, of which the most important are shown in Fig. 1.1.

Of course, predictors need not be statistically independent of each other. To visualize the statistical interdependencies among them, Fig. 1.2 shows a Bayesian network (BN) fit to the data (using the default hill-climbing (HC) algorithm in the *bnlearn* package in R), with *Latitude* and *Longitude* constrained to have only outward-pointing arrows and *DeathRate100k* constrained to have only inward-pointing arrows, to facilitate possible intuitive causal interpretations of the arrows leaving or entering these three nodes. (Presumably, latitude and longitude are not caused by anything else, and death does cause any of the other variables.) However, in general the arrows only signify statistical dependencies between variables, and not necessarily causal relationships. (Chapter 9 discusses principles of BN-learning and interpretation of BNs in more detail, and Chap. 2 provides a less technical introduction.)

For example, an arrow between PM2.5 and percent Hispanic (*X2000.2016AveragePM25* and *PCT_HISP*) does not suggest that either causes the other: it simply reflects that counties with higher percentages of Hispanic populations tend

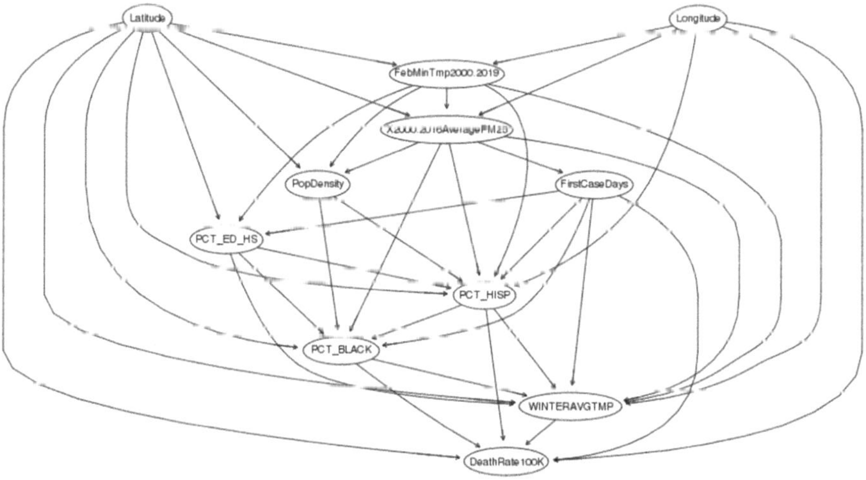

Fig. 1.2 Bayesian network for COVID-19 mortality (deaths per 100,000 people) showing statistical dependencies among variables. An arrow between two variables indicates that they are informative about each other (i.e., not statistically independent)

to also have higher PM2.5 levels. However, if variables depend on their direct causes, then absence of an arrow between two variables corresponds to absence of empirical evidence in the BN that either directly causes the other. COVID-19 mortality in Fig. 1.2 is shown as depending directly on latitude and longitude (which are presumably surrogates for other biologically effective causes), as well as on time since first case in a county (*FirstCaseDays*), average winter temperature, and ethnic composition (*PCT_BLACK* and *PCT_HISP*). Figure 1.3 shows an analogous BN for COVID-19 case rate, which depends directly on latitude and longitude, ethnic composition (*PCT_BLACK* and *PCT_HISP*), time since first case in a county (*FirstCaseDays*), and education (*PCT_ED_HS*). Bayesian network learning is a relatively new technique for exploring and visualizing direct and indirect dependencies among variables. As an alternative, Fig. 1.4 shows a classification and regression tree (CART) tree for COVID-19 mortality. The CART algorithm (implemented in the *rpart* package in R) recursively partitions counties into clusters with significantly different COVID-19 mortality rates, based on the results of binary tests ("splits"), such as whether *Longitude* <−75.61 (yes = left branch, no = right branch). For example, the counties with *Longitude* <−75.61, *PCT_BLACK* < 0.2636, and time since first case <44.5 days have an average COVID-19 mortality rate of less than 3 per 100,000 (2.436, although 3 significant digits is spurious precision), compared to a rate over 50 times greater (148.7 per 100,000) for counties further to the East with high population densities and longer times since first cases. Although CART trees are subject to residual confounding due their binary splits of continuous variables and are not very robust, in the sense that fitting them to multiple random samples from the same data set often produces different trees (e.g., with *WINTERAVGTMP* in some and *FebMinTmp2000.2019* in others), they provide a relatively simple, well-established nonparametric technique for exploring significant predictors of a selected dependent variable such as *DeathRate100K*. The

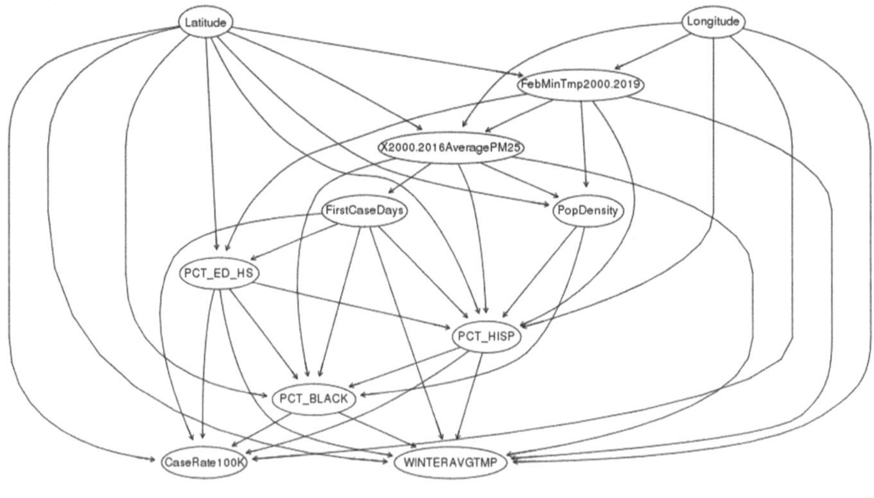

Fig. 1.3 Bayesian network for COVID-19 case rate (cases reported per 100,000 people)

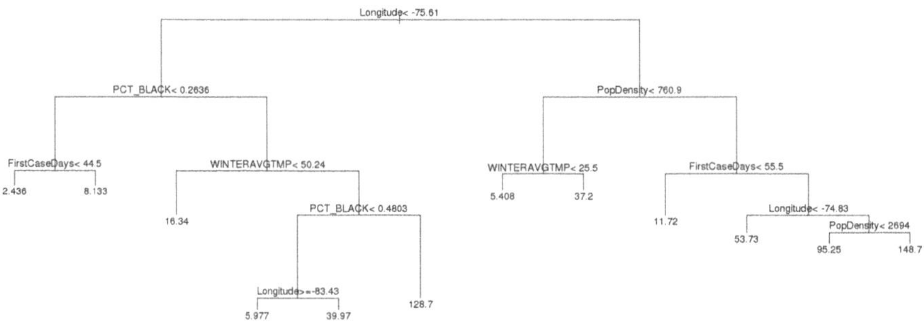

Fig. 1.4 A classification and regression tree (CART) tree for COVID-19 mortality (*DeathRate100K*). The tree was generated by the *rpart* package in R

predictors identified in Fig. 1.4 are *Longitude*, *PCT_BLACK*, *WINTERAVGTMP*, *FirstCaseDays*, and *PopDensity*.

Although we regard Figs. 1.2, 1.3, and 1.4 as only exploratory visualizations, they highlight the importance of confounders such as *Longitude* in understanding associations between PM2.5 and COVID-19 mortality. The usual way to control for measured confounders in regression modeling is to "adjust" for them by including them as predictors on the right side of a regression equation. For example, a multiple linear regression model that includes all of the variables identified in Figs. 1.2 and 1.4 on which COVID-19 mortality rate might directly depend, and that further hypothesizes a dependence on historical average ambient PM2.5 exposure concentration levels (*X2000.2016AveragePM25*), would posit an equation of the form in Eq. (1.1).

$$\begin{aligned} E\left(DeathRate100K\right) = {} & Intercept + b_1 * X2000.2016AveragePM25 \\ & + b_2 * PCT\_BLACK + b_3 * PCT\_HISP + b_4 \\ & * Latitude + b_5 * Longitude + b_6 * FirstCaseDays \\ & + b_7 * WINTERAVGTMP + b_8 * PopDensity \end{aligned} \tag{1.1}$$

For simplicity, Eq. (1.1) follows Wu et al. (2020) in assuming that risk depends on a weighted sum of terms on the right side, ignoring interaction terms (e.g., that increasing PM2.5 should not increase death rates if population density = 0); consequences of this modeling choice are discussed later. Fitting Eq. (1.1) to the data set via ordinary least squares (OLS) regression yields Table 1.1.

The regression coefficient for past estimated average ambient PM2.5 exposure concentration (denoted by *2000.2016AveragePM25* in Tables 1.1 and 1.2) is negative and not significantly different from 0 ($p = 0.87$), consistent with Fig. 1.2. However, regression modeling allows modelers to select variables to include in the model, which can drive the results that get published (Dominici et al. 2014). For example, dropping *Longitude* from the regression model yields Table 1.2. Now the regression coefficient for PM2.5 is positive instead of negative, and it is highly

Table 1.1 Multiple linear regression model for COVID-19 mortality rate

	Regression summary for dependent variable: DeathRate 100K, R = 0.39594357, $R^2 = 0.16$, Adjusted $R^2 = 0.16$					
N = 3009	b*	Std. err. of b*	b	Std. err. of b	t(3000)	p-value
Intercept			−15.81	9.28	−1.7	0.0885
2000-2016AveragePM25	−0.00	0.028	−0.04	0.23	−0.2	0.8718
PCT_HISP	0.09	0.021	13.17	3.19	4.1	0.0000
PCT_BLACK	0.28	0.021	40.75	2.99	13.6	0.0000
PopDensity	0.04	0.018	0.00	0.00	2.2	0.0261
WINTERAVGTMP	0.09	0.034	0.17	0.06	2.7	0.0063
FirstCaseDays	0.17	0.018	0.21	0.02	9.6	0.0000
Latitude	0.18	0.037	0.77	0.16	4.8	0.0000
Longitude	0.15	0.025	0.27	0.05	5.9	0.0000

The columns give standardized regression coefficients (*b**) and their standard errors; unstandardized regression coefficients (*b*) and their standard errors; and t-test values and significance levels (p-values) for each coefficient

Table 1.2 Multiple linear regression model for COVID-19 mortality rate with longitude omitted

	Regression summary for dependent variable: DeathRate100K R = 0.38341345, $R^2 = 0.147$, Adjusted $R^2 = 0.145$					
N = 3009	b*	Std. err. of b*	b	Std. err. of b	t(3001)	p-value
Intercept			−38.08	8.52	−4.5	0.0000
2000-2016AveragePM25	0.09	0.02	0.77	0.19	4.0	0.0001
PCT_HISP	0.06	0.02	9.05	3.13	2.9	0.0038
PCT_BLACK	0.29	0.02	41.08	3.00	13.7	0.0000
PopDensity	0.04	0.02	0.00	0.00	2.5	0.0109
WINTERAVGTMP	0.04	0.03	0.08	0.06	1.3	0.1876
FirstCaseDays	0.17	0.02	0.21	0.02	9.6	0.0000
Latitude	0.15	0.04	0.63	0.16	4.0	0.0001

significant ($p = 0.000053$) instead of non-significant. In effect, PM2.5 acts as a partial surrogate for longitude for predicting COVID-19 mortality risk, so that omitting longitude induces PM2.5 to enter with a significant positive coefficient. Figure 1.5 suggests why: both PM2.5 and COVID-19 mortality rates tend to be higher in the East than in the West. Interpreting the positive regression coefficient for PM2.5 in Table 1.2 as evidence that an increase in PM2.5 increases PM2.5 mortality risk, as in Wu et al. (2020) would be mistaken: it is only evidence that the modelers omitted longitude from the model.

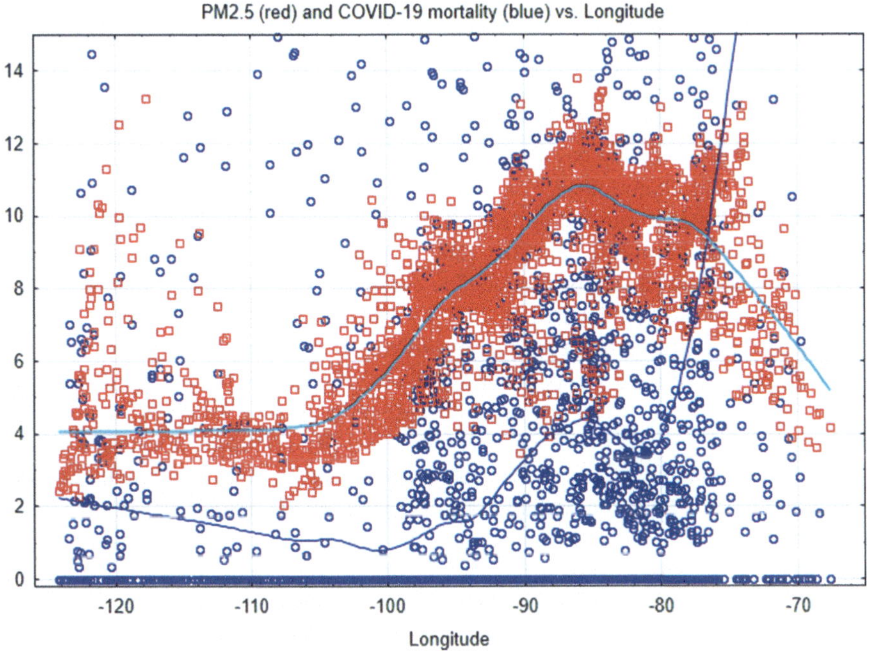

Fig. 1.5 Scatter plots of average estimated historical PM2.5 concentrations (in micrograms per cubic meter) (red squares) and COVID-19 deaths per 100,000 (blue circles) vs. *longitude*

Positive Regression Coefficients Created by Model Specification Error and Other Causes

More generally, there are many reasons that PM2.5 might have a significant positive regression coefficient that do *not* imply that increasing PM2.5 would increase risk. As a simple conceptual example, suppose that *PopDensityLog* is a confounder of the PM2.5-Risk association, and that the structural equations describing the causal relationships among these variables are as follows:

$$E(Risk) = PopDensityLog^2 \tag{1.2}$$

$$PM2.5 = PopDensityLog^2 \tag{1.3}$$

(In such structural equations, the values of the dependent variables on the left are causally determined by the values of the variables on the right: if the right-hand variables are exogenously changed, then the left-hand variables will change to make the equality hold.) Then fitting the following multiple linear regression model (1.4) to data generated by Eqs. (1.2) and (1.3) for a large number of counties with different population densities would yield the optimized parameter estimates $b_0 = 0$, $b_1 = 0$, $b_2 = 1$.

$$E(Risk) = b_0 + b_1 * PopDensityLog + b_2 * PM2.5 \tag{1.4}$$

Thus, regression identifies a significant positive coefficient for *PM2.5*, and not for the confounder *PopDensityLog*, because doing so minimizes prediction error (MSE). But this coefficient has no relevance for determining how or whether changing *PM2.5* would change *Risk*. A claim that such a regression analysis had "controlled for" potential confounding from *PopDensityLog* by including it in regression model (1.4), and yet still found that *PM2.5* increased risk, would be deluded. A judgment that such an analysis provides evidence that increasing PM2.5 levels increases risk would be mistaken.

As a less hypothetical example, suppose we create a synthetic data set that is identical to the real one, except for the addition of a new *Risk* variable defined as $Risk = PopDensityLog^2$. In other words, we artificially create a variable that we know is determined only by population density, via the nonlinear formula $Risk = (log(population\ density))^2$. (This example is suggested by Fig. 1.6, which shows a scatter plot of COVID-19 deaths per 100,000 against *PopDensityLog*.) Fitting a multiple linear regression model to the data with this artificial *Risk* variable as the dependent variable yields the results in Table 1.3. All but one of the predictors, including PM2.5 (*2000-2016AveragePM25*), have highly statistically significant positive regression coefficients, even though, by construction, *Risk* does not depend on anything other than *PopDensityLog*. The reason is that the multiple linear regression model's assumption that risk depends only on a weighted sum of the predictors is false. As illustrated in Fig. 1.6 for the real risk variable (*DeathRate100k*), risk varies nonlinearly with *PopDensityLog*. The mistaken modeling assumption of linearity is sufficient to induce many other predictors to enter the regression model with significant positive coefficients, because including them helps to reduce the mean squared prediction error due to model misspecification. Again, interpreting such a positive regression coefficient for exposure as evidence that reducing exposure would reduce risk is mistaken, even though no confounders have been omitted in this example. Instead, positive regression coefficients are only evidence that the assumed regression model does not describe the data. Chapter 2 develops further the point that model specification error often accounts for positive regression coefficients.

Nonparametric methods help to avoid these difficulties. Figure 1.7 shows a CART tree for the same example as in Table 1.3. In this tree, as also in a nonparametric Bayesian network fit to the same data, the only predictor of $PopDensityLog^2$ is found to be *PopDensityLog*.

More generally, PM2.5 could have a significant positive regression coefficient as a predictor of COVID-19 mortality risk for any or all of the following reasons, as discussed further in Chap. 2.

- Model specification errors, e.g., if mortality rate is assumed to depend on a weighted sum of variables, but in fact its dependence is better described by a model with nonlinearities or interaction terms, as in Table 1.3.

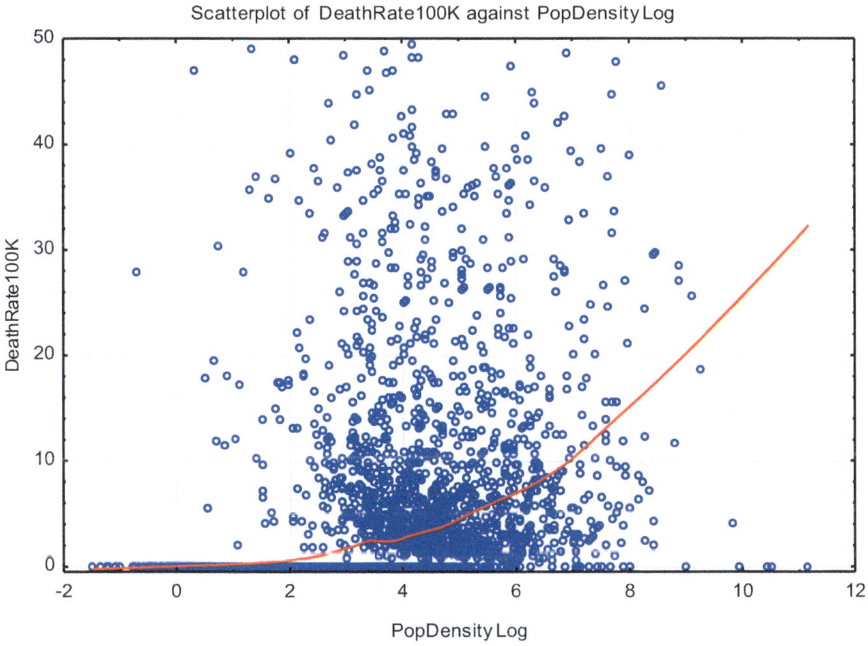

Fig. 1.6 Scatter plot of COVID-19 deaths per 100,000 (*DeathRate100k*) against *PopDensityLog*. A non-parametric (lowess) smoothing regression curve is superimposed to aid visual interpretation

Table 1.3 A multiple linear regression model for the dependent variable *PopDensityLog*²

	Regression summary for dependent variable: Risk = PopDensityLog², R^2 = 0.80040234, R^2 = 0.641, Adjusted R^2 = 0.640					
N = 3009	b*	Std. err. of b*	b	Std. err. of b	t(3000)	p-value
Intercept			−47.30	4.19	−11.3	0.000000
2000-2016AveragePM25	0.35	0.02	1.99	0.11	18.9	0.000000
PCT_BLACK	−0.01	0.01	−1.32	1.35	−1.0	0.326789
PCT_HISP	0.16	0.01	17.09	1.44	11.9	0.000000
Latitude	0.37	0.02	1.09	0.07	15.0	0.000000
Longitude	0.17	0.02	0.21	0.02	10.1	0.000000
WINTERAVGTMP	0.22	0.02	0.28	0.03	9.9	0.000000
FirstCaseDays	0.39	0.01	0.33	0.01	33.5	0.000000
PopDensity	0.38	0.01	0.00	0.00	33.1	0.000000

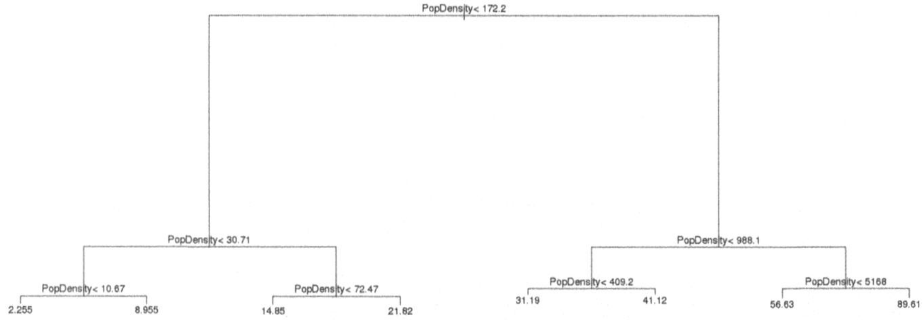

Fig. 1.7 A CART tree model for the dependent variable *PopDensityLog*[2]

- Omitted confounders, as in the example of PM2.5 and COVID_19 mortality risk depending on latitude and longitude (independently of other factors, as shown in Fig. 1.2), if these factors are omitted;
- Measurement errors in explanatory variables, e.g., if PM2.5 is correlated with other variables that are measured or estimated with error, so that including PM2.5 in the regression reduces prediction error due to uncertainties about those variables;
- Residual confounding, e.g., if older people tend to live in more polluted areas and, independently, to have higher mortality rates, but age is only measured in wide categories such as "% of people aged 65 or older";
- Use of surrogate variables, e.g., "Average winter temperature" since 2000, rather than more causally relevant variables such as low temperatures in the months of COVID-19 in 2020;
- Unmodeled interactions or dependencies among variables, e.g., if PM2.5 modifies or is modified by variables such as humidity and temperature that affect respiratory illnesses and COVID-19 mortality;

A positive regression coefficient explained by one or more of these sources does not provide evidence that reducing PM2.5 would reduce mortality risk.

Conclusion: Regression Models and Judgment Should Complement Science, not Substitute for it

We do not conclude from the foregoing considerations that PM2.5 does not increase risk of COVID-19 mortality; perhaps it does. Rather, we conclude that a positive regression coefficient *per se* does not provide useful evidence about the matter. Such regression coefficients are easily produced by modeling choices (Dominici et al. 2014), but lack clear causal interpretation. A judgment that such evidence provides reason to worry—that, in the words of a *New York Times* headline, "New Research Links Air Pollution to Higher Coronavirus Death Rates" (Friedman 2020)—is simply misleading: the "links" provided by positive regression coefficients are statistical links, not causal ones, and they may signify only that longitude was omitted, or that a linear form was assumed, or that different predictors are correlated with each other and estimated with errors, or that continuous variables have been categorized, and so forth. (All of these are true of the model of Wu et al. (2020) behind the headline.) This need not mean that the conclusion is false, but it does mean that the conclusion is not implied by the data and regression analyses from which it is said to be derived.

North (2020) wrote that "An established paradigm for interpreting epidemiological evidence causally, used by the US EPA, based on considerations proposed in 1965 by British epidemiologist Sir Austin Bradford Hill, is being challenged by another paradigm based on statistical procedures to distinguish between association and causation." Perhaps the most fundamental prescription of the causal paradigm is to recognize that statistical "links" such as positive regression coefficients (or relative risks greater than 1, or positive attributable risks and burdens of disease, and so forth) are neither more nor less than indicators of statistical association, which do not necessarily or usually provide relevant evidence about causation (Pearl 2009). Modern methods such as the Bayesian network in Fig. 1.2 can help to discover what evidence a data set does provide that some variables depend, directly or indirectly, on others. For example, Fig. 1.2 suggests that latitude and longitude have direct effects on COVID-19 mortality risk (meaning, effects in addition to those mediated by the other variables in Fig. 1.2). This discovery might not have been anticipated intuitively by an investigator, leading to the omission of these confounders, as in Wu et al. (2020). Such computer-assisted discoveries from data may assist, but not replace, the scientific work of formulating testable predictions about whether and how much changes in some variables affect changes in others, and then testing these predictions against new data and reporting the results. If COVID-19 mortality risk appears to be conditionally independent of PM2.5 in non-parametric analyses with adequate power to detect even relatively small effects (Figs. 1.2 and 1.4), then parametric regression modeling that imposes assumptions on the data sufficient to create a positive regression coefficient for PM2.5 (Tables 1.2 and 1.3) should not be construed as evidence that changing PM2.5 would change COVID-19 mortality risk. But neither should it preclude a search for alternative hypotheses, backed by empirical testing, that better explain the observations. For example, for the same average PM2.5 concentration over the past 20 years, do counties with constant or increasing PM2.5 levels over time have significantly earlier first dates of COVID-19 mortalities than counties with PM2.5 levels that decreased over time? Figure 1.2 leaves open this possibility, and additional research might pursue it further.

The question of what scientific hypotheses are worth investigating further is surely a proper matter for expert judgment. Interpretation of regression coefficients as evidence that reducing exposure would reduce risk is not. Thus, we conclude that empirical testing of predictive generalizations against data should not be skipped in favor of applying judgment to regression coefficients to draw policy-relevant causal conclusions. Regression coefficients simply do not provide the information needed to determine—or to make sound judgments about—whether or to what extent they are likely to be causal (Table 1.3). Judgment that seeks to bridge the gap between association and causation based on positive regression coefficients (e.g., Cromar and Ewart 2016) is akin to a Rorschach test: an expert may perceive evidence for causation in such coefficients, and may even use them to quantify health benefits to be expected from reducing exposure, but the perceived evidence and expectations of health benefits are solely in the mind of the expert, neither supported nor refuted by the regression coefficients themselves. The causal paradigm proposes that traditional scientific method, while often more time-consuming and difficult than applying judgment to regression models, is a far more reliable guide to determining whether and to what extent interventions that change exposure will cause risk to change. Verifying scientific models and predictions for PM2.5 and COVID-19 might take impractically long, and meanwhile decisions must be made and risks managed despite scientific uncertainties about causation. But we reiterate that unwarranted causal interpretation of statistical associations and regression coefficients in a WoE framework should not be substituted for sound science. A technically gullible press and policy makers should not be distracted by pre-scientific claims about health effects from PM2.5 based on judgment and regression modeling in the absence of traditional scientific method and careful evaluation of the predictive validity of such claims (Chap. 19). Risk analysis and the public interest can and should be better served by adhering to the principles of sound science articulated in the Introduction and to the principles of sound causal inference referred to by North (2020) and discussed in several chapters of this book, especially Chaps. 9 and 18.

The remainder of this book develops and illustrates these themes in more detail. Although regression modeling assisted by WoE judgments does not usually suffice to provide a sound or reliable basis for determining whether or to what extent reducing exposure would reduce risk, other techniques such as simulation modeling, Bayesian networks, and causal modeling using nonparametric machine-learning methods can do much to improve the risk analyst's ability to predict the effectiveness of interventions, including reductions in exposure concentrations or changes in time patterns of exposure, in changing the health risks of exposed public or occupational populations. Chapter 2 introduces these methods.

Appendix 1: Data

We collected data from many sources, including most of those cited by Wu et al. (2020), but with alternate authoritative sources for temperature, humidity, and cases/deaths data. We used more recent data for PM2.5, demographics, temperatures, and cases/deaths, and added further sources or fields. For example, we collected USDA county level economic characterizations along with various county attributes compiled by the UC Berkeley Yu Group (2020). Table 1.4 summarizes data sources and variables. Data building was accomplished using python scripts. The full data set can be downloaded from http://cox-associates.com/CausalAnalytics/; it is the file "covidpm25.xlsx" (Table 1.5).

Table 1.4 Data Sources and variable overview

Data category	Source	Comments
PM2.5	Pm2.5 annual average data from the Atmospheric Composition Analysis Group (http://fizz.phys.dal.ca/~atmos/martin). 0.01° × 0.01° grid resolution PM2.5 prediction in mcg/m^3	We averaged across grid cells in each county, and produced a 2000–2016 average, as well as separate values for each year 2000–2018
County boundaries	U. S. Census https://www.census.gov/geographies/mapping-files/time-series/geo/tiger-line-file.html	Used to provide county boundaries for PM2.5 attribution, land area for popdensity, and centroids lat/lons
Demographic	U.S. Census Bureau online API. 2018 ACS5 (American Community Survey 5-year data ending in 2018). County level data. https://www.census.gov/data/developers/data-sets/acs-5year.html	List of variables in table below
Temperatures	NOAA. County level annual data ftp://ftp.ncdc.noaa.gov/pub/data/cirs/climdiv/ Average: climdiv-tmpccy-v1.0.0-20200504' Min: climdiv-tmincy-v1.0.0-20200504' Max: climdiv-tmaxcy-v1.0.0-20200504' Description: county-readme.txt	We averaged across years 2000–2019 for long term averages by month. We also extracted monthly averages for Jan–Apr 2020
Humidity	Humidity averages by U.S. weather station (city) through 2018. https://www1.ncdc.noaa.gov/pub/data/ccd-data/relhum18.dat. City lat/lons from https://simplemaps.com/data/us-cities	For each county centroid, the humidity data from the closest (based on lat/lon coordinates) weather station is obtained
Hospital beds	Hospital level data with county identifier from Homeland Infrastructure Foundation-Level Data (HIFLD) https://hifld-geoplatform.opendata.arcgis.com/datasets/hospitals as of 10/7/2019	Aggregated hospitals over counties. Converted to beds per 100K population. Log version also
Mitigation policies	State level governmental COVID-19 policies compiled byRaifman et al. Boston University School of Public Health, COVID-19 United States state policy database (www.tinyurl.com/statepolicies)	Used to compute days since stay-at-home order and days since closure of non-essential businesses (from 5/11/2020)
Behavioral	County level data from Robert Wood Johnson (2020) https://www.countyhealthrankings.org/	Smoking, obesity, and overall health
County population	https://www2.census.gov/programs-surveys/popest/datasets/2010-2018/counties/totals/	Used to scale various variables

(continued)

Table 1.4 (continued)

Data category	Source	Comments
County attributes	Selected variables from the COVID Severity Forecasting project. UC Berkeley Departments of Statistics, EECS led by Professor Bin Yu https://github.com/Yu-Group/covid19-severity-prediction. See also https://www.stat.berkeley.edu/~binyu/ps/papers2020/covid19_paper.pdf	List of variables in table below
Economic characteristics	County level data from USDA—https://www.ers.usda.gov/data-products	3 county coding schemes described in table below
Outcomes (deaths, cases, days since first case)	Cumulative values by date downloaded from https://github.com/nytimes/covid-19-data and are as of 5/11/2020	5/11 values for cases and deaths extracted. Converted to per 100K. Days since first case computed by using first case date

Table 1.5 Additional variable details

Variable	Category	Description
PCT_POVERTY	Demographic	% below poverty
PCT_OWNEDHOM	Demographic	% owning home
PCT_ED_HS	Demographic	% with high school education
PCT_BLACK	Demographic	% black
PCT_HISP	Demographic	% hispanic
MED_INCOMERATIO	Demographic	Median income, converted to ratio relative to mean over counties
MED_HOMERATIO	Demographic	Median home value, converted to ratio relative to mean over counties
PCT_65PLUS	Demographic	% 65+ years
PCT_45TO64	Demographic	% 45–64 years
PCT_15TO44	Demographic	% 15–44 years
Rural-urban_ContinuumCode_2013	Economic characteristics	1–9 code indicating county degree of urbanization. https://www.ers.usda.gov/data-products/rural-urban-continuum-codes. Created binary column for each level
Urban_Influence_Code_2013	Economic characteristics	1–12 code indicating county degree of urban influence. https://www.ers.usda.gov/data-products/urban-influence-codes. Created binary column for each level
Economic_typology_2015	Economic characteristics	1–6 code indicating county economic condition. https://www.ers.usda.gov/data-products/county-typology-codes. Created binary column for each level

(continued)

Table 1.5 (continued)

Variable	Category	Description
PopDensity[Log]	County population	2018 population estimate divided by land area (square miles) from shape files. Log version also
CensusRegionName	County attributes	Created binary column for each level
CensusDivisionName	County attributes	Created binary column for each level
StateName	County attributes	Created binary column for each level
dem_to_rep_ratio	County attributes	Ratio of registered democrats to republicans in county
#ICU_beds100K[Log]	County attributes	Number of ICU beds per 100K population. Log version also

References

Bareinboim E, Pearl J (2013) Causal transportability with limited experiments. In: Proceedings of the 27th AAAI conference on artificial intelligence, pp 95–101

Bashir MF, Ma BJ, Bilal A, et al. Correlation between environmental pollution indicators and COVID-19 pandemic: a brief study in Californian context. Environ Res. 2020;187:109652. https://doi.org/10.1016/j.envres.2020.109652.

Burns J, Boogaard H, Polus S, Pfadenhauer LM, Rohwer AC, van Erp AM, Turley R, Rehfuess E. Interventions to reduce ambient particulate matter air pollution and their effect on health. Cochrane Database Syst Rev. 2019;5:CD010919. https://doi.org/10.1002/14651858.CD010919.pub2.

Burns J, Boogaard H, Polus S, Pfadenhauer LM, Rohwer AC, van Erp AM, Turley R, Rehfuess E. Interventions to reduce ambient air pollution and their effects on health: an abridged Cochrane systematic review. Environ Int. 2020;135:105400. https://doi.org/10.1016/j.envint.2019.105400.

Campbell DT, Stanley JC. Experimental and quasi-experimental designs for research. Boston: Houghton Mifflin Company; 1963.

Clancy L, Goodman P, Sinclair H, Dockery DW. Effect of air-pollution control on death rates in Dublin, Ireland: an intervention study. Lancet. 2002;360(9341):1210–4.

Clean Air Scientific Advisory Committee (CASAC) (2019) CASAC review of the EPA's policy assessment for the review of the National Ambient Air Quality Standards for Particulate Matter (External Review Draft—September 2019). https://yosemite.epa.gov/sab%5Csabproduct.nsf/E2F6C71737201612852584D20069DFB1/$File/EPA-CASAC-20-001.pdf

Cox LA Jr. Do causal concentration-response functions exist? A critical review of associational and causal relations between fine particulate matter and mortality. Crit Rev Toxicol. 2017;47(7):603–31. https://doi.org/10.1080/10408444.2017.1311838.

Cox LA Jr. Modernizing the Bradford Hill criteria for assessing causal relationships in observational data. Crit Rev Toxicol. 2018;15:1–31. https://doi.org/10.1080/10408444.2018.1518404.

Cromar K, Ewart G. Concentration-response associations used to estimate public health benefits of less pollution are not valid causal predictive models. Ann Am Thorac Soc. 2016;13(12):2281. https://doi.org/10.1513/AnnalsATS.201610-754LE.

Dockery DW, Rich DQ, Goodman PG, Clancy L, Ohman-Strickland P, George P, Kotlov T, HEI Health Review Committee. Effect of air pollution control on mortality and hospital admissions in Ireland. Res Rep Health Eff Inst. 2013;176:3–109.

Dominici F, Greenstone M, Sunstein CR. Science and regulation. Particulate matter matters. Science. 2014;344(6181):257–9. https://doi.org/10.1126/science.1247348.

Drugmond D (2020) EPA clean air panel chair dismisses his oil industry ties, slams Harvard study on air pollution and COVID risks. https://www.desmogblog.com/2020/05/18/epa-clean-air-committee-cox-harvard-air-pollution-covid

Friedman L (2020). https://www.nytimes.com/2020/04/07/climate/air-pollution-coronavirus-covid.html

Goldman GT, Dominici F (2019) Don't abandon evidence and process on air pollution policy. Science. https://doi.org/10.1126/science.aaw9460

Goodman GE, Thornquist MD, Balmes J, Cullen MR, Meyskens FL Jr, Omenn GS, Valanis B, Williams JH Jr. The beta-carotene and retinol efficacy trial: incidence of lung cancer and cardiovascular disease mortality during 6-year follow-up after stopping beta-carotene and retinol supplements. J Natl Cancer Inst. 2004;96(23):1743–50.

Henneman LR, Liu C, Mulholland JA, Russell AG. Evaluating the effectiveness of air quality regulations: a review of accountability studies and frameworks. J Air Waste Manag Assoc. 2017;67(2):144–72. https://doi.org/10.1080/10962247.2016.1242518.

Jiang Y, Wu XJ, Guan YJ. Effect of ambient air pollutants and meteorological variables on COVID-19 incidence. Infect Control Hosp Epidemiol. 2020;222:1–11. https://doi.org/10.1017/ice.2020.222.

Kahneman D. Thinking, fast and slow. New York: Farrar, Straus, and Giroux; 2011.

Krimsky S. The weight of scientific evidence in policy and law. Am J Public Health. 2005;95(S1):129–36.

Linkov I, Massey O, Keisler J, Rusyn I, Hartung T. From "weight of evidence" to quantitative data integration using multicriteria decision analysis and Bayesian methods. ALTEX. 2015;32(1):3–8. https://doi.org/10.14573/altex.1412231.

North DW. Commentary on "Should health risks of air pollution be studied scientifically?" by Louis Anthony Cox, Jr. Global Epidemiol. 2020;2:100021. https://doi.org/10.1016/j.gloepi.2020.100021.

Omenn GS, Goodman GE, Thornquist MD, Balmes J, Cullen MR, Glass A, Keogh JP, Meyskens FL Jr, Valanis B, Williams JH Jr, Barnhart S, Cherniack MG, Brodkin CA, Hammar S. Risk factors for lung cancer and for intervention effects in CARET, the Beta-Carotene and Retinol Efficacy Trial. J Natl Cancer Inst. 1996;88(21):1550–9.

Pearl J. Causal inference in statistics: an overview. Stat Surveys. 2009;3:96–146.

Peters J, Bühlmann P, Meinshausen N. Causal inference by using invariant prediction: identification and confidence intervals. J R Stat Soc Ser B. 2016;78(5):947–1012.

Schwartz S, Gatto NM, Campbell UB. Transportabilty and causal generalization. Epidemiology. 2011;22(5):745–6.

Spirtes P. Introduction to causal inference. J Mach Learn Res. 2010;11:1643–62.

Stenlund T, Lidén E, Andersson K, Garvill J, Nordin S. Annoyance and health symptoms and their influencing factors: a population-based air pollution intervention study. Public Health. 2009;123(4):339–45. https://doi.org/10.1016/j.puhe.2008.12.021.

Textor J, van der Zander B, Gilthorpe MS, Liskiewicz M, Ellison GT. Robust causal inference using directed acyclic graphs: the R package 'dagitty'. Int J Epidemiol. 2016;45(6):1887–94.

Tollefson J (2019) Air pollution science under siege at US environment agency Top EPA adviser attacks agency decision-making ahead of major review of air pollution standards. Nature. https://www.nature.com/articles/d41586-019-00937-w

Triantafillou S, Tsamardinos I. Constraint-based causal discovery from multiple interventions over overlapping variable sets. J Mach Learn Res. 2015;16:2147–205.

Wu X, Nethery RC, Sabath BM, Braun D, Dominic F (2020) Exposure to air pollution and COVID-19 mortality in the United States: a nationwide cross-sectional study. medRxiv 2020.04.05.20054502

Zhang L, Zhang Z, Wang C, et al. Int J Disaster Risk Sci. 2017;8:445. https://doi.org/10.1007/s13753-017-0149-2.

Zigler CM, Dominici F. Point: clarifying policy evidence with potential-outcomes thinking--beyond exposure-response estimation in air pollution epidemiology. Am J Epidemiol. 2014;180(12):1133–40. https://doi.org/10.1093/aje/kwu263.

Chapter 2
Modeling Nonlinear Dose-Response Functions: Regression, Simulation, and Causal Networks

Introduction

Recent advances in understanding of biological mechanisms and adverse outcome pathways for many exposure-related diseases show that certain common mechanisms involve thresholds and nonlinearities in biological exposure concentration-response (C-R) functions. These range from ultrasensitive molecular switches in signaling pathways, to assembly and activation of inflammasomes, to rupture of lysosomes and pyroptosis of cells. Realistic dose-response modeling and risk analysis must confront the reality of nonlinear C-R functions. This chapter reviews several challenges for traditional statistical regression modeling of C-R functions with thresholds and nonlinearities, together with methods for overcoming them. As mentioned in Chap. 1, statistically significantly positive exposure-response regression coefficients can arise from many non-causal sources such as model specification errors, incompletely controlled confounding, exposure estimation errors, attribution of interactions to factors, associations among explanatory variables, or coincident historical trends. This chapter discusses and illustrates these sources of positive C-R regression coefficients, explaining why the unadjusted regression coefficients do not necessarily predict how or whether reducing exposure would reduce risk. It discusses statistical options for controlling for such threats, and proposes that causal Bayesian networks and dynamic simulation models can be valuable complements to nonparametric regression modeling for assessing causally interpretable nonlinear C-R functions and understanding how time patterns of exposures affect risk. These approaches are promising for extending the great advances made in statistical C-R modeling methods in recent decades to clarify how to design regulations that are more causally effective in protecting human health.

Nonlinearities in exposure concentration-response (C-R) functions can wreak havoc on traditional statistical risk modeling developed for linear no-threshold (LNT) modeling assumptions. Nonlinearity in an agent's causation of a health

L. A. Cox Jr., *Quantitative Risk Analysis of Air Pollution Health Effects*, International Series in Operations Research & Management Science 299,
https://doi.org/10.1007/978-3-030-57358-4_2

endpoint implies that no single slope coefficient necessarily characterizes the change in risk from a given change in exposure. Nonlinearities in interactions of the agent with other factors (such as co-exposures, co-morbidities, or covariates that modify the agent's effect) and dependencies among these factors, complicate the interpretation and estimation of slope factors or of entire concentration-response (C-R) curves that seek to quantify how a health endpoint depends on exposure. At a minimum, it becomes necessary to specify what is assumed about the levels of other factors, and about how (if at all) they change when exposure is changed—for example, how changing one component of a mix of pollutants changes other components that affect the same health endpoint. More generally, in both linear and nonlinear models, failing to characterize causal pathways other than those leading directly from the agent to the effect, such as indirect (mediated) pathways, or exposure-response associations due to common causes (e.g., confounders) or to common effects (e.g., sample selection criteria), can make it difficult or impossible to determine how changing exposure would change response probabilities. The possibility of nonlinearity exacerbates C-R estimation problems if high-order interactions among factors must be considered, and no small number of parameters in a simple model form can be assumed to adequately represent the data-generating process.

Yet, nonlinear C-R functions are prevalent in practice. They require new ways to carry out each of the health risk assessment steps of hazard identification, dose-response modeling, exposure assessment, risk characterization, and uncertainty characterization. They also have strong implications for how to communicate risk accurately, and for how to manage risk effectively. This chapter reviews challenges for risk analysis posed by nonlinearity, and discusses constructive methods to meet these challenges using current techniques of data science and causal analytics. It focuses on techniques for using epidemiological data—typically, exposure, response, and covariate variables measured in a population over time—to identify and quantify health risks caused by exposures when underlying individual-level causal biological C-R functions are nonlinear. As an organizing framework, the chapter largely follows the practice of regulatory occupational risk assessment recently described by the National Institute for Occupational Safety and Health (NIOSH 2020) in a clear, thoughtful exposition of key principles and practices of current regulatory risk assessment. The following sections consider how to extend and apply these principles to nonlinear C-R functions with possible interactions and confounding, using simple examples to illustrate and clarify the main technical issues.

Why Does Nonlinearity Matter?

Advances in biological understanding of adverse outcome pathways and mechanisms have identified many sources of strong nonlinearities and thresholds in C-R functions. Examples include ultrasensitive molecular switches in key signaling pathways (Bogen 2019); positive feedback, cooperativity and bistability in regulatory networks and in dynamic processes such as assembly and activation of inflammasomes; discontinuous changes (e.g., rupture of lysosomes, ion fluxes, loss of organelle or cell membrane integrity, onset of pyroptosis); and saturation or depletion of protective (homeostasis-preserving) resources such as antioxidant pools in target cells and tissues (Cox Jr. 2018a). At the population level, important nonlinearities have been found in statistical models of associations between air pollutant concentrations, temperature, and mortality rates (Armstrong 2006; Li et al. 2019; Thurston and Ito 2001). The point of departure for the following sections is the need for realistic risk analyses to use data analysis methods that are appropriate for such nonlinearities. Traditional regression modeling relating exposures to risks may give misleading results when the underlying C-R functions are nonlinear. Each of the following sections describes challenges posed by nonlinearity and then discusses data analysis techniques for overcoming these challenges.

Hazard Identification

The central question of *hazard identification* is whether exposure to a substance causes increased risk of adverse health effects in at least some members of the exposed population. "Risk" for an individual is the probability that an adverse effect occurs in a given time interval. (Equivalently, it can be expressed as an age-specific hazard function, giving the expected rate of occurrence per unit time, given that it has not already occurred. Formulas from survival data analysis allow probabilities of occurrence by a given time or age, or within a specified interval, to be calculated from the age-specific hazard function, and *vice versa*.) NIOSH (2020) describes hazard identification as "the systematic process for assessing the weight of evidence on whether an agent of interest causes an adverse effect in exposed workers. The findings from hazard identification are characteristic descriptions and information on the exposures of interest, any important cofactors (e.g., other risk factors, moderating factors, mediating factors, or confounders); modes and mechanisms of action; and conditions (e.g., pre-existing diseases) under which changes in exposures change the probabilities or timing of adverse effects." Thus, we will consider that hazard identification addresses the following questions:

- "Risk *of* what?"—the adverse effect(s) of interest;
- "Risk *from* what?"—the source of risk, i.e., the hazard, of interest;
- "Risk *to* whom?"—the exposed population of interest;
- "Risk *under what conditions*?"—the context of conditions, such as co-exposures, co-morbidities, and sociodemographic covariates, under which risk is assessed; and
- "Risk *via what mechanisms*?—the causal mechanisms or pathways by which effects of changes in exposure to the hazard are transmitted to changes in risks of adverse health effects.

Hazard identification should clarify these defining elements of a risk—its source, target, effects, and mechanisms—in enough detail to uniquely specify the risk being assessed and to support quantitative assessment of how the conditional probability (or hazard rate) for occurrence of the effect in a stated interval changes in response to changes in exposure, given the values of other causally relevant variables.

Challenges for Regression-Based Hazard Identification

A common approach to hazard identification based on epidemiological data is to fit a regression model to data on estimated exposure levels and response rates (e.g., average numbers of mortalities or morbidities per person-year) and to test the null hypothesis that the regression coefficient relating exposure to response rate is 0. Rejection of this null hypothesis supports weight-of-evidence (WoE) determinations of a causal relationship between exposure and response, especially if plausible confounders have been controlled for by including them as predictor in the regression model. (Similarly for risk ratios, confident rejection of the null hypothesis that a risk ratio is 1 (no difference) in risk between more-exposed and less-exposed people in statistical analyses of relative risks, standardized mortality ratios, odds ratios in logistic regression models, or hazard ratios in Cox proportional hazards regression models, is usually interpreted as evidence supporting a causal relationship in weight-of-evidence (WoE) determinations, especially if chance, confounding, and biases have been ruled out as plausible explanations.) Against this practice is a fundamental objection that regression models quantify associations rather than causal impacts. Specifically, they quantify the conditional expected (average) value of the dependent variable, given the observed values of predictors, but this is different from answering the causal question of how or whether changing one or more of the predictors, such as exposure concentration, would change the dependent variable (Pearl 2009; Pearl and Mackenzie 2018).

In practice, as discussed in NIOSH (2020), many statistical issues also complicate the interpretation of regression models and of non-zero regression coefficients. These challenges include model misspecification errors, exposure estimation errors, omitted (unobserved, latent) variables, missing data, inter-individual heterogeneity and variability in causal dose-response functions, correlations and statistical dependencies among predictors, attribution of interactions, internal validity of study designs and conclusions (i.e., do the causal conclusions or interpretations follow from the data collected), and generalization and external validity of conclusions (i.e., their applicability in contexts other than those of the original studies). Large technical literatures in statistics and data science have developed to address these issues. Constructive approaches have matured enough to create practical algorithms and software.

Table 2.1 summarizes some key approaches and provides links for practitioners to R packages (vetted, documented, and maintained via the CRAN repository, https://cran.r-project.org/), where technical details and documentation can be found.

Table 2.1 Statistical techniques for commonly encountered data imperfections

Data/study imperfection	Examples of appropriate techniques and software
Model misspecification errors; unknown shapes of exposure-response dependencies	Flexible nonparametric models (e.g., MARS, https://cran.r-project.org/web/packages/earth/earth.pdf) and deep learning; non-parametric model ensembles (e.g., random forest, https://cran.r-project.org/web/packages/randomForest/randomForest.pdf) and superlearning (https://rdrr.io/cran/SuperLearner/f/vignettes/Guide-to-SuperLearner.Rmd) for model combination
Exposure estimation errors and errors in estimated or measured covariates (explanatory variables)	Errors-in-variables methods (e.g., the MMC package in R, https://cran.r-project.org/web/packages/mmc/mmc.pdf; see also https://www.jstatsoft.org/article/view/v048i02, https://cran.r-project.org/web/packages/GLSME/GLSME.pdf, https://arxiv.org/pdf/1510.07123.pdf)
Omitted variables; unobserved or unmeasured risk factors, confounders, and modifiers	Latent variable techniques and finite mixture distribution modeling methods (e.g., www.jstatsoft.org/article/view/v011i08; https://www.jstatsoft.org/article/view/v048i02; PROC CALIS in SAS)
Missing data values	Multiple imputation algorithms (e.g., MICE, https://cran.r-project.org/web/packages/mice/mice.pdf); data augmentation and EM (expectation-maximization) algorithms
Inter-individual heterogeneity and variability in causal exposure-response curves	Finite mixture distribution modeling, clustering, individual conditional expectation methods (e.g., https://cran.r-project.org/web/packages/ICEbox/ICEbox.pdf)
Correlated or interdependent explanatory variables	Probabilistic graphical methods, e.g., Bayesian networks (https://cran.r-project.org/web/packages/bnlearn/bnlearn.pdf; https://cran.r-project.org/web/packages/CompareCausalNetworks/index.html)
Interactions among risk factors or other explanatory variables	Nonparametric detection, estimation, and visualization of interactions (https://rdrr.io/cran/npIntFactRep/; https://rdrr.io/cran/npregfast/)
Uncertain internal validity (soundness of causal inferences)	Use quasi-experiment designs (or randomization and design of experiments where possible) to control for standard threats to internal validity, e.g., using PlanOut and PlanAlyzer software (https://hci.stanford.edu/publications/2014/planout/planout-www2014.pdf; https://dl.acm.org/doi/pdf/10.1145/3360608)
Uncertain external validity (generalizability of findings)	Multisite causal mediation analysis (https://cran.r-project.org/web/packages/MultisiteMediation/index.html), Bayesian evidence synthesis and hierarchical meta-analysis (https://cran.r-project.org/web/packages/jarbes/index.html)

Despite this progress, most of these methods are not yet widely used in regulatory risk assessment, which, as described in Chap. 1, often relies instead on the judgments of analysts to try to deal with limitations in study designs and data (NIOSH 2020). These difficulties can be exacerbated by nonlinearities, as illustrated in the following paragraphs.

Significant Regression Coefficients Arising from Trends and from Omitted Confounders

Statistically significant C-R associations and non-zero regression coefficients linking exposure to response probability can arise from many sources, even in the absence of a causal relationship between them. For example, if two variables, such as exposure and risk, follow statistically independent random walks, then regressing one against the other will usually produce a statistically significant regression coefficient between them, even though neither causes the other (Yule 1926). Likewise, coincidental historical trends can induce C-R associations without causation.

In ordinary least-squares (OLS) regression, a regression coefficient for exposure will differ significantly from zero if conditioning on exposure significantly reduces the mean squared error (MSE) of the values predicted by the regression model. This can happen for many reasons, even if the predicted variable does not depend on exposure. Perhaps the best known reason is that a *confounder*—a variable that makes both exposure and the response more likely when it is present (or, more generally, that shifts both their cumulative distribution functions rightward)—can induce a positive regression coefficient for exposure in a regression model that includes exposure but not the confounder. For example, suppose that cigarette smoking, exposure indicators such as blood concentrations of heavy metals (e.g., lead or cadmium), and response indicators such as age-specific mortality or morbidity are all mutually positively correlated in a data set. Then a regression model that omitted smoking could show positive regression coefficients for the exposure indicators, whether or not response risk depends directly on them, if they also act as surrogates for smoking, which directly affects risk. In this case, smoking would be a confounder for the estimated exposure-response association. In current practice, it is perhaps unlikely that such an obvious confounder would be omitted, unless the data were unavailable. However, fully controlling for effects of confounders can be surprisingly difficult, especially when linear models cannot be assumed.

The standard way to deal with a measured confounder is to include it as a predictor in the regression model. The estimate of the coefficient for exposure is then said to have been "controlled" or "adjusted" for the confounder. However, this tactic often fails to fully control for effects of confounding, for reasons discussed in the following sections on measurement errors, model specification errors, residual confounding, surrogate variables, variable selection, competing explanations, and attribution of joint effects. For example, indicators of smoking such as self-reported pack years and cotinine levels, are often imperfectly accurate (Hsieh et al. 2011). Residual effect of confounding might then still contribute to a positive regression coefficient for exposure. Even more challenging is the problem of omitted confounders (also called latent confounders or unobserved confounders)—that is, confounders that are not included in a regression model, perhaps because they were not measured. A useful current practice is to quantify how strong the effects of omitted confounders on exposure and risk would have to be to explain away the estimated effect of exposure on response. If the required effect sizes are much larger than those for measured confounders, then this suggests that any omitted confounder(s) would have to be stronger than the measured ones to provide a plausible alternative explanation for the estimated exposure-response association.

Significant Regression Coefficients Arising from Measurement Errors in Confounders

If a confounder is measured or estimated with some error, then including it on the right side of a regression model will typically not fully control for it, and exposure will still have a significant positive regression coefficient in large data sets. As a simple hypothetical illustrative example to clarify concepts, suppose that, unbeknownst to the risk modeler, the true relationship between a measure of health risk, R, and past pack-years of smoking, S, is the LNT structural equation $R = 0.01*S$; and that an exposure variable X (such as concentration of a metal in blood or urine) is also related to S by the equation $X = 0.02*S^{1/2}$; thus, S confounds the association between X and R. Consider what will happen if a multiple linear regression model is fit to data on R, S, and X if the estimated values of S and X values are unbiased but have uniformly distributed estimation errors. Specifically, suppose that estimated values are uniformly distributed between zero and twice their corresponding true values. Fitting the regression model $\text{E}(R \mid S, X) = b_0 + b_S S + b_X X$ to a simulated data set with 1000 cases having S values independently uniformly distributed between 0 and 1 yields the estimated regression model $\text{E}(R \mid S, X) = 0.002 + 0.0035*S + 0.094*X$. The intercept and both regression coefficients are significantly greater than 0 ($p < 0.00001$). By contrast, the correct causal relationship with accurately measured variables would be $\text{E}(R \mid S, X) = 0.01*S$ (or, $0.01*S + 0*X$). Thus, measurement error has induced a significant positive exposure coefficient ($b_X = 0.094$) for X, even though the regression model included the confounder S on its right side. Intuitively, the reason is that measurements of X provide useful information for reducing the mean squared prediction error when R is predicted from S alone, because of the measurement error in S. However, the positive regression coefficient does not represent a dependence of risk on exposure.

As another example, if two variables *exposure* and *risk* have a common cause such as *income*, but neither *exposure* nor *risk* causes the other (so that the relevant probabilistic graph model is *risk* ← *income* → *exposure*), and if variables are measured or estimated with error, then the multiple linear regression model

$$E\left(risk \mid, exposure \mid, income\right) = a * exposure + b * income,$$

may still show a significant regression coefficient, *a*, for *exposure*. Indeed, if the measurement error for *income* is large and that for *exposure* is small, regression modeling will conclude that *b* is not significantly different from 0 but that *a* is. Attempting to control for the confounder *income* by including it in the regression model fails if conditioning on the measured values of *exposure* helps to reduce prediction error for *risk* by reducing the effects of measurement error (essentially because measuring *exposure* reduces uncertainty about the true value of *income*).

More generally, regression selects values of regression coefficients to reduce the mean squared error of predicted values. Thus, it can give a significant coefficient to *exposure* if doing so reduces the contribution of measurement error to prediction error, whether or not exposure makes a causal contribution to *risk*. This is true for both linear and nonlinear models. A somewhat analogous phenomenon, examined next, arises specifically from unmodeled nonlinearity in exposure-response relationships.

Significant Regression Coefficients Arising from Model Specification Errors

Regression models can fail to fully control for confounding, even if explanatory variables, including potential confounders, are measured without errors, if the models assume a shape (e.g., linear) for the relationship between explanatory and dependent variables that differs from their empirical relationship. This is the problem of *model specification error* or model misspecification: the specified model form does not perfectly describe the data. As a simple hypothetical example, if $R = S^3$ and $X = S^2$, where S is uniformly distributed between 0 and 1, then fitting the (incorrectly specified) multiple linear model $E(R \mid S, X) = b_0 + b_S S + b_X X$ to a simulated data set of size $N = 1000$ cases produces a least-squares fit of $E(R \mid S, X) = 0.05 + 1.5{*}X - 0.60{*}S$, with an R^2 value of 0.99 and all coefficients and the intercept significantly different from zero ($p < 0.00001$). Although R depends only on the confounder S and not on exposure X, controlling for S by including it on the right side of the regression model does not fully control for its confounding effects, or preclude a statistically significant positive regression coefficient for X. Intuitively, the reason is that the shape of the assumed model (in this case, linear) is not correct, and including X and well as S helps to reduce prediction errors due to model specification error. However, although conditioning on X as well as S leads to an excellent fit as assessed by the R^2 value of 0.99, neither this high R^2 nor the small p value of the regression coefficient for X indicates that risk depends on X as well as on S. Analogous examples arise for dichotomous outcomes such as mortality. If $R = 1$ whenever $S > 0.5$, with S uniformly distributed between 0 and 1, and R is otherwise 0; and if X is uniformly distributed between 0 and 1 if $S > 0.5$, and is otherwise 0 (e.g., if X is an exposure marker that is only formed when $S > 0.5$), then the regression coefficient for X will be significantly positive (and larger than that for S) in a logistic regression model for R with X and S as predictors, even though R does not depend on X.

Model misspecification is often present even when goodness-of-fit tests do not reject the specified model form in favor of specified alternatives. A practical illustration, developed in detail in Chap. 11, concerns hazard identification of the metal molybdenum (*Mo*) as a hazard that might cause decreased testosterone (*T*) in men. For example, Lewis and Meeker (2015) presented a multiple linear regression of the dependent variable *log(T)* against the predictors *log(Mo)*, *BMI* (body mass index), and others, in order to estimate the statistical effect of *log(Mo)* on *log(T)* while controlling for potential confounding by *BMI*. (For simplicity, we focus on these three variables; the full model also included age and other potential confounders.) The multiple linear regression model showed statistically significant negative regression coefficients for both *log(Mo)* and *BMI*, i.e., greater values of each are associated with smaller values of *T* (and *log(T)*), given the value of the other. This led to the tentative identification of *Mo* as a reproductive hazard. As the authors state, "In adjusted analyses where metals were modeled as a continuous variable, we found significant inverse associations between urinary molybdenum and serum copper and serum testosterone.…These findings add to the limited human evidence that exposure to molybdenum and other metals is associated with altered testosterone in men, which may have important implications for male health."

Accounting for nonlinearity changes this conclusion. Figure 2.1, which is discussed further in Chap. 11, shows scatter plots of *log(T)* (upper, blue scatter plot) and *log(Mo)* (lower, red scatter plot) against *BMI* in kg/m^2 with linear regression lines and nonparametric (smoothing) regression curves superimposed on the data so that departures from linearity can be seen easily. (The data are for men aged 18–55

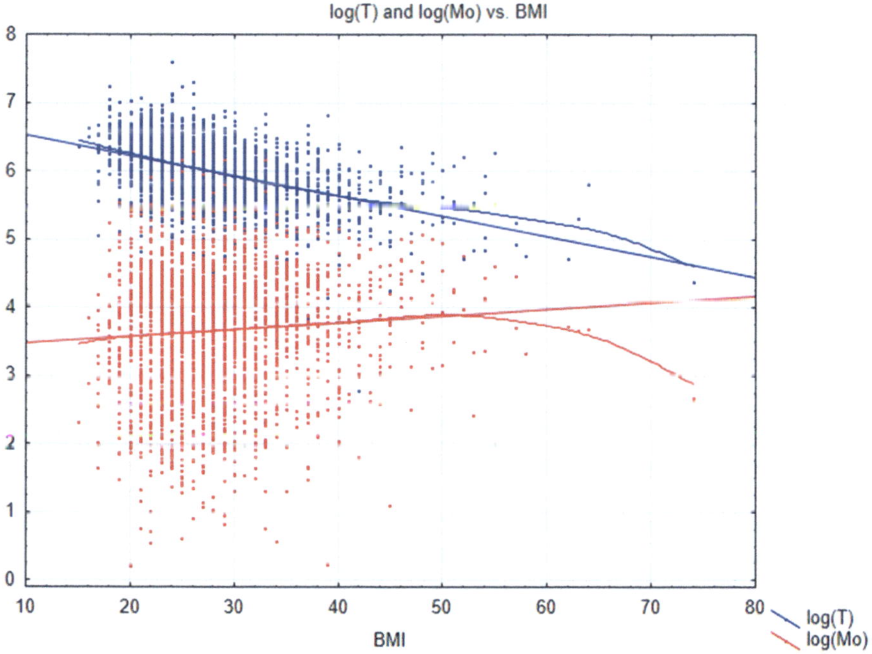

Fig. 2.1 Linear and nonlinear (smoothing) regressions for *log(T)* and *log(Mo)* as functions of *BMI*. The nonlinear curves are fit to the data using locally weighted scatterplot smoothing (LOWESS)

from the National Health and Nutrition Examination Survey (NHANES) for 2011–2016.) Although the nonparametric regression curves are close to the linear regression lines for BMI between 25 and 35, deviations from linearity occur at relatively high and low values of BMI. These nonlinearities allow the mean squared error (MSE) of *log(T)* values predicted from *BMI* using the regression line to be reduced by including *log(Mo)* as an additional predictor. (Subtracting a multiple of *log(Mo)* from the linearly predicted value of *log(T)* reduces the MSE caused by the departures from linearity.) A significant regression coefficient for *log(Mo)* arises because including *log(Mo)* as a predictor of *log(T)* reduces the prediction error due to model misspecification. It does not provide evidence about whether reducing *Mo* would increase *T*. (Analogously, if $Y = Z^2$ and $X = Z^{0.5}$, where *Z* is uniformly distributed between 0 and 2, then a multiple linear regression model for predicting *Y* from both *X* and *Z* has a lower prediction error (MSE) than one for predicting *Y* from Z alone, even though only *Z* and not *X* determines the value of *Y*.) Restricting the range of BMI values considered to the interval from 25 to 35, where the linear and nonparametric regression models nearly coincide, makes *log(Mo)* no longer a significant predictor of *log(T)* in the linear multiple regression model, suggesting that fully controlling for confounding by BMI eliminates the association between *log(Mo)* and *log(T)*. A significant regression coefficient that does not reflect a causal relationship, or that is eliminated by fully controlling for confounding, does not provide a sound basis for causal inference for hazard identification.

In summary, analyzing data with regression models that do not perfectly describe the data-generating process can lead to statistically significant regression coefficients even in the absence of any causal relationship. Much as a significant regression coefficient for exposure can simply indicate that including exposure reduces prediction errors due to measurement errors, it may also indicate that including exposure reduces prediction errors due to model misspecification. Neither has any necessary implications for hazard identification.

Significant Regression Coefficients Arising from Residual Confounding

It is common to refer to a confounder as having been controlled, or adjusted for, when a variable representing it has been included on the right side of a regression model, even if not all relevant information about it has been captured. For example, a variable such as "smoking status" with possible values of current, former, or never (or, even more simply, a binary value such as 1 for "has smoked at least 100 cigarettes in life" and 0 otherwise) might be used to "adjust" for smoking. Yet, such a summary variable leaves much quantitative information about smoking intensity and duration unaccounted for. Likewise, a regression model which adjusts for age by including a binary indicator such as 1 for over 65 years old, else 0, by including 5-year or 10-year age categories, leaves more detailed information about age unrepresented. Such omitted information about a confounder may induce a positive regression coefficient for exposure, even if risk does not depend on exposure; this is the problem of *residual confounding*. For example, if $Risk = (Age/100)^2$ for $0 < Age < 100$ years; $Exposure = Age$; and the correctly specified regression model $E(Risk \mid Exposure, Age) = b_0 + b_X{*}Exposure + b_A{*}Age^2$ is fit to 1000 randomly generated cases with *age* uniformly distributed between 0 and 100, then, in the absence of measurement error and sampling error, the result is as expected: $E(Risk \mid Exposure, Age) = (age/100)^2$. However, if age is rounded to the nearest decade, then fitting the same model yields $E(Risk \mid Exposure, Age) = -0.03 + 0.002{*}Exposure + 0.00008{*}Age^2$. The intercept and both regression coefficients are statistically significantly different from 0 ($p < 0.000001$), even though risk depends only on *Age* and not on *Exposure*. Intuitively, the reason is that conditioning on *Exposure* reduces the mean squared prediction error for *Risk* due to the limited accuracy of measurement of *Age*, by providing information about the precise value of *Age* within each age category formed by rounding age to the nearest decade. This additional information reduces prediction error.

Use of a few wide categories to code continuous variables is still fairly common in practice, despite decades of admonishment from methodologists (e.g., Streiner 2002; Naggara et al. 2011); hence the threat of residual confounding often has practical importance (Groenwold et al. 2013). For example, a recent study of the association between blood lead level (BLL) and age-specific mortality rate (Lanphear et al. 2018) used 3 categories for body mass index (BMI); 3 for self-reported smoking status (never, former, current); 2 for cotinine in blood serum (above or below 10 ng/mL); 2 for alcohol consumption; 3 for physical activity; 3 for cadmium; 2 for household income; 2 for hypertension; and 3 for a healthy eating index that runs from 1 to 100. Similarly, a regression analysis of the negative association between molybdenum (*Mo*) and testosterone (*T*) (Lewis and Meeker 2015) used 2 categories for BMI (below 25 kg/m^2 or not); 3 for income; and 2 for smoking. In such studies, the use of only a few categories for each of many potential confounders leaves unclear the extent to which reported exposure-response associations reflect residual confounding. For example, Fig. 2.1 shows that BMI is associated with *Mo* and *T* both above and below 25 kg/m^2, so use of a dichotomous BMI variable leaves this remaining (residual) confounding unaccounted for. The possibility of residual confounding does not necessarily imply that qualitative conclusions about exposure-response associations would change if residual confounding were better controlled, but it leaves open the question of how much they would change. To avoid this unnecessary uncertainty, it suffices to treat continuous variables as continuous, rather than artificially dichotomizing or categorizing them (Streiner 2002; Naggara et al. 2011).

Surrogate Variables

Similar to residual confounding, controlling for a confounder by including in the regression model a surrogate variable that is correlated with it does not fully eliminate its confounding effects. For example, including self-reported pack-years of smoking and/or measured blood levels of cotinine in a regression model as surrogates for smoking does not fully control for the confounding effects of smoking if exposure provides additional information about smoking (and hence helps to reduce mean squared prediction error for a health effect caused by smoking) even after other indicators of smoking have been included as predictors (Hsieh et al. 2011). Similarly, including county population density (average people per square mile) in a regression model for effects of PM2.5 on COVID-19 mortality risk to control for the possibility that more densely populated areas might have both higher PM2.5 pollution levels and higher COVID-19 mortality risk does not fully eliminate this source of potential confounding if local population densities differ from the county average.

Variable Selection

When predictors are correlated, including some on the right side of a regression model may prevent others from having a coefficient significantly different from zero. By choosing different subsets of other variables to include on the right side, modelers may affect the size of the regression coefficient for exposure, and, in some cases, even whether it is positive or negative (with each being significantly different from zero) (Dominici et al. 2014). In such cases, the results of the regression modeling are *model-dependent*: they reflect particular modeling choices rather than facts about the world. As a trivial example, suppose that the causal relationships among *Age*, *Exposure*, and *Risk* are described by the structural equations *E*(*Risk* | *Age*, *Exposure*) = *Age* - *Exposure* and *Exposure* = 0.5**Age*. Then the regression coefficient for *Exposure* is−1 if both *Age* and *Exposure* are included as predictors, but is +1 if only *Exposure* is included as a predictor. Including only *Exposure* is more parsimonious, and, once it has been selected, including *Age* does not improve predictive accuracy; thus *E*(*Risk* | *Age*, *Exposure*) = *Exposure* would be the preferred model by these criteria, even though the regression coefficient of +1 for *Exposure* reveals nothing about how *Risk* would change if *Exposure* were changed.

Significant Regression Coefficients Arising from Competing Explanations

If healthy workers are more likely to move away from a factory town each year than unhealthy workers, then workers in a birth cohort who have stayed the longest and gained the most cumulative exposure to the factory town environment will be disproportionately unhealthy compared to workers in the same birth cohort who have moved away. This creates a positive association (reflected in a positive regression coefficient) between cumulative exposure and risk of poor health, even if exposure *per se* does not affect health. If poor health is a predictor of increased risk for some diseases (e.g., cancers or heart diseases) in old age, then retired workers with the high cumulative exposures will be more likely to develop such diseases because of underlying poor health, even if exposure has no causal impact on health. Another example of a non-causal explanation that does not involve underlying health status would be if workers stay in a certain occupation only if they are poor or have low exposures (or both). If poverty causes increased health risks but exposure has no direct causal effect on health risks, then a study of workers who have stayed in the occupation may find that those with low exposures are less likely to be poor (since low exposures provide an alternative explanation for the choice to stay) and hence have lower average health risks. If poverty is not measured, but exposure and health effects are, there will be a positive association between exposure and risk (mediated by the unmeasured variable poverty), even if exposure does not affect risk. These examples illustrate the fact that statistically significant positive regression coefficients between observed levels of exposure and risk need not imply that changing exposure would change risk, or that exposure is a contributing cause of risk.

Significant Regression Coefficients Arising from Attribution of Joint Effects

Suppose that a disease occurs in a worker if and only if the sum of three exposures A, B, and C exceeds 15, where the three exposures, perhaps corresponding to concentrations of three pollutants, are independent random variables uniformly distributed between 0 and 2, between 4 and 8, and between 5 and 9, respectively. Then a generalized linear regression model, such as logistic regression, will assign significantly positive regression coefficients to each of the three exposures if there are a large enough number of observations (each consisting of values of A, B, and C for an individual). This remains true even if the data are modified to set A = 0 whenever B + C < 15 (e.g., if pollutant A is formed only by sufficiently high levels of pollutants B and C). Yet, in this case, A makes no contribution to risk. Whether a disease occurs depends only on the values of B and C. Thus, regression can create a significant positive regression coefficients for an exposure as a predictor of risk by (mis)*attributing* part of the joint effect of multiple variables to it, even if has no causal impact on risk. (Special techniques that deal more consistently with attribution of risk in the presence of joint causes, such as Shapley regression, avoid this problem, but are seldom used in regulatory risk assessment.) Finally, if the distribution of either B or C (or both) is changed to a bimodal distribution that is equally likely to be 0 or 20, then the regression coefficient for A becomes zero (or, rather, is not significantly different from 0 in large data sets). Thus, whether regression analysis provides evidence that A has a significant positive regression coefficient depends on the distributions of other variables, rather than only on the causal biological effect (if any) of A itself. Hazard identification based on whether a regression coefficient is significantly different from zero may therefore be misleading when multiple risk factors interact in contributing to disease causation.

Some Alternatives to Regression for Hazard Identification

Although thoughtless interpretation of regression coefficients can be misleading, numerous regression diagnostic plots and tests (e.g., Q-Q plots, homoscedasticity tests, all-subsets regression plots) have been developed to help assess how well a regression model describes the data to which is being applied. Flexible nonparametric regression models (using techniques such as locally estimated scatterplot smoothing (LOESS), locally weighted scatterplot smoothing (LOWESS), or splines) can also show where empirical relationships depart from parametric (e.g., linear or generalized linear) modeling assumptions, as in Fig. 2.1. Such techniques avoid the risk of overfitting inherent in many parametric models with (e.g., high-order polynomial regression models) by fitting simple low-order models to data in the neighborhood of each point of estimation (NIST 2013). For large data sets with many potential predictors, there are many alternatives to regression modeling; here we mention some of them, deferring to a large technical literature and recent surveys (see Chap. 9 and its references) for details. Non-parametric alternatives to regression for predicting risk from exposure and other variables include classification and regression tree (CART) analysis, which seeks to partition records into subsets with significantly different values of the dependent variable by conditioning on values (or ranges of values) of predictors; and random forest model ensembles that estimate *partial dependence plots* (PDPs) (see Chap. 9). A PDP shows how the average predicted value of the dependent variable (e.g., risk) changes as an explanatory variable of interest (e.g., exposure) is varied over its range, holding the values all other variables fixed at their observed levels in the data set. (This corresponds roughly to what epidemiologists term the natural direct effect of exposure on risk. Each predicted value is averaged over many CART trees fit to different subsets of the data.)

In reality, however, changing a single explanatory variable, such as the exposure concentration of a pollutant in ambient air, might cause the values of multiple other variables to change, making it unrealistic to hold their values fixed in assessing total effects on the dependent variable. To deal with this situation, causal Bayesian network (BN) propagate changes in the values of input variables (such as exposure) to changes in the conditional probability distributions of variables that depend on them. The full methodology for BN learning and inference from data is quite detailed; we refer the reader to Nagarajan et al. (2013) and Chap. 9 and its appendices for details. An arrow joins two variables in a BN if they are found to be dependent (i.e., mutually informative about each other, so that the null hypothesis of conditional independence—that the conditional probability distribution for one is the same for all values of the other—can be confidently rejected, e.g., using nonparametric tests for independence; see Nagarajan et al. 2013). If effects depend on their direct causes, then a BN provides evidence that *X* might be a direct cause of *Y* if and only if they are linked by an arrow. (Figure 2.2, discussed in the next section, is an example of a BN.) If *Y* is conditionally independent of *X* given the values of other variables (such as common causes), so that there is no arrow between them, then the BN provides no evidence that *X* might be a direct cause of *Y*.

A BN that includes exposure and risk variables, as well as other variables such as potential confounders and modifiers, can be used for hazard identification. Such a BN shows that exposure is a hazard, i.e., a potential direct cause of risk, if and only if exposure and risk are joined by an arrow. (The specific set of other variables that should be conditioned on—including confounders or common causes, but not common effects, of exposure and risk—to obtain an unbiased estimate of the effect of exposure on risk constitutes what is called an *adjustment set*. Adjustment sets can be calculated from BNs by modern causal analysis algorithms (see Chap. 9). This solves the problem of variable selection that often bedevils regression modeling. It does so by identifying minimal sufficient subsets of variables—the "adjustment sets"—to condition on, i.e., to include in a (perhaps nonparametric) regression model, to estimate direct and total effects of one variable on another while avoiding biases created by failure to condition on common ancestors, or by conditioning on common descendants (Textor et al. 2016).) Some causal analysis algorithms attempt to orient the arrows between variables in a BN to reflect the flow of causality (and information) between variables, but even without such orientation of arrows, the structure of a BN is useful for hazard identification insofar as it shows whether risk is found to depend on exposure after adjusting for other variables.

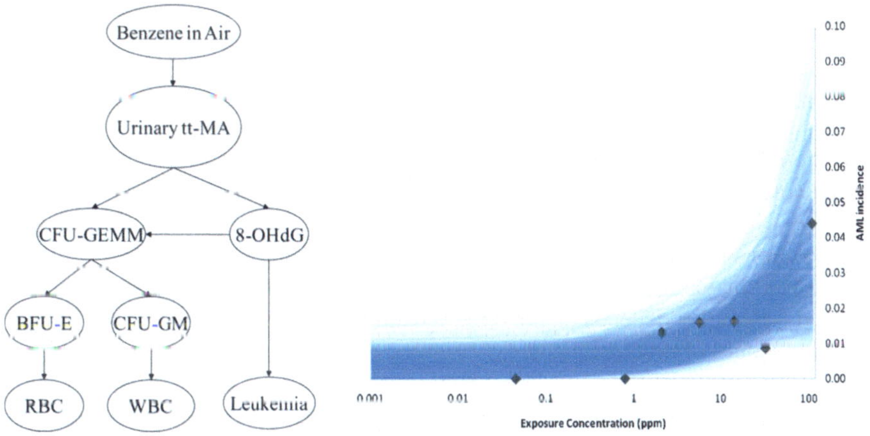

Fig. 2.2 Bayesian network (BN) model structure (left) and predictions (right). Abbreviations: *8-OHdG* 8-hydroxyguanosine (a biomarker of oxidative stress), *CFU-GEMM* colony-forming unit-granulocyte, erythrocyte, monocyte, megakaryocyte (a precursor to RBCs and WBCs), *BFU-E* burst-forming unit-erythroid (a RBC precursor cell type), *CFU-GM* colony forming unit—granulocyte-macrophage (a WBC precursor), *RBC* red blood cell count, *ttMA* trans,trans muconic acid, *WBC* white blood cell count. Diamonds on right indicate observed data. Blue curves are random samples from the uncertain dose-response relationship (*Source*: Hack et al. 2010)

BN analysis generalizes regression analysis in the following ways: (a) it models dependences among all variables in a data set, instead of only for a single dependent variable; (b) it quantifies dependences using nonparametric methods such as conditional probability tables or trees, which easily accommodate nonlinearities and complex interactions if all variables are discrete (although regression models are sometimes used for continuous variables); and (c) it quantifies the full conditional probability distribution of each variable, for any set of observed values for any subset of other variables, rather than only quantifying the conditional expected value of a single dependent variable given observed values for all other variables. BNs are therefore well suited for hazard identification in complex causal networks of many interacting variables, using the criterion that health effects should depend on exposures to hazards that directly cause them. Recent developments also address the challenge of synthesizing evidence across multiple studies—the important problem of *external validity* (or "transportability" or generalizability) of causal conclusions drawn from specific studies to other settings and conditions (Cox Jr. 2018a). BNs have also been used as an alternative to regression for quantitative risk assessment and dose-response modeling, as discussed next.

Dose-Response Modeling

Challenges for Regression-Based Dose-Response Modeling

While hazard identification provides a qualitative determination about whether there is evidence that exposure to a substance causes increased risk of an adverse health response, dose-response modeling quantifies how risk varies with exposure, typically using regression models. As stated by NIOSH (2020), "NIOSH generally obtains dose-response estimates via statistical models constructed to provide the conditional expectation of the dependent variable (the adverse effect) given one or more explanatory variables, but at least including the variable describing the agent exposure of interest. Model input data stem from toxicologic and/or epidemiologic investigations identified and assessed in hazard identification." A regression coefficient for estimated exposure is interpreted as providing information about how risk of an adverse effect depends on the observed value of exposure, conditioning on the observed values of other explanatory variables.

The limitations of regression modeling discussed in the previous section also apply to dose-response modeling (also called exposure-response modeling and exposure concentration-response (C-R) modeling) using epidemiological data. Regression describes associations among *observed* values of variables. These may not reveal how or whether risk would *change* if exposure were changed (with or without holding values of other variables fixed) (Pearl 2009). As already discussed, a significantly positive regression coefficient for exposure may arise simply because conditioning on measured values of exposure reduces the prediction error (MSE) from model specification errors or from measurement errors in other predictors. It may arise from attribution of joint effects, or from competing explanations, or from coincident historical trends. Dose-response models derived by regression modeling of epidemiological data reflect these non-causal sources of association, as well as any causal contributions. Hence, they cannot necessarily be used to predict by how much (or whether) a reduction in exposure would reduce risk. Although they are commonly used for this purpose in current regulatory risk assessments (NIOSH 2020), such use is not necessarily sound unless these various non-causal contributions to regression relationships are identified and corrected for.

The next two sections describe two alternatives to regression analysis that focus more explicitly on causality: causal Bayesian networks and dynamic simulation modeling. A causal BN provides a high-level description of how the conditional probability distribution of response varies with exposure and other variables, allowing Bayesian inference from observations on biomarkers or other variables in the network (Hack et al. 2010). Dynamic simulation models use systems of differential equations and algebraic formulas to model (a) flows of chemicals and metabolites among tissues; (b) internal doses (concentrations of toxic metabolites in target organs and tissues) over time, and (c) resulting rates of cell death and proliferation and transitions of cells among various states (e.g., normal, pre-malignant, and malignant) over time (Cox Jr. 2020). The following two sections highlight the main ideas of these methods as they apply to dose-response modeling, relegating their (extensive) mathematical details to the references.

Bayesian Networks for Dose-Response Modeling

Chapter 12 discusses Bayesian network model for benzene; here, we introduce some of the key ideas. The left side of Fig. 2.2, from Hack et al. 2010), shows the structure of a Bayesian network (BN) model for quantifying conditional probabilities of some variables (e.g., various metabolites, markers and acute myeloid leukemia (AML) (the *Leukemia* node at the lower right)), given observed or assumed values of other variables, including concentration of benzene in air (averaged over a worker's years of exposure). The conditional probability distribution of each variable (node in the network) depends on the values of the variables that point into it. This BN was constructed manually based on a detailed literature review of candidate markers and outcomes and predictive relationships among them, given limitations in measurement techniques (Hack et al. 2010). The right side of Fig. 2.2 shows random samples drawn (via Monte Carlo uncertainty analysis) from the BN-predicted dose-response function describing the conditional probability for AML given air benzene exposure concentrations. These curves are constructed from the conditional probability tables of the BN, as follows. Each value for benzene concentration (ppm) in inhaled air determines a conditional probability distribution for the benzene metabolite urinary tt-MA (trans-trans muconic acid). The value of tt-MA, in turn, determines a conditional probability distribution for 8-hydroxyguanosine

(8-OHdG), a marker of oxidative stress in lymphocytes. Finally, the value of 8-OHdG determines a conditional probability for AML. For each benzene concentration, a value of tt-MA is sampled from the conditional distribution of tt-MA, given the ppm of benzene; then a value for 8-OHdG is drawn from its conditional distribution given the sampled value of tt-MA; and finally this value is used to determine the conditional probability for AML given the sampled value of 8-OHdG. The needed conditional probabilities constitute the quantitative part of the BN model; they are estimated from data collected in multiple studies. Repeating this Monte Carlo sampling many times and averaging the results yields an estimate of the dose-response curve giving the conditional probability of AML for each value of air benzene concentration. Other potentially causally relevant variables (e.g., p-benzoquinone, inflammation status, age of patient, co-exposures and co-morbidities, and so forth) are "marginalized out" of the BN model in Fig. 2.2, meaning that they can still implicitly affect the probability of leukemia, but are not explicitly shown or conditioned on in calculating the sample dose-response curves for conditional probability of AML given ppm of benzene, shown on the right side of the diagram.

The uncertainty reflected in the band of dose-response functions (blue curves) on the right of Fig. 2.2 might in principle be reduced by conditioning on additional causally relevant variables, data permitting. Other investigators (or automated machine-learning programs for learning BNs directly from data) might select additional markers and perhaps get narrower uncertainty bands. In addition, durations of exposure and uncertainty in exposure concentrations could be included in refined models. However, the BN analysis in Fig. 2.2 already suffices to indicate both that benzene is a hazard for increased risk of AML (hazard identification) and also that predicted AML risk is not clearly increased above background at low exposure concentrations (e.g., <0.1 ppm) and reaches a prevalence of almost 0.02 at a concentration of 10 ppm for the exposure conditions and durations in the studies used to build the BN model. This quantitative dose-response information can help inform risk management decisions. In addition, the structure of the BN on the left side of Fig. 2.2 shows that measuring hematological parameters such as WBC and RBC, in addition to air benzene concentrations, can help to infer AML risks when the marker 8-OHdG is not directly measured.

Promising as BNs appear to be, there are several reservations and questions about them that should be addressed, and some important recent progress in addressing them. Key developments are as follows (Cox Jr. 2018a):

- *How are connections among variables to be determined?* Allowing complex networks of relationships raises the threat of "model shopping"—that is, of investigators searching among complex models to find those that support prior beliefs or desired conclusions. To help guard against this, software packages that learn Bayesian network structures from data using several different algorithms (e.g., R packages such as *bnlearn* and *CompareCausalNetworks*) are becoming increasingly popular (Chap. 9). The main idea of these algorithms is to use tests for statistical properties (such as conditional independence tests) to determine which

variables each variable is found to depend on, even after conditioning on other variables. Direct and indirect effects of one variable on another (e.g., of exposure on risk of adverse response) are identified by conditioning on appropriate adjustment sets calculated from the BN structure. This approach avoids the asymmetry inherent in regression models that specify one variable as the "dependent" variable and other variables as "independent" variables to explain it. Rather, it seeks to discover what each variable depends on (among the variables in a data set; extensions to latent variables are at the frontiers of current research). This symmetric approach (sometimes called "causal discovery," in contrast to testing prespecified causal hypotheses about what might cause a dependent variable), together with algorithms for specifying adjustment sets to condition on in estimating direct and indirect effects (thereby addressing the problem of variable selection in a principled way) help to reduce the potential for model shopping and confirmation bias in modeling.
- *What functional forms should be assumed* for dependence relationships among variables, and *how should interactions among variables be modeled*? A common approach to addressing both questions is to use a nonparametric *conditional probability table* (CPT) or CART tree at each node of a BN, i.e., for each variable, to describe how its conditional probability distribution depends on the values of the variables that point into it (its direct "parents"), if any. This nonparametric approach allows arbitrary nonlinear interactions among variables to be estimated and described in a uniform framework while avoiding the necessity of assuming any particular parametric model form.

At its best, BN technology may help investigators avoid using data and modeling choices to support or refute particular hypotheses (e.g., searching for a statistical model to show that exposure to an agent is significantly associated with a health effect, or for an alternative model in which it is not) and to instead support more dispassionate discovery of stable dependencies among variables, with clearer distinctions drawn between direct and indirect (mediated) causal effects. But much remains to be done to achieve this goal. Despite impressive recent technical progress, current BN-learning algorithms remain limited by their lack of common sense. (They are typically better at determining whether variables are conditionally independent than at identifying the directions of dependences between them, and may require human users to specify constraints such as that death is a possible effect but not a possible cause of other variables, or that sex, age, and ethnicity are possible causes but not possible effects of other variables.) They are prone to identify false-positive links between variables if confounders are omitted, or if discretization of continuous variables to form CPTs (e.g., using deciles of continuous variables as levels, or using CART trees) leaves substantial residual confounding. Validation that CPTs express stable causal relationships that hold across situations typically requires collecting data from multiple studies. Recent advances in theory and software for causal analysis and interpretation (e.g., the *InvariantCausalPrediction* and *CompareCausalNetworks* R packages) have started to address these and other challenges, including detecting and modeling hidden (latent, unobserved) variables based on otherwise unexplained correlations between observed variables), but these developments have as yet had little impact on regulatory risk assessment.

Dynamic Simulation for Dose-Response Modeling

An alternative to regression or BN modeling of epidemiological data for dose-response modeling, developed in greater detail in Chaps. 3, 4, 5, and 6, is to seek to understand biological causal mechanisms and to model them well enough to simulate the effects of exposures on risk. This typically involves integrating pharmacokinetics, which convert administered to internal doses; pharmacodynamics, which model effects of internal doses (e.g., changes in behaviors and transition rates) in target organs, tissues, and cell populations; and disease models, such as two-stage or multistage clonal expansion (TSCE or MSCE) models of carcinogenesis (Chaps. 4 and 5), which model the development of diseases over time. Causal dynamic processes leading from exposure to health effects can be simulated by linking pharmacokinetic, pharmacodynamic, and disease process simulation submodels, provided that sufficient knowledge is available to create them and sufficient data are available to populate them with realistic values for their parameters and functions, such as flow rate and stochastic transition rate coefficients. Chapters 4, 5, and 6 discuss such modeling in greater depth.

Dynamic simulation models enrich dose-response modeling by showing how time patterns of exposure concentrations affect risk over time. For example, which exposure pattern in each of the following pairs poses a higher risk (as measured, for example, by lifetime probability of a disease, or age-specific hazard function)?

- 1 ppm for 8 h/day vs. 8 ppm for 1 h/day
- Exposures on Monday and Friday each week (with none on other days) vs. the same exposures on Monday and Tuesday each week (with none on other days)
- Weekly exposures for 52 weeks/year, or twice those exposures for the first 26 weeks of each year only, or twice those exposures administered on alternating weeks throughout the year.
- Occupational exposures from ages 20–35 years or the same exposures from ages 35–50 years.

Many regression models applied in regulatory risk analyses use exposure metrics, such as cumulative ppm-years of exposure, that do not distinguish among time patterns of exposure, but both experimental evidence (e.g., from stop-exposure experiments) and dynamic simulation models show that they can have very different effects on risk.

Chapter 3 provides a relatively accessible introduction to dynamic simulation modeling using diagrams, and Chaps. 4, 5, and 6 provide more detailed development and examples. Figure 2.3, which is discussed further in Chap. 3, highlights key ideas via a notional diagram, without specific units on its axes, showing typical time patterns for how the internal dose (e.g., concentration of a toxic metabolite in a target organ or tissue) on the vertical axis varies over time (on the horizontal axis) for three different time patterns of dose administration (curves 1, 2, and 3). For an inhalation hazard, the administered dose rate corresponds to concentration of a substance in air. Curve 1 administers a certain concentration for a certain duration in each cycle of a repeated exposure pattern; curve 2 administers twice the concentration for half the duration; and curve 3 administers 3 times the concentration for 1/3 the duration, in each consecutive cycle. The vertical axis shows how concentration in a typical compartment varies with these time patterns of exposure. Specific versions of such model-predicted curves have been developed and validated for many chemicals using physiologically-based pharmacokinetic (PBPK) models. For purposes of dose-response modeling, an important feature is that administering higher concentrations for shorter durations produces higher peak internal concentrations than lower concentrations for longer durations, for the same total amount delivered (e.g., for the same ppm-hours per week). This implies that curve 3 can activate responses that curve 1 would not. If a response has an internal dose concentration threshold (denoted by *T* in Fig. 2.3), as in the examples mentioned in the Introduction, then repeated high-concentration, short-duration exposures are more dangerous than administering the same average amounts per unit time more gradually.

The disproportionate risk from shorter, more concentrated dose administration in each cycle in Fig. 2.3 holds even if pharmacokinetics are linear, so that concentrations in internal compartments are proportional to administered concentrations. But pharmacokinetics are often nonlinear at relevant exposure concentrations. For example, doubling the administered concentration of a substance in air can more than double its internal concentrations in tissues over time if clearance mechanisms are saturable, following Michaelis-Menten kinetics. Figure 2.4, which is discussed further in Chap. 3, illustrates the effects of such nonlinearity, motivated by a model for accumulation of asbestos fibers in lung and mesothelial tissue (Chap. 5). The vertical axis represents the amount of internal dose (e.g., fibers) added to a target tissue per unit time for different administered concentrations in air (horizontal axis),

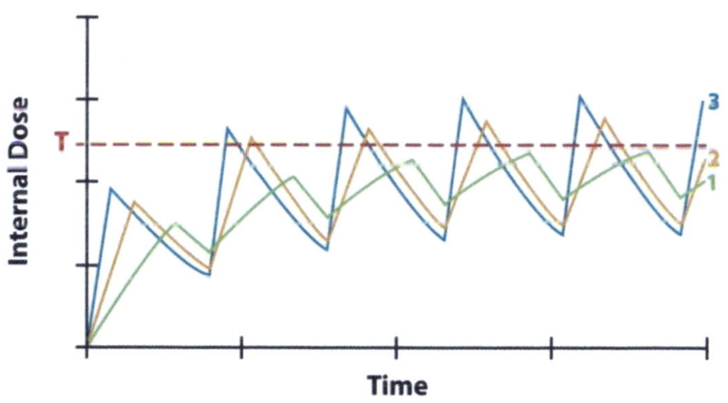

Fig. 2.3 Administering the same total amount of exposure (e.g., 100 ppm-hours/week) in different time patterns changes the maximum internal dose received

assuming sustained exposure to a constant concentration. The axes are scaled so that 1 on the horizontal axis represents the base case exposure scenario, e.g., what a typical worker might receive. The vertical axis shows the increase in cumulative internal dose (e.g., fibers lodged in target tissues) per unit time, and the curve plots this for different levels of administered concentration on the horizontal axis. Doubling exposure concentration from 1 to 2 increases the rate at which cumulative internal dose increases from about 1 to about 14, reflecting slower clearance from the lungs (e.g., via macrophages and the mucociliary escalator) and hence a higher fraction of inhaled dose (fibers) translocating to target tissues. By contrast, doubling exposure concentration from 0.1 to 0.2 makes relatively little difference, as most of the administered dose is cleared at such low concentrations.

Similar reasoning sheds light on the question of whether a worker suffers greater risk if exposed to a hazardous substance on two consecutive days each week (e.g., Monday and Tuesday) or on two days separated by a longer recovery interval (e.g., Monday and Friday). If the first day's exposure reduces clearance (or depletes protective resources such as antioxidant pools), then another exposure on the next day may create a greater internal dose than it would if it instead occurred a few days later, when clearance capacity has recovered. If so, then exposures on consecutive days in each cycle (e.g., in a work week) will produce more internal dose than the same exposures administered with longer recovery periods in between. The timing of internal doses can also interact with response dynamics, e.g., by delaying cell division during and following a first exposure until internal doses decline enough for mitosis to proceed safely; if the cells thus synchronized and undergoing mitosis at approximately the same time, receive a second dose part way through it, the cytotoxic effect may be far larger than if the second dose had occurred earlier or later.

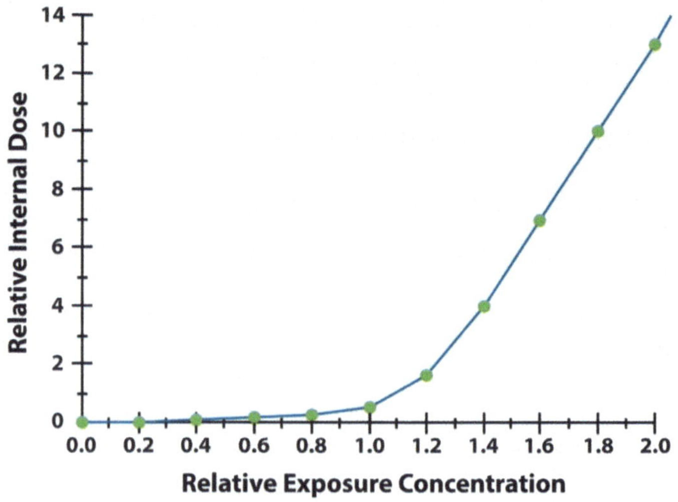

Fig. 2.4 Higher administered concentrations that reduce clearance rates are disproportionately efficient in producing internal concentrations

These examples reinforce the key point that *timing matters*: for many substances, the same average cumulative exposure per unit time can have very different effects on risk of adverse response, depending on how it is distributed over time. Chapter 6 discusses this point further.

In Fig. 2.3, whether internal doses in a target tissue become high enough to trigger an adverse response depends on details of timing that are lost when exposure metrics such as ppm-years or ppm-hours/day, averaged over some interval, from a week to a working lifetime, are used to summarize individual exposure histories. When response thresholds or other nonlinearities in dose-response functions are important, accurate risk prediction may require more detailed descriptions of exposure histories than are captured in the exposure metrics used in regression-based risk modeling. Conversely, when only these exposure metrics are available, it may be impossible to predict risk from them with useful accuracy. Dynamic simulation models provide rich opportunities to study how collecting and analyzing data on time patterns of exposure, and regulating the timing of exposures as well as permitted levels of exposures, can protect worker health in ways that regulating summary measures of exposure alone does not.

Exposure Assessment

NIOSH (2020) notes that, "In environmental risk assessments, exposure assessment is considered a separate step for assessing the likelihood of exposure for estimating population risks and/or disease burden. In contrast, NIOSH risk assessments… estimate the risks to a hypothetical working population from a known exposure. Although exposure probabilities are not typically calculated, dose-response analyses include exposure information; therefore, NIOSH systematically assesses the availability, magnitude, and validity of exposure data used in relevant studies as a part of hazard identification and applies this information, as applicable, in the dose-response assessment." The dose-response modeling considerations in the previous section imply that unmodeled uncertainty and variability in exposures may greatly affect risk estimates. It may do so in the following ways:

- If higher concentrations are disproportionately dangerous, as in Figs. 2.3 and 2.4, then an estimated exposure concentration with symmetrically distributed (e.g., normally distributed) estimation error will appear to be more dangerous (i.e., to cause higher risk) than the same concentration measured without error.
- Similarly, if exposure concentrations have some variance around their TWA means, then the risk caused by a given mean concentration may be much greater than it would be if the variance were zero. Indeed, risk may depend as much or more on the variance (and also on the autocorrelation structure, if consecutive high concentrations are disproportionately dangerous) than they depend on the mean concentration. Therefore, regulatory standards that address the mean (and

possibly occasional excursions above it) but that do not consider variance and autocorrelation may neglect key drivers of risk.
- Ignoring uncertainty and variability in exposures in regression modeling can make nonlinear dose-response relationships, including ones with sharp thresholds where risk jumps from a low level below a critical exposure concentration threshold to a high level above it, appear to be smooth, S-shaped, curves that are approximately linear at low doses (Rhomberg et al. 2011a, b; Cox Jr. 2018a). The reason is that exposure concentrations closer to the threshold are more likely to be mis-estimated as being on the wrong side of it than are concentrations further from it.

The apparent smoothness of an estimated exposure-response function when the underlying biological dose-response function has a discontinuous jump stems from the smoothness of the error distribution of estimated exposure values around true exposure values: at estimated exposure concentrations further below the threshold, the probability is less that a response will occur (because the probability is less that the true exposure is above the threshold). A practical consequence is that estimated exposure-response functions that appear to be smooth and approximately linear at low concentrations, with no evidence of a threshold, should not necessarily be interpreted as evidence that there is not a threshold. If errors in exposure estimates create this appearance whether or not there is a threshold, then the appearance does not provide evidence for or against a threshold. Similar caveats hold for other nonlinearities less extreme than thresholds: exposure estimation errors distort (and typically flatten and linearize) estimated exposure-response relationships (Rhomberg et al. 2011a, b). In simple linear regression models with exposure as the only predictor, ignoring measurement error biases the estimated slope (or potency) of the exposure-response line toward zero. By contrast, in nonlinear and multiple-predictor regression models, the bias can go in either direction. If higher exposures are disproportionately dangerous, as in Fig. 2.4, then neglecting errors in estimated exposures leads to over-estimates of risks at low exposures (since some high-exposure risks are misattributed to lower estimated exposures) and to under-estimates of risks at high exposures (since some low-exposure risks are misattributed to higher exposure levels). Thus, nonlinearity implies that measurement error need not attenuate estimated effects of exposure on risk, as in simple linear regression models, but rather exaggerates it at low concentrations and attenuates it at high concentrations.

Current practice in regulatory risk assessment often uses best estimates of exposures, e.g., estimates reconstructed from job exposure matrices for occupational risks, or from microsimulation models of individual movements and exposures for public health risks. Using best estimates of exposure without quantifying or correcting for effects of exposure estimation errors can lead to substantial biases in estimated exposure-response functions, as just discussed, and as examined further in Chap. 16 for air pollution. Fortunately, as mentioned in Table 2.1, a variety of "errors-in-variables" statistical methods have been developed to correct for the distorting effects of measurement or estimation errors in exposure and other predictors. Computational Bayesian methods infer the shape of the exposure-response function by treating the true exposure as an unobserved quantity on which the estimated exposure depends. Other techniques (such as instrumental variables and repeated measurement methods) use observations on other variables, or repeated observations of the same variables, to help estimate the true shape of the exposure-response function when exposures are estimated with errors. Thus, the distortions in estimated exposure-response functions described in this section can often be avoided by careful design of data collection and analysis of data.

Risk Characterization, Uncertainty Characterization, and Risk Communication

NIOSH (2020) explains its approach to risk characterization as follows.

> The final step in NIOSH risk assessment is risk characterization. It is the translation of information from hazard identification and dose-response assessment into a basis, completely or in part, for recommendations on limiting workplace exposure. The framework of NIOSH risk characterization centers on a choice between two distinct approaches, based primarily on the evidence supporting the absence or presence of an impairment threshold. For effects with a response threshold, NIOSH typically [develops] an estimate of a safe dose. Here the term safe implies that excess risk at this exposure level is absent or negligible. …When effects appear to be without a response threshold, NIOSH obtains quantitative estimates of low-dose risk by model-based extrapolation of the risk at doses below the observed data.

Nonlinear dose-response relationships raise the possibility that neither of these two options—a threshold model, or model-based extrapolation of risk below the observed data range—describes the true dose-response relationship. Figure 2.4 illustrates the problem. There is no threshold in this curve, but neither can risk at low concentrations be confidently extrapolated from observations at higher concentrations (e.g., the nearly linear segment of the curve to the right). When linearity cannot be assumed, extrapolation is an unreliable guide because of the variety of possible shapes for nonlinear functions below the observed data range.

The admirable goal of using risk characterization to translate hazard identification and dose-response assessment into a basis for recommendations to limit workplace exposures to protect worker health is also threatened if any or all of the following previously discussed conditions hold:

- **Model form misspecified**. The assessed dose-response function describes *statistical effects* on risk attributed to exposure in multivariate modeling of nonlinear interactions; but not *causal relationships* revealing how reducing exposure (with or without holding other factors fixed) would affect risk (Pearl 2009). Statistical effects may arise from non-causal sources such as departure of nonlinear dose-response functions from assumed linearity (Fig. 2.1). Using flexible nonparametric regression methods can reduce the threat of model specification errors and incomplete control of confounding, as illustrated in Fig. 2.1.

- **Ignored exposure dynamics**. The assessed dose-response function predicts conditional expected values of risk indicators from exposure metrics that ignore essential details of the time pattern (Fig. 2.3), variability, and autocorrelation of exposure time series. Dynamic simulation risk models can clarify how changes in exposure affect changes in risk over time (Fig. 2.3).
- **Ignored exposure estimation errors**. The assessed dose-response function estimates probabilities of adverse effects at different estimated dose (or exposure) levels, but errors in exposure estimates distort the shape of this function, e.g., making a nonlinear or threshold dose-response function appear to be approximately linear no-threshold, exaggerating risks at exposures below the observed data range. Errors-in-variables methods can help to avoid such distortions due to measurement and estimation errors.
- **Non-causal explanations**. More generally, the assessed dose-response functions describes association but not causation, whether due to nonlinearity or other explanations. For example, if higher exposure concentrations are significantly positively associated with higher risks, but this is explained by the fact that both are declining over time, or that poorer areas tend to have higher values of both variables, or because lower-risk people are more likely to move away before they are sampled, then the assessed dose-response association does not necessarily predict whether, or to what extent, reducing workplace exposure would change risk. Bayesian network models (Fig. 2.2) can help to clarify causal pathways and competing explanations for observed exposure-response associations.

In all these cases, dose-response functions estimated by regression without making the suggested corrections cannot necessarily be used to predict whether or to what extent a change in exposure would change risk. This undermines attempts to use them to provide a rational, causally effective basis for recommendations to limit workplace exposures to protect worker health. Failures of workplace exposure standards to reduce some exposure-associated risks may reflect this lack of causally effective regulations, as discussed further in Chap. 6.

To inform more effective regulation and policy-making, uncertainty characterization and risk communication must convey more than estimates of slopes and confidence intervals for dose-response functions at and below current exposure levels. They should also convey any significant uncertainty about whether reducing exposure will reduce risk, as described by the estimated dose-response function, or whether the function instead describes risks attributed to or associated with different levels of exposure, but not necessarily preventable by reducing it. Dose-response functions assessed by regression modeling applied to epidemiological data, and referring only to associations and attributed risks, do not address how much (if at all) reducing exposure would reduce risk. Hence, they do not tell policy makers what they need to know to take effective action to protect health based on quantitative evaluation of risk reductions expected from limiting exposures. This uncertainty about the causal relevance of estimated dose-response relationships, and of risk characterizations based on them, is seldom clearly communicated to policymakers. Yet is often more relevant for well-informed and causally effective decision-making than the widths of confidence intervals for estimated slope factors (Pearl 2009).

In practice, as discussed in Chap. 19, risk assessments often present policy-makers with point estimates and confidence or uncertainty intervals for risk estimates using units such as "premature deaths prevented per unit of reduction in exposure concentration" or "statistical lives (or life-years or quality-adjusted life-years) saved per unit reduction in cumulative exposure." These estimated risks per unit of exposure seldom warn policy-makers (or other recipients) that "per" here only signifies that an estimated burden of mortality or morbidity attributed to exposure has been divided by an estimated exposure amount, and that it has no necessary implication that reducing the amount of exposure would reduce the burden of mortalities or morbidities attributed to it. Uncertainty about the causal interpretation of estimated dose-response models cannot be communicated or characterized by presenting confidence intervals (or resampling or Bayesian uncertainty intervals) and sensitivity analyses. It requires explicitly discussing whether the dose-response relationship and risk characterizations being presented reflect association and causal attribution, or changes in risk caused by changes in exposure. Conflating these two very different concepts risks misinforming risk management decision and policy makers about the expected consequences caused by interventions (Pearl 2009).

Failure to clearly distinguish in charactering and communicating risk between risks associated with and attributed to exposure and risk preventable by reducing exposure leads to what might be called the *supralinearity paradox*: substances that have been studied for decades and subjected to increasingly tight regulations are often estimated to be more dangerous (potent) in newer regression-based studies than in older ones, leading to speculations that lower concentrations have been discovered to be disproportionately dangerous compared to higher concentrations (Hornung and Lanphear 2014). A different possible explanation is that causally ineffective reductions in exposures do not reduce the risks attributed to them as originally predicted by regression models (e.g., because imperfectly controlled confounding or nonlinearities or interactions created regression coefficients that do not represent the causal effect on risk per unit reduction in exposure). Attributing undiminished risk to substantially lowered exposures increases the regression coefficient linking exposure to risk. This can create the appearance of higher risk-per-unit exposure at lower concentrations (supralinearity) simply because the exposure-risk association attributes risk to exposures that do not cause it.

Figure 2.5, based on an example discussed further in Chap. 10, illustrates some of these points. It fits three univariate models—linear, quadratic, and nonparametric smoothing regression curves—to data on blood lead concentration levels (x axis) and mortality during follow-up for nonsmokers in the NHANES data set. (Logistic regression would also seem appropriate a priori, but does not fit the data well, as the conditional expected value of the mortality rate is not a sigmoid function of concentration.) The data are used only to illustrate some conceptual and methodological points, so we forego discussion of details of the data set (or of alternative regression models, such as proportional hazards, that would better address survival data); the blood lead example is discussed further in Chap. 10. The

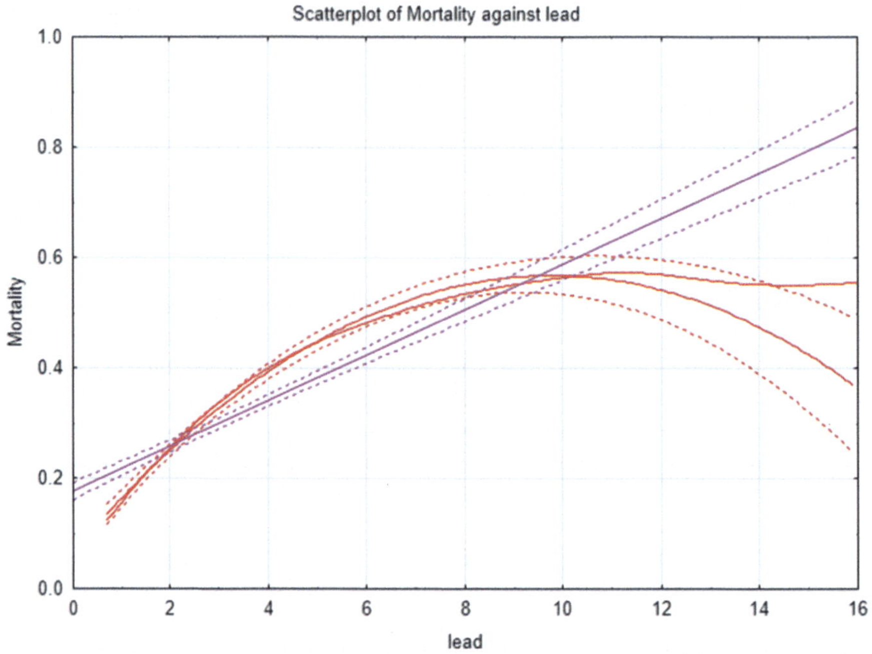

Fig. 2.5 Mortality during follow-up vs. blood lead level (μg/dL) for non-smokers in NHANES data

linear and quadratic models have 95% confidence bands (shown as dashed curves). The nonparametric regression curve is approximately flat at the right edge of the plot, where the linear model slopes up and the quadratic model slopes down.

Figure 2.5 illustrates several points about risk and uncertainty characterization. First, confidence intervals do not convey model uncertainty. For most of the range of values, the 95% confidence intervals for the linear and quadratic model specifications do not overlap. The best-fitting non-parametric model does not fall within either set of confidence bands at the right side of the graph. Although model diagnostics and goodness-of-fit tests can help select the best-fitting parametric model (e.g., quadratic rather than linear), they do not fix the fact that even the best-fitting model in a parametric class of models may not fit the data very well. Confidence bands do not communicate this aspect of model uncertainty: the confidence intervals in Fig. 2.5 are narrow compared to the range of predictions from different models. For this reason, sensitivity analysis (discussed later) is often used to assess the extent to which predictions from different models differ (NIOSH 2020). This leaves open the question of how to reconcile (or select or combine) risk predictions from different models, as well as the question of whether any of the different models considered make accurate predictions.

Second, without a thorough discussion of confounding, model specification errors, nonlinearities, interactions, and residual confounding, it is impossible for a policy maker to determine from the results of such models—for example, from the linear model and its confidence bands extrapolated below the range of data—how or whether reducing blood lead levels (BLL) would affect mortality risk. For example, if the strong positive association between BLL and mortality risk were entirely explained by incompletely controlled confounding by age (i.e., older people having higher BLL and higher mortality probabilities, with a nonlinear increase in mortality risk at older ages (and higher BLL levels) that is not entirely corrected for including age as a predictor in a regression model), then the models in Fig. 2.5 would have no implications for how much or whether reducing lead concentrations would reduce risk. That statistical risk models do not necessarily describe how or whether reducing exposure would reduce risk, even if they fit observed data well and have tight confidence limits, is well understood by many epidemiologists. It is not always as clear to risk managers and regulatory risk assessors, many of whom seek to use quantitative estimates of low-dose risk obtained from regression-based dose-response models to inform decisions about how to limit exposures in order to protect health (NIOSH 2020), without understanding that regression-based dose-response (or exposure-response) curves are usually not appropriate for this purpose because they do not reveal how changing exposures would change risks (Pearl 2009). Risk and uncertainty characterization and communication of risk assessment results to decision-makers should clarify such uncertainty about consequences of interventions in order to inform rational and effective decision-making. More constructively, presenting results from validated causal Bayesian network risk models (Pearl 2009; Hack et al. 2010; Cox Jr. 2018a) and dynamic simulation risk models (Cox Jr. 2020) can show how interventions, such as reducing exposure, are predicted to change the probability distributions of outcomes such as disease and mortality risks.

LNT Rationales: Additivity to Background, Population Heterogeneity, and Upper-Bound Estimates

Regulatory risk assessment has long embraced three principles that help to explain why LNT assumptions are often used in practice, despite reservations about possible nonlinearity such as those described in previous sections (e.g., Crump 2017; Rhomberg et al. 2011a, b). The first is an assumption that incremental cancer risks from exposures to chemicals (and radiation) add to background processes, leading to effective linearity of the exposure-related dose-response or C-R curves. The second is that population heterogeneity tends to linearize population C-R functions, even if individual-level ones have thresholds or other nonlinearities. The third is that statistical upper confidence limits on LNT functions are appropriately conservative or health-protective in the face of uncertainty about the true C-R functions. All three principles have been debated for years. However, some recent examinations of the additivity-to-background assumption have concluded that it is often unjustified on both mathematical and biological grounds, in part because mechanisms of exposure-related harm, such as activation of inflammasomes, differ from mechanisms of background (spontaneous) harm in many cases, leading to genuine thresholds or threshold-like nonlinearities for incremental risks caused by exposures (Bogen 2016; Bogen 2019; Calabrese 2018; Cox Jr. 2018b). Likewise, careful explication of the population heterogeneity argument has concluded that it broadens C-R uncertainty bands, but does not linearize nonlinear functions and should not be construed as implying linear upper confidence limits at low exposure concentrations (Rhomberg et al. 2011a, b; Crump 2017; Bogen 2016). Specifically, if thresholds or sharp nonlinearities hold for individuals, a combination of uncertainty and variability in C-R functions, together with exposure estimation error, may create a range of values at the population level over which estimated population risk decreases gradually with decreasing exposure concentrations, but this does not in general imply that linear extrapolation down to zero is justified, even if it is deemed prudent to use upper uncertainty limits (Cox Jr. 2018a; Bogen 2019). As illustrated on the right side of Fig. 2.2, Monte Carlo uncertainty analysis can lead to wide uncertainty bands that are nonetheless nonlinear. We conclude that, in general, nonlinear models deserve careful uncertainty analysis, and that simplifying LNT assumptions are not necessarily well justified even if there are positive background rates, population heterogeneity, and a desire to be conservative by using upper uncertainty distribution quantiles for purposes of regulation. Monte Carlo uncertainty analysis provides a constructive approach for characterizing uncertainty in nonlinear models, including Bayesian networks and dynamic simulation models, if enough is known to model uncertain quantities as random variables.

Use of Sensitivity Analysis and Scientific Judgment

Scientific judgment is often invoked as a way to address limitations in knowledge and data used in risk assessment. For example, NIOSH (2020) states that "Limitations in available data often require scientific judgment in order to fill gaps in model specifications. Risk assessors identify and characterize these judgements by conducting additional analyses to test plausible alternative assumptions, examine the robustness of main analyses, and improve transparency in the risk assessment process. These alternative analyses comprise sensitivity and modeling uncertainty analyses. NIOSH defines sensitivity analysis as a study of the uncertainty in estimates from the mathematical model that can be apportioned to uncertainties in its inputs." Sensitivity analyses inform decision-makers about how predictions depend on uncertain modeling assumptions, for example, by showing the different risk predictions made by different models (e.g., the linear and quadratic models in Fig. 2.5). However, they do not identify whether any of the alternatives is close to the truth. Moreover, they are limited to studying how changes in assumed inputs affect outputs, but they do not address uncertainty about what the outputs mean, such as whether estimated C-R functions reflect association or causation (or both).

Similarly, NIOSH (2020) adds that "Because different model specifications can lead to different estimates, a key step in dose-response analysis is model selection. Clearly, it is preferable to base model selection on biologic plausibility, although a strong advantage of one model among several plausible models is rarely evident. Furthermore, data from most studies are imperfect and potentially incomplete. In lieu of available statistical techniques and algorithms designed to deal with data imperfections, the risk assessor may have to rely on assumptions based on scientific judgment." However, empirical studies of the performance of scientific judgment under uncertainty indicate that it (like other human judgments under uncertainty) is subject to heuristics and biases that make it a very unreliable guide to truth (Kahneman 2011). Using statistical techniques and algorithms usually outperforms expert judgment, in a variety of domains (Tetlock and Gardner 2015). Moreover, the previous sections have illustrated that the specific technical challenges arising in risk assessment can be quite subtle and challenging, making it difficult to see how scientific judgment alone can address them usefully. Scientific judgment used in lieu of available statistical techniques and algorithms might be hard pressed to yield useful assessments and corrections for residual confounding due to slight departures from linearity (Fig. 2.1); or to reliably assess whether a repeated exposure will produce internal doses that exceed a threshold (Fig. 2.3); or to adjust correctly for nonlinearities in extrapolating from high to low exposure concentrations (Fig. 2.4); or to predict how correcting for exposure measurement errors and for variance and autocorrelations in exposure time series would change the shapes of estimated C-R functions; or to estimate the results of averaging the risk predictions from an ensemble of models while varying exposure and holding other variables fixed, as in a partial dependence plot. These are not types of tasks that judgment excels at. Scientific judgments about causality are notoriously error-prone (Kahneman 2011), and C-R regression curves and confidence intervals do not offer even the most discerning judgment the empirical information needed to determine how or whether changing exposure would change risk (Fig. 2.5, Pearl 2009). Judgments organized, tested, and validated using causal Bayesian network appear to be promising for addressing causal questions (Pearl 2009) and for quantifying causal C-R functions (Fig. 2.2). However, testing and validating such models with data makes heavy use of statistical methods and algorithms (such as conditional independence tests, calculation of adjustment sets, estimation of conditional probability tables of functions, and inference of causal exposure-response relations controlling for the variables in adjustment sets) (Cox Jr. 2018b).

At the same time, it is undeniable that formulating useful models of pharmacokinetics, pharmacodynamics, and disease processes requires considerable scientific judgment. Understanding of relevant biology is needed to decide what details can (and should) be safely omitted or averaged over, what simplifying assumptions should be made to reduce complex reality to a manageable model with parameters that can be estimated from data, and how compartments and flow rates or transition rates should be specified and parameterized. Thus, scientific judgment is often essential in identifying appropriate model structures and simplifications, and in specifying what needs to be estimated from data, e.g., exposure-dependent transition rates or flow rates among model compartments. Statistical algorithms can then be applied to appropriate data (if it is available) to estimate these parameters. Thus, we conclude that scientific judgment should not be used in lieu of statistical methods and algorithms, but rather should be used in combination with them. For example, human understanding and judgment of plausible causal pathways and mechanisms, together with statistical models of conditional probabilities and of pharmacokinetics, pharmacodynamics, and disease processes, can contribute to developing, testing, and refining causal Bayesian networks (Hack et al. 2010) or dynamic simulation models (Chaps. 3, 4, 5, and 6). Such data-informed models, in turn, can contribute to credible risk estimates and causal C-R models for informing risk management decisions.

Discussion and Conclusions: Risk Management and Risk Assessment Implications of Nonlinearity

The aspects of nonlinear C-R function estimation and interpretation discussed in previous sections have clear implications for more effective risk management, as well as for more informative risk assessment. One is the importance of including time patterns of exposure, such as times between successive high-concentration exposures, in characterizing and regulating risks for some substances. Reducing time-weighted average (TWA) exposure concentrations may not succeed in reducing risk if times between consecutive high exposure concentrations are not also controlled (Fig. 2.3). Likewise, regulations that limit variability around mean concentrations, and autocorrelations in exposure histories on different time scales, may be much more effective in controlling risks for some substances that measures that only reduce time-weighted average (TWA) mean concentrations. When C-R functions are nonlinear, as in Fig. 2.4, an exposure history that randomly fluctuates between 0 and 2 units of concentration, with a mean of 1 unit of concentration, may be far more dangerous than a constant (zero-variance) exposure to 1 unit of concentration, even though they have identical TWA values. Regulations based on C-R functions that do not model and correct for errors in exposure estimates may be ineffective if the distorted shape of the estimated C-R function (e.g., a low-dose linear curve) does not accurately represent the true shape of the underlying causal dose-response function (e.g., a threshold or other nonlinear function). Thus, the practical design of regulations that better protect worker health by addressing these aspects of exposure appears to be a worthwhile topic for further applied research. Dynamic simulation modeling can help to understand effects of proposed regulations on risk, accounting for realistic variability and autocorrelations in exposure time series; time patterns of mean exposure concentrations within working weeks and over lifetimes; inter-individual variability in exposure response parameters; and exposure estimation errors and uncertainty (Cox Jr. 2020). However, developing and validating dynamic simulation models of exposures, pharmacokinetics, pharmacodynamics, and disease processes requires considerable research compared to regression modeling of estimated exposure-response data.

Recognizing the limitations and challenges of regression modeling for nonlinear C-R functions does not imply that regression modeling should not be used in risk assessment, characterization, and communication; but only that it should be used and interpreted carefully, and in conjunction with other techniques that better address causality. Regression modeling should be used with errors-in-variables corrections if there are errors in variables; with nonparametric model ensembles such as random forest if there is uncertainty about model forms and specification errors; with latent variable techniques if there might be unobserved factors that modify observed responses; and so forth (Table 2.1). Well-specified regression models can be highly useful for estimating conditional probability and conditional expected value relationships between a variable and its direct causes, if these are known and suitable adjustment sets have been identified. However, regression *per se* does not address the causal interpretation of the C-R curves that it produces (Pearl 2009). Judgment is not usually a viable substitute for more formal analytic methods (Kahneman 2011; Tetlock and Gardner 2015); conversely, even the best current formal analytic methods cannot replace the scientific judgment and knowledge needed to create dynamic simulation models and other detailed causal models of complex biological processes. Even strong and statistically significant associations observed in well-designed and controlled health impact assessment studies do not necessarily translate into valid causal relationships between the studied variables. Furthermore, statistical significance does not necessarily imply practical or clinical significance. These limitations are relevant for both linear and nonlinear models.

When regression modeling is used to estimate C-R functions, additional data-driven techniques are usually needed to establish valid causal interpretations of these curves, and, in particular, to clarify the extent to which changes in exposures will bring about changes in risk (Pearl 2009). Difficulty in demonstrating direct causation is not the same as evidence against such causation, and finding alternative ways to explain apparent associations of agents and effects should not necessarily be taken as evidence that the agent does not contribute to causing the effect. Rather, causal analysis clarifies whether a data set provides evidence of a dependence of effects on exposure that is not fully explained away by standard alternative explanations such as confounding, model specification error, and measurement error. If so, then it may be prudent to treat the dependence (at least provisionally) as being causal. If not—for example, if effects are conditionally independent of exposure to an agent, given the values of measured confounders, or given plausible assumptions about effects of unmeasured or residual confounding—then additional evidence is needed to warrant a hazard identification conclusion that exposure to the agent increases probability of the effect.

There has been remarkable progress in statistical and machine learning methods for dose-response modeling in recent decades. Most of the techniques we have mentioned for extending regression modeling to meet challenges arising from realistic imperfections in relevant data and knowledge, from the statistical methods in Table 2.1 to machine learning methods such as random forest partial dependence plots, were not widely taught or practiced by earlier generations of risk analysts; many did not yet exist. Now they are increasingly being included in state-of-the-art risk analysis practice and guidance (NIOSH 2020). We suggest that, in addition to these methods, causal Bayesian networks (Fig. 2.2) (Pearl 2009, Hack et al. 2010, Chap. 9) and dynamic simulation modeling of nonlinear C-R functions (Figs. 2.3 and 2.4, Chaps. 3, 4, 5, and 6) appear promising for further improving the realism of C-R modeling. These techniques typically require more detailed analysis than regression modeling alone, but the extra effort may be rewarded by increased success in designing exposure regulations that more effectively protect human health.

References

Armstrong B. Models for the relationship between ambient temperature and daily mortality. Epidemiology. 2006;17(6):624–31. https://doi.org/10.1097/01.ede.0000239732.50999.8f.

Bogen KT. Linear-No-Threshold default assumptions for noncancer and nongenotoxic cancer risks: a mathematical and biological critique. Risk Anal. 2016;36(3):589–604. https://doi.org/10.1111/risa.12460.

Bogen KT. Inflammation as a cancer co-initiator: new mechanistic model predicts low/negligible risk at noninflammatory carcinogen doses. Dose-Response. 2019;17(2):1559325819847834. https://doi.org/10.1177/1559325819847834.

Calabrese EJ. The additive to background assumption in cancer risk assessment: a reappraisal. Environ Res. 2018;166:175–204. https://doi.org/10.1016/j.envres.2018.05.015.

Cox LA Jr. Effects of exposure estimation errors on estimated exposure-response relations for PM2.5. Environ Res. 2018a;164:636–46. https://doi.org/10.1016/j.envres.2018.03.038.

Cox LA Jr. Modernizing the Bradford Hill criteria for assessing causal relationships in observational data. Crit Rev Toxicol. 2018b;48(8):682–712.

Cox LA Jr. Nonlinear dose-time-response functions and health-protective exposure limits for inflammation-mediated diseases. Environ Res. 2020;182:109026. https://doi.org/10.1016/j.envres.2019.109026.

Crump KS. Bogen's critique of linear-no-threshold default assumptions. Risk Anal. 2017;37(10):1802–7. https://doi.org/10.1111/risa.12748.

Dominici F, Greenstone M, Sunstein CR. Science and regulation. Particulate matter matters. Science. 2014;344(6181):257–9. https://doi.org/10.1126/science.1247348.

Groenwold RH, Klungel OH, Altman DG, van der Graaf Y, Hoes AW, Moons KG, PROTECT WP2 (Pharmacoepidemiological Research on Outcomes of Therapeutics by a European Consortium, Work Programme. Adjustment for continuous confounders: an example of how to prevent residual confounding. CMAJ. 2013;185(5):401–6. https://doi.org/10.1503/cmaj.120592.

Hack CE, Haber LT, Maier A, Shulte P, Fowler B, Lotz WG, Savage RE Jr. A Bayesian network model for biomarker-based dose response. Risk Anal. 2010;30(7):1037–51. https://doi.org/10.1111/j.1539-6924.2010.01413.x.

Hornung RW, Lanphear BP. The supralinear dose-response for environmental toxicants: a statistical artifact? Clin Toxicol. 2014;52(2):88–90. https://doi.org/10.3109/15563650.2013.878946.

Hsieh SJ, Ware LB, Eisner MD, Yu L, Jacob P 3rd, Havel C, Goniewicz ML, Matthay MA, Benowitz NL, Calfee CS. Biomarkers increase detection of active smoking and secondhand smoke exposure in critically ill patients. Crit Care Med. 2011;39(1):40–5. https://doi.org/10.1097/CCM.0b013e3181fa4196.

Kahneman D. Thinking, fast and slow. New York: Farrar, Straus and Giroux; 2011.

Lanphear BP, Rauch S, Auinger P, Allen RW, Hornung RW. Low-level lead exposure and mortality in US adults: a population-based cohort study. Lancet Public Health. 2018;3(4):e177–84. https://doi.org/10.1016/S2468-2667(18)30025-2.

Lewis RC, Meeker JD. Biomarkers of exposure to molybdenum and other metals in relation to testosterone among men from the United States National Health and Nutrition Examination Survey 2011-2012. Fertil Steril. 2015;103(1):172–8.

Li T, Guo Y, Liu Y, et al. Estimating mortality burden attributable to short-term $PM_{2.5}$ exposure: a national observational study in China. Environ Int. 2019;125:245–51. https://doi.org/10.1016/j.envint.2019.01.073.

Nagarajan R, Scutari M, Lebre S. Bayesian Networks in R with applications in systems biology. New York: Springer; 2013.

Naggara O, Raymond J, Guilbert F, Roy D, Weill A, Altman DG. Analysis by categorizing or dichotomizing continuous variables is inadvisable: an example from the natural history of unruptured aneurysms. AJNR Am J Neuroradiol. 2011;32(3):437–40. https://doi.org/10.3174/ajnr.A2425.

NIOSH (2020) Current intelligence bulletin 69: NIOSH practices in occupational risk assessment. By Daniels RD, Gilbert SJ, Kuppusamy SP, Kuempel ED, Park RM, Pandalai SP, Smith RJ, Wheeler MW, Whittaker C, Schulte PA. Cincinnati, OH: U.S. Department of Health and Human Services, Centers for Disease Control and Prevention, National Institute for Occupational Safety and Health. DHHS (NIOSH) Publication No. 2020-106 (revised 03/2020). https://doi.org/10.26616/NIOSHPUB2020106revised032020

NIST (2013) NIST/SEMATECH e-handbook of statistical methods. https://www.itl.nist.gov/div898/handbook/pmd/section1/pmd144.htm

Pearl J. Causal inference in statistics: an overview. Stat Surveys. 2009;3:96–146.

Pearl J, Mackenzie D. The book of why: the new science of cause and effect. New York: Basic Books; 2018.

Rhomberg LR, Chandalia JK, Long CM, Goodman JE. Measurement error in environmental epidemiology and the shape of exposure-response curves. Crit Rev Toxicol. 2011a;41(8):651–71.

Rhomberg LR, Goodman JE, Haber LT, Dourson M, Andersen ME, Klaunig JE, Meek B, Price PS, McClellan RO, Cohen SM. Linear low-dose extrapolation for noncancer heath effects is the exception, not the rule. Crit Rev Toxicol. 2011b;41(1):1–19. https://doi.org/10.3109/10408444.2010.536524.

Streiner DL. Breaking up is hard to do: the heartbreak of dichotomizing continuous data. Can J Psychiatr. 2002;47(3):262–6.

Tetlock PE, Gardner D. Superforecasting: the art and science of prediction. New York: Penguin; 2015.

Textor J, van der Zander B, Gilthorpe MS, Liskiewicz M, Ellison GT. Robust causal inference using directed acyclic graphs: the R package 'dagitty'. Int J Epidemiol. 2016;45(6):1887–94.

Thurston GD, Ito K. Epidemiological studies of acute ozone exposures and mortality. J Expo Anal Environ Epidemiol. 2001;11(4):286–94. https://doi.org/10.1038/sj.jea.7500169.

Yule GU. Why do we sometimes get nonsense-correlations between time series? -- A study in sampling and the nature of time-series. J R Stat Soc. 1926;89(1):1–63. https://doi.org/10.2307/2341482.

Chapter 3
Simulating Exposure-Related Health Effects: Basic Ideas

Introduction

This is the first of four chapters that explain and apply the dynamic simulation approach to health risk assessment modeling introduced in Chap. 2. The exposition in this chapter is motivated by an observation made by toxicologist Julie Goodman of Gradient, who coauthored the article on which this chapter is based (Cox Jr et al. 2020): most of the key ideas of the dynamic simulation approach can be explained through relatively simple diagrams, without going through the underlying equations. This chapter undertakes such a high-level diagrammatic exposition. It returns to two of the diagrams used in Chap. 2 to introduce simulation modeling, and adds further discussion and diagrams to help understand and visualize the dynamic responses of internal doses to different time patterns of exposure. Chapters 4, 5, and 6 and their appendices then cover the same material in much greater depth, providing the underlying systems of equations used to model biological responses to exposures to inhaled respirable crystalline silica (Chap. 4) and asbestos (Chap. 5) and highlighting the importance of nonlinearities in these responses for risk assessment and risk management (Chap. 6).

Chronic Inflammation and Biological Roles of the Inflammasome in Exposure-Associated Diseases

This section introduces the crucial roles of chronic inflammation in many exposure-related diseases. The widespread significance of chronic inflammation has recently been recognized in the pathogenesis of an astonishingly diverse set of exposure-related and autoimmune diseases, including Alzheimer's, asbestosis, bacterial infections, various cancers (e.g., lung, colon, mesothelioma), gout, heart disease, kidney

L. A. Cox Jr., *Quantitative Risk Analysis of Air Pollution Health Effects*,
International Series in Operations Research & Management Science 299,
https://doi.org/10.1007/978-3-030-57358-4_3

disease, rheumatism, silicosis, and various viral illnesses (e.g., influenza, Zika) (Kaplan et al. 2019). Underlying many of these diseases is a common mechanism that is triggered by sufficiently high and prolonged exposures to substances that threaten cellular integrity and homeostasis. These exposures cause protein filaments to be secreted by the cell nucleus and to self-assemble, together with other proteins (specifically, an adaptor protein called ASC, standing for "apoptosis-associated speck-like protein containing a C-terminal caspase recruitment domain," and pro-caspase-1), in the cytosol of macrophages (and some other cells) into *inflammasomes* (Jo et al. 2016). Inflammasomes, which were discovered in 2002, are a family of pro-inflammatory signaling complexes (i.e., multi-protein oligomers) that orchestrate inflammatory responses in many tissues. A specific inflammasome, the NLRP3 inflammasome, has been implicated in all of the above diseases. The NLRP3 inflammasome and its dynamic responses to exposure play a key role in the models developed and discussed in Chaps. 4, 5, and 6. Here, we focus on the main ideas, reserving additional biological details for later chapters and their references.

The NLRP3 inflammasome is now known to be activated by some important occupational carcinogens and irritants, including asbestos and respirable crystalline silica, as well as by many pathogenic bacteria, some viruses, ionizing radiation, and endogenously produced uric acid crystals (Chaps. 4 and 5; Jo et al. 2016; Yang et al. 2019). Once activated, it broadcasts biochemical signals (cytokines) that stimulate inflammation in nearby cells. It also triggers a form of programmed cell death called *pyroptosis* that ends in the rupture of a cell's plasma membrane and the release of its toxic and pro-inflammatory contents into the extracellular environment, where they may stimulate inflammation and pyroptosis in nearby cells. Fortunately, inflammation is usually dampened by processes that deactivate and degrade inflammasomes. Thus, once the exposures that caused the inflammation cease, the acute inflammation normally resolves. However, sufficiently intense and prolonged inflammation can become self-sustaining, persisting and spreading even in the absence of continued exposure. This is the condition of *chronic inflammation*. Chronic inflammation is accompanied by an increase in toxic products of inflammation, such as reactive oxygen species (ROS) and cathepsins released from lysosomes, which rupture in the process of pyroptosis. These toxic products act on cells of the inflamed tissue, sometimes for years or decades, damaging the tissue and increasing the risk of disease or reducing the time until disease onset (Bogen 2019). This broad outline of the process of chronic inflammation-mediated disease risks has been described in considerable detail for the NLRP3 inflammasome, although some important additional details remain to be elucidated.

A framework to conceptualize how the NLRP3 inflammasome initiates chronic inflammation can be described by an Adverse Outcome Pathway (AOP), in which a molecular initiating event (MIE; i.e., a molecular interaction between a xenobiotic, such as an inhaled particle, and a specific biomolecule, in this case a sensor or receptor that signals to the cell nucleus to upregulate production of proteins needed for the NLRP3 inflammasome) is followed by sequential key events and their relationships, ultimately leading to an adverse outcome (Aleksunes and Eaton 2019; Allen et al. 2016; Ankley et al. 2010). Ankley et al. (2010) stated that AOPs "may have gaps where details on the exact chain of events leading to adverse outcome are unknown, but enough information is available about the pathway to improve risk assessment decisions." Such is the case for the NLRP3 inflammasome. Its AOP includes the following steps:

- *Activation of the NLRP3 inflammasome.* Some proportion of alveolar macrophages (or other cells) are activated in response to sufficiently intense and prolonged exposures. The NLRP3 inflammasome has an *activation threshold*, meaning that brief and low-concentration exposures do not result in inflammation (Cox 2019; Yang et al. 2019). Thus, routinely encountered small quantities of dusts, fibers, bacteria, and other challenges can often be cleared (*e.g.*, *via* the mucociliary escalator and macrophages) without activating inflammasomes.
- *Transition from acute to chronic inflammation.* When the proportion of inflamed cells in a tissue is large enough, the release of pro-inflammatory signals and materials from cells that have undergone pyroptosis becomes sufficient to sustain permanent ongoing (i.e., chronic) inflammation that will not normally resolve on its own. More generally, a small stimulus does not cause a small inflammatory response due to a low level of activation of the NLRP3 inflammasome. Rather, it causes a zero response because it does not activate the NLRP3 inflammasome at all. Threshold mechanisms prevent adverse responses to routine small stimuli (although possibly diminishing some resilient capacity by depleting protective reserves). Such mechanisms include ultrasensitive molecular switches in key regulatory pathways (such as NF-κB); positive feedback, cooperativity and bistability in regulatory networks and oligomerization processes; discontinuous changes (e.g., rupture of lysosomes, ion fluxes, loss of membrane integrity); and homeostasis-preserving responses (e.g., production of antioxidants in response to oxidative stress in target tissues, cells, and organelles) that must be depleted (Bogen 2019; Cox 2018; Kelley et al. 2019).
- *Damage to chronically inflamed tissue.* The onset of chronic inflammation increases the exposure of cells in the inflamed tissue (such as alveolar epithelial cells or bronchiolar epithelial cells in the lung) to ROS and other toxic products, and thereby may increase their rates of cell death and turnover. However, clinically significant damage to the tissue may not occur unless and until its capacity to replace damaged or dead cells with healthy ones is strained or exhausted by continued damage from chronic inflammation.
- *Onset of disease.* Additional changes (e.g., fibrosis or scarring of tissue) may need to accumulate before inflammation and tissue damage lead to diagnosable disease states.

Figure 3.1 diagrams this simplified process.

Over the nearly two decades since the discovery of the inflammasome, a large and deep technical literature has developed that clarifies many mechanistic details

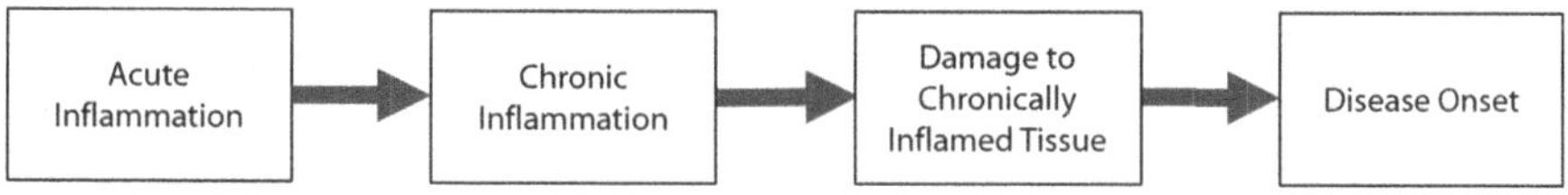

Fig. 3.1 Flow chart depicting the steps necessary for acute inflammation to result in disease onset

and poses sharply defined questions about others. For example, activation of the NLRP3 inflammasome requires sufficiently sustained and intense exposure to generate two signals separated in time (Jo et al. 2016; Yang et al. 2019). Signal 1 causes enough active NLRP3 protein filaments and other components to be secreted from the cell's nucleus (in response to signaling from receptors to the nucleus via the NF-κB pathway) into its cytoplasm so that the inflammasome can self-assemble, which occurs by a process of oligomerization. (The protein coded for by the NLRP3 gene in humans is called the NALP3 protein.) Signal 2 causes the assembled inflammasome to start producing mature pro-inflammatory cytokines (specifically, interleukins IL-1β and IL-18, which it produces by recruiting the enzyme caspase-1 and using it to cleave the protective caps from pro-IL-1β and pro-IL-18 in the cytosol). These events and their mechanisms have been elucidated in considerable detail (Cox 2018 and references therein). Briefly, the priming signal (Signal 1) typically comes from microbial molecular patterns or endogenous cytokines that activate transcription factor NF-κB, leading to upregulated production of NLRP3 and pro-interleukin-1β. The activation signal (Signal 2) can be generated by a variety of stimuli (e.g., extracellular ATP, RNA viruses, bacterial and fungal toxins, silica crystals, urea crystals, and asbestos fibers) that trigger fluxes in calcium and potassium ions, mitochondrial dysfunction, ROS production, and lysosomal membrane damage) (Kelley et al. 2019). On the other hand, the precise identity of Signal 2 remains a puzzle. It is associated with a host of closely interrelated events. These include potassium ion (K+) efflux, calcium ion (Ca2+) signaling, mitochondrial and cytoplasmic ROS production, inhibition of thioredoxin antioxidants, mitochondrial damage, lysosomal rupture induced by lysosome membrane permeabilization (LMP), and release of cathepsins from the lysosome (Yang et al. 2019).

The sequence of events mediated by the NLRP3 inflammasome is now understood well enough to propose useful AOP descriptions for use in risk assessment, even though some biochemical details of the key events have yet to be fully elucidated. The two initial steps of NLRP3 inflammasome activation and transition from acute to chronic inflammation, which are common to AOPs for many inflammation-mediated diseases, can be understood using a few relatively simple diagrams that summarize their main features.

Modeling the Time for Internal Doses to Reach Activation Thresholds

Internal Doses After Exposures to Constant Concentrations with Elimination

The existence of an activation threshold for the NLRP3 inflammasome has important practical implications for risk assessment. Figure 3.2 shows typical curves for the increases over time of concentrations of a pollutant in an internal compartment of a physiologically based pharmacokinetic (PBPK) model (the lung in this hypothetical case study, but applicable for any organ or tissue) for different concentrations in inhaled air. The exposure scenarios described in Fig. 3.2 are long-term exposures to constantly administered concentrations. Higher curves in Fig. 3.2 correspond to higher administered concentrations in inhaled air. The concentrations in the lung increase until inflow and outflow are equal; this dynamic equilibrium level can be seen as each line becomes approximately horizontal. (Concentration units on the y axis are nominal, since the same general pattern—an initial increase that levels off at an equilibrium value, for any given input concentration in air—holds for a wide variety of chemicals, e.g., respirable crystalline silica in alveolar macrophages, as described in Chap. 4). Similar curves hold for more complex PBPK models, in which metabolites rather than only parent compounds may be of interest. Time units (x axis) are also nominal; it is the height of the eventual dynamic equilibrium along the y axis that matters most.) When the pattern depicted in Fig. 3.2 holds, a key implication for risk assessment is that, at low enough exposure concentrations, the internal doses never reach the activation threshold (i.e., the horizontal line labeled T in Fig. 3.2, representing the smallest activation threshold for initiating an NLRP3-mediated inflammatory response). For example, as shown in Fig. 3.2, curves 1, 2, and 3 fail to reach the threshold (line T), meaning that chronic inflammation will never occur, regardless of the length of the exposure. By contrast, curves 4 and 5 reach the threshold after a short time, meaning chronic inflammation will occur.

If the activation threshold is different in different cells, or varies over time, then this line (T) may be thickened into a band or modeled with a distribution, but the point remains: sufficiently small concentrations in inhaled air will not activate the NLRP3 inflammasome when internal doses behave as in Fig. 3.2. Figure 3.2 also suggests that most exposure concentrations that cause inflammation do so relatively

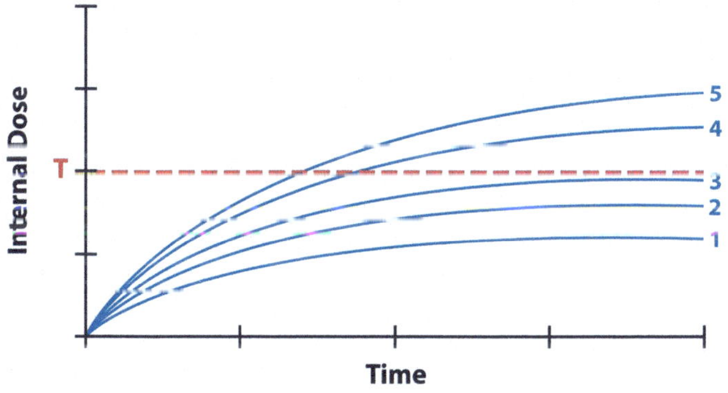

Fig. 3.2 Typical time courses for internal doses of a substance in the lung of a person exposed to a constant concentration in air. Higher curves correspond to higher administered concentrations in inhaled air. Regardless of exposure concentration, the internal dose reaches a maximum after a period of time and then remains constant as long as exposure continues at the same concentration. Line T is the threshold above which chronic inflammation will occur

quickly (typically in less than a year in biomathematical models for respirable crystalline silica; see Chap. 4). Portions of the internal dose curves above the activation threshold in Fig. 3.2 are not necessarily realistic because the onset of inflammation can change pharmacokinetics parameters (e.g., by degrading the performance of alveolar macrophages in clearing pollutants). However, the portions below or at the threshold model the time until an inflammatory response is triggered. This time is infinite if administered concentrations are low enough (curves 1–3), and limiting exposures to such concentrations is one way to protect against NLRP3 inflammasome-mediated disease risks. The approximate size of the threshold T in Fig. 3.2 can be estimated from PBPK models and toxicological data for some well-studied substances, such as respirable crystalline silica, but both interindividual variability in model parameters and variability in the physical, chemical, and biological properties of different samples of substances (e.g., due to differences in size, shape, and surface chemistry of crystals) create several-fold differences in T (Chap. 4).

Internal Doses After Exposures to Constant Concentrations Without Elimination

Figure 3.3 illustrates a different pattern for accumulation of internal dose if there is no elimination from the internal compartment. A practical example is the accumulation of extremely biopersistent mineral fibers embedded in lung or mesothelial tissue. The lines correspond to constant concentrations in inhaled air; steeper lines correspond to higher administered concentrations. The *y* axis shows the cumulative dose in the internal compartment (in this case, the lung) over time. Unlike Fig. 3.2, there is no clearance from the lung, so the internal dose steadily increases over time without ever leveling off at a final equilibrium value. However, if the administered concentration is small enough (i.e., lines 1, 2, and 3), the time required to reach a threshold level (line T) that activates the NLRP3 inflammasome will be more than a lifetime (the vertical green line). Such exposure concentrations may be regarded as virtually safe, in the sense that they will not cause inflammation during the course of an exposed worker's life.

The points made so far hold even in linear PBPK models with fixed (time-invariant) parameter values. In reality, however, high concentrations can saturate some clearance pathways and deplete protective resources (e.g., antioxidant pools), leading to nonlinear pharmacokinetics. The rates at which internal doses accumulate in the lung in Fig. 3.3 are not simply proportional to corresponding concentrations in inhaled air (as they would be for a linear PBPK model), but depend nonlinearly on the inhaled concentration. Fig. 3.4 makes this clearer by plotting the cumulative internal dose in an internal compartment (i.e., the lung) with no clearance over a fixed time interval for different exposure concentrations in air. (These values correspond to the different heights attained along the y axis by the time course curves in Fig. 3.3 after 100 time steps for exposure concentrations somewhat greater or less

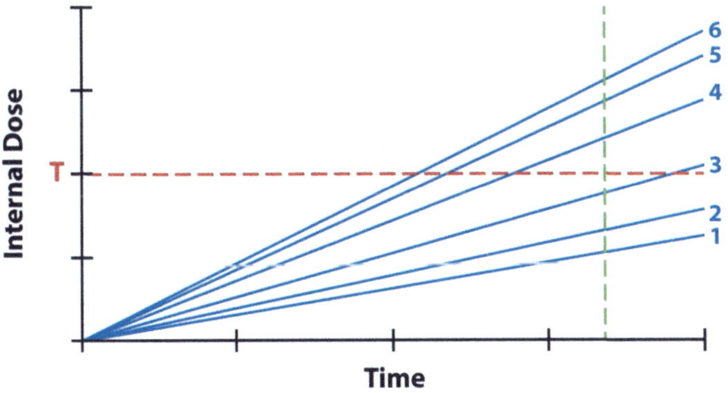

Fig. 3.3 Cumulative internal dose in the absence of elimination. Steeper lines correspond to higher administered concentrations in inhaled air. The internal dose continues to increase with time if exposure remains constant. T is the threshold above which chronic inflammation will occur. The vertical green line represents the average human lifetime

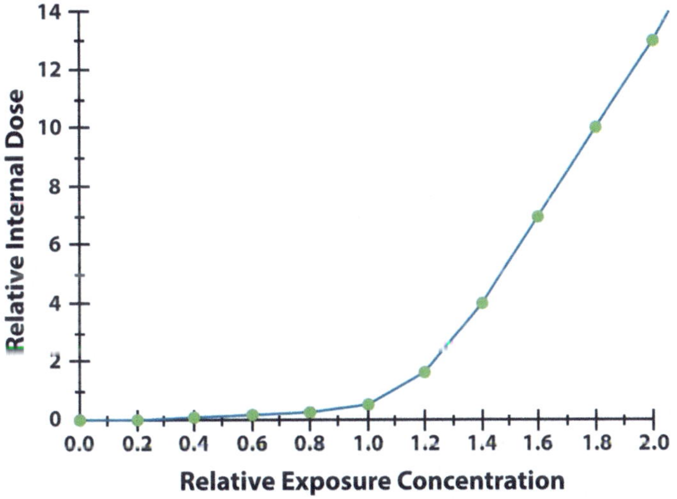

Fig. 3.4 Predicted relative accumulation (i.e., internal dose) in a target tissue over a given time interval *vs.* relative exposure concentration in inhaled air (where 1.0 is the exposure concentration for a typical worker). This demonstrates that the internal dose of an inhaled pollutant accumulates very slowly for small exposure concentrations, but much more quickly at higher concentrations

than the "normal" (base case) concentration, which is scaled to have a value of 1 on the horizontal axis in Fig. 3.4). Figure 3.4 shows a constant exposure concentration of 2 (i.e., twice the normal concentration) increases the accumulation rate of internal doses dramatically, by more than tenfold, instead of only by a factor of 2, as would be the case in a linear PBPK (or compartmental flow) model. The mathematical analysis leading to this nonlinearity is described in Chaps. 5 and 6, but its intuitive basis is that clearance of inhaled dose is impaired by high concentrations that, in effect, clog removal pathways, thus making accumulation of internal doses more rapid.

In Fig. 3.4, small relative exposure concentrations (x axis below about 0.8) cause relatively slow accumulations of internal doses, reflecting the body's ability to clear the substance from the body. At the elbow of the curve (between 0.8 and 1.2), the body's clearance mechanisms (e.g., mucociliary clearance of macrophages engulfing, or at least entangling, particles or fibers) start to saturate, and the substance builds up relatively quickly in target tissues. For example, for a constant exposure of 1, the body is able to clear almost all of the substance before it reaches the target tissue, but, at an exposure of 2, the body is unable to do so, leading to much more rapid accumulation of the internal dose in the target tissue (y axis). A striking implication is that an exposure concentration history that is 0 for half the time and 2 for the rest of the time can pose a much larger risk than a constant exposure concentration of 1, because the rate of accumulation of internal dose when the inhaled concentration is 2 is much more than twice the rate when the inhaled concentration is 1. Hence, the average concentration per unit time is not a sufficient basis for predicting risk—specifically, the time until internal dose in a target tissue first reaches an activation threshold.

Activation of Chronic Inflammation

Our discussion thus far has focused on keeping inhaled concentrations low enough to prevent internal doses from reaching an activation threshold for acute inflammation. A less demanding alternative is to prevent *chronic* inflammation, while allowing for the possibility of occasional transient, localized acute inflammatory responses. Recall from Fig. 3.2 that curves 1 through 3 never reached the activation threshold regardless of the length of the exposure duration. Such inflammatory responses that resolve spontaneously do not activate the AOP for chronic inflammation and the causal chain—involving long-term cell stress, death (pyroptosis), increased cell turnover, and increased cancer risk in the chronically inflamed tissue—is never completed. To be more specific about how much acute inflammation is needed before chronic inflammation occurs, we use the results of a mathematical model, presented in Chap. 4, and elaborated and applied in Chap. 5, in which uninflamed cells are more likely to become inflamed as the number of neighboring cells that are already inflamed increases. Figure 3.5 shows how inflammation (e.g., fraction of inflamed cells or sites) spontaneously evolves over time in the absence of further exposures, starting from different initial levels of inflammation at time 0. For example, at time 0, the curves below line T show varying levels of inflammation, but for all of them, the inflammation resolves over time and never becomes chronic. Although the two green curves begin at a similar level of inflammation, the curve that starts above the threshold (T) at time 0,rather than resolving, eventually increases to a new, higher equilibrium level (E), resulting in chronic inflammation. In fact, all initial levels of inflammation above threshold T, starting both above and below equalibrium, follow trajectories that lead to the chronic inflammation equilibrium (E).

The underlying model in Chap. 4 postulates that a large number of units, sites, or patches of tissue are susceptible to inflammation and that the probability per unit time that a currently uninflamed unit becomes inflamed depends on the fraction of its neighbors that are inflamed (i.e., undergoing pyroptosis). The uninflamed unit becomes inflamed if and only if its activation threshold is exceeded (for example, the lines above the threshold T in Fig. 3.5). Inflammation is also resolved at a certain rate, and this can keep inflammation in check if it starts at a low level (the lines below T in Fig. 3.5). The model quantifies the expected extent of inflammation via the fraction of inflamed units, as indicated by the height of the curve along the *y* axis, over time. According to this model, low initial levels of inflammation (below T) will be spontaneously extinguished over time in the absence of additional inflammatory exposures, while higher initial levels (above T) will spontaneously adjust toward a stable dynamic equillibrum level (indicated by E), at which the rate of resolution is balanced by the rate of new inflammation. Both this simple "tipping point" result, with T as the tipping point, and the existence of a sharp transition threshold are outcomes of analyses of underlying stochastic processes with the trajectories in Fig. 3.5 being generated by a mean-field approximation (Chap. 4). That the end result is simple, even though the underlying process is complex, is typical for such spatial stochastic processes (Page 2018).

The key aspect of Fig. 3.5 is that existence of activation thresholds for inflammation of individual units implies a threshold for the transition from self-resolving (acute) inflammation to self-sustaining (chronic) inflammation in the entire population of units. In other words, in this model, cell-level thresholds imply tissue-level

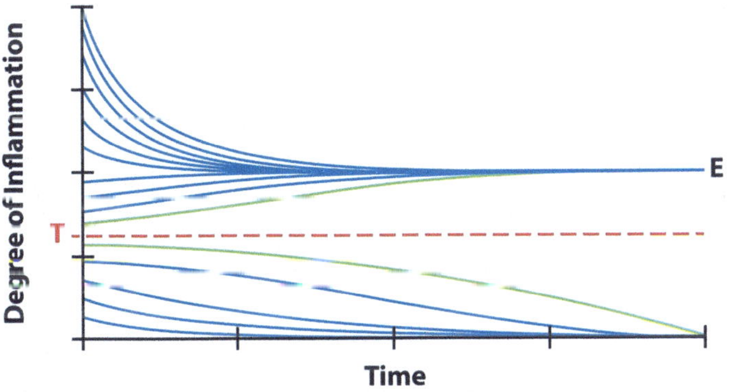

Fig. 3.5 A family of inflammation trajectories. There is a threshold (T) below which an inflammatory response will never result in chronic inflammation. At an inflammation level above this threshold, an inflammatory response will always lead to chronic inflammation. This occurs regardless of the cause of inflammation. T is the threshold above which chronic inflammation will occur. The stable dynamic equillibrum level is represented by E. The lower green line represents the highest degree of inflammation that will not result in chronic inflammation and the higher green line represents the lowest degree of inflammation that will result in chronic inflammation

thresholds. This is a characteristic of many models, in which the state of each individual cell or tissue (e.g., inflamed or not) depends on the states of its neighbors in such a way that transitions become easier or more likely as they become more common (Page 2018).

Internal Doses After Time-Varying Exposure Concentrations

While the foregoing discussion has focused on hypothetical constant-rate exposures over time, these scenarios have limited practical applications, particularly for occupational exposures. Regulators such as the United States Occupational Safety and Health Administration often assume lengthy exposures to constant concentrations (e.g., 45 years of an inhaled exposure to a certain concentration of a workplace chemical), but real-world occupational exposures typically involve considerable variability (e.g., long stretches of relatively low concentrations punctuated by occasional episodes of higher concentrations). Although realistic exposure histories depend on specific substances and exposure scenarios, Fig. 3.6 shows the time courses of internal doses in an internal compartment (the lung) for the same model as Fig. 3.2 (a linear PBPK model) for three different scenarios (see Chap. 4 for details of the PBPK model and parameter values). The same cumulative exposure (e.g., a given number of ppm-hours) is experienced by a worker each week, but it is spread over different numbers of days, from 5 days (curve 1, in green) to 1 day (curve 3, in blue). The worker is not exposed on weekends and the exposure pattern repeats consistently each week. Under these conditions, more concentrated exposures lead to higher internal peak concentrations, but the effect is modest: peak concentrations differ by less than a factor of 1.5 across these exposure scenarios.

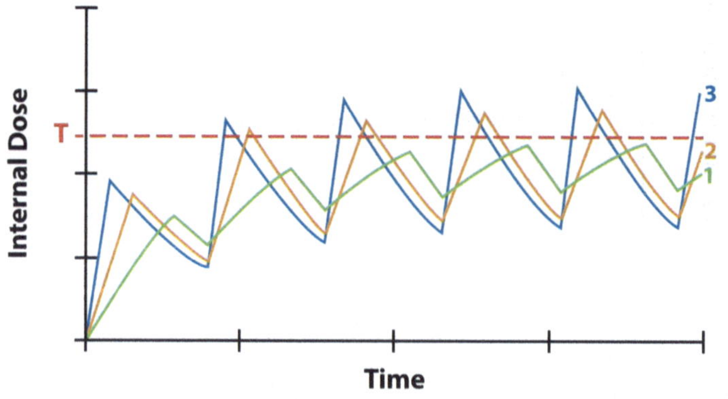

Fig. 3.6 Time courses of concentrations in an internal dose (i.e., the lung) for three different exposure profiles giving the same cumulative exposure per week. The total cumulative exposure each week is constant, but in curve 1 (green), the exposure is spread over 5 days each week, in curve 2 (orange), it occurs over 4 days each week, and in curve 3 (blue), the exposure occurs over 1 day each week. T is the threshold above which chronic inflammation will occur

For the nonlinear model in Figs. 3.3 and 3.4, the analogous effect is much largerbecause clearance is satuarable: different ways of allocating the same total exposure per week can change the resulting internal dose by more than tenfold, changing a virtually safe concentration (e.g., one that would not cause chronic inflammation over a duration of 120 years) to an unsafe one (e.g., causing chronic inflammation within 12 years), even though the average concentration per unit time is the same. The intuitive explanation is that high concentrations narrowly spaced in time (e.g., on Monday and Tuesday) are more hazardous than the same high concentrations more widely separated in time (e.g., on Monday and Friday) because the first high concentration partly saturates, and hence reduces, clearance, making the exposures that immediately follow more more effective in producing higher internal doses.

In the hypothetical example in Fig. 3.6, it is clear that having the exposure spread out over 5 days each week will never result in lung concentrations above the threshold (T). However, in every other situation, lung concentrations exceed the threshold every week. This again demonstrates how exposures spread out over a longer time period will not necessarily lead to chronic inflammation, but if this occurs over a shorter time period, it may.

Discussion

As has been extensively discussed in the literature, a mode-of-action (MoA) analysis describes a biologically plausible series of events (i.e., the functional or anatomical changes at a cellular level) that results from exposure to a particular substance or class of substances (Aleksunes and Eaton 2019). Meek et al. (2014) noted that "[t]he mode of action framework is based…on the premise that any human health effect caused by exposure to an exogenous substance can be described by a series of causally linked biochemical or biological key events that result in a pathological or other disease outcome." An AOP framework incorporates information from mechanistic and MoA analyses, but because the AOP framework describes a toxicity pathway and an adverse outcome resulting from different types of exposures, it is more broadly applicable to multiple classes of chemicals and to an entire population. Thus, AOPs augmented with insights from quantitative modeling can be useful in risk assessment because, once an exposure initiates the key events on the AOP, one can follow the trajectory of the AOP.

This chapter has illustrated how the existence of an activation threshold below which inflammation is not triggered contrasts with familiar probabilistic dose-response models and stochastic models of carcinogenesis in which even arbitrarily small exposures create some (albeit possibly very small) increased risk. The existence of such thresholds for chronic inflammation is well supported by current understandings of multiple biological mechanisms. Figures 3.2, 3.3, and 3.6 highlight the importance of exposure duration, as well as exposure concentration, in determining whether and when an adverse response will occur; this is examined further in Chap. 6. In these diagrams, chronic inflammation occurs if and only if internal dose reaches a sufficiently high level (denoted by T) within an exposed individual's lifetime. On a much shorter time scale, as previously noted, NLRP3 inflammasomes are activated only if concentrations remain high enough for long enough to generate both Signals 1 and 2 (Jo et al. 2016; Yang et al. 2019). Previous work on receptor-mediated AOPs has also demonstrated the importance of sustained activation in causing adverse events, showing that simplified idealizations in which occurrence of an MIE triggers a sequence of changes leading to adverse outcomes are often too simple: how long the MIEs last, and how widely-spread they are in the target tissue, may also be important in determining the probability of an adverse outcome (Becker et al. 2015; Cox 2019) (Fig. 3.7).

The NF-κB pathway that transduces damage signals to the macrophage nucleus to upregulate transcription of NLRP3 protein filaments acts as an ultrasensitive switch. If too few receptors are bound, this signaling pathway is not activated, and the NLRP3 inflammasome will not be formed. The assembly of the NLRP3 inflammasome involves oligomerization and exhibits cooperativity (the addition of polymer units becomes more energetically favorable as the process proceeds), which has characteristic all-or-nothing kinetics. The rupture of lysosomes (subsequent to induction of LMP) does not occur unless concentrations of toxic contents in the lysosome exceed a threshold (Cox 2018; Kelley et al. 2019). The induction of pyroptosis occurs only when caspase-1 activation passes a critical threshold (Bergsbaken et al. 2009).The NLRP3 inflammasome has an activation threshold – activation requires sustained exposures high enough to generate Signals 1 and 2 close enough together so that activation can proceed.

Thus, it appears realistic that sufficiently low exposure concentrations do not increase the risk of NLRP3 inflammasome-mediated diseases because they do not activate the NLRP3 inflammasome. This warrants focusing on how long it will take (if ever) for internal doses of substances to become high enough to activate the NLRP3 inflammasome (or to initiate consequent chronic inflammation). The simple diagrams we have presented provide a framework for answering that question. The main implications for risk assessment, discussed further in Chap. 6, are as follows:

1. Safe and virtually safe exposure concentrations can be defined as those that do not activate NLRP3 inflammasomes or chronic inflammation during a worker's lifetime.
2. Exposure metrics based on cumulative exposure per unit time are insufficient for predicting risks, since risks and their determinants (e.g., time to onset of chronic inflammation, which increases disease risks) depend on the time patterns of administered concentrations.
3. In particular, high concentrations spaced closely in time can be disproportionately risky compared to the same high concentrations more widely separated in time, even if the average cumulative dose per unit time is the same. Many current occupational safety standards do not address this important aspect of exposure.
4. Randomly fluctuating exposure concentrations can be substantially more dangerous than constant exposure concentrations with the same cumulative dose per unit time if high fluctuations do disproportionately more harm than low fluctuations prevent (Fig. 3.4).

Fig. 3.7 Adverse outcome pathway for NLRP3-induced chronic inflammatory diseases

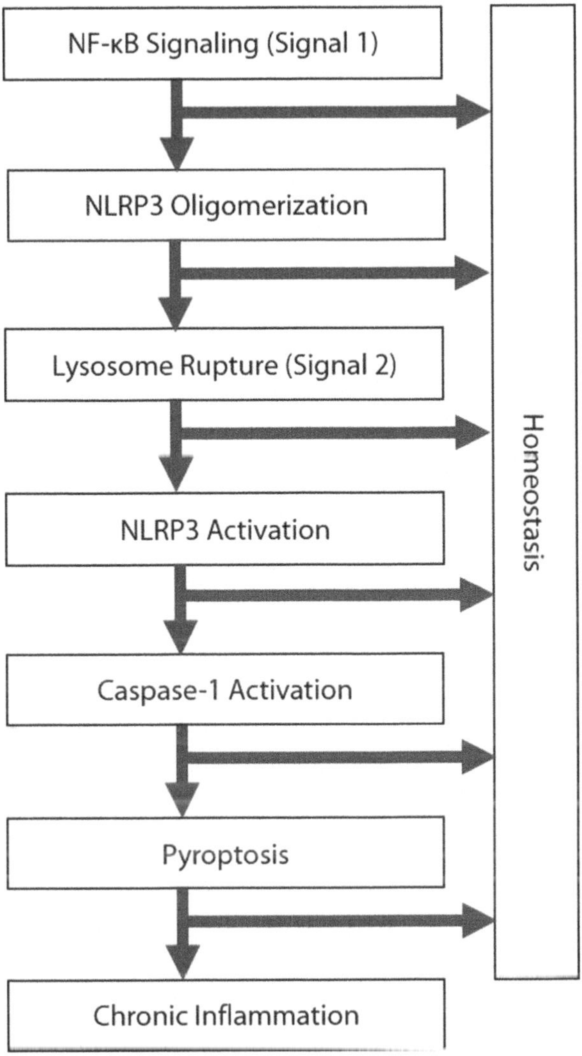

Conclusions

Chronic inflammation has emerged as an increasingly recognized MoA for many exposure-related pathologies in the past two decades. This chapter has presented a simple graphical description of the main features of internal dose time courses that are important when pharmacodynamics are governed by an activation threshold, using the NLRP3 inflammasome activation AOP to illustrate general principles. Similar considerations hold for other AOPs that are rate-limited by processes with activation thresholds (e.g., ultra-sensitive molecular switches, cooperativity in assembly of signaling complexes, programmed cell death). The risk analysis implications of AOPs with threshold or threshold-like pharmacodynamic responses include the need to consider how cumulative dose per unit time is distributed over time and the possibility that safe, or virtually safe, exposure concentrations can be defined for such processes. The next three chapters develop these points in more detail. They point toward an improved approach to risk management in which biologically realistic responses to exposure patterns, rather than aggregate statistical models of average responses to average exposure concentrations, play a greater role in informing standards and practices to better protect worker health.

References

Aleksunes LM, Eaton DL. Principles of toxicology. In: Klaassen CD, editor. Casarett and Doull's toxicology: the basic science of poisons. 9th ed. New York: McGraw-Hill Education; 2019. p. 25–64.

Allen TEH, Goodman JM, Gutsell S, Russell PJ. A history of the molecular initiating event. Chem Res Toxicol. 2016;29(12):2060–70. https://doi.org/10.1021/acs.chemrestox.6b00341.

Ankley GT, Bennett RS, Erickson RJ, Hoff DJ, Hornung MW, Johnson RD, Mount DR, Nichols JW, Russom CL, Schmieder PK, Serrano JA, Tietge JE, Villeneuve DL. Adverse outcome pathways: a conceptual framework to support ecotoxicology research and risk assessment. Environ Toxicol Chem. 2010;29(3):730–41.

Becker RA, Patlewicz G, Simon TW, Rowlands JC, Budinsky RA. The adverse outcome pathway for rodent liver tumor promotion by sustained activation of the aryl hydrocarbon receptor. Regul Toxicol Pharmacol. 2015;73(1):172–90. https://doi.org/10.1016/j.yrtph.2015.06.015.

Bergsbaken T, Fink SL, Cookson BT. Pyroptosis: host cell death and inflammation. Nat Rev Microbiol. 2009;7(2):99–109. https://doi.org/10.1038/nrmicro2070.

Bogen K. Inflammation as a cancer co-initiator: new mechanistic model predicts low/negligible risk at noninflammatory carcinogen doses. Dose-Response. 2019;17(2):1–12. https://doi.org/10.1177/1559325819847834.

Cox LA (2018) Biological mechanisms of non-linear dose-response for respirable mineral fibers. Toxicol Appl Pharmacol. https://doi.org/10.1016/j.taap.2018.06.016

Cox LAT. Risk analysis implications of dose-response thresholds for NLRP3 inflammasome-mediated diseases: Respirable crystalline silica and lung cancer as an example. Dose-Response. 2019;17(2):1–21. https://doi.org/10.1177/1559325819836900.

Cox LA Jr, Goodman JE, Engel AM. Chronic inflammation, adverse outcome pathways, and risk assessment: a diagrammatic exposition. Regul Toxicol Pharmacol. 2020;114:104663. https://doi.org/10.1016/j.yrtph.2020.104663.

Jo EK, Kim JK, Shin DM, Sasakawa C. Molecular mechanisms regulating NLRP3 inflammasome activation. Cell Mol Immunol. 2016;13(2):148–59. https://doi.org/10.1038/cmi.2015.95.

Kaplan BLF, Sulentic CEW, Haggerty HG, Holsapple MP, Kaminski NE. Toxic responses of the immune system. In: Klaassen CD, editor. Casarett and Doull's toxicology: the basic science of poisons. 9th ed. New York: McGraw-Hill Education; 2019. p. 633–717.

Kelley N, Jeltema D, Duan Y, He Y. The NLRP3 inflammasome: an overview of mechanisms of activation and regulation. Int J Mol Sci. 2019;20(13):3328. https://doi.org/10.3390/ijms20133328.

Meek ME, Boobis A, Cote I, Dellarco V, Fotakis G, Munn S, Seed J, Vickers C. New developments in the evolution and application of the WHO/IPCS framework on mode of action/species concordance analysis. J Appl Toxicol. 2014;34(1):1–18. https://doi.org/10.1002/jat.2949.

Page SE. The model thinker: what you need to know to make data work for you. New York: Basic Books; 2018.

Yang Y, Wang H, Kouadir M, Song H, Shi F. Recent advances in the mechanisms of NLRP3 inflammasome activation and its inhibitors. Cell Death Dis. 2019;10(2):128. https://doi.org/10.1038/s41419-019-1413-8.

Chapter 4
Case Study: Occupational Health Risks from Crystalline Silica

Introduction

Chapter 3 pointed out that chronic inflammation mediates an extraordinarily wide range of diseases. Recent progress in understanding intracellular inflammasome assembly, priming, activation, cytokine signaling, and interactions with mitochondrial reactive oxygen species (ROS), lysosome disruption, cell death, and prion-like polymerization and spread of inflammasomes among cells, has potentially profound implications for dose-response modeling. This chapter and Chaps. 5 and 6 further discuss biological mechanisms of exposure concentration and duration thresholds for NLRP3 inflammasome-mediated inflammatory responses, and develop and apply simple biomathematical models of the onset of exposure-related tissue-level chronic inflammation and resulting disease risks. This chapter focuses on respirable crystalline silica (RCS) and lung cancer risk as an example. It proposes an inflammation-mediated two-stage clonal expansion (I-TSCE) model of RCS-induced lung cancer that explains why relatively low estimated concentrations of RCS (e.g., <1 mg/m^3) do not increase lung cancer risk and why even high occupational concentrations increase risk only modestly (typically RR < 2). The model of chronic inflammation implies a dose-response threshold for excess cancer risk, in contrast to traditional linear no-threshold (LNT) assumptions. If this implication is correct, then concentrations of crystalline silica (or amphibole asbestos fibers, or other environmental challenges that act via the NLRP3 inflammasome) below the threshold do not cause chronic inflammation and resulting elevated risks of inflammation-mediated diseases.

L. A. Cox Jr., *Quantitative Risk Analysis of Air Pollution Health Effects*, International Series in Operations Research & Management Science 299,
https://doi.org/10.1007/978-3-030-57358-4_4

Setting: Linear No-Threshold (LNT) vs. Threshold Dose-Response Models

As the field of quantitative risk assessment (QRA) for carcinogens has matured over the past 40 years, several fundamental questions have remained unresolved. One concerns the realism of the linear-no-threshold (LNT) assumption or hypothesis that exposure-related excess risk of cancer is proportional to exposure concentrations and durations, at least at relatively low levels of exposure. Despite its convenience and widespread use in regulatory risk assessments, enthusiasts of biologically-based risk assessment and mode-of-action (MoA) considerations have long criticized the LNT assumption as being unrealistic for important non-genotoxic and genotoxic carcinogens (Bolt et al. 2004; Fukushima et al. 2005; Lutz and Kopp-Schneider 1999). Quantitative risk assessments using two-stage and multi-stage clonal expansion (TSCE and MSCE) models of carcinogenesis indicate that many carcinogens increase cancer risks by increasing the net birth rates of altered (e.g., "initiated" or "pre-malignant") cells, rather than (or in addition to) hastening mutations by acting directly on DNA. Exposure-related increases in proliferation ("clonal expansion" or "promotion") of the pool of cells at risk of being transformed into malignant cells by a mutation can increase cancer risk even if exposure does not increase the mutation rate itself. Dose-response thresholds arise if compensating proliferation of cells in response to unrepaired cytotoxic damage (e.g., regenerative hyperplasia subsequent to oxidative damage) occurs only at doses high enough to cause such damage, perhaps after depleting pools of protective resources such as antioxidants or saturating capacity for repair of oxidative damage (Fukushima et al. 2005; Lutz 1998; Butterworth et al. 1992). This might be called the *cytotoxicity/cell proliferation* model of thresholds in carcinogen dose-response relations. It can also lead to U-shaped or J-shaped dose-response relations (hormesis) if low exposure levels upregulate production of antioxidants or other protective resources that become depleted or saturated at higher doses (Calabrese and Baldwin 1998; Lutz 1998; Conolly et al. 2005; Bogen 2017). Thresholds and hormesis have been discussed for important carcinogens including arsenic (Snow et al. 2005), chloroform (Wolf and Butterworth 1997), 1,4-dichlorobenzene (Butterworth et al. 2007), diesel exhaust, ionizing radiation, dioxin (TCDD) (Lutz 1998), saccharin (Cohen 1995; Whysner and Williams 1996) and many others (Bogen 2017; Calabrese and Baldwin 1998).

For lung carcinogenesis caused by inhaled particles or fibers, compensating proliferation of relevant cell populations (possibly including alveolar type II epithelial cells, bronchial basal cells, Clara cells, pulmonary neuroendocrine cells (PNECs), and bronchioalveolar stem cells at bronchioalveolar duct junctions) depends not only on direct damage from particles, but also on resulting releases of reactive oxygen species (ROS) from damaged cells and inflammation of the alveolar epithelium and the lung environment (Castranova et al. 2002; Porter et al. 2006; Borm et al. 2018). ROS is known to act as a second messenger affecting stem cell proliferation for self-renewal and differentiation in many systems (Paul et al. 2014; Ren et al. 2015). In the lung, ROS signaling is involved in inflammation and cytotoxicity caused by quartz dusts and other particles, with response thresholds for measures of cytotoxicity and inflammation (e.g., tumor necrosis factor (TNF) alpha) that depend on the physicochemical properties of the dust (Bruch et al. 2004). Experiments in rats show an exposure threshold above which lungs become overburdened and clearance of even low-solubility and relatively low toxicity particles by alveolar macrophages (AMs) becomes impaired (Borm et al. 2015). However, the rat lung appears to be unique in this regard, so it is widely considered that there are no clear implications for human carcinogenicity of rat lung carcinogenicity that depends on such overburdening of the lung (Warheit et al. 2016).

More recently, a different type of threshold has been elucidated for inflammatory responses in lungs exposed to mineral particles and mineral fibers. Cell-level inflammatory responses to a wide variety of environmental dangers, including inhaled respirable crystalline silica (RCS) and asbestos fibers, are mediated by the NOD-like receptor protein 3 (NLRP3) inflammasome introduced in Chap. 3. This is a supramolecular signaling complex that assembles in the cytosol of alveolar macrophages, alveolar epithelial cells, and other lung cells and coordinates releases of inflammatory cytokines in response to a wide range of stimuli (Sayan and Mossman 2016). The complex contains a pattern recognition receptor that recognizes a variety of pathogen-associated and danger-associated molecular patterns—specifically, a nucleotide-binding oligomerization domain (NOD)-like receptor (NLR), from which the name "NOD-like receptor protein 3" (NLRP3) is taken, although it is also often used to abbreviate longer and more descriptive names for the same complex, e.g., "nucleotide-binding domain, leucine-rich repeat region-containing family, pyrin domain-containing-3" (Abderrazak et al. 2015). As mentioned in Chap. 3, the NLRP3 inflammasome has been implicated in the pathogenesis of exposure-related fibrosis, silicosis, asbestosis, lung cancer, mesothelioma, and many other diseases, from viral and bacterial infections to type 2 diabetes, heart disease, Alzheimer's and Parkinson's diseases, atherosclerosis, and arthritis, gout, and other autoimmune diseases (Coll et al. 2016). It senses damage signals and disruption of cellular homeostasis and responds by producing the potent pro-inflammatory cytokines interleukin (IL) 1 beta (IL-1β) and IL-18 and inducing pyroptosis, a highly inflammatory form of programmed cell death that leads to swelling and lysis (rupture) of the cell membrane and release to the extracellular environment of its inflammatory contents—including IL-1β, IL-18, and activated NLRP3 inflammasomes (Sayan and Mossman 2016; Peeters et al. 2013; Luna-Gomes et al. 2015; Bergsbaken et al. 2009; Baroja-Mazo et al. 2014). The NLRP3 inflammasome provides important portions of a previously missing link between exposure-related ROS-associated oxidative damage within the cell and the spread of inflammation between cells that creates chronic inflammation of tissue, a high-ROS tissue environment, and increased risks of inflammation-mediated diseases (Abais et al. 2015; Harijith et al. 2014).

The remainder of this chapter focuses on RCS-lung cancer dose-response relations as a specific example, but similar biological points hold more generally for other inflammatory agents and inflammation-mediated diseases. The main points in the following sections are as follows.

- Contrary to the LNT model, priming, assembly and activation of the NLRP3 inflammasome in response to inhaled RCS, asbestos, and other agents involve clear dose-response thresholds. Both concentration and duration of exposure have thresholds: neither high concentrations for too short a period nor prolonged exposures to too low a concentration can activate the NLR3 inflammasome. Exposures that do not generate sufficiently high concentrations of RCS particles in alveolar macrophages (AMs) for sufficiently prolonged durations do not activate their NLRP3-mediated responses. We review several different biological mechanisms that prevent NLRP inflammasome activation in response to brief or low-concentration exposures.
- Dose-response thresholds at the level of intracellular NLRP3 inflammasome activation imply tissue-level dose-response thresholds for inflammation, and specifically for onset of chronic, irreversible inflammation mediated by activation of NLRP3 inflammasomes. We present a simple mathematical model linking individual cell-level thresholds for inflammation to resulting tissue-level thresholds for self-sustaining chronic inflammation in a population of cells.
- These dose-response thresholds hold also for induction of chronic inflammation-mediated diseases such as silicosis, fibrosis, or lung cancers in workers exposed to high and prolonged concentrations of RCS. We illustrate this point by embedding the tissue-level inflammation model between a previously developed physiologically-based pharmacokinetic (PBPK) front end model for RCS and a two-stage clonal expansion (TSCE) model of lung carcinogenesis.
- Realistic estimates of inter-individual variability in dose-response thresholds suggest that populations of people (e.g., occupationally exposed populations) also have exposure concentration and duration thresholds below which exposures do not increase risks of NLRP3 inflammasome-mediated diseases.

If dose-response thresholds exist, they dramatically change both the information requirements for quantitative risk assessment (QRA) and the nature of risk management calculations that compare marginal benefits to marginal costs of incremental reductions in allowed exposure concentrations. This chapter concludes by briefly examining some of the risk assessment and risk management implications of exposure thresholds for inflammation-mediated diseases, and Chaps. 5 and 6 examine these implications further.

Biological Thresholds for NLRP3 Inflammasome Priming, Assembly, and Activation

This section reviews and synthesizes aspects of NLRP3 inflammasome biology related to dose-response thresholds for triggering NLRP3 activation and induction of inflammatory responses to stressors such as RCS. The goal here is qualitative: to review biological grounds for concluding that dose-response thresholds exist at the level of individual cells. The next section explains how cell-level thresholds create tissue-level thresholds for exposure-induced chronic inflammation. Quantitative assessment of how large this chronic inflammation dose-response threshold is for RCS, and how variable it is among individuals, are addressed subsequently.

The NLRP3 inflammasome is a multi-protein complex that forms in alveolar macrophages (AMs) and other cells in response to detection of sufficiently strong damage-associated signals or bacterial proteins by toll-like receptors (TLR) (especially TLR4) on the cell's surface or by other pattern recognition receptors. In response, signaling cascades transmitted via NFκB pathways to the cell's nucleus induce it to upregulate production ofpro-IL1β and of NLRP3 protein filaments, secreting them into the cytosol, thus "priming" the cell for assembly and activation of the NLRP3 inflammasome (Lopez-Castejon et al. 2013). Two distinct thresholds, one for NFκB signaling and one for MAPK activation, provide an ultrasensitive, switch-like response to TLR4-induced inflammation, preventing production of inflammatory cytokines by macrophages in response to weak TLR4 stimulation so that low levels of damage do not trigger inflammatory responses (Gottschalk et al. 2016). Priming also increases deubiquitination of NLRP3, allowing post-translational as well as transcriptional regulation of active NLRP3 protein balance (He et al. 2016; Hauenstein et al. 2015; Lopez-Castejon et al. 2013). If the concentration of protein becomes high enough and damage signals continue to be received, Pyrin domain-containing sensors on the filaments lose their autoinhibition (e.g., via deubiquitination) and begin a process of orderly nucleated polymerization (assembly of successive units) of ASC and oligomerization (assembly of different types of protein units into a complex) to create a fully functional, wheel-shaped NLRP3 inflammasome in the cytosol (He et al. 2016; Hauenstein et al. 2015). The completed inflammasome has a Pyrin domain-containing sensor in its NLRP3 component, a caspase-recruitment domain (CARD) effector component, and an adaptor molecule linking them: the apoptosis-associated speck-like protein containing a CARD (ASC). The assembly process is cooperative: once begun, addition of further units becomes more favorable, leading to a characteristic all-or-nothing response for inflammasome assembly (Lu and Wu 2015; Hauenstein et al. 2015). Conversely, if damage signals are too rare or too brief, the critical concentration of uninhibited NLRP3 sensors and protein filaments required to initiate cooperative oligomerization is not reached, and assembly and activation of NLRP3 inflammasomes then do not occur. Thus, cooperative assembly in response to sustained damage signaling creates dose-response thresholds for NLRP3 inflammasome assembly and NLRP3-mediated inflammatory responses. In general, nucleated cooperative polymerization exhibits a sharp concentration threshold for the constituent subunits (e.g., ASC): below a critical concentration, they cannot self-assemble into polymers, but above it, they do. This has long been understood on biophysics grounds (e.g., Miraldi et al. 2008). Assembly of the NLRP3 inflammasome illustrates this general principle.

Activation of the assembled inflammasome requires a second signal, referred to as "signal 2," that again appears to operate in part by deubiquitinating NLRP3. (In contrast to the mitochondrial ROS (mROS)-dependent deubiquitination involved in priming, deubiquitination induced by the second signal is mROS independent, indicating that distinct regulatory mechanisms are involved (Lopez-Castejon et al.

2013; Bednash and Mallampalli 2016).) Although the precise molecular identity of signal 2 is uncertain, the closely correlated events of potassium ion (K+) efflux, calcium ion (Ca2+) signaling, high mitochondrial and cytoplasmic ROS production, inhibition of antioxidant thioredoxins (Harijith et al. 2014), mitochondrial damage, and lysosomal rupture following lysosome membrane permeabilization (LMP), leading to release of cathepsins from the lysosome (Boya and Kroemer 2008), have all been proposed to play critical and interrelated roles; these are reviewed in more detail elsewhere (Cox Jr. 2018b; Katsnelson et al. 2016). LMP and rupture of the lysosome appears to be a central pathway for NLRP3 inflammasome activation specifically for silica and other respirable particles (Jessop et al. 2017). Lysosome rupture is of course an intrinsically threshold phenomenon: it occurs only when intralysosomal concentrations of cell damage products become too high. Moreover, the fact that signal 2 must be received after the inflammasome has assembled and before it is deactivated and removed via ubiquitination and autophagy (Shi et al. 2012) implies a temporal threshold for activation: substantial damage to the cell must be sustained for long enough for priming, assembly, and activation to take place in a cell, i.e., on a time scale of hours.

The activated NLRP3 inflammasome recruits caspases, including caspase-1, which cleaves the protective protein ends from pro-IL-1β and pro-IL-18, forming mature active IL-1β and IL-18 that are secreted by the AM and exert a number of potent pro-inflammatory effects via the cytokine and chemokine networks, including recruiting phagocytes (neutrophils and macrophages) to engulf and devour the sources of inflammation. Caspase-1 also causes mitochondrial damage and disassembly, production of mitochondrial ROS (mROS), and pyroptosis, leading to the swelling and bursting (lysis) of the inflamed cell and release of its inflammatory contents, fueling further inflammatory signaling to nearby cells. The release of ASC specks and NLRP3 inflammasomes into extracellular spaces facilitates ROS activation and spread of inflammation to nearby cells, as the released components continue to promote IL-1β maturation and broadcast inflammatory danger signals to neighboring cells, which can respond by becoming inflamed themselves (Baroja-Mazo et al. 2014). The release and subsequent uptake of ASC specks also spreads inflammation among macrophages via a prion-like transmission mechanism: when phagocytized by macrophages, the previously released ASC specks induce lysosomal damage, nucleation of ASC oligomerization and fiber growth, and IL-1β activation in their new hosts, initiating a new round of inflammatory responses, pyroptosis and lysis, and release of the specks to continue the process (Franklin et al. 2014).Inflammation is like a contagious disease in the population of cells: inflamed neighbors increase the probability that a currently uninflamed cell will become inflamed.

At the tissue level, if inflammation remains unresolved due to prolonged damage signaling and continued activation of NLRP inflammasomes, then it is plausible that the inflammatory response ceases to serve the health-preserving role of restoring homeostasis for local cells and tissue and instead crosses over to a pathological process of harmful unresolved inflammation: in response to chronic exposure to NLRP3-activating conditions, "the homeostatic threshold of NLRP3 may be surpassed, resulting in excessive and detrimental inflammation, and ultimately disease. …NLRP3 becomes pathologic either because of too strong a response or because of the overactivation of macrophages due to the chronic presence of excessive amounts of [noxious] substances" (Coll et al. 2016). A simple mathematical model is proposed next to help illuminate how chronic high exposures can cause such a transition from homeostatic to pathological inflammation.

Modeling the Transition from Transient to Chronic Inflammation

The main point of this section is that inflammation dose-response thresholds for individual cells imply inflammation dose-response thresholds for cell populations and tissues. As mentioned in Chap. 3, this conclusion holds in various models of inflammation. It is robust to many variations and uncertainties about the details of the inflammatory process, but follows mainly from the key assumption that the rate (probability per unit time) at which a cell or area of tissue becomes inflamed increases when its neighbors are inflamed. The spread of inflammation in a population of cells is then somewhat analogous to the spread of an infectious epidemic in a spatially distributed population of people. Dose-response thresholds for self-sustaining inflammation, analogous to epidemic thresholds, arise for a wide range of alternative assumptions about transmission among neighbors. The modeling in this section also seeks to clarify how sustained exposure can drive a transition from homeostatic, transient, self-resolving inflammation to pathological, self-sustaining, chronic inflammation.

Several mathematical and computational models have previously been proposed for lung tissue inflammation in response to inhaled microbial, viral, and particulate challenges (Smith et al. 2011; Camberlein et al. 2015). Both ordinary differential equation (ODE) models (Smith et al. 2011) and agent-based models (ABMs) (Brown et al. 2011) have been used to simulate the dynamics of lung cell populations and their behaviors (e.g., activation, functioning, and removal of alveolar macrophages) over time and within different lung compartments or locations. However, no model of which we are aware has attempted to provide a realistically detailed account of the many variations in individual cell states or the complex interactions among cells within and between different cell populations, and nor shall we. Rather, models of lung inflammation typically aggregate many cell populations into a few representative types or compartments and study simplified version of their interactions to illuminate their qualitative and quantitative dynamics. ABMs (and partial differential equations models) can consider locations as well as time, but typically model changes in the states of cells at specified locations as depending on the states of neighboring cells via very simple rules or functions. Current biological knowledge is probably too limited to permit much more detailed and realistic modeling; for example, the precise identity of the "Signal 2" activating primed NLRP3 inflammasomes is not yet known, and many other details of inflammasome regulation both within and among cells (e.g., which specific ubiquitins and deubiquitins are involved and how they vary over time and location within the cytosol) are still being elucidated. Despite these knowledge gaps, we propose that a simple model that borrows ideas from both ODE (compartmental) and ABM modeling can shed useful light on how exposure to RCS or other NLRP3-activating substances can shift responses from homeostatic inflammation (self-limiting inflammation that resolves itself after exposure ceases) to pathological chronic inflammation that sustains itself even in the absence of further exposure.

Consider a population of units, each of which is either "inflamed" or "not inflamed" at any moment. For example, we might focus on AMs as the units and declare a macrophage to be inflamed if it contains at least one activated NLRP3 inflammasome. Or, since there are many other cell types (e.g., monocytes, neutrophils, dendritic cells, etc.) involved in lung inflammation, and knowledge and data to support detailed modeling of their behaviors, movements, and interactions are not readily available, the units might instead be interpreted as referring to locations or "patches" within the lung, with each patch being either inflamed or not at any moment. (Patches can be envisioned as small contiguous spatial areas that collectively tessellate the tissue, analogous to elements in a finite-element analysis grid. A patch might be defined as being inflamed if it harbors activated AMs, or has ASC specks or activated inflammasomes in its extracellular spaces, or has concentrations of IL-1β significantly above normal background levels, or meets other operational criteria for inflammation.) We can afford such useful ambiguity about the precise interpretation of the model because the same equations describe a wide range of phenomena and can be interpreted in different ways without affecting the validity of resulting conclusions about the spread of inflammation.

For notation, Let $P(t)$ denote the fraction of the population (of cells or patches) that is inflamed at time t; thus, $1 - P(t)$ is the fraction that is not inflamed then. When time is not being emphasized, these quantities can be denoted more briefly as P and $1 - P$. Assume that inflamed cells are replaced by uninflamed ones at a rate of d replacements per inflamed cell per unit time. The parameter d represents the combined effects of inflammation-resolving mechanisms such as ubiquitination of active NLRP3 inflammasomes, autophagy by autophagosomes formed in response to, and partly co-localized with, inflammasomes (Shi et al. 2012), removal of macrophages (perhaps with burdens of RCS particles that have not been successfully phagocytosized) via the mucociliary escalator, and pyroptosis and replacement of inflamed cells that have died (e.g., via pyroptosis) or been removed by uninflamed ones. (Of course, modeling these clearance processes by a single rate constant is an over simplification, as they have their own detailed dynamics. For example, particle overload in the rat lung compromises the efficiency of phagocytosis, corresponding to a decrease in d (Porter et al. 2004; Borm et al. 2018). However, we shall focus on lower exposures and long averaging times, for which the simplified approximate description of clearance of inflamed cells by a single rate constant yields useful insights into inflammation dynamics on a time scale of months to years or decades, despite its undoubted inaccuracies on short time scales.) Conversely, uninflamed cells become inflamed at a rate that depends on the states (inflamed or not) of other cells. The flow of cell from uninflamed to inflamed at any moment is proportional to $1 - P$ and to the strength of proinflammatory signals that these currently uninflamed cells are receiving. Mean-field theory (Wang et al. 2016) suggests approximating this signal strength by some increasing function of *P*, which we will denote by *f*. Internal doses of RCS particles or other poorly soluble particles can also cause AMs to become inflamed as phagocytosis fails to dissolve the particles and they prime and activate NLRP3 inflammasomes. Letting *c* denote the fractional rate at which uninflamed cells become directly inflamed by internal dose, the mass-balance ODE describing the evolution of *P* over time has the following form:

$$dP / dt = f(P) * (1 - P) + c * (1 - P) - d * P. \tag{4.1}$$

All of the parameters and *P* itself may vary over time and in response to dose. To focus at first on the dynamic response of the lung to previous inflammatory exposures, we will consider scenarios in which a history of exposure has led to some starting level of inflammation, denoted *P*(0) (with time 0 defined as the time at which we start tracking the inflammation process) and with no further exposure, i.e., $c = 0$, over the time interval for which response is modeled. For such exposure scenarios, Eq. (4.1) simplifies to Eq. (4.1a).

$$dP / dt = f(P) * (1 - P) - d * P. \tag{4.1a}$$

To obtain an explicit model that supports simulation of time courses, it is necessary to specify the function *f*. ABM simulations have each cell or patch (i.e., each "agent" or unit) make state transitions (e.g., from uninflamed to inflamed) at each moment based on the current states of its neighbors. Although mean-field approximations and lumped-parameter ODEs such as Eq. (4.1a) do not explicitly identify the neighbors of individuals units, a stochastic ABM with discrete time steps would typically assume that each unit has *N* neighbors and that each uninflamed unit becomes inflamed if and only if at least *k* of its *N* neighbors are inflamed. The crucial assumption that each individual cell (or other unit) has a dose-response threshold for becoming inflamed is reflected in the threshold parameter *k*. (In reality, cells are heterogeneous and *k* may be different for different cells; this and other complexities are discussed subsequently.) When the fraction of units that are inflamed is *P*, the probability that a randomly sampled uninflamed unit will have at least *k* inflamed neighbors, and hence become inflamed itself, is determined from the binomial cumulative distribution function (CDF) with parameters *k*, *N*, and *P*: using the notation of the R statistical programming language, it is simply 1 *pbinom*(*k*, *N*, *P*), where *pbinom*(*k*, *N*, *P*) denotes the CDF for the binomial distribution with these parameters. An explicit formula for *f*(*P*) in discrete time, given the parameters *K* and *N*, is thus:

$$f(P) = 1 - pbinom(k, N, P) \tag{4.2}$$

For example, if each patch has $N = 4$ neighbors and if it becomes inflamed if and only if at least 3 of them are inflamed, then the fractional inflammation rate per unit time will be:

$$f(P) = P^N + N * P^{N-1}(1-P) = P^4 + 4P^3(1-P)$$

This is just the probability that the number of inflamed neighbors is 4 or 3. In continuous time, assuming a mean field approximation with averaging over relevant time scales, $f(P)$ refers to a fractional rate per unit time (which may exceed 1) rather than to a discrete probability, and Eq. (4.2) is replaced by Eq. (4.3):

$$f(P) = a * (1 - pbinom(k, N, P)). \tag{4.3}$$

Here, the parameter a converts from probability per discrete time step to fractional rate per unit of continuous time. It can be interpreted as a potency or intensity parameter reflecting how quickly a cell becomes inflamed once its inflammation dose-response threshold has been exceeded. The mean-field approximation ignores detailed stochastic fluctuations and considers only average values; thus, it is entirely possible that some cells with high probabilities of becoming inflamed during some interval won't, while others with lower probabilities will; also, the local percolation of inflamed cells through parts of a population may take more or less time than average-case calculations reveal, with local patches of inflammation spreading or extinguishing randomly. Nonetheless, mean-field approximation gives useful analytic insight into drivers of the long-run dynamic behavior of the system, as discussed next, even though it is not the right tool for analyzing short-term stochastic fluctuations or variability around average-case behaviors (Wang et al. 2016). Substituting Eq. (4.3) into the ODE model (4.1a) yields Eq. (4.4) as the final ODE model governing the evolution of inflammation P over time in the absence of further contributions from internal doses.

$$dP / dt = a * (1 - pbinom(k, N, P)) * (1 - P) - d * P. \tag{4.4}$$

The qualitative nature of possible solutions (i.e., time courses of P) for this model are of greatest interest, rather than specific numerical values. One qualitative question is whether there are any equilibrium values of P, corresponding to levels of inflammation that will sustain themselves if they are reached. $P = 0$ is a trivial example of such an equilibrium—if the system starts with no inflamed cells and no inflammatory exposures, it remains with no inflamed cells—but the more interesting qualitative questions remain of how many non-zero equilibria there are (if any), their relative sizes, their local stability, and how they are approached (if at all) from any starting degree of inflammation, $P(0)$. Answers can be gained by recognizing that, at equilibrium, $dP/dt = 0$ and hence the ODE in Eq. (4.4) reduces to the algebraic equation $a*(1 - pbinom(k, N, P))*(1 - P) - d*P = 0$, which can be rearranged as Eq. (4.5) if P is an equilibrium value strictly between 0 and 1 for the fraction of inflamed units.

$$P/(1-P)=(a/d)(1-pbinom(k,N,P)) \tag{4.5}$$

As P is increased from 0 toward 1, the left side of Eq. (4.5) considered as a function of P increases from 0 toward infinity and the right side increases from 0 to a/d. Equilibrium levels of P occur wherever these two curves cross, so that Eq. (4.5) is satisfied. Since the right side of Eq. (4.5) approaches a horizontal line at height a/d and the left side approaches a vertical asymptote at $P = 1$, if the left side is ever less than the right side, then there must exist at least one non-zero equilibrium value of P at which the left curve crosses the right curve from below. Such an equilibrium is locally stable, since $P/(1\text{-}P)$ as a function of P crosses $(a/d)(1 - pbinom(k, N, P))$ from below only if $P/(1\text{-}P) < (a/d)(1 - pbinom(k, N, P))$, which can be rearranged as $a^*(1 - pbinom(k, N, P))^*(1 - P) - d^*P > 0$, or (by Eq. (4.4)) as $dP/dt > 0$. In other words, P is increasing at P values to the left (and, conversely, is decreasing at P values to the right) of the intersection point, making it a stable equilibrium. At the origin, where $P = P/(1\text{-}P) = (a/d)(1 - \text{pbinom}(k, N, P)) = 0$, the function $P/(1\text{-}P)$ on the left side of (4.5) has slope 1 and the function $(a/d)(1 - \text{pbinom}(k, N, P))$ on the right side of (5) has slope 0, so $dP/dt < 0$ at the origin, implying that $P = 0$ is a locally stable equilibrium. Figure 4.1 diagrams both of these functions of P (P is on the horizontal axis) and shows their intersections for the specific combination of values of the model parameters $a/d = 2$, $k = 2$, $N = 6$.

Although different choices of these parameter values change the exact form of the S-shaped curve for $(a/d)(1 - pbinom(k, N, P))$ in Fig. 4.1, the following qualitative properties do not depend on details of the curve or the precise numerical values of its parameters.

1. The origin is a stable equilibrium (since $P/(1 - P) > (a/d)(1 - pbinom(k, N, P))$, so $dP/dt < 0$). In other words, small levels of inflammation tend to resolve themselves.
2. There is a threshold level of inflammation, corresponding to the unstable equilibrium intersection point T in Fig. 4.1, below which the inflammation level P will decrease to 0 over time (self-resolving transient inflammation). In the jargon of dynamical systems theory, the "basin of attraction" for the healthy equilibrium point $P = 0$ extends from 0 to T. From any initial inflammation level $0 < P(0) < T$, the values of $P(t)$ decrease toward zero as time t increases. If the vertical scaling factor a/d is small enough so that the two curves only intersect at $P = 0$, then T becomes effectively infinite and $P = 0$ is the only equilibrium and is globally stable. Otherwise, if the threshold T is finite (between 0 and 1), a second stable equilibrium exist, as described next.
3. There is a unique non-zero stable equilibrium level of inflammation, corresponding to intersection point E in Fig. 4.1. From any initial inflammation level to the right of T, the inflammation level P will move toward E over time (self-

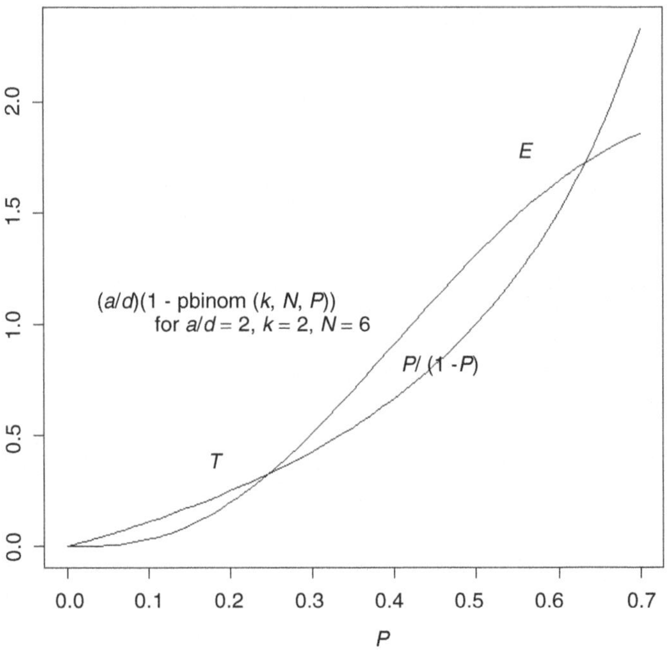

Fig. 4.1 Left and right sides of Eq. (4.5) as functions of the inflammation level P on the horizontal axis. T is an unstable equilibrium. The origin and E are stable equilibria

sustaining chronic inflammation). The basin of attraction for E extends from $P = T$ to $P = 1$.

Figure 4.2 shows explicit trajectories (time courses) for $P(t)$ starting from different initial levels of inflammation, $P(0)$, on the vertical axis. (For variety, this is illustrated for an inflammation process described by parameters $k = 2$, $N = 3$, $d = 0.2$, $a = 0.4$, although the qualitative properties of interest do not depend on these specific numerical values.) The vertical axis is scaled in percentages, running from 0% to 100% of units inflamed. Time is on the horizontal axis, with units reciprocal to those of the fractional rate parameters a and d; if they are in fraction per day, then time units would be days; if they are in fraction per month, then the time units would be months, and so on. Since the qualitative nature of the trajectories is of central interest, we leave the time units unspecified for now. For sufficiently small initial inflammation levels (below T on the vertical axis), all trajectories lead down to zero, although the times for them to reach any given level become longer and longer as the initial inflammation fraction increases toward T. At a starting value of $P(0) = T$, the time to decline becomes infinite (since T is an equilibrium level, albeit an unstable one). For initial inflammation levels greater than T, the trajectories instead approach a self-sustaining higher equilibrium value, $E = 50\%$ on the vertical axis, with the approach taking less time from initial inflammation levels closer to E.

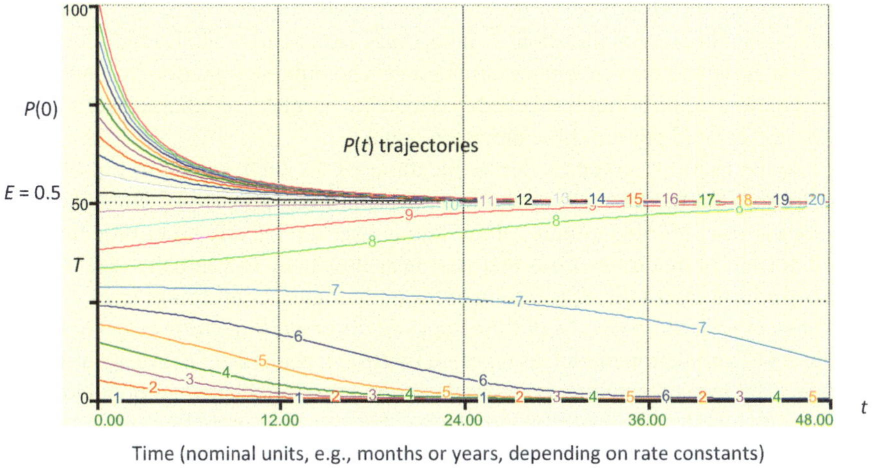

Fig. 4.2 A family of $P(t)$ inflammation trajectories starting from different initial inflammation levels on the vertical axis (numbers show percent of units inflamed). Time is on the horizontal axis. Time units are left unspecified here; they are the inverse of the units for the rate parameters a and d. For these trajectories, $k = 2$, $N = 3$, $d = 0.2$, $a = 0.4$

If internal dose is constant and contributes an additional fractional rate of inflammation c per unit time by inhaled particles interacting directly with as-yet uninflamed cells, then Eqs. (4.4) and (4.5) generalize to Eqs. (4.6) and (4.7), respectively.

$$dP / dt = \left(a * \left(1 - pbinom\left(k, N, P\right)\right) + c\right) * \left(1 - P\right) - d * P \tag{4.6}$$

$$P / \left(1 - P\right) = \left(a / d\right)\left(1 - pbinom\left(k, N, P\right)\right) + c / d \tag{4.7}$$

The qualitative analysis is almost identical to that already given for the case $c = 0$, but now continued exposure shifts the S-shaped curve in Fig. 4.1 upward by c/d (which shifts T down and E up and moves the stable homeostatic equilibrium from 0 to a positive value).

Figures 4.1 and 4.2 show how sufficiently large exposures in this model can eventually shift inflammation from self-limiting to chronic by pushing the system from the basin of attraction for $P = 0$ (below threshold T) to the basin of attraction for $P = E$ (above threshold T). The system dynamics described by Eq. (4.4) exhibit bistability: exposure histories that do not inflame more than the threshold fraction T of units lead to transient inflammations that eventually resolve themselves in the absence of continued exposure, returning to zero units inflamed, but inflammation above this threshold level leads to irreversible chronic inflammation that settles at the new, higher level E. Thus, spread of inflammation among cells provides a different route to bistability than positive feedback loops, a well-known source of bistability in biological control systems that are also important in RCS-induced lung inflammation (Cox Jr. 2011). Rather than a feedback loop, a tipping point (similar to an epidemic threshold) better describes the bistability here: once the proportion of inflamed cells exceeds threshold T, it becomes easier (more likely) for other cells to become inflamed than to remain uninflamed. The inflamed fraction expands until a new dynamic equilibrium is reached in which the numbers of inflamed cells being created and cleared per unit time are again equal.

It may be useful to compare the implications of the ODE model in Eq. (4.6) to those from ABM models. Although the analysis in Figs. 4.1 and 4.2 show that Eq. (4.6) describes a bistable system, observations over a relatively short time horizon would not reveal that trajectories that start near threshold T eventually reach either of two levels, homeostatic equilibrium or the pathological equilibrium level, E. This is consistent with observations of three types of observed trajectories reported for a simple ABM simulation model, as reported by Brown et al. (2011): "An agent-based model (ABM) was created to examine the response of an abstracted population of inflammatory cells (nominally macrophages, but possibly including other inflammatory cells such as lymphocytes) and cells involved in remodeling (nominally fibroblasts) to particulate exposure. .... The model yielded three distinct states that were equated with (1) self-resolving inflammation and a return to baseline, (2) a pro-inflammatory process of localized tissue damage and fibrosis, and (3) elevated pro- and anti-inflammatory cytokines, persistent tissue damage, and fibrosis outcomes. Experimental results consistent with these predicted states were observed in histology sections of lung tissue from mice exposed to particulate matter." If the ABM model is consistent with our ODE model in Eq. (4.6), then extending the lengths of these ABM would show that their state 2 ("a pro-inflammatory process of localized tissue damage and fibrosis") eventually progresses to either state 1 or state 3.

Although the particular combination of ODE, ABM, and mean-field theory simplifications in Eq. (4.6) for modeling the transition from acute to chronic inflammation is new to the best of our knowledge, the finding that there is a threshold below which self-sustaining inflammation does not occur and above which it does has many analogies and generalizations in the biomathematical theory of epidemics and in percolation theory for spatial stochastic processes. The existence of such a threshold is robust to many modeling details. It occurs in a wide variety of stochastic models that allow for realistic complexities such as heterogeneity in the thresholds for different units, or heterogeneous and time-varying numbers of neighbors for different units (Neri et al. 2011; Kenah and Robins 2007; Volz and Meyers 2009). The key feature driving the occurrence of bistability and a threshold between two basins of attraction for extinction vs. self-sustaining activation of units in such models is that a unit is more likely to become activated when more of its neighbors are activated. This interdependence is realistic for phenomena ranging from spread of forest fires to epidemics, and we believe that it is also realistic for NLRP3-mediated spread of pyroptosis and lung inflammation.

To summarize this section, the inflammation model in Eq. (4.6) takes as input an initial level of inflammation, $P(0)$, and produces as output a time course of subsequent inflammation levels, $P(t)$, as illustrated in Fig. 4.4, and as previously described in Chap. 3. In the long run, initial levels of inflammation above a threshold T lead to a high chronic level of inflammation, *E*, as a stable equilibrium. By contrast, initial levels of inflammation below the threshold lead to resolution and zero inflammation in the absence of further exposure. Although the time required to approach the long-run level (0 or *E*) can be arbitrarily long if the initial level is arbitrarily close to the threshold *T*, it is shorter for starting values far from *T*.

A PBPK Model for Internal Doses of Respirable Crystalline Silica (RCS)

To extend a model of lung inflammation to provide a dose-response model for RCS and lung cancer (or for other agents and diseases) it is necessary to embed the inflammation model between a front-end model describing how dose affects inflammation and a back-end model describing how inflammation affects lung cancer or other disease risks. This section describes a front-end PBPK model that takes RCS exposure histories as inputs and produces resulting time courses of inflammation levels as outputs. A later section develops a lung cancer response model that takes time courses of inflammation levels as inputs and produces resulting time courses of lung cancer risk as outputs.

Physiologically-based pharmacokinetic (PBPK) models have previously been developed for RCS exposures and measures of resulting AM and neutrophil inflammation in the lungs of exposed rats and humans (Kuempel et al. 2001; Tran et al. 2001, 2002). Although these models preceded the elucidation of the role of the NLRP3 inflammasome, they provide useful descriptions of how RCS exposures affect resulting time courses of inflammation markers in the lung, consistent with considerable data from rat experiments and limited data in humans (Tran et al. 2001).

Figure 4.3 shows the structure of the main part of the PBPK model for RCS described and partly validated by Tran et al. (2001). This figure uses standard visual conventions for systems dynamics modeling environments: rectangular boxes represent compartments ("stocks"), thick arrows represent flows among compartments, thin arrows represent information about dependencies, circles represent constants or calculated quantities, and dashed circles represent quantities input or calculated from outside the part of the model shown in the diagram. All quantities are allowed to be time-varying. Thus, Fig. 4.3 describes a dynamic system in which dose (specified exogenously as a time series of concentrations of RCS in inhaled air) causes a flow of crystalline silica particles to the alveolar surface of lung; X_1 denotes the mass in mg of alveolar silica. (Surface area of deposited silica particles is a crucial aspect of internal dose, as discussed later.) From the alveolar surface, RCS is redistributed to alveolar macrophages (X_2 = mg of silica burden in AMs), the interstitium (X_3 = mg of silica in the interstitium), and lymph nodes (X_4 = mg of silica in the lymph nodes). RCS in AMs can be cleared by the mucociliary escalator or returned to the alveolar surface (e.g., as is now understood, following pyroptosis of an AM).

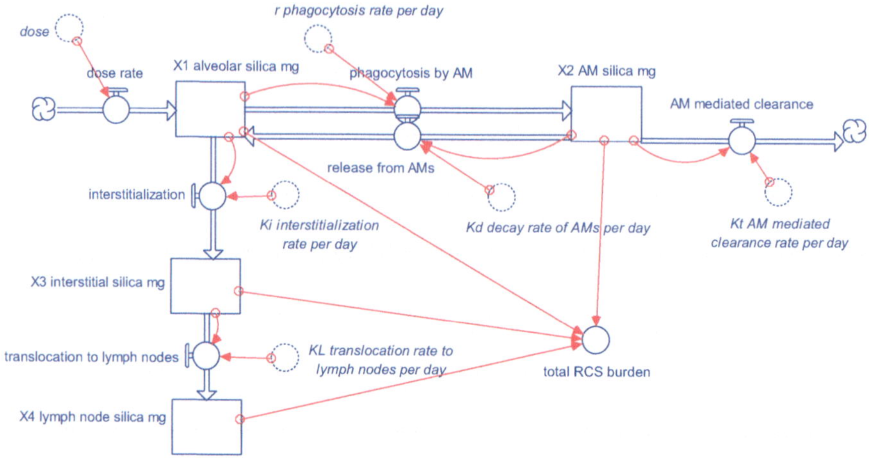

Fig. 4.3 Systems dynamics model diagram for a basic RCS PBPK model, adapted from Tran et al. (2001). The diagram was created using the *iThink* software, www.iseesystems.com/

RCS in the interstitium is eventually translocated to the lymph nodes, although the process is relatively slow.

Table 4.1 summarizes the ODEs and rate constants describing these time-varying flows of RCS among the PBPK model compartments for low levels of exposure. The model described in Fig. 4.3 and Table 4.1 can be downloaded as an *iThink* file ("RCS_PBPK_small.itm") from http://cox-associates.com/CausalAnalytics/. The full PBPK model developed by Tran et al. is considerably more complicated, in part because, at higher exposure levels, phagocytosis becomes impaired and several of the fractional rate constants in Table 4.1 (and in additional modules describing dose-dependent recruitment of macrophages and neutrophils, eventual fibrosis, and changes in oxidants and markers such as NF-κB) become dependent on the silica burden in AMs, denoted by X_2. As explained by Tran et al. (2001), "The AM-mediated defense of the alveolar region is effective (i.e. phagocytosis and clearance of silica particles by AMs are unimpaired) as long as the silica burden inside AMs (X_2) remain below a critical burden level (m_{crit}). Once m_{crit} is reached clearance quickly becomes impaired and an inflammatory reaction is switched on." The "inflammatory reaction" that is switched on in this model is onset of acute (neutrophil-dominated) inflammation, rather than the possible subsequent onset of chronic (macrophage-and lymphocyte-dominated) inflammation (Moldoveanu et al. 2009), which is our main focus. Thus, the model and parameter values in Table 4.1 describe the distribution of RCS at exposures low enough that they do not trigger acute inflammation, as defined by a significant increase in alveolar neutrophils. Tran et al. (2001) found it necessary to include the threshold parameter m_{crit} to fit data from experiments in rats with doses high enough to trigger acute inflammation of the lung, even in the absence of lung overload. They estimated its value for humans as $m_{crit} = 146.8$ by scaling from the values estimated from rate data based on mg/g of lung weight.

To understand how the time until acute inflammation is initiated depends on exposure concentrations, Fig. 4.4 shows the trajectories for accumulation of silica loads in AMs (measured in mg, X_2 on the vertical axis)over time (measured in days on the horizontal axis) for different exposure concentrations, as predicted by the PBPK model in Table 4.1. As just discussed, this model ignores the accelerated rate of accumulation of RCS that occurs once the load exceeds m_{crit}, and therefore the trajectories in Fig. 4.4 are only meant to be descriptively accurate for X_2 values on the vertical axis that are no greater than $m_{crit} = 146.8$. (Although, as discussed next, we do not believe that the part of the Tran et al. PBPK model that describes what happens after inflammation is triggered is correctly specified, it is still useful for quantitatively estimating RCS accumulation until that point, and hence for quantifying the time for X_2 to reach m_{crit}, i.e., the time until acute inflammation is triggered.) The concentrations are varied from 0.08 mg/m^3 (curve 1, the lowest curve in Fig. 4.4) to 0.18 mg/m^3 (curve 11, the highest curve) in increments of 0.01 mg/m^3.

The main results from the PBPK modeling, as shown in Fig. 4.4, are as follows:

1. Exposures to sufficiently low concentrations of RCS in inhaled air (below about 0.13 mg/m^3) do not trigger inflammation: the AM load accumulation curves flatten out below m_{crit}.
2. Most exposure concentrations that are high enough to trigger acute inflammation (i.e., that cause load curves that eventually exceed $m_{crit} = 146.8$) do so relatively

Table 4.1 Human lung RCS PBPK model and parameter values, from Tran et al. (2001)

Compartments
• X_1 = mg of alveolar silica
• X_2 = mg of silica burden in AMs
• X_3 = mg of silica in the interstitium
• X_4 = mg of silica in the lymph nodes
Flow equations and parameter values for $X_2 < m_{crit}$
• $dX_1/dt = dose - rX_1 - k_iX_1 + k_dX_2$; $r = 0.9660$, $k_i = 0.5289$, $k_d = 0.03$
• $dX_2/dt = rX_1 - k_tX_2 - k_dX_2$; $r = 0.9660$, $k_t = 0.0036$, $k_d = 0.03$
• $dX_3/dt = k_iX_1 - k_lX$; $k_i = 0.5289$, $k_l = 9.40$
• $dX_4/dt = k_lX_3$; $k_l = 0.001$
These fractional rate parameters (all in units of fraction per day) are interpreted as follows
• r = phagocytosis rate = 0.9660 per day
• k_i = interstitialization rate= 0.5289 per day
• k_d = AM decay rate= 0.03 per day
• k_t = AM clearance rate= 0.0036 per day
• k_l = translocation rate to lymph nodes = 0.001 per day
Internal dose model
• *dose* = (concentration of RCS in inhaled air, mg/m^3)*(deposited fraction)*(liters of air inhaled per minute)*(days per week exposed/7)*(hours per day exposed)
• = 0.10*0.32*13.5*0.714*8*1 = 2.47 in base case scenario of Tran et al. (2001). (Although Tran et al. refer to "days per week," they clearly mean "fraction of days per week exposed," i.e., (days per week exposed)/7. The value 0.714 = 5/7)

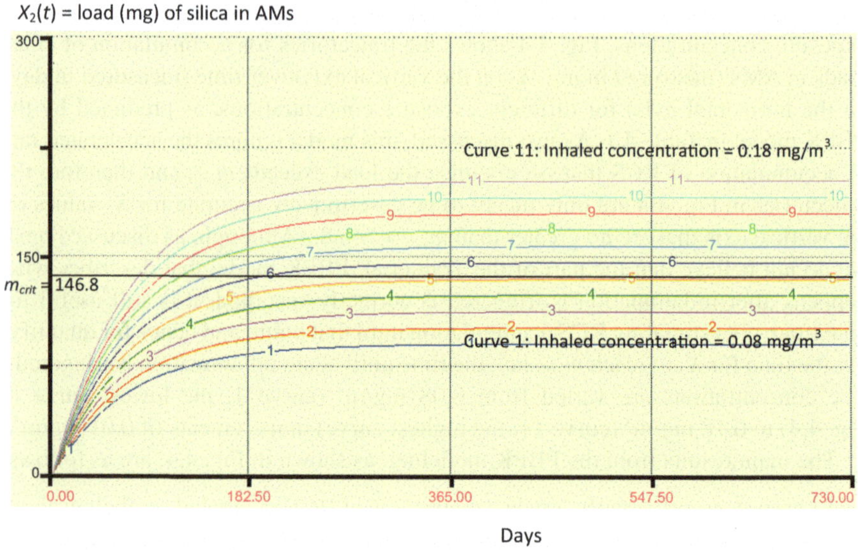

Fig. 4.4 Predicted time courses of silica load (mg) in AMs for 11 different exposure concentrations, 0.08 mg/m³ (curve 1) through 0.18 mg/m³ (curve 11), increments of 0.01 mg/m³. Predictions are made using the PBPK model in Table 4.1. The threshold level m_{crit} = 146.8 mg for acute inflammation is reached only for exposure concentrations >0.13 mg/m³ (curve 7). Most exposures that achieve this threshold do so within about 3–6 months

quickly, within about 3–6 months, as shown by the times where the curves for $X_2(t)$ cross the horizontal line $X_2 = m_{crit}$.

3. For prolonged exposures to constant concentrations, the entire PBPK model acts as a multiplier, yielding a steady-state AM RCS load in mg equal to 1115.1 times the inhaled concentration of RCS in mg/m³. Thus, a threshold for the minimum concentration of RCS that can initiate acute inflammation when m_{crit} = 146.8 can be estimated as follows: m_{crit}/1115.1 = 146.8/1115.1 = 0.13 mg/m³.
4. Even concentrations that are high enough to trigger acute inflammation (above about 0.13 mg/m³) do so only if exposure lasts for long enough (typically 3–6 months) for the curves to rise to exceed m_{crit}. Thus, there is a minimum duration of exposure necessary for an exposure concentration greater than 0.13 mg/m³ to trigger acute inflammation. This agrees with experimental data in rodents (Porter et al. 2004).

The threshold value m_{crit} is clearly similar to threshold T in Figs. 4.1 and 4.2, insofar as both represent a critical threshold for inflammation. However, there are the following crucial differences between them. First, m_{crit} is for onset of acute (neutrophil) inflammation as a response to internal dose (AM load) of RCS, while T is for onset of chronic inflammation as a response to any stimulus that increases the level of inflammation. Second, although m_{crit} is a threshold for dose-dependent parameter values, it does not separate the basins of attraction for two different stable equilibria. No matter how detailed, complex, and well-validated for goodness-of-fit in describing experimental time course data it may be, *any* PBPK model in which parameter values depend only on RCS loads and inhaled RCS is eventually cleared implies that, after cessation of exposure, all parameters will eventually return to their initial levels. This contrasts with the bistability in Figs. 4.1 and 4.2, where crossing threshold T implies that the system will move toward the new pathological equilibrium E and will never return to its initial (homeostatic) equilibrium, even if exposure ceases. Empirically, "Human epidemiologic studies have found that silicosis may develop or progress even after occupational exposure has ended, suggesting that there is a threshold lung burden above which silica-induced pulmonary disease progresses without further exposure," and similar thresholds have been found for chronic inflammation and fibrosis in RCS-exposed rats at exposure concentrations and durations that do not cause lung overload (Porter et al. 2004; Borm et al. 2018). Thus, bistability appears to be an important feature of real-world exposure-response data for RCS. The PBPK model in Table 4.1 does not allow for bistability, so its modeling of inflammation after the critical threshold m_{crit} has been reached must be extended, e.g., in the direction of Eq. (4.6), to better describe results of stop-exposure experiments and the continued progression of inflammation mediated diseases after exposure ends.

The PBPK model in Table 4.1 was developed before the roles of the NLRP3 inflammasome and the importance of self-sustaining inflammation and bistability were well understood. Its focus on the load of RCS crystals in AMs (X_2) as the key driver for an inflammation threshold was an important and useful contribution, although in hindsight, it is probably not phagocytosis (or attempted phagocytosis) of RCS crystals by AMs *per se* but rather activation of NLRP3 inflammasomes in AMs in response to RCS crystals that is the rate-limiting step for onset of RCS-induced NLRP3 inflammasome-mediated inflammation (Pollard 2016). Despite their limitations for describing the time course of acute and chronic inflammation (below T and above T, respectively) *after* inflammation has been initiated, PBPK models such as the one in Table 4.1 are highly useful for relating concentrations of RCS in inhaled air to time courses of RCS loads in various compartments before inflammation begins and for predicting when and whether these loads reach specified levels such as m_{crit}.

Once acute inflammation has been initiated, what happens next? The most usual outcome is that the acute, neutrophil-dominated inflammation runs its course and then resolves itself via a number of well-characterized mechanisms involving apoptosis and efferocytosis (Robb et al. 2016). Porter et al. (2004) expressed concern that "relatively low silica exposures may pose a serious health risk because silica-induced disease, once initiated even at low threshold lung burdens, can progress even in the absence of further exposure" but noted that "However, the mechanism(s) responsible for the progression of silica-induced pulmonary disease, despite decreasing silica lung burden [after exposure ceases], remains to be established." Figure 4.1 suggests a more reassuring answer, assuming that the mechanism of self-sustaining disease progression involves activation and propagation of ASCs and NLRP3 inflammasomes among cells (especially, AMs), as well as silica particle-induced LMP and pyroptosis (or necrosis) of macrophages. A sustained constant internal dose at a level high enough to induce inflammation (represented by $c > 0$ in Eq. (4.6)) shifts the S-shaped curve in Fig. 4.1 upward (by an amount c/d). This, in turn, shifts rightward the left-most intersection between the two curves in Fig. 4.1, which is at the origin when $c = 0$, creating a locally stable positive equilibrium level of inflammation. Because this is a stable equilibrium, nothing else happens as long as internal dose is not further increased: the new, low-level inflammation persists as long as internal dose continues to create local inflammation (e.g., by RCS particles interacting with AMs and causing them to undergo inflammation and pyroptosis). Discontinuing exposure (so that c returns to 0) will cause the inflammation to return to 0. Only if exposure is increased and sustained so that inflammation exceeds threshold T will the system move into the basin of attraction for chronic inflammation. In short, because the left-most (homeostatic) equilibrium in Fig. 4.1 is locally stable, inflammation will not spontaneously progress when exposure ceases, but will subside back down to 0. Self-sustaining inflammation and progression of inflammation-mediated diseases is only triggered by exposures that lead to inflammation greater than the threshold T. Low exposure concentrations of RCS (e.g., below about 0.13 mg/m^3 in Fig. 4.4) do not even trigger acute inflammation, much less chronic inflammation, and thus, according to the model in Fig. 4.1, they cannot initiate inflammation-mediated diseases that will then progress even after exposure ceases.

Sensitivity to Interindividual Variability and Physical and Chemical Properties of RCS

The following qualitative modeling conclusions are robust to scientific and modeling uncertainties and inter-individual variability, in the sense that they do not depend on the specific parameter values or details of the models used:

1. Prolonged exposure to a relatively low constant concentration of RCS leads to a steady-state level in each PBPK compartment that is approximately proportional to the inhaled concentration of RCS (e.g., the steady-state burden of RCS in AMs, in mg, is about 1115.1 times the concentration of RCS in inhaled air in mg/m^3.) This direct proportionality is a consequence of the fact that internal pools of RCS increase until the amount cleared per unit time just balances the amount added per unit time from further inhalation of RCS. Hence, it holds independent of the details of the PBPK model used (e.g., the number of compartments, how they are connected, and the fractional flow rate parameters governing rates of flow among them), although of course the specific numerical value of the constant of proportionality is sensitive to these details. This part of the model satisfies LNT assumptions.
2. As a corollary, only exposure concentrations in excess of a certain threshold, maintained for at least a certain minimum duration, can cause the size of a given pool to reach or exceed a specified threshold size (e.g., in Fig. 4.4, only exposure

concentrations greater than about 0.13 mg/m^3 cause X_2 to eventually exceed m_{crit}, triggering neutrophil inflammation).

Because these properties do not depend on details of the models used, interindividual variability and uncertainty in parameter values and variations in physicochemical details of the RCS particles to which individuals are exposed do not change the conclusion that thresholds exist. However, such variations can alter the numerical values of the thresholds for different individuals and compositions of RCS. It is worthwhile to consider the plausible extent of resulting variability in threshold values.

Many studies have established that toxicity of specific quartz dusts is most strongly correlated with properties that affect the surface area (rather than the mass or volume) of the particles interacting with AMs and other lung cells. The specific quartz dust assumed in Table 4.2 is DQ12. Other quartz dusts vary widely in their

Table 4.2 TSCE model equations, functions, and parameter values for RCS-induced lung cancer

Flows among compartments:
• $d(Normal)/dt = 0 = -a_1*Normal + replacement$ $a_1 = 1.4*10^{-7}$
– $replacement = a_1*Normal$
• $d(Initiated)/dt = a_1*Normal + (g - a_2)*Initiated$ $g = 0.075 + (1 + (1 - exp(-0.373E)))$
• $d(Malignant)/dt = a_2*Initiated$ $a_2 = 1.4*10^{-7}$
• $h(t)$ = age-specific hazard function for lung cancer ≈ expected number of new malignant cells formed per year = $a_2*Initiated$. (McCarthy et al. assume a 5-year lag between formation of a malignant cell and mortality; we ignore this latency period, following Zeka et al. (2011) for RCS-specific lung cancers)
Interpretations of model parameters and variables
• a_1 = fractional rate for first mutation ("initiation"), in units of expected fraction of normal cells initiated per year. The number of normal stem cells at risk of mutation is assumed to be $Normal$ = 1E7 for purposes of estimating a_1. (Only the product $a_1*Normal$ affects model predictions.) The value $a_1 = 1.4*10^{-7}$ is taken from McCarthy et al. (2012), Table 4.1
• G = net proliferation rate of initiated (i.e., premalignant) cells = clonal expansion rate for initiated cells, in units of expected new initiated cells per year per initiated cell. The zero-exposure baseline value $g = 0.075$ is taken from McCarthy et al. (2012), Table 4.1. For RCS-exposed workers in the diatomaceous earth mining and processing industry, Zeka et al. (2011) estimate that RCS exposure increase g by a factor of $1 + (1 - exp(-0.373E))$, where E = RCS concentration, if RCS acts as a promoter. This exposure-dependent multiplier ranges from 1 when $E = 0$ to 2 as E becomes large
• a_2 = fractional rate for second mutation ("progression"), in units of expected fraction of initiated cells that become malignant per year. The value $a_1 = 1.4*10^{-7}$ is taken from McCarthy et al. (2012), Table 4.1
• $h(t)$ is approximately the probability per year of a malignant cell being formed (assuming negligible latency period, as in Zeka et al. 2011), given that none has been formed so far. When this number is small, it is well approximated by the expected number of new malignant cells formed per year
• $P(t) = 1 - \exp\left(-\int_0^t h(s)\,ds\right) = 1 - exp(-Malignant(t))$ = probability of lung cancer by time t

potencies for eliciting various inflammatory responses and triggering LMP and activation of the NLRP3 inflammasome in AMs based on factors such as polymorphic type; freshly fractured vs. aged surfaces; particle size, shape, and surface area; and presence of iron or other substances (Clouter et al. 2001; Meldrum and Howden 2002; Bruch et al. 2004; Kusaka et al. 2014; Pavan et al. 2014; Rabolli et al. 2016; Turci et al. 2016). For example, Clouter et al. (2001) compared inflammatory responses in rats exposed to workplace samples of RCS to rats exposed to DQ12 and reported that "The workplace samples did not cause inflammation at any dose or time point [but] DQ12 quartz caused marked inflammatory responses, as measured by an increased number of neutrophils in the lungs of instilled animals for both time points and doses." Similarly, experiments that identified the cytokine TNF-α as an inflammation marker that is especially sensitive to differences in RCS physicochemical properties concluded that "Apart from quartz samples with an activity close to that of DQ12, one also finds examples with a very low activity. … Threshold effects in the range of 15 to ≥ 120 μm can be demonstrated" for the relatively sensitive TNF-α responses to different quartz dusts in *in vitro* tests (Bruch et al. 2004). Thus, among RCS dusts with sufficient potency to activate an acute inflammatory response, an approximate 8-fold range (120/15 = 8) for the acute inflammation threshold m_{crit} based on variability in RCS properties may be plausible, with the threshold for DQ12 being in the upper part of this range. Reciprocally, the minimum RCS concentration needed to elicit an acute inflammatory response might also have an approximately an 8-fold range among samples capable of causing inflammation. A dust with 1/8 the inflammatory potency of DQ12 would imply an RCS concentration threshold for triggering acute inflammation of just over 1 mg/m^3 (= 8*0.13 mg/m^3) for the PBPK model in Table 4.1.

Even for identical exposures to the same quartz dust, different individuals can form different burdens of RCS in AMs, reflecting differences in their individual PBPK or biochemical parameter values. This widens the range of RCS concentration thresholds that might trigger acute inflammation. Based on a Bayesian uncertainty and variability analysis that compared respirable dust loads found in lungs of autopsied coal miners in the US and UK to amounts predicted using a three-compartment PBPK model similar to the one in Table 4.1, Sweeney et al. (2013) concluded that "On average, the model predictions were within a factor of just less than 2 of the experimentally measured amounts of dust in lungs and lymph nodes. Almost all of the predictions for [individual] miners were within tenfold of the measured values." This analysis ignored exposure estimation errors and uncertainties by treating individual exposure concentration estimates as if they were known to be accurate measurements of true exposure concentrations. The authors also note that actual exposure concentrations might be higher than the estimated concentrations. Thus, unmodeled errors in exposure concentration estimates contribute an unquantified amount to the differences between predicted and measured values for individual miners. Even with these uncertainties and omissions, it appears that inter-individual variability usually contributes no more than one order of magnitude difference between PBPK model-predicted and actual values for lung burdens. The same may be plausible for RCS in AMs in the RCS PBPK model in Table 4.1.

Mamuya et al. (2006) quantified exposure estimation errors for respirable dust and quartz dust (RCS) for individual workers in a coal mine by comparing the exposure estimates based on job teams to actual measurements based on personal dust sampling. They found that "[R]atios of the 97.5th and 2.5th percentiles of the between-worker distribution of respirable dust exposure were relatively low, varying between 1.0 and 22.5 in the 8 job teams, while the analogous within-worker distribution varied between 2.2 and 3902." For both respirable dust and quartz dust (RCS) specifically, 10% of the workers had actual cumulative exposures more than 15 times higher than the estimated median cumulative exposures. Thus, it appears that most or all of the apparent interindividual variability in RCS levels reported by Sweeney et al. might be explained by unmodeled variability in individual cumulative exposures around their estimated values. Comparing averaged individual values to model predictions can reduce the effects of this unmodeled estimation error in individual exposures. The finding that "On average, the model predictions were within a factor of just less than 2 of the experimentally measured amounts of dust in lungs" (Sweeney et al. 2013) then suggests that a factor of about 2 may account for most interindividual variability in the PBPK model converting concentrations in air to accumulation of silica in lungs (or, perhaps, in AMs). If so, then since the base model in Table 4.1 predicts a nominal steady-state AM RCS load in mg equal to about 1115 times the inhaled concentration of RCS in mg/m^3, different individuals might have loads from about 1115/2 = 557 to 1115.1*2 = 2230 times the air concentration of RCS.

Inflammation-Mediated Promotion of Lung Cancer in a Two-Stage Clonal Expansion Model

It has recently become well understood that chronic inflammation plays an essential role in promoting RCS-induced lung cancer (Borm et al. 2018; Freire et al. 2013) and other inflammation mediated exposure-related lung cancers, in part by promoting cell proliferation and tumor formation (Schetter et al. 2010; Vineis et al. 2010). However, to our knowledge, this qualitative understanding of the role of chronic inflammation has not yet been incorporated into quantitative mathematical models of carcinogenesis such as the two stage clonal expansion (TSCE) model (Vineis et al. 2010; Zeka et al. 2011 for silica- and asbestos-associated lung cancer; Zaballa and Eidemüller 2016 for joint carcinogenicity of radon and silica). The purpose of this section is to develop a TSCE model for RCS-induced lung cancers that incorporates the promoting role of chronic inflammation.

Figure 4.5 shows the structure of a TSCE model for carcinogenesis using the same systems dynamics graphical notation as for the PBPK model in Fig. 4.3. Table 4.2 lists the corresponding system of ODEs, which are determined directly from the diagram. Parameter values are drawn from the literature, as follows: $a_1 = a_2 = 1.4 \times 10^{-7}$ is taken from McCarthy et al. (2012), Table 4.1, for lung cancer

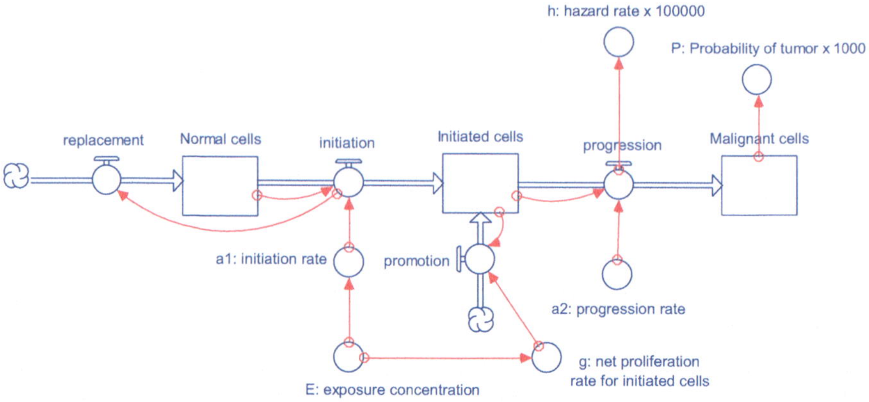

Fig. 4.5 Structure of a two-stage clonal expansion (TSCE) model of carcinogenesis (Zeka et al. 2011; McCarthy et al. 2012)

initiation and promotion rates, respectively, in non-smokers. $g = g_0*(1 + (1 - exp(-0.373E)))$ is the estimated functional dependency between exposure E and net proliferation rate g for initiated cells specified for RCS and lung cancer by Zeka et al. (2011) under the assumption that RCS exposure acts only as a promoter, increasing g but not the other parameters. Here, g_0 is the baseline (zero-exposure) net proliferation rate for initiated cells, estimated as $g_0 = 0.075$ by McCarthy et al. (2012), Table 4.1.

TSCE models, also called MVK (Moolgavkar-Venzon-Knudson) models after the researchers who pioneered them, have been extensively developed and well described in many existing sources (e.g., Zeka et al. 2011; McCarthy et al. 2012), along with careful discussions of how well their deterministic equations approximate the underlying stochastic processes. For our purposes, their most important features are that they provide excellent fits to many data sets, as illustrated in Figure 4.6 for the American Cancer Society (ACS) CPS II data set for lung cancer hazard functions for non-smoking men and women; and they can be used to predict the effects on lung cancer of exposures with different modes of action, such as increasing the first mutation rate (often called "initiator" carcinogens in the older literature, especially for skin cancer), increasing the net proliferation rate of initiated cells, e.g., by increasing their division rates or slowing their apoptosis rates ("promoter" carcinogens); or increasing the second mutation rate ("progressor" carcinogens).

Of course, some carcinogens can act via multiple modes of action, represented by making two or more of the TSCE rate parameters a_1, a_2, and g functions of exposure. However, a recent review of mechanistic information for RCS-induced genotoxicity suggests that a much higher dose of RCS is required to reach a threshold for genotoxic effects than to reach a threshold for persistent inflammation in the rat lung, by a factor of 10 to 20-fold (2–4 mg vs. 0.2 mg), and that carcinogenic effects of inhaled RCS are driven by inflammation (Borm et al. 2018), consistent with

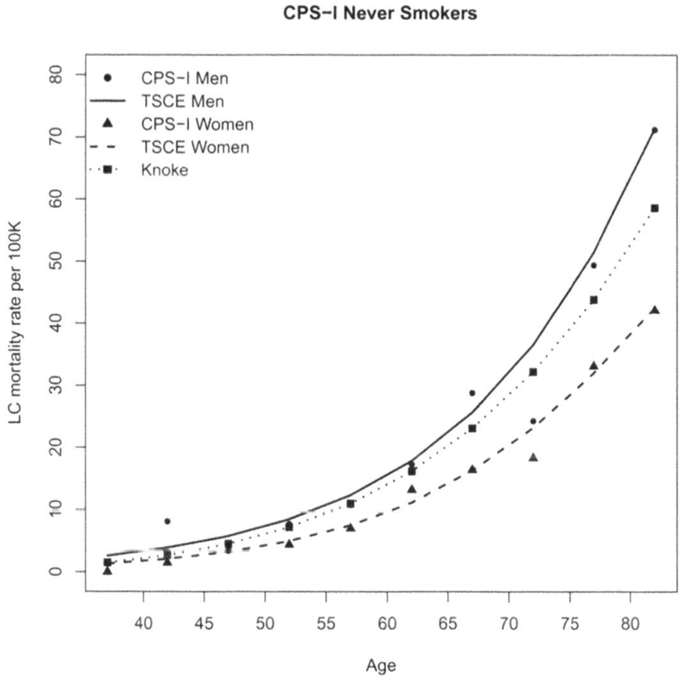

Fig. 4.6 Comparison of observed data (points) and TSCE model predictions of age-specific hazard functions (top curve for men, bottom curve for women) in the ACS CPS I data set for non-smokers. ("Knoke" is an additional, non-TSCE, model.) (*Source:* Reproduced with permission, from Fig. 2 of McCarthy et al. 2012)

earlier findings (Freire et al. 2013). We therefore use the TSCE model in Table 4.2, in which exposure increases the proliferation of initiated cells by increasing the net proliferation rate, g, but has no significant direct effect on the mutation rates a_1 or a_2.

Lung cancer typically develops on a time scale of decades. On this time scale, exposure affects lung cancer risk primarily by inducing chronic inflammation of the lung, which then increases one or more parameters in the TSCE model. From this perspective, exposure-related increases depend mainly on (a) When chronic inflammation begins; and (b) How much it increases age-specific hazard functions. We propose that the relation between exposure and risk on the time scale of decades can be usefully approximated as a jump in TSCE model parameters from their baseline levels to new, higher levels following chronic inflammation. This contrasts with the usual modeling assumption that risk increases smoothly with exposure as in the TSCE model in Table 4.2 (Zeka et al. 2011). However, the following considerations suggest that a discontinuous jump may be more realistic than continuous dependence of risk on exposure. The underlying epidemiological data (Checkoway et al. 1997), shown in Fig. 4.7, do not indicate that lung cancer risk is increased by exposure levels below the highest category ($\geq$5 mg/m^3-years); to the contrary, they appear more consistent with a U-shaped (hormetic) relationship. As stated by Zeka

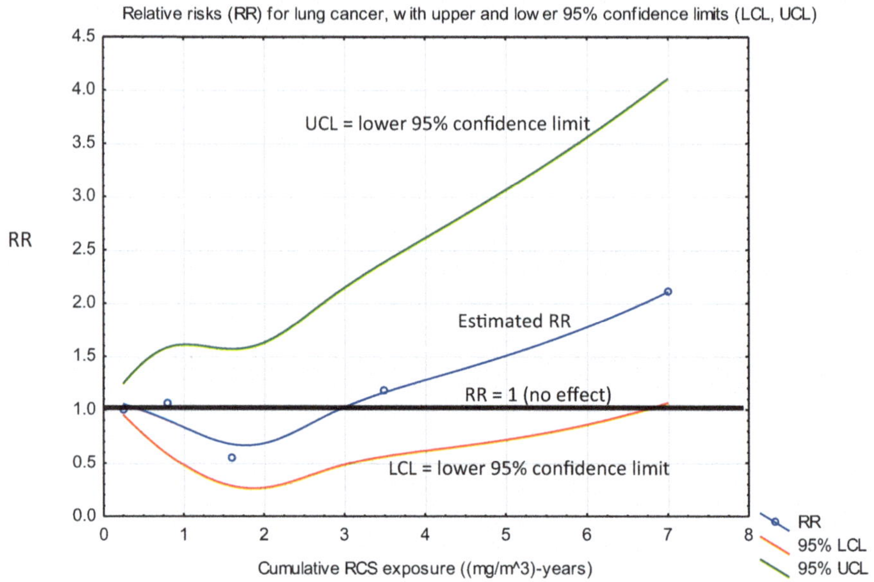

Fig. 4.7 Relative risk (RR) of lung cancer (middle curve) for different levels of cumulative RCS exposure, with estimated upper and lower 95% confidence limits (top and bottom curves). The data points shown use the midpoint for each of the first four cumulative exposure interval (< 0.5, 0.5–1.1, 1.1–2.1, 2.1–5 mg/m^3) considered by Checkoway et al. (1997). By repeating the pattern of doubling the previous midpoint, 7 mg/m^3 is used to represent the highest cumulative exposure interval (≥5 mg/m^3). *Source*: Data from Table 6 of Checkoway et al. (1997)

et al. (2011), "excess mortality based on 77 lung cancer deaths was found in association with cumulative crystalline silica equal or greater than 5 mg/m^3-years (RR = 2.11, 95% CI 1.07–4.11…)". (Median duration of employment was 5.54 years in this cohort, and mean cumulative exposure to RCS was 2.16 mg/m^3-years, suggesting RCS exposure concentrations on the order of 2.16/5.54 ≈ 0.4 mg/m^3.) Even for the highest cumulative exposure category, the increase in lung cancer risk is barely significant (lower 95% confidence limit just above the no-effect line RR = 1), without adjustment for multiple testing. On the other hand, the TSCE model with continuous dependence of model parameter g on cumulative exposure to RCS substantially over-predicts observed relative risks at high exposure concentrations. Using the model in Table 4.2, for example, the relative risk of lung cancer compared to background increases smoothly from 1 (no effect) to more than 7 as exposure concentration E increases from 0 to 2 mg/m^3. But real-world relative risks seldom exceed 2–3, and are commonly less than 2 even for the highest exposure categories. A recent meta-analysis of ECS-lung cancer associations found that "workers in the mining industry had the highest risk of lung cancer with a pooled SMR [standardized mortality ratio] of 1.48 (95% CI 1.18–1.86). …The positive association between silica and lung cancer became weaker with increasing quality of the included studies, from 2.56 (95% CI 1.57–4.19) among SMR studies with an NOS

[Newcastle-Ottawa scale for study quality] score of 1–3 to 1.24 (95% CI 1.01–1.52) in those with an NOS score of 7–9" (Poinen-Rughooputh et al. 2016). Even emphysema, corresponding to severe and prolonged inflammation of the lung, has been estimated to increase risk of lung cancer by less than twofold (Mayne et al. 1999). Thus, the assumption that g is a smoothly increasing function of RCS concentration appears to overstate relative risks, both at low exposures where there is no clear evidence of increased risk in the underlying data (Fig. 4.7) and at high exposures where the assumption implies relative risks much greater than those observed in numerous epidemiological studies of RCS and lung cancer

If it is assumed instead that g increases from its base value of 0.075 to a somewhat higher value such as 0.09 when and if occupational exposure triggers chronic inflammation, and then remains at this increased level (corresponding to the stable right-most equilibrium, E, in Fig. 4.1), then these disagreements with data disappear. No excess risk occurs at exposure concentrations too low to trigger inflammation, but relative risks remain in the observed range even for high exposure concentrations. Figure 4.8 illustrates the increase in age-specific risk (hazard rate) of lung cancer mortality for workers with and without occupational exposure that induces chronic inflammation of their lungs at age 50, assuming that the effect of the inflammation is to increase g by 20%, from 0.075 to 0.09. (In Fig. 4.8, ages are shifted leftward by 5 years compared to those in Fig. 4.6 because, following Zeka et al. (2011), we model the lag (latency) period between creation of a malignant cell and observed lung cancer as negligible. Assuming a positive lag before lung cancer mortality would shift the age axis rightward by the amount of the lag, which is 5 years in Fig. 4.6.) Figure 4.9 repeats this exercise for ages at onset of chronic inflammation ranging from 30 years to 80 years in 5-year increments. The corresponding relative risks of lung cancer mortality for exposed workers range from 1.00 (no excess risk) if inflammation is avoided to 1.6 if it begins at age 30. This

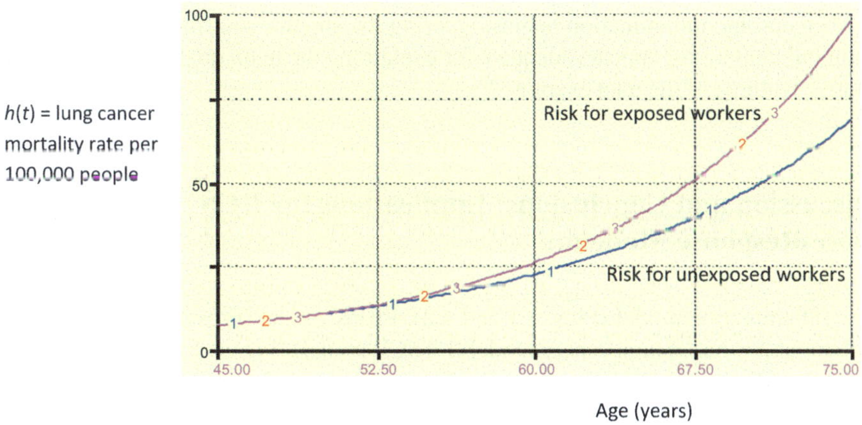

Fig. 4.8 Age-specific hazard rates for lung cancer mortality with (upper curve) and without (lower curve) onset of chronic inflammation at age 50, assuming that the effect of inflammation on cancer risk is to increase g from 0.075 to 0.09

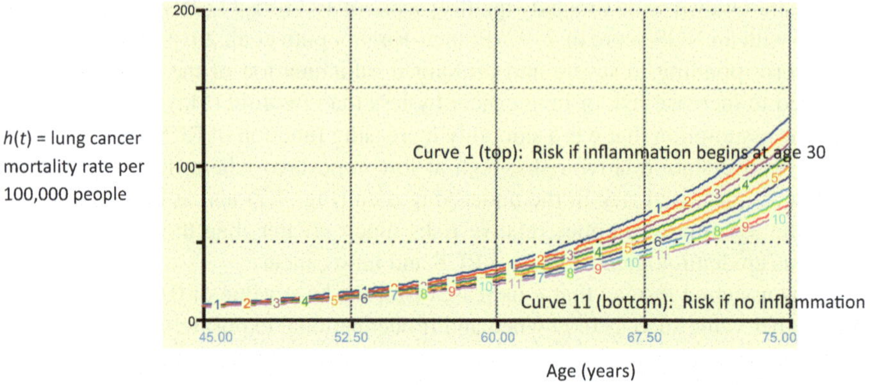

Fig. 4.9 Sensitivity analysis of increase in lung cancer mortality hazard rates with age vs. age at which chronic inflammation begins (curve 1 = age 30, curve 11 = age 80, 5-year increments). Corresponding relative risks for lifetime probability of tumor vary from RR = 1 (no excess risk) for inflammation onset age = 80 to RR = 1.6 for inflammation onset age = 30 years

range is consistent with observed relative risks in populations occupationally exposed to RCS (Poinen-Rughooputh et al. 2016). Thus, the proposed inflammation-mediated TSCE (I-TSCE) model with the background promotion proliferation rate g being increased by about 20% following onset of chronic inflammation, and no other effects of RCS exposure on lung cancer, appears to provide a satisfactory fit to epidemiological data showing no clear increase in risk for less-exposed workers and very limited increases in relative risks even in the highest exposure categories. Of course, transient inflammation might also increase risk while it lasts, if it increases the proliferation rate g for initiated cells; data permitting, a refined analysis could model such transient increases in g. However, on a time scale of decades, simply approximating the effect of exposure on risk as a one-time increase in the value of g when chronic inflammation begins, ignoring the smaller contributions from transient inflammations, appears adequate for explaining the main qualitative and quantitative features of the data examined.

Discussion and Conclusions: Implications for RCS Dose-Response Modeling

The inflammation model (Figs. 4.1 and 4.2), PBPK model (Figs. 4.3 and 4.4), and inflammation-based lung cancer model (I-TSCE model) (Figs. 4.5, 4.6, 4.8 and 4.9) have the following strong implications for exposure modeling and dose-response models. First, *cumulative exposure (in units such as mg/m^3-years) is not an appropriate exposure metric*, despite its widespread use in practice (e.g., Checkoway et al. 1997; Zeka et al. 2011). One reason is that it does not discriminate between many years of exposure to a concentration below the inflammation threshold (e.g., at 0.75*T in Fig. 4.1) and half as many year of exposure to twice the concentration, which may be well above the threshold (e.g., at 1.5*T in Fig. 4.1). A second reason is that, once chronic inflammation has been initiated, continued exposure plays no further role, at least in the simple I-TSCE model we have described. This modeling assumption may be too extreme for substances that exert toxic or immunosuppressive effects that modify the course and effects of inflammation. But the general point remains that exposure contributes to cancer risk differently before and after onset of inflammation. Standard exposure metrics such as cumulative exposure do not take this into account. Doing so raises challenges for statistical inference, since when and whether chronic inflammation begins for each individual is not typically recorded in industrial hygiene data sets. Thus, the times at which I-TSCE model parameters change are latent (unobserved) variables. For practical risk analysis, however, simple sensitivity analyses and bounds on risks, similar to Fig. 4.9, can help to quantify plausible ranges of exposure effect sizes on risks.

Second, *dose-response thresholds for inflammatory changes at the micro level of individual cells and organelles imply dose-response thresholds for inflammation in cell populations and tissues.* We have reviewed how the existence of dose-response thresholds for AM inflammatory responses to RCS and other NLRP3 inflammasome-activating substances at the sub-cellular level is now firmly established on biological grounds, e.g., from ultrasensitive switching in signaling via NF-κB and MAPK pathways to the nucleus to secrete NLRP3 protein filaments into the cytosol; switch-like cooperative polymerization of ASC and cooperative oligomerization in assembly of the NLRP3 inflammasome once concentrations in the cytosol exceed critical threshold levels; LMP and lysosome membrane rupture and release of cathepsins into the cytosol; receipt of signal 2 to activate the NLRP3 inflammasome; and pyroptosis and release of ASCs and activated NLRP3 inflammasomes into extracellular spaces, where they are taken up by, and activate inflammation in, nearby cells. We argued that such intracellular threshold mechanisms, which have been increasingly well elucidated by recent biochemical and biophysical research, create (probably heterogeneous) thresholds for activating AMs. These cell-level thresholds, together with transmission of inflammatory signals and elements (ASCs and inflammasomes) to neighboring cells, imply a corresponding exposure concentration threshold for triggering spread of self-sustaining chronic inflammation in cell populations or tissues (specifically, the administered concentration in inhaled air that produces a steady-state internal dose of RCS particles interacting with AMs in Fig. 4.4 corresponding to inflammation level T in Fig. 4.1). These concentration thresholds for causing chronic inflammation, in turn, imply concentration thresholds for increasing lung cancer risk in the I-TSCE model. Thus, the existence of dose-response thresholds for activating NLRP3-mediated inflammation at the intracellular level ultimately implies dose-response thresholds for lung cancer caused by NLRP3-activating agents such as RCS or asbestos.

Qualitatively, because the existence of thresholds in the inflammation and PBPK models does not depend on the precise numerical values of the various parameters involved, the fact that PBPK parameters and RCS properties vary around their nominal values (e.g., the values shown for DQ12 quartz dust in Table 4.1) does not alter the conclusion that *each individual has a threshold concentration for RCS exposures below which they do not increase risk of chronic inflammation or inflammation-mediated lung diseases* such as fibrosis, silicosis, or lung cancer. Interindividual variability and heterogeneity in the inflammation, PBPK, and I-TSCE models do not smooth out the population dose-response function to create an aggregate population dose-response function that satisfies LNT assumptions. Rather, the existence of ranges of values for the steady-state levels of RCS in AMs in Fig. 4.4 for any given RCS concentration in inhaled air, together with ranges of values for the inflammation thresholds in Fig. 4.1 and the resulting times of onset of chronic inflammation in Fig. 4.9, simply implies that there is an overall range of values for the RCS exposure concentration thresholds that increase risk of lung cancer in different individuals, as well as a range of values for the resulting increase in risk (e.g., from RR = 1 to RR = 1.6 in Fig. 4.9). The lower end of the RCS concentration range for increasing risk in different individuals is a population exposure concentration threshold below which population risk is not increased. Thus, it is ultimately intracellular thresholds for priming, assembling, and activating NLRP3 inflammasomes via signals 1 and 2 that prevent sufficiently small (or brief) exposure concentrations from increasing risks of inflammation and resulting diseases in exposed populations, and hence prevent LNT-like population dose-response relations from emerging from aggregation of individual dose-response relations.

Quantitatively, past PBPK models suggest possible lower bounds on RCS concentrations that can trigger acute (neutrophil-dominated) inflammation in rat lungs. These can be scaled to estimate corresponding thresholds in humans (e.g., at around 0.13 mg/m^3 in Fig. 4.4 for DQ12 quartz dust, and about 1 mg/m^3 for a dust with about 1/8 the inflammatory potency of DQ12). Such acute inflammation responses are part of normal, healthy, self-resolving inflammation as a means of responding to and clearing RCS. Corresponding thresholds for triggering chronic (macrophage-dominated) inflammation—the numerical value of T in Fig. 4.1—cannot be easily derived for humans from acute inflammation data in rats. If no clear excess risk of lung cancer is found at cumulative exposures less than about 5 mg/m^3-years (Fig. 4.7) among workers with a median exposure duration of 5.54 years, then average exposure concentrations below about 5/5.54, or about 1 mg/m^3, may not increase lung cancer risk under past exposure conditions. Such averages do not reveal the concentrations required to cause chronic inflammation and increased lung cancer risk, however, since some individual RCS workers under past exposure conditions might have accumulated exposures one or more orders of magnitude greater than the estimated values for the job groups to which they belong (Mamuya et al. 2006). If the excess risks observed among workers come primarily from those with the highest exposure concentrations, then the threshold RCS concentrations for causing chronic inflammation and excess risk of lung cancer could also be one or more orders of magnitude higher than calculations based on estimated values suggest, e.g., more than 10 mg/m^3 instead of less than 1 mg/m^3.Almost all past risk assessments for RCS have ignored such exposure estimation error in fitting risk models, which can make even exposure-response relations with clearly defined, unique thresholds mistakenly appear to satisfy LNT assumptions (Cox Jr. 2018a; Rhomberg et al. 2011). To better estimate the exposure concentrations and durations needed for RCS to induce chronic inflammation, it will be essential to model individual exposure uncertainty and variability in future RCS risk assessments.

For quantitative risk assessment, uncertainty about true exposures may dominate other sources of uncertainty and variability in estimating the excess risk of lung cancer created by a given estimated exposure history. An estimated average exposure concentration such as 0.1 mg/m^3 for years or decades might occasionally produce much higher exposures for an individual worker (Mamuya et al. 2006). If enough such anomalously high exposure events happen to occur closely enough together in time to trigger chronic inflammation, then lung cancer risk may be permanently increased, even though the threshold would never have been crossed had the variance around the mean concentration been smaller. Study designs and risk assessments that pair observed lung cancer mortality rates with estimated average cumulative exposure levels based on job exposure matrices or other dose reconstruction efforts, ignoring errors and interindividual variability around the estimated exposures (e.g., Checkoway et al. 1997), mistakenly attribute to the estimated average exposure concentration adverse effects that arise from such unmodeled variability. Hence they can greatly over-estimate the (possibly zero) risks from low concentrations (Cox Jr. 2018a). To obtain more realistic guidance for protecting worker health, the PBPK model in Table 4.1 can calculate numerically how the distribution of daily exposure concentrations affects the probability distribution of the time to cross an internal dose threshold such as m_{crit}. The I-TSCE model (Fig. 4.9) quantifies the consequences for lung cancer risk of initiating chronic inflammation at different ages. Data are not yet available to quantify how many times greater the RCS concentration needed to trigger chronic inflammation is than the concentration required to trigger acute inflammation, but RCS exposure histories that avoid letting internal doses of RCS in macrophages cross the acute inflammation threshold m_{crit} will also avoid initiating chronic inflammation.

The analysis presented in this chapter has several limitations. The modeling is deliberately very simplified; as noted in several places, detailed transients and stochastic fluctuations are ignored, and smoothed (mean-field) approximations on relatively long time scales are used throughout. This keeps the data requirements and modeling tractable, but at the cost of omitting potentially important inter-individual variability of individual risks around the average-case values. Stochastic simulation or agent-based models could provide much more information about such variability than the deterministic ODE modeling and sensitivity analyses presented here (e.g., Fig. 4.9). In addition, the proposed model has not been thoroughly validated. Although it appears to explain some aggregate epidemiological data fairly well, and although the PBPK and TSCE sub-models have been studied and validated as providing useful descriptions of other data sets, the integrated I-TSCE model with inflammation is, to our knowledge, new. It has yet to be applied and validated using diverse data sets (e.g., for asbestos-induced lung cancer and perhaps mesothelioma) in addition to RCS; thus, at this stage, we regard it as a promising proposed approach, but in need of further application and validation. Finally, it is possible to model mechanisms of RCS-induced lung cancer without modeling the NLRP3 inflammasome-mediated inflammation mechanism emphasized here. For example, a previous model (Cox Jr. 2011) describes an inflammatory mode of action in which prolonged exposure to high RCS concentrations activates overlapping positive feedback loops involving recruitment and activation of alveolar macrophages and neutrophils, increasing levels of reactive oxygen species (ROS) and nitrogen species, increased levels of pro-inflammatory cytokines, and eventual increases in apoptosis and compensating proliferation of the alveolar epithelium. That model, which has considerable biological support, describes the sources and consequences of chronic inflammation in terms of feedback loops among cell populations and signals, without specifying how events within individual cells create and propagate inflammation. The NLRP3 inflammasome biology discussed earlier supplies these missing details, explaining how sustained RCS exposure leads to signaling to the AM cell nucleus, secretion of NLRP3 protein to the cytosol, intracellular assembly and activation of the NLRP3 inflammasome, secretion of mature IL-1β and IL-18, upregulation of other pro-inflammatory cytokines, recruitment of neutrophils and macrophages, increases in ROS, eventual lysosome rupture, pyroptosis, and propagation of activated NLRP3 and ASC specks between cells. Although the models are consistent, it is not necessary to understand or model the NLRP3-mediated events within and between cells to describe the interdependencies and positive feedback loops among cell populations and the chemokine and cytokine signals that they generate. The model in this chapter explains why sustained high exposures eventually lead to an inflammation "tipping point" beyond which inflammation spontaneously spreads (Figs. 4.1 and 4.2), but dose-response models that simply accept this as an empirical fact and that do not seek to derive it from cell-level thresholds could be

Although this chapter has focused on respirable crystalline silica (RCS) as a case study, the NLRP3 inflammasome mediates inflammatory responses to many other agents, from viruses and bacteria (Coll et al. 2016) to radiation (Liu et al. 2017) to mineral fibers such as asbestos (Sayan and Mossman 2016), as discussed in Chaps. 5 and 6. The bistability illustrated in Figs. 4.1 and 4.2 does not depend on which particular substance activates the NLRP3 inflammasome. Rather, it reflects the fact that NLRP3 inflammasome assembly and activation in some cells, followed by pyroptosis and discharge of NLRP3 inflammasomes and ASC specks to the intracellular environment, makes induction of this same process in nearby cells more likely. In this sense, NLRP3-mediated inflammation is analogous to a contagious condition in a population (of cells); as in many models for contagious processes and epidemics, a threshold exists below which the contagion process moves toward a low-prevalence equilibrium and above which it moves toward a higher-prevalence equilibrium level for the contagious condition (i.e.,NLRP3-mediated inflammation). This aspect of the model may be applicable to many substances other than RCS that act via NLRP3 inflammasome-mediated inflammation of various tissues.

In summary, as highlighted in Chaps. 2 and 3 and discussed further in Chap. 6, the existence of dose-response thresholds for NLRP3 inflammasome-mediated diseases emphasizes the need and opportunity to protect worker health by reducing high exposures that can trigger chronic inflammation. This may not affect average exposure concentrations much (or, perhaps, at all) but it is probably essential to reduce risk of inflammation-mediated diseases such as RCS-associated lung cancer.

References

Abais JM, Xia M, Zhang Y, Boini KM, Li PL. Redox regulation of NLRP3 inflammasomes: ROS as trigger or effector? Antioxid Redox Signal. 2015;22(13):1111–29. https://doi.org/10.1089/ars.2014.5994.

Abderrazak A, Syrovets T, Couchie D, El Hadri K, Friguet B, Simmet T, Rouis M. NLRP3 inflammasome: from a danger signal sensor to a regulatory node of oxidative stress and inflammatory diseases. Redox Biol. 2015;4:296–307. https://doi.org/10.1016/j.redox.2015.01.

Baroja-Mazo A, Martín-Sánchez F, Gomez AI, Martínez CM, Amores-Iniesta J, Compan V, Barberà-Cremades M, Yagüe J, Ruiz-Ortiz E, Antón J, Buján S, Couillin I, Brough D, Arostegui JI, Pelegrín P. The NLRP3 inflammasome is released as a particulate danger signal that amplifies the inflammatory response. Nat Immunol. 2014;15(8):738–48. https://doi.org/10.1038/ni.2919.

Bednash JS, Mallampalli RK. Regulation of inflammasomes by ubiquitination. Cell Mol Immunol. 2016;13(6):722–8. https://doi.org/10.1038/cmi.2016.15.

Bergsbaken T, Fink SL, Cookson BT. Pyroptosis: host cell death and inflammation. Nat Rev Microbiol. 2009;7(2):99–109. https://doi.org/10.1038/nrmicro2070.

Bogen KT. Low-dose dose-response for in vitro Nrf2-ARE activation in human HepG2 Cells. Dose-Response. 2017;15(2):1559325817699696. https://doi.org/10.1177/1559325817699696.

Bolt HM, Foth H, Hengstler JG, Degen GH. Carcinogenicity categorization of chemicals-new aspects to be considered in a European perspective. Toxicol Lett. 2004;151(1):29–41.

Borm P, Cassee FR, Oberdörster G. Lung particle overload: old school -new insights? Part Fibre Toxicol. 2015;12:10. https://doi.org/10.1186/s12989-015-0086-4.

Borm PJA, Fowler P, Kirkland D. An updated review of the genotoxicity of respirable crystalline silica. Part Fibre Toxicol. 2018;15(1):23. https://doi.org/10.1186/s12989-018-0259-z.

Boya P, Kroemer G. Lysosomal membrane permeabilization in cell death. Oncogene. 2008;27(50):6131–51. https://doi.org/10.1038/onc.2008.310.

Brown BN, Price IM, Toapanta FR, DeAlmeida DR, Wiley CA, Ross TM, Oury TD, Vodovotz Y. An agent-based model of inflammation and fibrosis following particulate exposure in the lung. Math Biosci. 2011;231(2):186–96. https://doi.org/10.1016/j.mbs.2011.03.005.

Bruch J, Rehn S, Rehn B, Borm PJ, Fubini B. Variation of biological responses to different respirable quartz flours determined by a vector model. Int J Hyg Environ Health. 2004;207(3):203–16.

Butterworth BE, Popp JA, Conolly RB, Goldsworthy TL. Chemically induced cell proliferation in carcinogenesis. IARC Sci Publ. 1992;116:279–305.

Butterworth BE, Aylward LL, Hays SM. A mechanism-based cancer risk assessment for 1,4-dichlorobenzene. Regul Toxicol Pharmacol. 2007;49(2):138–48.

Calabrese EJ, Baldwin LA. Can the concept of hormesis be generalized to carcinogenesis? Regul Toxicol Pharmacol. 1998;28(3):230–41.

Camberlein E, Cohen JM, José R, Hyams CJ, Callard R, Chimalapati S, Yuste J, Edwards LA, Marshall H, van Rooijen N, Noursadeghi M, Brown JS. Importance of bacterial replication and alveolar macrophage-independent clearance mechanisms during early lung infection with Streptococcus pneumoniae. Infect Immun. 2015;83(3):1181–9. https://doi.org/10.1128/IAI.02788-14.

Castranova V, Porter D, Millecchia L, Ma JY, Hubbs AF, Teass A. Effect of inhaled crystalline silica in a rat model: time course of pulmonary reactions. Mol Cell Biochem. 2002;234-235(1-2):177–84.

Checkoway H, Heyer NJ, Seixas NS, Welp EA, Demers PA, Hughes JM, Weill H. Dose-response associations of silica with nonmalignant respiratory disease and lung cancer mortality in the diatomaceous earth industry. Am J Epidemiol. 1997;145(8):680–8.

Clouter, A, Brown, D, Höhr, D. Inflammatory effects of respirable quartz collected in workplaces versus standard DQ12 quartz: particle surface correlates. Toxicol Sci. 2001;63(1):90–98.

Cohen SM. Role of cell proliferation in regenerative and neoplastic disease. Toxicol Lett. 1995;82:15–21.

Coll RC, O'Neill L, Schroder K. Questions and controversies in innate immune research: what is the physiological role of NLRP3? Cell Death Dis. 2016;2:16019. https://doi.org/10.1038/cddiscovery.2016.19.

Conolly RB, Gaylor DW, Lutz WK. Population variability in biological adaptive responses to DNA damage and the shapes of carcinogen dose-response curves. Toxicol Appl Pharmacol. 2005;207(2 Suppl):570–5.

Cox LA Jr. An exposure-response threshold for lung diseases and lung cancer caused by crystalline silica. Risk Anal. 2011;31(10):1543–60. https://doi.org/10.1111/j.1539-6924.2011.01610.x.

Cox LA Jr. Effects of exposure estimation errors on estimated exposure-response relations for PM2.5. Environ Res. 2018a;164:636–46. https://doi.org/10.1016/j.envres.2018.03.038.

Cox LA Jr. (2018b) Biological mechanisms of non-linear dose-response for respirable mineral fibers. Toxicol Appl Pharmacol. https://doi.org/10.1016/j.taap.2018.06.016.

Franklin BS, Bossaller L, De Nardo D, Ratter JM, Stutz A, Engels G, Brenker C, Nordhoff M, Mirandola SR, Al-Amoudi A, Mangan MS, Zimmer S, Monks BG, Fricke M, Schmidt RE, Espevik T, Jones B, Jarnicki AG, Hansbro PM, Busto P, Marshak-Rothstein A, Hornemann S, Aguzzi A, Kastenmüller W, Latz E. The adaptor ASC has extracellular and 'prionoid' activities that propagate inflammation. Nat Immunol. 2014;15(8):727–37. https://doi.org/10.1038/ni.2913.

Freire J, Ajona D, de Biurrun G, Agorreta J, Segura V, Guruceaga E, Bleau AM, Pio R, Blanco D, Montuenga LM. Silica-induced chronic inflammation promotes lung carcinogenesis in the context of an immunosuppressive microenvironment. Neoplasia. 2013;15(8):913–24.

Fukushima S, Kinoshita A, Puatanachokchai R, Kushida M, Wanibuchi H, Morimura K. Hormesis and dose-response-mediated mechanisms in carcinogenesis: evidence for a threshold in carcinogenicity of non-genotoxic carcinogens. Carcinogenesis. 2005;26(11):1835–45.

Gottschalk RA, Martins AJ, Angermann BR, Dutta B, Ng CE, Uderhardt S, Tsang JS, Fraser ID, Meier-Schellersheim M, Germain RN. Distinct NF-κB and MAPK activation thresholds uncouple steady-state microbe sensing from anti-pathogen inflammatory responses. Cell Syst. 2016;2(6):378–90. https://doi.org/10.1016/j.cels.2016.04.016.

Harijith A, Ebenezer DL, Natarajan V. Reactive oxygen species at the crossroads of inflammasome and inflammation. Front Physiol. 2014;5:352. https://doi.org/10.3389/fphys.2014.00352.

Hauenstein AV, Zhang L, Wu H. The hierarchical structural architecture of inflammasomes, supramolecular inflammatory machines. Curr Opin Struct Biol. 2015;31:75–83. https://doi.org/10.1016/j.sbi.2015.03.014.

He Q, Fu Y, Tian D, Yan W. The contrasting roles of inflammasomes in cancer. Am J Cancer Res. 2018;8(4):566–583. Published 2018 Apr 1.

Jessop F, Hamilton RF Jr, Rhoderick JF, Fletcher P, Holian A. Phagolysosome acidification is required for silica and engineered nanoparticle-induced lysosome membrane permeabilization and resultant NLRP3 inflammasome activity. Toxicol Appl Pharmacol. 2017;318:58–68. https://doi.org/10.1016/j.taap.2017.01.012.

Katsnelson MA, Lozada-Soto KM, Russo HM, Miller BA, Dubyak GR. NLRP3 inflammasome signaling is activated by low-level lysosome disruption but inhibited by extensive lysosome disruption: roles for K+ efflux and Ca2+ influx. Am J Physiol Cell Physiol. 2016;311(1):C83–C100. https://doi.org/10.1152/ajpcell.00298.2015.

Kenah E, Robins JM. Network-based analysis of stochastic SIR epidemic models with random and proportionate mixing. J Theor Biol. 2007;249(4):706–22.

Kuempel ED, Tran CL, Bailer AJ, Porter DW, Hubbs AF, Castranova V. Biological and statistical approaches to predicting human lung cancer risk from silica. J Environ Pathol Toxicol Oncol. 2001;20(Suppl 1):15–32.

Kusaka T, Nakayama M, Nakamura K, Ishimiya M, Furusawa E, Ogasawara K. Effect of silica particle size on macrophage inflammatory responses. PLoS One. 2014;9(3):e92634. https://doi.org/10.1371/journal.pone.0092634.

Liu YG, Chen JK, Zhang ZT, Ma XJ, Chen YC, Du XM, Liu H, Zong Y, Lu GC. NLRP3 inflammasome activation mediates radiation-induced pyroptosis in bone marrow-derived macrophages. Cell Death Dis. 2017;8(2):e2579. https://doi.org/10.1038/cddis.2016.460.

Lopez-Castejon G, Luheshi NM, Compan V, High S, Whitehead RC, Flitsch S, Kirov A, Prudovsky I, Swanton E, Brough D. Deubiquitinases regulate the activity of caspase-1 and interleukin-1β secretion via assembly of the inflammasome. J Biol Chem. 2013;288(4):2721–33. https://doi.org/10.1074/jbc.M112.422238.

Lu A, Wu H. Structural mechanisms of inflammasome assembly. FEBS J. 2015;282(3):435–44. https://doi.org/10.1111/febs.13133.

Luna-Gomes T, Santana PT, Coutinho-Silva R. Silica-induced inflammasome activation in macrophages: role of ATP and P2X7 receptor. Immunobiology. 2015;220(9):1101–6. https://doi.org/10.1016/j.imbio.2015.05.004.

Lutz WK. Dose-response relationships in chemical carcinogenesis: superposition of different mechanisms of action, resulting in linear-nonlinear curves, practical thresholds, J-shapes. Mutat Res. 1998;405(2):117–24.

Lutz WK, Kopp-Schneider A. Threshold dose response for tumor induction by genotoxic carcinogens modeled via cell-cycle delay. Toxicol Sci. 1999;49(1):110–5.

Mamuya SH, Bråtveit M, Mwaiselage J, Moen BE. Variability of exposure and estimation of cumulative exposure in a manually operated coal mine. Ann Occup Hyg. 2006;50(7):737–45.

Mayne ST, Buenconsejo J, Janerich DT. Previous lung disease and risk of lung cancer among men and women nonsmokers. Am J Epidemiol. 1999;149(1):13–20.

McCarthy WJ, Meza R, Jeon J, Moolgavkar SH. Chapter 6: lung cancer in never smokers: epidemiology and risk prediction models. Risk Anal. 2012;32(Suppl 1):S69–84. https://doi.org/10.1111/j.1539-6924.2012.01768.x.

Meldrum M, Howden P. Crystalline silica: variability in fibrogenic potency. Ann Occup Hyg. 2002;46:27–30.

Miraldi ER, Thomas PJ, Romberg L. Allosteric models for cooperative polymerization of linear polymers. Biophys J. 2008;95(5):2470–86. https://doi.org/10.1529/biophysj.107.126219.

Moldoveanu B, Otmishi P, Jani P, Walker J, Sarmiento X, Guardiola J, Saad M, Yu J. Inflammatory mechanisms in the lung. J Inflamm Res. 2009;2:1–11.

Neri FM, Pérez-Reche FJ, Taraskin SN, Gilligan CA. Heterogeneity in susceptible-infected-removed (SIR) epidemics on lattices. J R Soc Interface. 2011;8(55):201–9. https://doi.org/10.1098/rsif.2010.0325.

Paul MK, Bisht B, Darmawan DO, Chiou R, Ha VL, Wallace WD, Chon AT, Hegab AE, Grogan T, Elashoff DA, Alva-Ornelas JA, Gomperts BN. Dynamic changes in intracellular ROS levels regulate airway basal stem cell homeostasis through Nrf2-dependent Notch signaling. Cell Stem Cell. 2014;15(2):199–214. https://doi.org/10.1016/j.stem.2014.05.009.

Pavan C, Rabolli V, Tomatis M, Fubini B, Lison D. Why does the hemolytic activity of silica predict its pro-inflammatory activity? Part Fibre Toxicol. 2014;11:76. https://doi.org/10.1186/s12989-014-0076-y.

Peeters PM, Perkins TN, Wouters EF, Mossman BT, Reynaert NL. Silica induces NLRP3 inflammasome activation in human lung epithelial cells. Part Fibre Toxicol. 2013;10:3.

Poinen-Rughooputh S, Rughooputh MS, Guo Y, Rong Y, Chen W. Occupational exposure to silica dust and risk of lung cancer: an updated meta-analysis of epidemiological studies. BMC Public Health. 2016;16(1):1137.

Pollard KM. Silica, silicosis, and autoimmunity. Front Immunol. 2016;7:97. https://doi.org/10.3389/fimmu.2016.00097.

Porter DW, Hubbs AF, Mercer R, Robinson VA, Ramsey D, McLaurin J, Khan A, Battelli L, Brumbaugh K, Teass A, Castranova V. Progression of lung inflammation and damage in rats after cessation of silica inhalation. Toxicol Sci. 2004;79(2):370–80.

Porter DW, Millecchia LL, Willard P, Robinson VA, Ramsey D, McLaurin J, Khan A, Brumbaugh K, Beighley CM, Teass A, Castranova V. Nitric oxide and reactive oxygen species production causes progressive damage in rats after cessation of silica inhalation. Toxicol Sci. 2006; 90(1):188–97.

Rabolli V, Lison D, Huaux F. The complex cascade of cellular events governing inflammasome activation and IL-1β processing in response to inhaled particles. Part Fibre Toxicol. 2016;13(1):40. https://doi.org/10.1186/s12989-016-0150-8.

Ren F, Wang K, Zhang T, Jiang J, Nice EC, Huang C. New insights into redox regulation of stem cell self-renewal and differentiation. Biochim Biophys Acta. 2015;1850(8):1518–26. https://doi.org/10.1016/j.bbagen.2015.02.017.

Rhomberg LR, Chandalia JK, Long CM, Goodman JE. Measurement error in environmental epidemiology and the shape of exposure-response curves. Crit Rev Toxicol. 2011;41(8):651–71. https://doi.org/10.3109/10408444.2011.563420.

Robb CT, Regan KH, Dorward DA, Rossi AG. Key mechanisms governing resolution of lung inflammation. Semin Immunopathol. 2016;38(4):425–48. https://doi.org/10.1007/s00281-016-0560-6.

Sayan M, Mossman BT. The NLRP3 inflammasome in pathogenic particle and fibre-associated lung inflammation and diseases. Part Fibre Toxicol. 2016;13(1):51. https://doi.org/10.1186/s12989-016-016.

Schetter AJ, Heegaard NH, Harris CC. Inflammation and cancer: interweaving microRNA, free radical, cytokine and p53 pathways. Carcinogenesis. 2010;31(1):37–49. https://doi.org/10.1093/carcin/bgp272.

Shi CS, Shenderov K, Huang NN, Kabat J, Abu-Asab M, Fitzgerald KA, Sher A, Kehrl JH. Activation of autophagy by inflammatory signals limits IL-1β production by targeting ubiquitinated inflammasomes for destruction. Nat Immunol. 2012;13(3):255–63. https://doi.org/10.1038/ni.2215.

Smith AM, McCullers JA, Adler FR. Mathematical model of a three-stage innate immune response to a pneumococcal lung infection. J Theor Biol. 2011;276(1):106–16. https://doi.org/10.1016/j.jtbi.2011.01.052.

Snow ET, Sykora P, Durham TR, Klein CB. Arsenic, mode of action at biologically plausible low doses: what are the implications for low dose cancer risk? Toxicol Appl Pharmacol. 2005;207(2 Suppl):557–64.

Sweeney LM, Parker A, Haber LT, Tran CL, Kuempel ED. Application of Markov chain Monte Carlo analysis to biomathematical modeling of respirable dust in US and UK coal miners. Regul Toxicol Pharmacol. 2013;66(1):47–58. https://doi.org/10.1016/j.yrtph.2013.02.003.

Tran CL, Graham M, Buchanan D (2001) A biomathematical model for rodent and human lung describing exposure, dose, and response to inhaled silica. institute of occupational medicine technical memorandum. Available at: http://www.iom-world.org/pubs/IOM_TM0104.pdf

Tran CL, Kuempel ED, Castranova V. A rat lung model of exposure, dose and response to inhaled silica. Ann Occup Hyg. 2002;46(Supplement 1):14.

Turci F, Pavan C, Leinardi R, Tomatis M, Pastero L, Garry D, Anguissola S, Lison D, Fubini B. Revisiting the paradigm of silica pathogenicity with synthetic quartz crystals: the role of crystallinity and surface disorder. Part Fibre Toxicol. 2016;13(1):32. https://doi.org/10.1186/s12989-016-0136-6.

Vineis P, Schatzkin A, Potter JD. Models of carcinogenesis: an overview. Carcinogenesis. 2010; 31(10):1703–9. https://doi.org/10.1093/carcin/bgq087.

Volz E, Meyers LA. Epidemic thresholds in dynamic contact networks. J R Soc Interface. 2009;6(32):233–41. https://doi.org/10.1098/rsif.2008.0218.

Wang W, Liu Q-H, Zhong L-F, Tang M, Gao H, Stanley HE. Predicting the epidemic threshold of the susceptible-infected-recovered model. Sci Rep. 2016;6:24676. https://doi.org/10.1038/srep24676.

Warheit DB, Kreiling R, Levy LS. Relevance of the rat lung tumor response to particle overload for human risk assessment - update and interpretation of new data since ILSI 2000. Toxicology. 2016;374:42–59. https://doi.org/10.1016/j.tox.2016.11.013.

Whysner J, Williams GM. Saccharin mechanistic data and risk assessment: urine composition, enhanced cell proliferation, and tumor promotion. Pharmacol Ther. 1996;71(1-2):225–52.

Wolf DC, Butterworth BE. Risk assessment of inhaled chloroform based on its mode of action. Toxicol Pathol. 1997;25(1):49–52.

Zaballa I, Eidemüller M. Mechanistic study on lung cancer mortality after radon exposure in the Wismut cohort supports important role of clonal expansion in lung carcinogenesis. Radiat Environ Biophys. 2016;55(3):299–315. https://doi.org/10.1007/s00411-016-0659-0.

Zeka A, Gore R, Kriebel D. The two-stage clonal expansion model in occupational cancer epidemiology: results from three cohort studies. Occup Environ Med. 2011;68(8):618–24. https://doi.org/10.1136/oem.2009.053983.

Chapter 5
Case Study: Health Risks from Asbestos Exposures

Introduction

Can a single fiber of amphibole asbestos increase the risk of lung cancer or malignant mesothelioma (MM)? Traditional linear no-threshold (LNT) risk assessment assumptions imply that the answer is yes: there is no safe exposure level. This chapter draws on recent scientific progress in inflammation biology, especially elucidation of the activation thresholds for NLRP3 inflammasomes and resulting chronic inflammation as discussed in Chaps. 3 and 4, to model dose-response relationships for malignant mesothelioma and lung cancer risks caused by asbestos exposures. Similar to the model for respirable crystalline silica in Chap. 4, the modeling in this chapter integrates a physiologically based pharmacokinetics (PBPK) front end with inflammation-driven two-stage clonal expansion (I-TSCE) models of carcinogenesis to describe how exposure leads to chronic inflammation, which in turn promotes carcinogenesis. Together, the combined PBPK and I-TSCE modeling predict that there are practical thresholds for exposure concentration below which asbestos exposure does not cause chronic inflammation in less than a lifetime, and therefore does not increase chronic inflammation-dependent cancer risks. Quantitative examples using model parameter estimates drawn from the literature suggest that practical thresholds may be within about a factor of 2 of some past exposure levels for some workers. The I-TSCE modeling framework presented here explains some previous puzzling aspects of asbestos epidemiology, such as why age at first exposure is a better predictor of lifetime MM risk than exposure duration.

As discussed in Chap. 3, the type of dynamic simulation modeling illustrated here can be a valuable tool for risk analysts when LNT assumptions are not justified due to inflammation response thresholds mediating dose-response relationships. To make this chapter relatively self-contained, the followings sections summarize relevant biological background, and the appendices to this chapter summarize and

L. A. Cox Jr., *Quantitative Risk Analysis of Air Pollution Health Effects*,
International Series in Operations Research & Management Science 299,
https://doi.org/10.1007/978-3-030-57358-4_5

extend key aspects of the mathematical model from Chap. 4, and adapt it for asbestos.

Is There a Threshold for Asbestos Carcinogenicity?

A question of great scientific and practical interest—and one that has long excited the passions of litigators favoring various answers—is whether a single asbestos fiber lodged in the lung or in the pleural or peritoneal mesothelium can increase risk of lung cancer (LC) or of malignant mesothelioma (MM), respectively. Whatever the answer is, what is the explanation for it? If there is a lowest concentration below which exposures do not increase risk of MM or LC and above which they do, then traditional one-hit and linear no-threshold (LNT) risk modeling assumptions would not hold, raising the challenge of identifying causal biological mechanisms that explain how high concentrations can cause malignancies even if lower concentrations do not. Such scientific questions and challenges cannot be answered confidently solely by fitting empirical curves to epidemiological data or by studying rat cells *in vitro*. Epidemiological data alone typically lack accurate fiber counts and are inconclusive about risks at very low concentrations, and *in vitro* techniques do not provide accurate estimates of biologically effective doses (e.g., numbers of fibers accumulated in alveolar epithelial tissue or in mesothelial tissue over time) that occur under realistic exposure scenarios. A different approach is needed to make progress beyond the studies of previous decades.

This chapter builds on the recent advances in understanding of inflammation biology mentioned in Chaps. 3 and 4 to seek new traction for understanding shapes of dose-response functions at low exposure concentrations for substances and diseases with exposure-induced chronic inflammation as a main (rate-limiting) mode of action for increasing disease risk. Asbestos fibers, like crystalline silica and other mineral fibers and particles, are now to known to activate the NLRP3 inflammasome, the inflammation-coordinating platform introduced in Chaps. 3 and 4. The low-dose biological activity and toxicology of these compounds depends in part on the fact that there is an activation threshold for NLRP3 inflammasome activation (Chap. 4). Inflammasome activation by inhaled asbestos fibers has been reported *in vivo* in mouse inhalation models exposed to chrysotile asbestos for 9 days (Dostert et al. 2008), but, to our knowledge, human dose-response data for NLRP3 inflammasome activation *in vivo* by amphibole asbestos fibers are not available. This chapter proposes that partial progress can be made by combining models of asbestos fiber pharmacokinetics and inflammation in humans, and positing that chronic inflammation increases cancer risk (Bogen 2019), even though precise quantitative dose-response modeling is not yet possible (and is likely to depend heavily on physical and chemical properties of fibers, which vary greatly among asbestos samples).

Advances in science can have policy and risk management implications, and those that affect current regulatory or litigation practice invite scrutiny and challenge. The main focus of this chapter is on science-grounded risk assessment modeling, and not on potential policy implications, which are relegated to a discussion section toward the end, along with a discussion of some past debates in the literature.

Recent Progress in Elucidating the Roles of the NLRP3 Inflammasome and Chronic Inflammation

This section provides a brief introduction to the relevant biological background; readers who have already read Chaps. 3 and 4 may skip to the next section without loss. As explained in Chap. 3, over the past two decades, a wide variety of diseases have been discovered to involve inflammation, mediated by the NLRP3 inflammasome. This is a wheel-shaped multiprotein signaling complex. It self-assembles in the cytosol of macrophages and other myeloid cells, including monocytes and dendritic cells, and some epithelial cells with mucosal surfaces, in response to sustained, sufficiently intense danger signals generated by receptors that recognize bacterial pathogens, mineral fibers or particles, organic crystals (e.g., of uric acid for gout, or cholesterol for atherosclerosis), or other threats requiring clearance via an inflammatory response. Once assembled and activated, the NLRP3 inflammasome coordinates pro-inflammatory signaling (by using the enzyme caspase-1 to cleave the protective ends from, and thereby activate, interleukin cytokines IL-1β and IL-18). "NLRP3" stands for "nucleotide-binding oligomerization domain (NOD)-like receptor protein 3" or "nucleotide-binding domain, leucine-rich repeat region containing family, pyrin domain-containing-3," referring to the components of the inflammasome oligomer, which links damage sensors (receptors in the cytosol, built from the leucine-rich repeat domain) to effectors that recruit pro-caspase-1 via a caspase recruitment domain (CARD). Increased release of intracellular and mitochondrial reactive oxygen species (ROS) is followed by lysosomal membrane permeabilization (LMP), leakage of the lysosome's toxic contents into the cell cytosol, programmed cell death and rupture of the cell's plasma membrane—the process of *pyroptosis* (Heid et al. 2013; Jo et al. 2016; Swanson et al. 2019). Pyroptosis, in turn, releases the pro inflammatory contents of ruptured cells, including their activated NLRP3 inflammasomes, into the extracellular environment, where they can help to induce inflammation in nearby cells. Despite this potential to spread, inflammation usually resolves itself as regulatory mechanisms (involving phosphorylation and ubiquitination and degradation of NLRP3 protein filaments and inhibitory signals from other cells, e.g., mesenchymal stem/stromal cells) halt continued assembly and activation of the NLRP3 inflammasomes (Latz et al. 2013; Oh et al. 2014; Song and Li 2018; Tezcan et al. 2019; Yang et al. 2019). If continued exposure overcomes these safeguards, however, then chronic inflammation ensues, leading to a high-ROS microenvironment, increased oxidative stress and damage, and increased risk of cancer and other diseases in the chronically inflamed tissue.

This general mechanism of NLRP3 inflammasome-mediated chronic inflammation has recently been implicated in a wide variety of diseases, including Alzheimer's disease, arthritis, atherosclerosis, bacterial infections, cardiovascular diseases, chronic obstructive pulmonary disease (COPD), colorectal cancer and other cancers, diabetes, gout, inflammatory bowel disease, kidney diseases, lung cancer, malignant mesothelioma, obesity, periodontis, pulmonary fibrosis, prion diseases, respiratory diseases including silicosis, stroke, and various viral infections (Abderrazak et al. 2015; He et al. 2018; Hutton et al. 2016; Kadariya et al. 2016; Karasawa and Takahashi 2017; Karki et al. 2017; Kuriakose and Kanneganti 2017; Macpherson et al. 2017; Sayan and Mossman 2016; Shao et al. 2019; Thompson et al. 2017; Zhiyu et al. 2016; Zhou et al. 2016). The NLRP3 inflammasome is activated by sufficiently high and prolonged exposures to a variety of agents, including mineral fibers and particles in the lung (e.g., asbestos fibers and respirable crystalline silica) and asbestos fibers in the mesothelium (Sayan and Mossman 2016; Thompson et al. 2017); long, needle-like carbon nanotubes (Palomäki et al. 2011); infectious bacteria (Tezcan et al. 2019); uric acid crystals (Yang et al. 2019); and some viruses, including influenza virus (Kuriakose and Kanneganti 2017). That so many different stimuli and exposure-related diseases are mediated by the same causal mechanisms, involving activation of the NLRP3 inflammasome, pyroptosis, and chronic inflammation of target tissue, provides an opportunity to study general properties of dose-response functions for these pathogenic processes, rather than viewing each individual exposure-response pair as a separate modeling challenge.

The relevance of NLRP3 inflammasome activation (Sayan and Mossman 2016) and chronic inflammation (Carbone et al. 2012) specifically for causation of MM and LC by asbestos have been elucidated very recently. Asbestos-induced priming (via upregulated transcription of pro-IL-1β and NLRP3 protein filaments and post-translational modification of inflammasome components) and activation (via oligomerization) of the NLRP3 inflammasome triggers an autocrine feedback loop (via the IL-1 receptor) in human mesothelial cells targeted in pleural fibrosis and carcinogenesis (Hillegass et al. 2013), and chronic inflammation is considered a key mediator in development of MM and LC based on several lines of evidence (Carbone et al. 2012; He et al. 2018; Huang et al. 2019; Miller et al. 2014; Moossavi et al. 2018). NLRP3 inflammasome-mediated chronic inflammation appears to be involved in the origin of MM, creating an environment of chronic IL-1β signaling in human mesothelial cells and promoting mesothelial-to-fibroblastic transition (Thompson et al. 2017). Similarly, in the lung, NLRP3-dependent increases in IL-1β and IL-18 promote epithelial-to-mesenchymal transition (EMT) and lung cancer progression (He et al. 2018). Chronic inflammation also increases the risk of lung cancer, while regular use of anti-inflammatory agents such as ibuprofen appears to reduce risk of lung cancer mortality in recent epidemiological studies (Vlachogianni et al. 2013; Bittoni et al. 2017). *Frustrated phagocytosis,* in which long (greater than 4 μm (Schinwald et al. 2012)), high aspect ratio, rigid, bio-persistent fibers cannot be successfully engulfed and destroyed by macrophages or other cells, but instead pierce the lysosome and trigger pyroptosis, is a common mechanism for induction of chronic inflammation, mesothelioma and lung carcinogenicity in rats; it has been demonstrated for carcinogenic multiwall carbon nanotubes and metallic nanowires as well as for amphibole asbestos fibers (Donaldson et al. 2013; Kane et al. 2018; Lee et al. 2018). Multiple studies have shown that asbestos fibers not only prime and activate the NLRP3 inflammasome *in vitro* in human mesothelial cell lines (Hillegass et al. 2013), and probably in macrophage cell lines (Li et al. 2012), but they also activate the NLRP3 inflammasome and increase pulmonary inflammation and inflammation-related malignant mesothelioma (MM) risk *in vivo* in murine models of asbestos inhalation and MM (Dostert et al. 2008; Kadariya et al. 2016). Although of course corresponding human experimental data are not available, these findings in cell lines and mice are generally coherent with limited evidence from human observational data suggesting that occupational asbestos exposures increase production of growth factors involved in early inflammatory responses prior to clinical manifestation of asbestos-induced diseases (Comar et al. 2016). Thus, although inhalation exposure levels of asbestos fibers that activate the NLRP3 inflammasome in humans have not been determined, it appears plausible that NLRP3 inflammasome activation and inflammation are relevant to *in vivo* responses to inhalation of asbestos fibers in mice and humans, as well as to *in vitro* responses at high doses in various cell lines. However, it should be noted that much of the experimental evidence that has clarified these mechanistic roles for the NLRP3 inflammasome consists of *in vitro* tests and limited *in vivo* inhalation experiments in mice, with relatively high concentrations and brief durations of exposure (e.g., 9 days) (Dostert et al. 2008), and that nothing like a clear, empirically validated dose-response model for NLRP3 inflammasomes in humans exposed to realistic levels of various types or mixtures of asbestos fibers is currently available.

A Risk Assessment Paradigm for NLRP3 Inflammasome-Mediated Carcinogenesis

The preceding advances in elucidating inflammasome biology and mechanisms of pathogenicity for asbestos fibers and similar nanomaterials have not yet been matched by corresponding advances in quantitative risk assessment (QRA) and dose-response modeling for substances that increase disease risks by causing chronic inflammation in target tissue. This mode of action appears to be sufficiently common and important to warrant its own dose-response modeling framework. The following sections propose such a framework and apply it to LC and MM caused by inhalation of amphibole asbestos fibers as important illustrations of inflammation-mediated risks.

Qualitative Aspects of NLRP3 Inflammasome-Mediated Dose-Response

The biology of the NLRP3 inflammasome implies the following qualitative properties of dose-response relationships for NLRP3 inflammasome-mediated diseases; ideally, these qualitative properties should be reflected in any quantitative dose-response models for NLRP3 inflammasome-mediated diseases.

- *NLRP3 inflammasomes have activation thresholds.* The NLRP3 inflammasome is not assembled and activated in a cell unless it receives sufficiently strong, sustained danger signals. Quantitatively, only one large NLRP3 inflammasome oligomer forms per cell; once formed and activated, it causes pyroptosis and destruction of the cell (Elliott and Sutterwala 2015). The following details are not essential for quantitative dose-response modeling, but the existence of activation thresholds, which may be modulated by continued exposure, is crucial for understanding inflammatory responses to asbestos and other hazards.
  - Assembly of the NLRP3 inflammasome requires both a priming signal and a subsequent second signal (Jo et al. 2016; Juliana et al. 2012). Priming up-regulates transcription and secretion of NLRP3 protein filaments from the nucleus into the cytosol (via an NF-κB signaling pathway from the danger-sensing receptors to the nucleus) and deubiquitinates NLRP3 protein already in the cytosol, freeing it for use in a subsequent oligomerization step.
  - Signal 2 (e.g., from binding of damage-associated or pathogen-associated molecular patterns (DAMPs or PAMPs) or cytokines to receptors) triggers self-assembly and activation of the NLRP3 inflammasome in the cytosol. Self-assembly takes place by polymerization and oligomerization of the protein filaments and other components, specifically the adaptor protein ASC (apoptosis-associated speck like protein containing a caspase recruitment domain (CARD)) bridging between damage sensor receptor proteins and pro-caspase-1. This self-assembly exhibits cooperativity, and hence has a switch-like, all-or-nothing character (Hauenstein et al. 2015).
  - This two-step design creates cell-level exposure concentration and duration thresholds for triggering assembly and activation of the NLRP3 inflammasome. This prevents inflammation in response to routine, brief encounters with bacteria or other stimuli. If a priming signal is not sufficiently intense for sufficiently long (i.e., it fails to bind enough receptors to sustain signals to the nucleus upregulating secretion of NLRP3 protein filaments into the cytosol), then not enough deubiquitinated NLRP3 protein filaments will accumulate in the cytosol for subsequent oligomerization and NLRP3 inflammasome self-assembly to proceed. Even if sufficient protein is available but the danger has passed, so that no signal 2 is received, the inflammasome will not form.
  - NLRP3 activation and pyroptosis depend on triggering events that also have exposure thresholds, such as sustained generation of intracellular and mitochondrial ROS, depletion of antioxidant pools, and lysosomal destabilization

and rupture. These events occur only if sufficient concentrations of toxic components accumulate, typically because damage (e.g., from oxidative stress) occurs too quickly for too long to be removed by homeostatic mechanisms. Sufficiently high and prolonged exposures to crystals of cholesterol, amyloid β, uric acid, silica, alum, and asbestos, all can damage lysosomes (leading to release of the protease cathepsin B into the cytosol) and activate the NLRP3 inflammasome in various tissues (Jo et al. 2016).

- *Inflammation and NLRP3 inflammasome activation can spread among nearby cells.* Inflammation of some of the cells in a location (e.g., a small patch of the alveolar epithelium, or of the pleural or peritoneal mesothelium) makes inflammation of other nearby cells more likely. Thus, at the cell population or tissue level, NLRP3 inflammasome-mediated inflammation acts somewhat like a contagious disease in a population: the more cells are inflamed (e.g., undergoing NLRP3 inflammasome activation and pyroptosis) in a given patch of tissue, the easier it is for neighboring cells to become inflamed. The biological reasons for this have been elucidated. When cells undergo pyroptosis, the ruptured cells release activated NLRP inflammasomes and ASC specks, as well as pro inflammatory and toxic contents, to the extracellular space. There they continue to broadcast pro-inflammatory cytokine signals; they may also be taken up by other macrophages, which then undergo pyroptosis themselves, perpetuating the inflammatory cycle (Baroja-Mazo et al. 2014).
- *Sustained tissue inflammation can cause a transition from acute to chronic inflammation.* Normally, inflammation resolves itself as dead cells and their debris are removed (via efferocytosis and phagocytosis) and NLRP3 proteins are ubiquitinated and degraded without being replaced (Grabiec and Hussell 2016; Robb et al. 2016; Tezcan et al. 2019; Yang et al. 2019). However, sustained frustrated phagocytosis and failure to resolve the acute phase of inflammation can lead to inflammation becoming chronic, i.e., self-sustaining rather than self-limiting. Failure to resolve acute inflammation in respiratory mucosa, and resulting oxidative damage and stress, contribute importantly to many lung diseases associated with chronic inflammation (Robb et al. 2016).
- *Chronic inflammation increases cancer risk.* Chronic inflammation increases lung cancer risk and promotes tumor progression in lung (and other) cancers (Cho et al. 2011; Gomes et al. 2014; Galani et al. 2019; Kumagai-Takei et al. 2018); it is believed to play an intrinsic role in the origin of MM (Thompson et al. 2017), although data are limited (Attanoos et al. 2018).

These qualitative aspects imply that a quantitative dose-response model for asbestos should describe the numbers of fibers reaching target tissue sites (e.g., lodging in alveolar epithelial tissue for LC or mesothelial tissue for MM); model the relationship between the density of these fibers in the target tissue and the onset of chronic inflammation; and estimate how chronic inflammation increases risk of cancer over time. These quantitative modeling aspects are discussed next.

Quantitative Dose-Response Modeling for Asbestos: Methods and Models

Three main sub-models are needed to apply the preceding qualitative biological insights to quantitative dose-response and risk modeling for specific substances such as asbestos fibers. A physiologically-based pharmacokinetics (PBPK) model is needed to convert exposure concentration histories (i.e., time series) of asbestos fibers in inhaled air to resulting time series of concentrations of delivered doses in target tissues. For asbestos, this delivered dose takes the form of fibers lodged or embedded in lung and mesothelial tissues for LC and MM, respectively. Next, an inflammation model is needed to describe the inflammatory response of the target tissue to the delivered doses over time, and especially the time until onset of chronic inflammation; for this, we reuse and adapt the model from Chap. 4. Finally, a quantitative model of carcinogenesis, such as a two-stage clonal expansion (TSCE) model, including effects of chronic inflammation, is used to describe resulting risk of malignancy (i.e., LC and MM) over time. We refer to the result as an inflammation-mediated TSCE (I-TSCE) model of age-specific and exposure-specific cancer risk. These three sub-models are developed next for asbestos. They follow the modeling framework developed in Chap. 4, linking PBPK, inflammation, and I-TSCE sub-models. This approach may also be useful for dose-response modeling of other exposures and illnesses with an NLRP3 inflammasome-mediated chronic inflammation mode of action.

PBPK Model for Asbestos Fibers

The PBPK model for asbestos is taken from Dostert et al. (2008) and is detailed in Appendix 1. Briefly, inhaled fibers can be cleared from the lung by alveolar macrophages (AMs), which can attach to fibers and then be removed by mucociliary clearance (ultimately being spat out or digested) even when phagocytosis is frustrated; or via the lymphatic system, from which they may reach the blood, and eventually be eliminated via the kidney. Fibers that are not cleared may lodge in lung tissue, other soft tissue, or the mesothelium. Both AM clearance and renal clearance are saturable processes with maximum possible rates. The recruitment of AMs to the lung is also a saturable process. The number of fibers in any compartment (free in lung, in lung tissue, bound to AMs, in the lymphatic system, blood, other soft tissue, pleural cavity, or the mesothelium) at any time is the cumulative difference between the number that have entered (via inhalation or translocation from other compartments) and the number that have been cleared from the compartment. Expressing this mass balance requirement as a system of ordinary differential equations (ODEs) completes the PBPK model. The model in Appendix 1 makes the simplifying assumption that all fibers are the same; otherwise, different subsets of fibers (e.g., with different lengths, aspect ratios, and stiffness) would move among compartments or become lodged in tissues at somewhat different rates.

Chronic Inflammation Model

Experimental evidence (mainly in rats) and epidemiological and clinical observations in humans have established that "Asbestos dose is a crucial determinant for triggering inflammation; high doses over short periods promote an acute neutrophil-predominant inflammation, whereas low doses over prolonged exposure periods promote alveolar macrophage (AM)-predominant chronic inflammation. Fiber dimensions are important because only very thin fibers (diameter <0.4 μm and length <10 μm) are respirable to the distal alveolar space; long fibers cannot be fully engulfed by AMs because they are biodurable. Phagocytosis of fibers is also limited by the size of the AMs (generally 14–21 μm)" (Liu et al. 2013). Consistent with the PBPK model, we will focus exclusively on hypothetical exposures to homogeneous (identical) amphibole fibers of appropriate length and shape to be carcinogenic, and will consider how the dose of these fibers triggers inflammation, without trying to model variations in potency by length, stiffness, or other parameters. As previously mentioned, in reality, quantitative dose-response relationships depend on physical and chemical properties of asbestos fibers, which vary greatly among asbestos samples. In particular, fibers of different lengths, stiffness, aspect ratios, and surface chemistry may have quite different biological activities and potencies in causing inflammation and other changes. These complexities are not taken into account in the modeling reported here, which is limited to simplified hypothetical exposure scenarios in which all fibers are assumed to be identical and able to cause inflammation and increased cancer risks.

Appendix 2 discusses the model for inflammation, adapted from the respirable crystalline silica model in Chap. 4. Although the reasoning is somewhat detailed, the conclusion is very simple: in this model, the activation thresholds for the NLRP3 inflammasomes in individual cells imply an exposure threshold (a critical cumulative number of fibers in target tissues) for causing chronic inflammation. Chronic inflammation in a target tissue (lung or mesothelium) begins when the cumulative number of fibers lodged or embedded in the tissue reaches a certain threshold number, F^*. This threshold level is plausibly on the order of hundreds of thousands to millions of long (>5 μm), thin fibers per gram dry weight of tissue, although uncertainty and interindividual variability make this estimate only a rough approximation. Appendix 2 provides the basis for this rough estimate. Cumulative exposure below this threshold may cause local acute inflammation around fibers embedded in tissue, but such inflammation is acute and self-limiting: it does not spread, and resolves itself in the absence of continued exposure (e.g., from biopersistent fibers in tissue). By contrast, cumulative exposure above threshold F^* leads to chronic inflammation, with pyroptosis increasing until a new, higher stable equilibrium level is reached. This level is self-sustaining and would not resolve itself even if exposure ceased. The key insight from the model in Appendix 2 is that, despite the complexity of inflammation processes in cell populations, an overall threshold exists for chronic inflammation in a variety of models, much as many models of contagious diseases in human populations predict epidemic thresholds below which disease spread is self-limiting and above which epidemics occur. In this sense chronic inflammation is analogous to a self-sustaining epidemic in a population. Modeling responses to mixtures of different fiber types with different properties and pharmacokinetics is beyond the intended scope of the current model, but one plausible consequence of exposure to heterogeneous fibers would be to replace the relatively sharp transition times discussed in Appendix 2 (Table 5.3) for onset of chronic inflammation with random times having variances that increase with the variance in inflammatory potencies of inhaled fibers. Equivalently, the threshold number of fibers that must be reached to trigger chronic inflammation, F^*, would become a random variable, reflecting the fact that a given number of fibers might have different pro-inflammatory effects, depending on the composition of the fibers. Quantitative sensitivity analysis of the time to initiate inflammation to variance in the potencies of inhaled fibers is beyond the scope of the homogeneous-fiber model developed here, but appears to be a worthwhile direction for future research.

As noted in Chap. 3, the existence of two stable equilibria separated by a threshold or tipping point (bistability) is a property of many models in which it becomes easier for units to switch states (e.g., from uninflamed to inflamed) as the fraction that have already done so increases (Page 2018). This property is implied by the fact that NLRP3 inflammasome activation for a cell becomes more likely to occur when the neighbors of a cell are already inflamed and undergoing pyroptosis, with their inflammasomes broadcasting pro-inflammatory signals, than when they are not. There are many potential complexities, e.g., in the dynamics of the spatial configurations of inflamed and uninflamed locations, but the modeling in Appendix 2 suggests that the large number of fibers needed to trigger chronic inflammation implies (with the help of results on sharp transitions in probabilistic systems) that the simple model of a threshold F^* below which chronic inflammation is not triggered, and above which it is, is probably a useful approximate description. This threshold value may vary substantially (e.g., by an order of magnitude) among individuals due to polymorphisms and other sources of interindividual variation (Kukkonen et al. 2014).

Carcinogenesis Models for Lung Cancer (LC) and Malignant Mesothelioma (MM)

Appendix 3 presents two-stage clonal expansion (TSCE) models for LC and MM caused by asbestos. Such models consist of a pool of normal stem cells, some of which may undergo an initiation mutation to become initiated (pre-malignant) cells. These cells can proliferate—the "clonal expansion" part of the TSCE framework, also called promotion—and some of them may undergo a second transformation ("progression") to yield fully malignant cells. A malignant cell eventually develops into a tumor, after some latency period, unless competing risks (e.g., death from other causes) interrupts the growth process. Traditional TSCE models for asbestos and other substances postulate that internal doses (i.e., fiber burdens in target tissues) can potentially affect the initiation, promotion, or progression rates. They fit parameter estimates to data (for parameter combinations that can be identified from observed exposure-response data, often with the help of assumed constraints such as that two parameters are equal) and test whether exposure effects on these reduced parameters are significantly different from zero. Ideally, the final TSCE model is then validated by testing whether it adequately describes exposure-response data not used in building the model.

Appendix 3 discusses that this standard approach to TSCE modeling must be modified for substances that act via chronic inflammation to increase cancer risk. For such substances, the onset of chronic inflammation leads to a change in the values of TSCE parameters, and exposures and internal doses affect risk only by changing the time until chronic inflammation occurs, at which point TSCE parameters change. For both LC and MM, the best supported effect of asbestos exposure is to increase the promotion rate, i.e., the net proliferation rate of initiated cells. In the lung, this is manifested as epithelial cell proliferation and hyperplasia of type II alveolar epithelia cells (Travis et al. 2002); in the mesothelium, it is associated with formation of pleural plaques, although these are not precursors of MM. We therefore focus on inflammation-mediated TSCE (I-TSCE) model in which the proliferation rate increases from its zero-dose base level to a new, higher level when chronic inflammation begins. The size of this increase in promotion rate and the other model parameters (initiation and progression rates) are chosen to match epidemiological data on the observed increases in age-specific cancer risk over time in exposed populations—e.g., with increases in absolute risk (lifetime probability) of MM from close to zero (about 1 case per 100,000 person-years in the general population) to almost 5% in heavily exposed populations (Carbone et al. 2012). (Gilham et al. 2016 suggest a 25% maximum rate, but this seems exceptional compared to most of the literature.)

This is only an approximate model, of course. For example, localized acute infections might slightly increase cancer risks (e.g., via increased mutations in response to reactive oxygen species) before chronic inflammation is fully established, and chronic inflammation has multiple effects on hastening the acquisition of hallmarks of carcinogenesis during carcinogenesis and tumor development (Liu et al. 2013), which are condensed into only two stages and a latency period in the I-TSCE model; moreover, the transition from acute to chronic inflammation, which we approximate as a step change when viewed on a time scale of decades, is not instantaneous in reality. However, these realistic details and transients probably have little quantitative impact on cancer risk over time compared to induction of chronic inflammation. Both human and animal data show that "asbestos carcinogenesis is linked to the chronic inflammatory process caused by the deposition of a sufficient number of asbestos fibers and the consequent release of pro-inflammatory molecules" (presumably as a result of NLRP3 inflammasome activation), with subsequent continued exposures (or lack of them) making no clear difference to risk in the decades after chronic inflammation is initiated (Carbone et al. 2012). Moreover, the hypothesis of chronic inflammation-mediated (or inflammation-driven) carcinogenesis explains several observations that might otherwise be puzzling, such as the limited increase in relative risks for LC even at very high lifetime exposures compared to the increases for much smaller exposures (Cox 2019). The following sections therefore explore the quantitative consequences of this model of chronic inflammation-driven carcinogenesis for asbestos-related LC and MM risks, while acknowledging that it is a simplification, in hopes that it nonetheless captures the main qualitative and quantitative features of cancer risks caused by carcinogenic asbestos exposures.

Results: Dose-Response and Age-Specific Risk Implications of the I-TSCE Model

Figure 5.1 shows how the risk of MM predicted by the I-TSCE model in Appendix 3 varies with age, from 0 to 90 years, if chronic inflammation of the mesothelium starts at different ages. The uppermost curve corresponds to onset of chronic inflammation at age 10; the lowest curve corresponds to onset of chronic inflammation at age 60; and intermediate curves correspond to 5-year increments in the age at start of chronic inflammation. The vertical axis shows risk, i.e., probability of at least one malignant cell having been formed, over time. (For these small numbers, this probability is indistinguishable from the expected number of malignant cells formed. Because the probability numbers are small, they are multiplied by 1000 on the vertical scale in Fig. 5.1; thus, 70 at the top of the scale refers to 70/1000, or 7% lifetime probability of tumor. MM is typically a very aggressive and fast-growing tumor type, so no distinction is made between the time of formation of a malignant (MM) cell and the time of its expression as a tumor, which in practice might follow a few years later. Finally, competing risks, effects of other risk factors, and co-exposures are ignored. In a real population, individual lifetimes may not extend to 90 years, and the probability of a malignant cell forming may be interrupted by death from other causes.)

The age at start of chronic inflammation makes an enormous difference in predicted lifetime risk of MM in Fig. 5.1, which is approximately 100 times greater for the top curve than for the bottom curve by age 90. Figure 5.2 shows the corresponding predictions for LC if chronic inflammation of the lung starts at different ages. Here, the timing of chronic inflammation makes a smaller difference, with the relative risk (RR) between the top and bottom curve being only about a factor of 2 rather than a factor of 100. This reflects the very different sensitivities of the lung and the mesothelium to exposures that act by reducing the time until onset of chronic inflammation. Figure 5.3 visualizes some of this information in a different form, comparing lifetime risk (for a 90-year life) for LC and MM as a function of age at the start of chronic inflammation in the target tissue (lung or mesothelium, respectively).

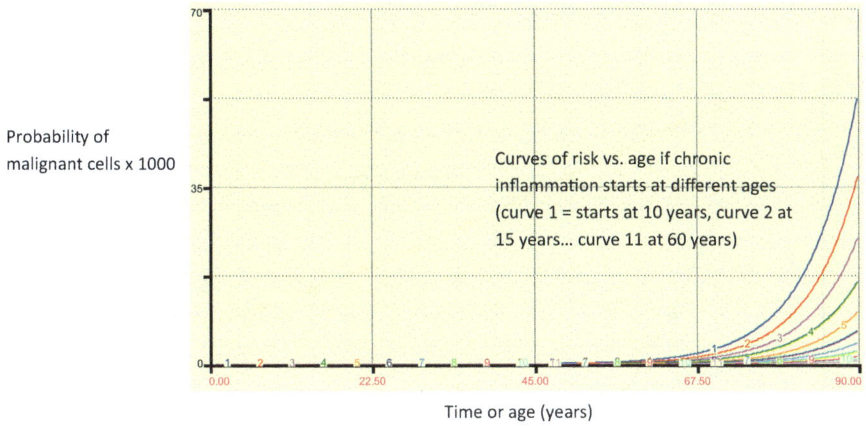

Fig. 5.1 Age-specific MM risk (probability of malignant cell) predicted by the I-TSCE model if chronic inflammation starts at different ages, from 10 years (top curve) to 60 years (bottom curve)

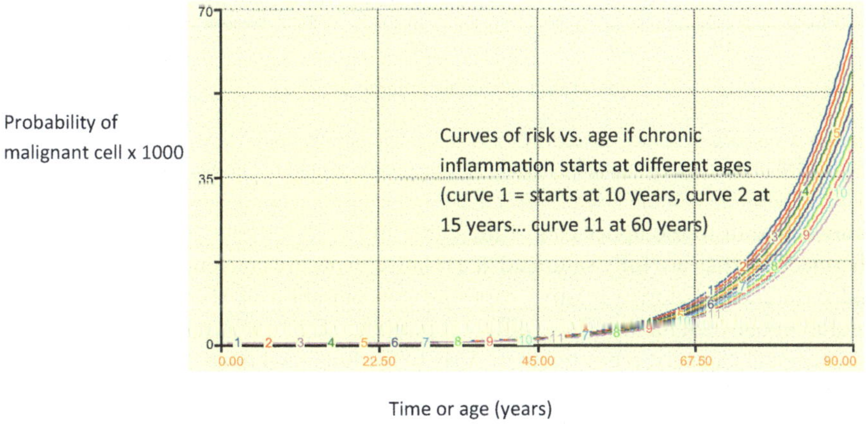

Fig. 5.2 Age-specific LC risk predicted by the I-TSCE model if chronic inflammation starts at different ages, from 10 years (top curve) to 60 years (bottom curve)

Changing exposures to avoid or postpone chronic inflammation is predicted to be far more effective in reducing MM risk than in reducing LC risk.

How long it takes a target tissue to accumulate sufficient fibers to initiate chronic inflammation depends on the ratio of the critical number of fibers, F^*, to the fiber deposition rate in the tissue. The fiber deposition rate, in turn, depends on the inhaled concentration of fibers and on the pharmacokinetics of their distribution and elimination. Figure 5.4 shows the relative rate of fiber deposition (vertical axis) as a function of the relative concentration of fibers in inhaled air (horizontal axis), as predicted by the asbestos PBPK model in Appendix 1. The model deals only with relative concentrations and rates, rather than with absolute fiber counts, but the

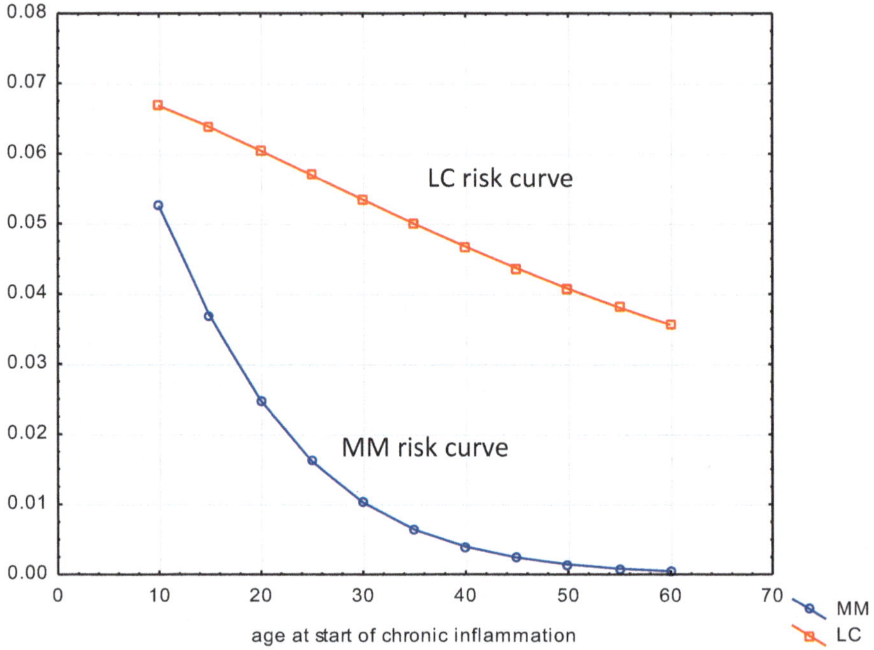

Fig. 5.3 Comparison of lung cancer (LC) and malignant mesothelioma (MM) lifetime risks (probability of malignant cell forming by age 90) vs. age (years) at start of chronic inflammation

straight line fit to the upper end of the curve in Fig. 5.4 shows that, on this scale, pharmacokinetics are fully saturated at a relative exposure concentration of about 3, and unsaturated below a relative exposure concentration of about 1, with partial saturation between them. (These different portions of the curve reflect saturation of AM removal capacity, recruitment of AMs to the lung, and renal clearance.)

Table 5.1 shows the corresponding numbers for relative fiber accumulations in lung and mesothelium (after 100 time steps) for different relative exposure concentrations in air (column 1) under both normal and inflamed pharmacokinetics (columns 2–3 and 4–5, respectively) predicted by the PBPK model in Appendix 1. We use only the normal pharmacokinetics, since our focus is on the time until chronic inflammation is initiated.

The maximum efficiency (steepest slope) of fiber deposition per unit concentration in air occurs at relative concentrations above 3 in Fig. 5.4. In this high-concentration range, each additional unit of exposure concentration produces a proportional increase in fiber deposition rate. However, this maximally efficient fiber deposition occurs only when all pharmacokinetic processes are saturated, and in this sense corresponds to very high exposure concentrations. By contrast, DeStefano et al. (2017) assume a relative concentration of 1 for studying asbestos pharmacokinetics. The relative exposure concentration value of 1 might be thought of as the level of exposure encountered by an industrially exposed worker with

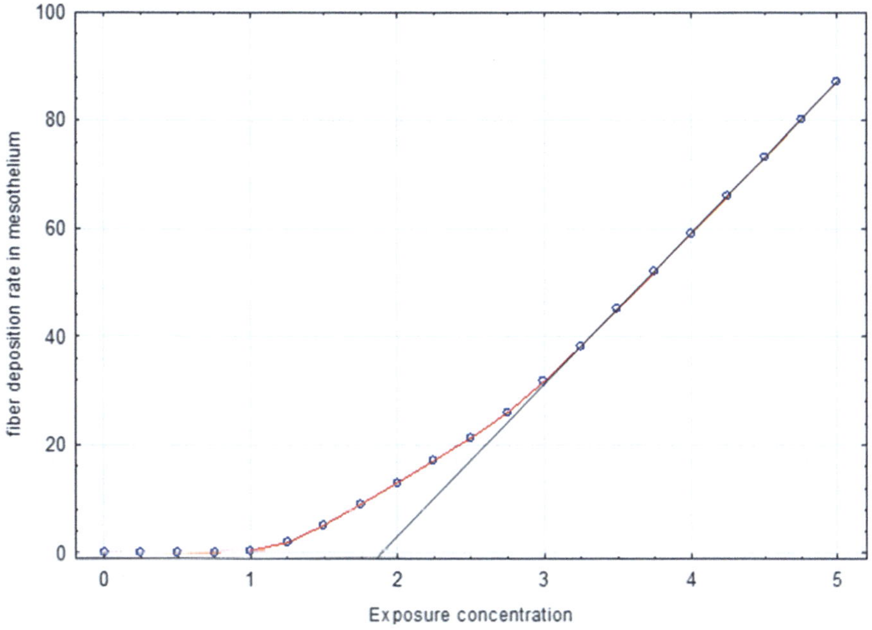

Fig. 5.4 PBPK model predictions of relative fiber deposition rate in mesothelium (vertical axis) for different relative exposure concentrations (horizontal axis)

fairly heavy exposure—high enough to increase the risk of MM, but not high enough to saturate clearance.

Figure 5.5 expands the portion of the mesothelium fiber deposition rate curve in Fig. 5.4 below the relative concentration of 1 to more clearly show its shape. It also shows a comparison curve for lung fiber deposition (upper curve). Both curves are sub-linear, indicating that higher concentrations produce disproportionately more fibers in both tissues; conversely, reductions in exposure concentrations yield greater-than-proportional reductions in fiber deposition rates and corresponding counts of fibers lodged in target tissues over time. For example, halving relative exposure concentration from 0.8 to 0.4 reduces relative fiber deposition rate by more than 4-fold, from 0.25 to 0.06 in the mesothelium and from 0.34 to 0.08 in the lung. This implies that the time to accumulate enough fibers in target tissues to trigger chronic inflammation more than quadruples. Such arithmetic suggests that reducing exposure concentrations can make the time until enough fibers accumulate to trigger chronic inflammation longer than a lifetime, although the required number of fibers—and hence the time needed for prolonged exposure to a given concentration to initiate chronic inflammation—probably varies several-fold among individuals. For example, Carbone et al. (2012) state that "The threshold limit above which asbestos and erionite will cause MM may vary among individuals due to genetics, exposure to cofactors, the type of mineral fiber inhaled, etc." Empirically, "prolonged continuous exposures to asbestos of at least 3 months and usually

Table 5.1 Relative fiber counts deposited in mesothelium and lung for different relative exposure concentrations of fibers in air

1. Exposure concentration	2. Fibers in mesothelium	3. Fibers in lung tissue	4. Fibers in mesothelium with inflammation	5. Fibers in lung with inflammation
0	0	0	0	0
0.2	0.02	0.03	0.02	0.04
0.4	0.06	0.08	0.05	0.12
0.6	0.12	0.17	0.11	0.25
0.8	0.25	0.34	0.22	0.5
1	0.55	0.77	0.48	1.12
1.2	1.55	2.15	1.35	3.12
1.4	3.85	5.33	3.35	7.73
1.6	6.77	9.38	5.89	13.61
1.8	9.87	13.68	8.59	19.84
2	13.06	18.09	11.36	26.22
2.2	16.31	22.59	14.16	32.7
2.4	19.64	27.21	17.01	39.27
2.6	23.14	32.06	19.91	45.97
2.8	27.07	37.5	22.93	52.94
3	31.74	43.97	26.19	60.47
3.2	36.95	51.19	29.87	68.96
3.4	42.38	58.72	33.94	78.36
3.6	47.92	66.38	38.21	88.23
3.8	53.49	74.11	42.58	98.31
4	59.1	81.88	46.99	108.51
4.2	64.72	89.67	51.44	118.77
4.4	70.36	97.48	55.9	129.06
4.6	76.01	105.3	60.37	139.39
4.8	81.66	113.13	64.85	149.73
5	87.32	120.97	69.33	160.09

These are relative rates (ratios), and hence are dimensionless. They assume homogeneous, identical fibers; see discussion in text

10 years were required to induce MM," although the MM often occurred decades later, with "the mean latency of MM from exposure [being] 44.9 years for crocidolite miners, 51 years for amosite miners, and 12.3 years for miners exposed to both, crocidolite and amosite," suggesting that "once a sufficient amount of asbestos or erionite has been inhaled, such as in a 6 year old child growing up in a village contaminated with erionite, they will develop MM." As a base scenario for purposes of illustration, suppose that at least 3 months of exposure to a very high concentration of carcinogenic asbestos fibers in air (corresponding to the pharmacokinetics-saturating level of 3 in Fig. 5.4) is needed to trigger chronic inflammation and increased MM risk. Then the relative fiber deposition numbers (or rates, as the two are proportional) in Table 5.1 imply that reducing relative exposure concentration

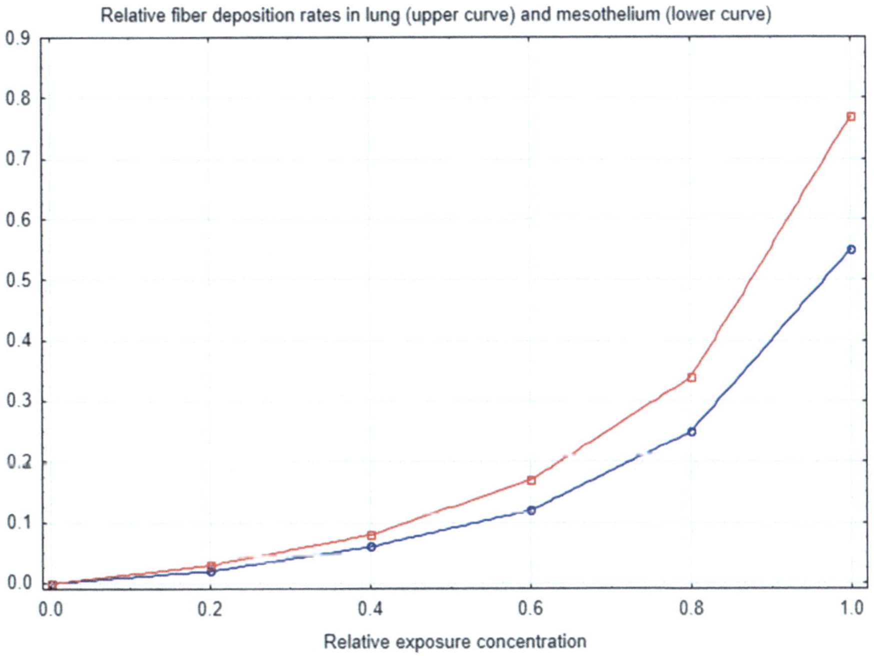

Fig. 5.5 PBPK model predictions of relative fiber deposition rate in mesothelium (lower curve) and lung (upper curve) for non-saturating exposure concentrations (horizontal axis)

from 3 to 1 would reduce the rate of fiber accumulation in the mesothelium by a factor of 0.55/31.74, and hence increase the time needed to initiate chronic inflammation and increase MM risk from a minimum of at least 3 months to (31.74/0.55)*3 = 173 months, or close to 15 years. Figure 5.3 shows that the effects on MM risk of postponing the onset of chronic inflammation by 15 years can be substantial. Similarly, reducing relative exposure concentration from 3 to 0.4 would increase the time to reach chronic inflammation in the mesothelium from at least 3 months to at least (31.74/0.06)*(3/12) = 132 years. For LC, the corresponding number is (43.97/0.08) * (3/12) = 137 years. Such an exposure level might be regarded as virtually safe, insofar as significantly increased relative risks of MM and LC caused by chronic inflammation would not be triggered in a lifetime of realistic length. This does not imply that the risk from exposure is exactly zero, since even localized inflammation around individual fibers might contribute to other sources of inflammation (e.g., from smoking, for LC) to cause localized oxidative damage, but such effects are probably *de minimis* compared to those from chronic inflammation.

In this example, the reduction in exposure concentrations needed to protect worker health is relatively small. If a relative exposure concentration of 1 is a usual exposure level for workers in come industry, and a relative exposure level of 0.4 is protective, then industrial hygiene practices that reduce exposure concentrations by less than one order of magnitude could make occupational exposures virtually safe, in the sense just described. Although this ignores important practical implementation challenges and realistic complexities, such as the variability in exposure concentrations around their mean levels and the difficulty of assuring that occasional very high spikes of concentration are not encountered, it suggests that efforts to reduce exposure concentrations for heavily exposed workers by a factor of twofold to threefold might have substantial health protection benefits for new workers, but that larger reductions would not be expected to cause further detectable reductions in MM and LC risks.

Discussion

Traditional risk assessment using the assumptions of the linear no threshold (LNT) model conceptualizes low-dose risks as being approximately proportional to exposure concentration and duration, with estimates of carcinogenic potency expressed in increased lifetime probability of malignancy per unit of cumulative exposure being a key result of risk assessment calculations. For example, a recent influential study of the relationship between lung burden of asbestos fibers and MM risk concluded that "Lifetime mesothelioma risk is approximately 0.02% per 1000 amphibole fibres per gram of dry lung tissue over a more than 100-fold range." Against this conceptualization of linear dose-response for asbestos is the long-standing, puzzling observation that, although time since first exposure and exposure concentration are both independently associated with pleural plaques (indicating chronic inflammation) and asbestosis, total duration of exposure has not been confirmed as a significant predictor of asbestos-related disease risk (Paris et al. 2008).

The previous sections have presented an alternative model of asbestos exposure-related risk that predicts and explains the limited relevance of total exposure duration. It is based on evidence that, following exposure to a sufficient number of fibers, "a long period elapses during which the asbestos present in the pleural tissues promotes a chronic inflammatory process that with time may favor the emergence of a malignant clone" in exposed susceptible individuals (Carbone et al. 2012). In this view, the key events in increasing risk of asbestos-induced MM or LC are (a) Accumulation of enough fibers in the target tissue to trigger chronic inflammation; and (b) Promotion of initiated (premalignant) stem cells by chronic inflammation, i.e., an increase in the net proliferation rate of these cells, and hence in the size of the pool of cells that may undergo additional transformations to become malignant. Therefore, duration of continued exposure following the onset of chronic inflammation has little relevance to lifetime risk: once chronic inflammation has begun, it drives increased risk, as described in the I-TSCE model.

Although the concept of a threshold number of fibers that is necessary and sufficient to induce chronic inflammation, regardless of future exposures, can be motivated by epidemiological observations (Carbone et al. 2012), it finds biological justification in recent studies showing activation thresholds for NLRP3 inflammasomes. In simple mathematical and computational models (Appendix 2), these activation thresholds for individual cells imply an exposure (fiber count) threshold for chronic inflammation of the population of cells in a tissue. For exposures below this tipping point threshold, only acute, self-limiting inflammation occurs. Above it, a transition to chronic inflammation occurs. If such a threshold exists, and if chronic inflammation drives subsequent detectable increases in cancer risk, then risk for any individual worker depends only on the early part of exposure and the resulting time until onset of chronic inflammation in target tissues. Total duration of exposure is therefore of limited relevance. (Continued exposure may, however, also play important roles in the development of tumors once a malignant cell has been formed. For MM, this process is unfortunately quick, measured in months to years instead of decades (Carbone et al. 2012), and is not modeled in our risk calculations, which focus on the cumulative probability or expected number of malignant cells formed over time.)

These considerations suggest a paradigm for NLRP3-inflammasome mediated, inflammation-driven cancer risk assessment in which the onset of chronic inflammation determines subsequent cancer risk. Quantitative modeling suggests that predicted MM risk is relatively sensitive to the timing of chronic inflammation compared to LC risk, with later onset producing substantially smaller lifetime risks for MM (Fig. 5.3). Moreover, the predicted timing of chronic inflammation is, in turn, sensitive to the concentration of fibers in inhaled air at low concentrations, with some reductions in exposure concentration producing much more than proportional increases in time to chronic inflammation, reflecting the greater efficiency of clearance at lower fiber concentrations (Figs. 5.4 and 5.5 and Table 5.1). Conversely, however, once exposure concentrations are small enough to avoid causing chronic inflammation during a lifetime, further reductions in concentration cannot achieve further reductions in chronic inflammation-mediated risk. In this sense, there is a practical threshold for excess lifetime risk created by asbestos exposures. These testable implications are quite different from those of the LNT model, and appear to be more consistent with epidemiological data suggesting that low levels of asbestos exposure do not lead to detectably elevated risks for LC or MM, while higher concentrations lead to increases in risk that are insensitive to total duration of exposure (Paris et al. 2008; Carbone et al. 2012).

Limitations and Robustness of the I-TSCE Model

This account of asbestos carcinogenicity has several limitations. Perhaps most importantly, it does not model interindividual variability in PBPK or pharmacodynamics parameters, even though it is well known that gene polymorphisms and other sources of interindividual variability may produce substantial differences in individual susceptibility to the inflammatory and carcinogenic effects of asbestos exposures (Kukkonen et al. 2011, 2014; Carbone et al. 2012). Our risk calculations have been illustrated for scenarios with estimated relative concentrations and response thresholds (F^*) using a single PBPK model and I-TSCE model (detailed in Appendices A and C), but exposed populations may have individuals with different PBPK and I-TSCE parameter values. Fortunately, sensitivity analysis is straightforward. If an individual has a chronic inflammation response threshold that is twice as great or half as great as assumed, then it will take twice as long or half as long, respectively, for a given exposure concentration to cause chronic inflammation. Likewise, if an individual's PBPK parameters result in twice or half as many fibers being delivered to target tissue per unit time, then the time to initiate chronic inflammation will be halved or doubled, respectively. Thus, a joint frequency distribution of PBPK and I-TSCE model parameters in a population leads to a distribution of the practical threshold concentration needed to protect against chronic inflammation and resulting increased inflammation-mediated cancer risks within a lifetime. These refinements do not change the conclusions that such a practical threshold exists for each individual, and that concentrations below the threshold for the most sensitive individual will be protective for all members of the population. However, the extent of interindividual variability in model parameters and resulting practical concentration thresholds deserves further study.

Similarly, we have avoided discussing the large, vexed, literatures on appropriate definitions of "asbestos" and "fiber," appropriate dose metrics for heterogeneous fibers, relationships among fiber physical and chemical properties and inflammatory or carcinogenic potency, and choices of definitions and analytic techniques for counting fibers in air and in tissues (Churg 1982). We bypassed these important but contentious issues by modeling homogeneous fibers and using relative concentrations in the mathematical modeling. Fibers with different shapes, lengths, stiffness, or chemical properties than those assumed in the PBPK model in Appendix 1 might have different values for the PBPK parameters, the chronic inflammation response threshold F^*, and effects on I-TSCE model parameters (e.g. the promotion rate) in Appendix 3. Likewise, co-exposures, co-morbidities, cigarette smoking, and other factors, as well as the chance spatial distribution of fibers and asbestos bodies within target tissues, might affect these model parameters.

Perhaps the greatest practical limitation of the modeling undertaken here is that, even if the underlying biological and mathematical theory (essentially, that activation thresholds for NLRP3 inflammasomes imply bistability and a threshold for chronic inflammation of tissue, which then increases cancer risk (Bogen 2019)) turns out to be a useful approximation, there is still no easy way to link the threshold in the model (F^*) to specific numerical values of fiber counts in real-world occupational settings. Fiber morphology and biological activity are very heterogeneous, so that fiber counts alone are not accurately predictive of risk in the absence of knowledge of other properties (length, aspect ratio, stiffness, chemical composition, etc.) This issue has been finessed in the modeling reported here by simply assuming homogeneous fibers capable of causing inflammation and increased cancer risk, and by using relative exposure concentrations, with "1" taken as a "normal" occupational exposure concentration (as in Fig. 5.4 and Table 5.1), but no specific corresponding fiber concentration has been identified. This follows the normalization in the asbestos PBPK model of DeStefano et al. (2017), as described in Appendix 1, but no careful or quantitative rationale is given. The normalization implies that normal exposure concentrations do not cause saturation of clearance rates, while concentrations about 3 times as great do (Fig. 5.4). This amounts to simply defining a "normal" exposure concentration as being about 1/3 of the clearance rate-saturating concentration (for constant exposures). No quantitative link to absolute fiber counts per cubic centimeter can be determined from this definition, and, indeed, any such quantitative link would have to specify fiber properties relating fiber counts to biological effects. To our knowledge, such quantitative dose-response models for humans are not currently available. The normalization approach still produces insights that we consider potentially useful for improving risk assessment and management—especially, that risk increases rapidly and disproportionately with exposure concentration over a relatively narrow range of relative concentrations (Fig. 5.4)—but estimating specific quantitative fiber concentrations to protect worker health based on these insights will require further research and modeling of dose-response potencies for different types of fibers and composition of fiber types for realistic occupational exposures.

Contributions and Relationships to Previous Research Literature

The model described here suggests that low concentrations of asbestos do not trigger chronic inflammation and resulting increases in cancer risk observed at higher exposure concentrations. This suggestion is perhaps unavoidably controversial. It conflicts with many epidemiological models that predict increased risk at all levels of exposure. For example, a recent Dutch study (Offermans et al. 2014) estimated asbestos exposures for 58,279 men from job-exposure matrices and concluded that a "multivariable adjusted model showed overall positive associations between all levels of asbestos exposure and mesothelioma, lung cancer, and laryngeal cancer." However, others have warned that, in this study as in others, "exact data on fibre concentration and type of asbestos are lacking. We suggest that occasional exposure to asbestos poses hardly any risk for the general population" (Baas and Burgers 2014). This pattern of conflicting and inconclusive interpretations maybe there is increased risk at low concentrations of fibers, but exposure estimates are uncertain and it is hard to be sure—has persisted in the literature for decades, notwithstanding repeated efforts to develop and defend more definitive conclusions. Thus, a recent review (Markowitz 2015) concluded that "No safe threshold for asbestos exposure has been established for lung cancer and mesothelioma. The validity of fiber-type risk assessments depends critically on the quality of exposure assessments, which vary considerably, leading to a high degree of uncertainty." Conversely, a recent review of toxicology and epidemiology for specific types of asbestos concluded that "Biopersistence and sub-chronic inhalation toxicology studies have shown that exposure to chrysotile at up to 5000 times the current threshold limit value (0.1 fibers/cm) produces no pathological response. …In contrast, short-term exposure to amphibole asbestos results quickly in the initiation of a pathological response in the lung and the pleural cavity" (Bernstein 2014).

More generally, in the litigious context for much asbestos research, decades of advances in understanding of the molecular and cell biology, toxicology, and pathogenesis of LC and other lung diseases caused by asbestos and other mineral fibers and particles have not led to general agreements on such fundamental questions such as what kinds of fibers or cleavage fragments (e.g. those with certain aspect ratios, lengths, etc.) increase risk of LC and MM; by how much different exposure histories increase these risks (i.e., the shape of exposure concentration-duration-response function); and how best to model quantitative exposure concentration-duration-response functions for LC and MM caused by asbestos exposures (Moolgavkar et al. 2014; Lacourt et al. 2017). While many scientists have opined that long, thin, stiff fibers are far more dangerous than cleavage fragments, and that chrysotile causes lower risk of MM than amosite or crocidolite (Gilham et al. 2016), others have opposed suggestions that some types of asbestos are less hazardous than others (e.g., Woitowitz and Baur 2018). There is no generally accepted dose-time-response model for MM or LC caused by inhalation exposures to amphibole asbestos fibers or cleavage fragments having different geometries and physiochemical properties, although many models have been proposed (Gilham et al. 2016). Understanding NLRP3 inflammasome activation adds useful biophysical mechanistic detail to inform this debate. Briefly, long, thin, stiff fibers can rupture lysosomes (in effect, "poking holes" in them from the inside) and activate the NLRP3 inflammasome, but shorter or more flexible fibers can't. This behavior can be demonstrated experimentally using other substances such as carbon nanotubes (Kane et al. 2018; Lee et al. 2018; Palomäki et al. 2011; Donaldson et al. 2013).

Earlier attempts to use reviews and meta-analyses to build agreement about risks to human health from different types and geometries of fibers and fragments did not resolve debates. For example, a 2004 paper argued as follows:

> A considerable body of evidence, gathered over the last 30 years, demonstrates that amphibole cleavage fragments do not show the same toxicity as their asbestiform analogues. Since there still continues to be confusion and controversy on this point, this review is aimed at resolving a major portion of this controversy. It has done so by bringing together the supporting mineralogical, animal and human evidence from many sources. These observations demonstrate that cleavage fragments and amphibole asbestos fibers have fundamentally different properties and these differences are biologically relevant. Indeed, the toxicity of respirable cleavage fragments is so much less than that of the fibrous amphiboles that by any reasonable measure they are not biologically harmful" (Ilgren 2004)

Hopes of settling a major portion of the controversy via synthesis and review of past findings are probably unrealistic. Resistance to conclusions of low or negligible risks for some types of exposure remains strong, and those who articulate such conclusions draw fire from those who disagree. Nonetheless, earlier papers explain why different substances, such as amphibole fibers and cleavage fragments, should be expected to have very different pathogenic potentials and potencies and why different exposure concentrations have different potencies (slope factors for risk vs. exposure concentration). For example,

- Browne (1991) placed the pathogenesis of asbestosis and asbestos-related LC and MM in the context of inflammation-mediated lung disease, with activated alveolar macrophages with incomplete phagocytosis of fibers releasing hazardous mediators that increase LC and MM risk. This work, conducted a decade before the discovery of the NLRP3 inflammasome, presented evidence for exposure concentration thresholds below which risk is not increased by asbestos fibers (estimated as about 5 fibers per ml-year for MM and about 50 fibers per ml-year for LC) and noted that fibers less than 5 μm long do not appear to increase risk of MM or LC. The risk models presented here can be viewed as adding to the qualitative insights of Browne (1991) more recently discovered biological details about the specific inflammatory mediators involved (e.g., the NLRP3 inflammasome, ASC particles, and IL-1β and IL-18), as well as a more quantitative risk assessment modeling approach.
- Berman and Crump (2008) fit different empirical models to data without trying to model the underlying biological dose-time-response causal relation. For both MM and LC, they conclude that the epidemiological data examined are sufficient to reject the null hypotheses that amphibole fibers and chrysotile fibers have equal potencies, consistent with the earlier conclusion of Ilgren (2004) and others. They also note that fibers less than 5 μm in length may not cause MM or LC and that there are unresolved differences in potency estimates across occupational studies that may reflect differences in fiber size distributions. Some of these conclusions are supported by evidence reviewed by Gamble and Gibbs (2008), who concluded from epidemiological data that "The cohorts exposed to non-asbestiform amphiboles had no excesses of lung cancer or mesothelioma.... The weight of evidence fully supports a conclusion that non-asbestiform amphiboles do not increase the risk of lung cancer or mesothelioma."
- Shukla et al. (2004) complement these epidemiological studies with animal data, presenting *in vitro* and *in vivo* experimental evidence of dose-response thresholds and no adverse effect levels based on gene expression profiles in rat mesothelial and lung epithelial cells (target cells for MM and LC, respectively) and living rats exposed to crocidolite and chrysotile asbestos.

Such findings support the plausibility of concentration thresholds in the dose-time-response relationships for carcinogenic amphibole asbestos fibers. To our knowledge, however, the present paper is the first to propose a quantitative model for how such thresholds can arise in the context of NLRP3 inflammasome-mediated inflammation.

Conclusions

This chapter has introduced an alternative to the LNT modeling framework that seems more appropriate for describing exposure-response relationships and risks for NLRP3 inflammasome-mediated, chronic inflammation-driven diseases. If the key qualitative aspects of this alternative model are correct, then there are practical thresholds below which exposures do not cause detectably increased lifetime risks of these diseases, and above which they do. The quantitative modeling we have discussed, although only based on illustrative scenarios and not addressing the distribution of individual dose-response parameters or thresholds in the population, suggests that such practical thresholds might well be within reach (perhaps within about a factor of 2 of some past exposure levels) for many workers, if they have not yet been achieved. Achieving them could eliminate all or most exposure-caused risk for many workers. More detail on the expected health-protective benefits of relatively modest further reductions in occupational exposure concentrations would require additional characterization of the variability of individual response thresholds in exposed populations and of the variability in potency of chronic inflammation in increasing promotion rates (and possibly other parameters) of the I-TSCE model.

Appendix 1: A Physiologically Based Pharmacokinetics (PBPK) Model for Asbestos Fibers in the Human Body

Figure 5.6 shows the structure of a systems dynamics model for the asbestos fiber PBPK model. The notation and model are adapted from DeStefano et al. (2017); details are given at www.tandfonline.com/doi/pdf/10.1080/17513758.2017.1355489 cf p. 368. For example, FL denotes free fibers in lung; ML denotes macrophages in lung that have not yet ingested a fiber; MI denotes macrophages that have ingested fibers (at least partly, as frustrated phagocytosis may block full ingestion) and await mucociliary clearance; FY, FB, FP, and FM denotes free fibers in the lymphatic system, blood, pleural cavity, and mesothelium, respectively, and AM denotes alveolar macrophages. These stocks, or compartments, of the fiber inventory are represented by rectangular boxes in Fig. 5.6. Flows among them, representing movement of fibers among compartments, are shown by thick arrows. Thin arrows represent information flows: the value of a variable is calculated from the values of the variables that point into it.

Fibers flow among these compartments at rates indicated by the constants (e.g., pLY for clearance rate of free fibers in the lung by lymphatic removal) or determined by Michaelis-Menten parameters for saturable removal kinetics in the kidney (where the removal rate of fibers from blood by the kidney is given by the formula $KB = k1*FB/(k2 + FB)$) and lung (where the rate of uptake of free fibers by alveolar macrophages is given by the formula $GLM = g1*FL*ML/(g2 + FL)$, expressing a maximum removal capacity of g1 fibers per unit time per macrophage). The recruitment rate of alveolar macrophages to the lung likewise follows saturable kinetics: it is given by the formula $GI = gI1*FL/(gI2 + FL)$ when the lung is inflamed. Multiplying compartment contents by outflow rates gives the values of the outflows at each moment, corresponding to the thick arrows between compartments in

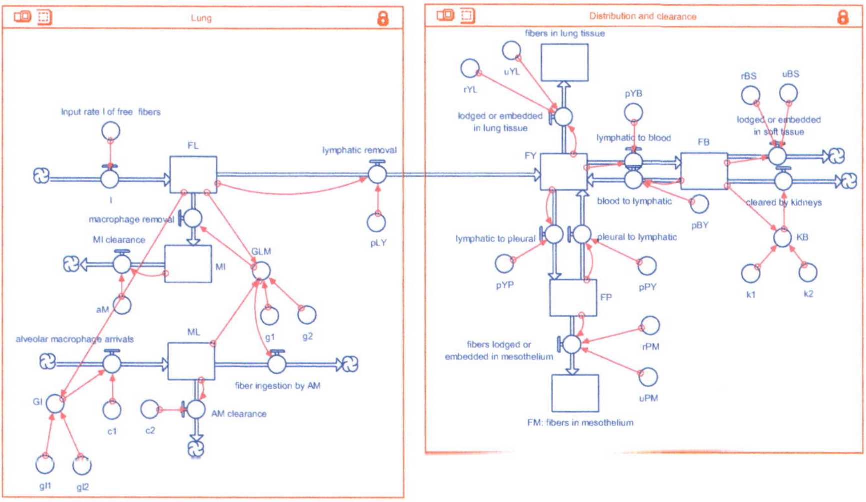

Fig. 5.6 Systems dynamics diagram for the PBPK model

Fig. 5.6. Summing the inflows and outflows for each compartment at each moment gives the rate of change of its contents, and these dynamics can be described by a system of ordinary differential equations (ODEs). DeStefano et al. (2017) give rough estimates for the parameter values and constants for both inflamed and normal tissues in humans, noting that the qualitative and quantitative dynamics of the system are determined primarily by the ratios of the different flow rates to each other, so that the exact numerical values, which in any case differ among individuals, are less important than the approximate values of their ratios. Table 5.2 lists the detailed equations and parameter and constant values for the PBPK model under normal conditions. DeStefano et al. (2017) (Table 5.1) also provide suggested modifications of these input values for fibrosis and chronic inflammation. However, our focus is on the normal case, as the transition to chronic inflammation locks in a high-ROS environment and changes the parameters of the TSCE model (Appendix 3), making the continued dynamics of fiber distribution less relevant for calculating cancer risks. Although experiments in rats exposed to many types of dusts and particles show that lungs become overburdened at high concentrations, impairing clearance of even low-solubility and relatively low toxicity particles by alveolar macrophages (AMs) (Borm et al. 2004, 2015), the rat lung appears to be unique in this regard, and relevance for human pathology is unclear (Warheit et al. 2016). The model in Table 5.2 focuses on PBPK dynamics for the healthy human lung prior to onset of chronic inflammation or fibrosis.

Table 5.2 Listing of PBPK model formulas and parameter values

FB(t) = FB(t - dt) + (lymphatic_to_blood - blood_to_lymphatic - lodged_or_embedded_in_soft_tissue - cleared_by_kideys) * dt
INIT FB = 0
INFLOWS:
lymphatic_to_blood = pYB*FY
OUTFLOWS:
blood_to_lymphatic = pBY*FB
lodged_or_embedded_in_soft_tissue = (rBS + uBS)*FB
cleared_by_kideys = KB
fibers_in_lung_tissue(t) = fibers_in_lung_tissue(t - dt) + (lodged_or_embedded_in_lung_tissue) * dt
INIT fibers_in_lung_tissue = 0
INFLOWS:
lodged_or_embedded_in_lung_tissue = (rYL + uYL)*FY
fibers_in_mesothelium(t) = fibers_in_mesothelium(t - dt) + (fibers_lodged_or_embedded_in_mesothelium) * dt
INIT fibers_in_mesothelium = 0
INFLOWS:
fibers_lodged_or_embedded_in_mesothelium = (rPM+ uPM)*FP
FL(t) = FL(t - dt) + (I - lymphatic_removal - macrophage_removal) * dt
INIT FL = 0
INFLOWS:
I = Input_rate_I_of_free__fibers
OUTFLOWS:
lymphatic_removal = FL*pLY
macrophage_removal = GLM
FP(t) = FP(t - dt) + (lymphatic_to_pleural - pleural_to_lymphatic - fibers_lodged_or_embedded_in_mesothelium) * dt
INIT FP = 0
INFLOWS:
lymphatic_to_pleural = pYP*FY
OUTFLOWS:
pleural_to_lymphatic = pPY*FP
fibers_lodged_or_embedded_in_mesothelium = (rPM+ uPM)*FP
FY(t) = FY(t - dt) + (lymphatic_removal + pleural_to_lymphatic + blood_to_lymphatic - lymphatic_to_blood - lymphatic_to_pleural - lodged_or_embedded_in_lung_tissue) * dt
INIT FY = 0
INFLOWS:
lymphatic_removal = FL*pLY
pleural_to_lymphatic = pPY*FP
blood_to_lymphatic = pBY*FB
OUTFLOWS:
lymphatic_to_blood = pYB*FY
lymphatic_to_pleural = pYP*FY

(continued)

Table 5.2 (continued)

lodged_or_embedded_in_lung_tissue = (rYL + uYL)*FY
MI(t) = MI(t - dt) + (macrophage_removal - MI_clearance) * dt
INIT MI = 0
INFLOWS:
macrophage_removal = GLM
OUTFLOWS:
MI_clearance = aM*MI
ML(t) = ML(t - dt) + (alveolar_macrophage_arrivals - AM_clearance - fiber_ingestion_by_AM) * dt
INIT ML = 3
INFLOWS:
alveolar_macrophage_arrivals = c1 + GI
OUTFLOWS:
AM_clearance = c2*ML
fiber_ingestion_by_AM = GLM
aM = 1
c1 = 0.5
c2 = 0.5
g1 = 2
g2 = 0.01
GI = gI1*FL/(gI2 + FL)
gI1 = 1
gI2 = 0.01
GLM = g1*FL*ML/(g2 + FL)
Input_rate_I_of_free__fibers = 0.1
k1 = 1
k2 = 0.01
KB = k1*FB/(k2 + FB)
pBY = 1
pLY = 1
pPY = 1
pYB = 2
pYP = 2
rBS = 0.25
rPM = 0.25
rYL = 1
uBS = 0.125
uPM = 0.125
uYL = 0.25

Appendix 2: A Model of the Transition from Acute to Chronic NLRP3-Inflammasome-Mediated Inflammation

The mathematical model for the transition from acute to chronic inflammation is adapted from the inflammation submodel for respirable crystalline silica and lung cancer described in Chap. 4. This appendix recapitulates key aspects of that model. In brief, exposure can cause uninflamed cells in target tissues to become inflamed (e.g., undergoing pyroptosis) either *directly*, as when a macrophage interacts with a fiber; or *indirectly* if pro-inflammatory materials from nearby inflamed cells exceed the cell's own NLRP3-inflammasome activation threshold and cause it to become inflamed. Although the spread of inflammation has been modeled and simulated in various ways, including deterministic ordinary differential equations (ODEs), stochastic (probabilistic) models, and agent-based models (ABMs), these multiple modeling approaches approaches converge on a conclusion that there is a tipping point, or threshold, for the fraction of inflamed cells (or sites, in a spatial model) (Gilroy and Lawrence 2008; Page 2018). Above this threshold, inflammation is self-sustaining, i.e., chronic, and will continue even if exposure stops; below, it is not, i.e., it remains acute and self-resolving once exposure stops.

This conclusion can be illustrated using a simple deterministic ODE model (as a mean-field approximation to an underlying probability model), as follows. Suppose that there are a large number of potential sites of inflammation (e.g., cells in a tissue); that the fraction of inflamed sites at time t is $P(t)$; that a randomly selected site that is not yet inflamed and that does not directly interact with a fiber becomes inflamed if and only if more than k of its neighbors are inflamed, where k reflects its activation threshold; that direct interaction with fibers can also cause an uninflamed site to become inflamed; and that inflamed sites are resolved (i.e., cease to be inflamed) at some rate (possibly zero, for sites where fibers are lodged). A simple model for the dynamics of such a system is as follows:

$$dP/dt = f(P)(1-P) + c(1-P) - dP \tag{5.1}$$

where P is the fraction of sites that are inflamed; $(1 - P)$ is the faction of sites that are not inflamed; c is the rate at which exposure directly causes uninflamed sites to become inflamed; d is the rate at which inflamed sites become uninflamed; and $f(P)$ is the rate at which uninflamed sites become inflamed when the fraction of inflamed sites is P. Particle overload in the rat lung compromises the efficiency of phagocytosis, corresponding to a decrease in d (Borm et al. 2015). However, we shall focus on lower exposures and long averaging times in humans, for which the simplified approximate description of clearance of inflamed cells by a single rate constant yields useful insights into inflammation dynamics on a time scale of months to years or decades, despite its undoubted inaccuracies on short time scales. If a site has N neighboring sites, then the probability that at least k of these N neighbors are inflamed if the fraction of inflamed sites is P can be modeled (using a mean field approximation for the average case) as $1 - pbinom(k, N, P)$, where $pbinom(k,$

N, P) denotes the cumulative probability that no more than k of the N neighbors are inflamed if each has probability P of being inflamed. The average rate at which the uninflamed site becomes inflamed due to proinflammatory effects from neighboring cells can be modeled as

$$f(P) = a * (1 - pbinom(k, N, P)) \tag{5.2}$$

The mathematical analysis in Chap. 4 established that Eq. (5.1) with $f(P)$ specified as in Eq. (5.2) has two stable equilibria, one at a low value of the inflammation fraction P (equal to zero if $c = 0$, e.g., in the absence of exposure) and the other at a higher value of P. Between them is an unstable equilibrium level—a tipping point threshold. Initial values of P below the threshold will decline to the lower equilibrium value, while initial values of P above the threshold will increase to the higher equilibrium value. Figure 5.7, the key points of which have been discussed in Chaps. 3 and 4, illustrates this behavior by showing $P(t)$ trajectories starting from different initial inflammation fractions $P(0)$ (expressed as percents, between 0 and 100) on the vertical axis. Time is on the horizontal axis. (The units of time are left unspecified, since the qualitative behavior shown does not depend on the choice of time units, which are the reciprocal units of the rate constants.) The trajectories in Fig. 5.7 assume that no further exposure occurs to cause direct inflammation, so that only indirect effects (i.e., inflammation caused by existing inflammation) are shown. In this case, the lower equilibrium value is $P = 0$ and the upper equilibrium value is $P = 0.5 = 50\%$, denoted by E on the vertical axis. More generally, if ongoing exposure

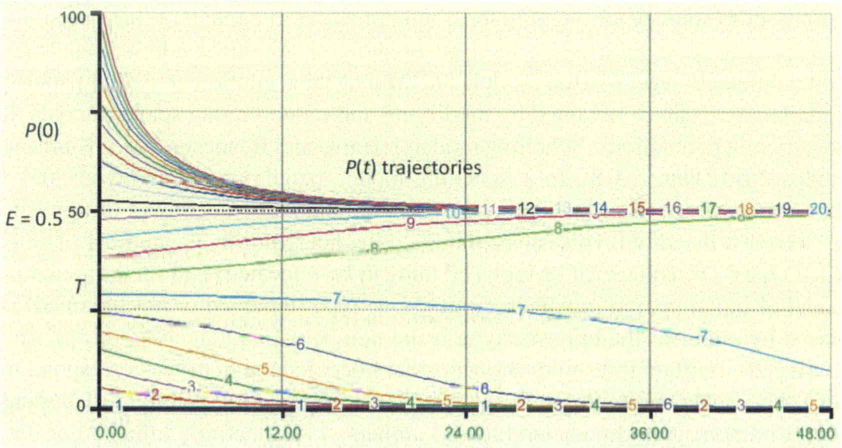

Fig. 5.7 A family of $P(t)$ inflammation trajectories starting from different initial inflammation levels on the vertical axis (numbers show percent of units inflamed). Time is on the horizontal axis. Time units are left unspecified here; they are the inverse of the units for the rate parameters a and d. For these trajectories, $k = 2$, $N = 3$, $d = 0.2$, and $a = 0.4$. Time (nominal units, e.g., months or years, depending on rate constants). For human inflammation, a time scale of weeks to months is plausible

is nonzero (as is plausibly the case for biopersistent fibers), then the lower equilibrium value increases with c, the potency of ongoing exposure. Trajectories that start below the threshold T correspond to acute inflammation (declining to the lower equilibrium level, which is 0 in the absence of further exposure), while trajectories that start above T correspond to chronic inflammation and approach level E. Of course, as Fig. 5.7 illustrates, the time between onset of chronic inflammation and convergence to the chronic inflammation equilibrium level of inflammation may be long for trajectories that start very close to the threshold T, where creation and resolution of inflamed units are in balance, but this is an unlikely circumstance, and in any case is unlikely to persist if exposure to fibers continues.

Such bistable, or tipping point, behavior is common in models where it becomes easier for unchanged units to change as the fraction of already-changed units increases (Page 2018). As previously noted (Cox 2019), "The existence of such a threshold is robust to many modeling details. It occurs in a wide variety of stochastic models that allow for realistic complexities such as heterogeneity in the thresholds for different units, or heterogeneous and time-varying numbers of neighbors for different units. The key feature driving the occurrence of bistability and a threshold between 2 basins of attraction for extinction versus self-sustaining activation of units in such models is that a unit is more likely to become activated when more of its neighbors are activated. This interdependence is realistic for phenomena ranging from spread of forest fires to epidemics, and we believe that it is also realistic for NLRP3- mediated spread of pyroptosis and lung inflammation."

For biopersistent fibers, as opposed to crystalline silica particles, the preceding model can be extended to consider the spatial distribution of fibers lodging in target tissues over time. If they are very biopersistent, such fibers may never be cleared, and may act as sources for ongoing local inflammation. If enough of them accumulate in proximity to each other, then all or most of the surrounding tissue may become inflamed. Agent-based models (ABMs) of the evolution of spatial patterns of inflammation can be obtained by modifying and reinterpreting spatial models of segregation in populations (Schelling models) (Hatna and Benenson 2015; Kaul and Ventikos 2015; Page 2018). In a Schelling model, people are of two types and a person moves away from a neighborhood if the number of neighbors of opposite type exceeds a threshold. This can be modified (by not requiring the number of units of each type to be conserved or replaced units to be relocated) and reinterpreted as a model of inflammation in which a unit of one type (inflamed or not inflamed) is replaced by a unit of the opposite type if the number of its neighbors of opposite type exceeds a certain threshold. Biopersistent fibers lodged in tissue correspond to people who never move. In such spatial models, there are typically still tipping points, so that neighborhoods tend to end up being either entirely inflamed or not. However, the final pattern of inflammation need not be all-or-nothing: inflamed and uninflamed regions may co-exist (Page 2018). This is consistent with the qualitative pattern in Fig. 5.7, in which an initial inflamed fraction greater than the tipping point T will spontaneously evolve to the new equilibrium fraction E.

Rather than seeking to develop an ABM or other model of the detailed dynamics of the spatial distribution of inflammation within a target tissue over time, we will make the simplifying approximation that the tissue can be modeled as a set of patches, within each of which a tipping point model for chronic inflammation, similar to the one in Fig. 5.7, holds. Each patch becomes chronically inflamed if and only if at least some critical number K of fibers are lodged in it; K is its tipping point, and once it is passed, inflammation spreads in the patch until some final spatial equilibrium configuration is reached. We do not model the details of the spatial configuration, but simply call the patch chronically inflamed (while acknowledging that the spatial density of inflammation may increase somewhat if further fibers arrive).

In this simplified patch approximation, a key question is how the fraction of chronically inflamed patches increases over time. The PBPK model in Appendix 1 can be interpreted as providing the average or expected cumulative number of fibers lodged in target tissues (lung and mesothelium) over time, but it provides no information about their spatial locations within the target tissues. However, the theory of Poisson processes implies that if the average arrival rate of fibers is λ fibers per patch unit time (with arrivals modeled as a Poisson process), then the random time for a patch to accumulate K fibers has a gamma distribution with parameters K and $1/\lambda$, implying a mean time of K/λ and a variance of K/λ^2. (These formulas can be generalized if the arrival rate is time-varying, so that the Poisson process is non-homogeneous.) For example, if a single lodged fiber were enough to chronically inflame a patch ($K = 1$) and the average arrival rate is $\lambda = 0.25$ lodged fibers per patch-year, then the mean time until the onset of chronic inflammation in the patch is $K/\lambda = 1/0.25 = 4$ years, and the standard deviation is $(K/\lambda^2)^{0.5} = K^{0.5}/\lambda = 4$ years. On the other hand, if $K = 8$ and $\lambda = 2$, then the mean time until the patch becomes inflamed is 4 years and the standard deviation is 1.4 years. If $K = 16$ and $\lambda = 4$, then the mean time until the patch becomes inflamed is 4 years and the standard deviation is 1 year. Table 5.3 summarizes these and other calculations showing how the standard deviation of the random time at which chronic inflammation begins decreases as K increases, holding the mean time fixed at 4 years. These numbers illustrate an important general principle: for any given mean time until the onset of chronic inflammation in a patch the more lodged fibers are needed to initiate chronic inflammation in the patch (i.e., the larger is K), the smaller is the standard deviation of the actual (random) time until onset of inflammation around its expected value. If

Table 5.3 Standard deviations for different pairs of K and λ values giving a mean time of years to initiation of chronic inflammation

K	λ	Mean years to chronic inflammation	Standard deviation = $K^{0.5}/\lambda$
1	0.25	4	4
4	1	4	2
8	2	4	1.4
16	4	4	1
100	25	4	0.4
10,000	2500	4	0.04
1,000,000	250,000	4	0.004

K is large, the time at which chronic inflammation begins will be very close to its expected value, K/λ. The coefficient of variation, i.e., the ratio of the standard deviation to the mean, of this random time, approaches zero. This follows from the theory of *sharp transitions* in random processes (Cox Jr. 2006; Motwani and Raghavan 1995). Therefore, even if there are many patches, all of them will tend to undergo the transition to chronic inflammation at around the same time if many fibers are required to initiate chronic inflammation in each patch. Since counts of asbestos fibers greater than 5 micrometers long in the lungs of patients with LC or MM are typically on the order of millions to hundreds of millions of fibers per gram of dry tissue weight (Gilham et al. 2016; Churg 1982; Tuomi et al. 1989; Rödelsperger et al. 2001), this large-number assumption for K seems plausible.

The net result of these modeling considerations is the very simple conclusion that chronic inflammation in a given patch of target tissue begins when the cumulative burden of fibers embedded in that patch reaches a certain threshold; moreover, this is likely to occur in multiple patches at approximately the same time. The exact time is a random variable, with a gamma distribution if the fiber arrival process is a Poisson process with constant intensity; and the exact spatial distribution of chronically inflamed locations, as revealed by ABM modeling, may be complex (Kaul and Ventikos 2015). But these details and complexities can be ignored if the goal is to integrate the inflammation model with the PBPK model, which only produces rough average fiber count estimates. Thus, the onset of chronic inflammation can be characterized, to a useful first-order approximation, as the time needed for a sufficient number of fibers to accumulate in target tissues to trigger the transition from acute to chronic inflammation. If this critical number of fibers is denoted by F^*, then the time until chronic inflammation is triggered will be approximately the time for the PBPK model, driven by the exposure history, to deliver F^* fibers to the target tissue.

A rough order-of-magnitude estimate for F^* can be derived by considering a worker who eventually develops MM or LC after a 40-year career of exposure to asbestos. Such a worker might accumulate roughly 1–10 million long (>5 μm), thin fibers per year per gram dry weight of lung tissue (Gilham et al. 2016; Churg 1982; Tuomi et al. 1989; Rödelsperger et al. 2001). If chronic inflammation is caused after roughly 4 years, i.e., 10% of the way into the 40-year exposure, then the cumulative number of fibers per gram dry weight of lung tissue that triggers chronic inflammation would be about $F^* = 100{,}000$ to 1 million. This simple estimate ignores sources of inter-individual variability, such as differences in pharmacokinetic parameters, or gene polymorphisms which might readily account for a several-fold difference in vulnerability to asbestos exposure, as measured by outcomes such as pleural thickening or signs of fibrosis (Kukkonen et al. 2011, 2014). It also ignores uncertainty about when chronic inflammation actually begins (e.g., at year 2 or 8 or 16 instead of 4). However, even with such uncertainty and variability, the simple baseline estimate of F^* as being between a few hundred thousand and a few million long, thin fibers per gram dry weight of lung tissue suggests that a single fiber lodged in lung or mesothelial tissue has effects that are orders of magnitude too small to trigger chronic inflammation.

Appendix 3: Two-Stage Clonal Expansion (TSCE) Models of Lung Cancer (LC) and Malignant Mesothelioma (MM) Risk

Both LC and MM, as well as numerous other malignancies, have been modeled using two-stage clonal expansion (TSCE) models of carcinogenesis in which normal stem cells undergo an initial mutation to become initiated (premalignant) cells; this pool of initiated cells proliferates via clonal expansion (promotion), i.e., increased net growth rate of initiated compared to normal stem cells; and initiated cells may then undergo a second mutation or transformation to yield fully malignant cells (Tan and Warren 2011; Zeka et al. 2011). Once a malignant cell is formed, it and its clonal progeny may acquire other hallmarks of carcinogenesis (e.g., angiogenesis, metastasis), yielding observed tumors after a latency period. Exposure to a carcinogen such as asbestos—or to the products of chronic inflammation caused by exposure, such as ROS—can increase cancer risk by increasing one or more of the first mutation rate, the second mutation rate, or the net proliferation rate of initiated cells.

Figure 5.8 shows the structure of a systems dynamics model for one proposed MM TSCE risk model, taken from Tan and Warren (2011). Table 5.4 lists the model formulas and parameter values. Similarly, Fig. 5.9 shows a TSCE systems dynamics diagram for LC risk, building on the baseline (no-exposure) model of McCarthy et al. (2012) with a step increase in the net proliferation rate of initiated (premalignant) stem cells when chronic inflammation begins. Table 5.5 list its equations.

These two models are essentially conceptually different. In Fig. 5.8, internal doses (tissue burdens) of asbestos fibers act directly on the initiation and promotion rates, causing them to increase by amounts that increase with the internal doses.

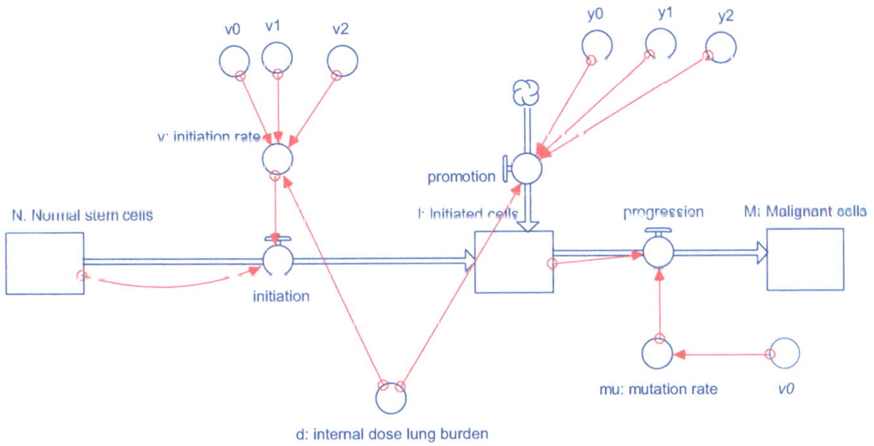

Fig. 5.8 Systems dynamics diagram for a TSCE model for MM caused by asbestos (Tan and Warren 2011)

Table 5.4 Listing of TSCE model formulas and parameter values for MM model in Fig. 5.8

I:_Initiated_cells(t) = I:_Initiated_cells(t - dt) + (initiation + promotion - progression) * dt
INIT I:_Initiated_cells = 0
INFLOWS:
initiation = N:_Normal_stem_cells*v:_initiation_rate
promotion = y0 + y1*d:_internal_dose_lung_burden^y2
OUTFLOWS:
progression = I:_Initiated_cells*mu:_mutation_rate
M:_Malignant_cells(t) = M:_Malignant_cells(t - dt) + (progression) * dt
INIT M:_Malignant_cells = 0
INFLOWS:
progression = I:_Initiated_cells*mu:_mutation_rate
N:_Normal_stem_cells(t) = N:_Normal_stem_cells(t - dt) + (- initiation) * dt
INIT N:_Normal_stem_cells = 1E5
OUTFLOWS:
initiation = N:_Normal_stem_cells*v:_initiation_rate
d:_internal_dose_lung_burden = 0*1E6
Malignant_cells_x_100000 = M:_Malignant_cells*100000
mu:_mutation_rate = v0
v0 = 2E-7
v1 = 0.007
v2 = 1
v:_initiation_rate = v0 + v1*d:_internal_dose_lung_burden^v2
y0 = 0.042
y1 = 1
y2 = 0.097

(The specific modeled dependences, taken from model T2 of Tan and Warren (2011), are as follows:

$$\text{Initiation mutation rate}: v = v_0 + v_1 d^{v2} = 2\text{E} - 7 + 0.007 * d$$

$$\text{Promotion rate}: y = y_0 + y_1 d^{y2} = 0.042 + 1 * d^{0.097}$$

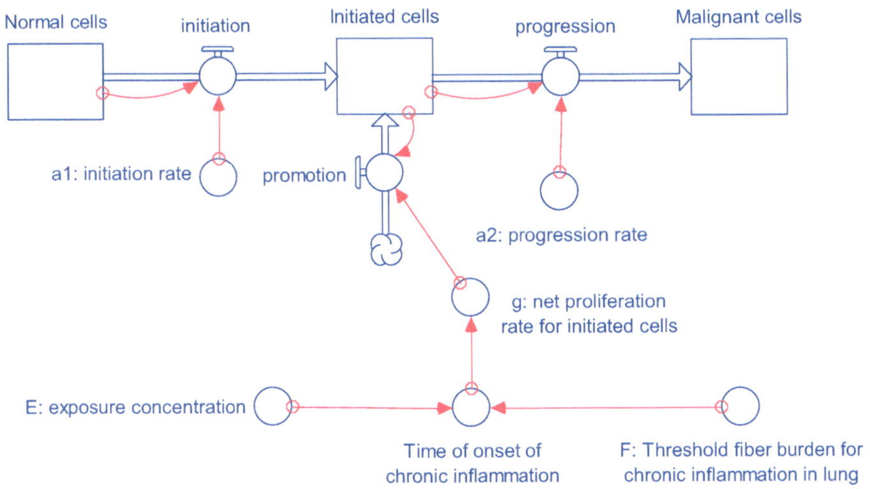

Fig. 5.9 Systems dynamics diagram for an I-TSCE model

$$\text{Progression rate}: mu = v_0$$

where d is a measure of internal dose of fibers.) By contrast, in Fig. 5.9, exposure and internal dose determine the time at which chronic inflammation begins; it is then the chronic inflammation state, rather than the internal dose of fibers *per se*, that increases subsequent cancer risks. (Figure 5.9 follows the notation in Chap. 4, with v in Fig. 5.8 renamed as $a1$, y as g, and mu as $a2$ for the initiation, promotion, and progression rates, respectively.)

Which approach better describes observations? They make different testable predictions. The model in Fig. 5.8 implies that risk increases with increasing accumulation of fibers throughout an exposed individual's life, but the model in Fig. 5.9 implies that, to the contrary, accumulation of fibers increases risk only by bringing about the onset of chronic inflammation, and becomes irrelevant thereafter. (Intermediate possibilities can be imagined, such as that continued fiber accumulation leads to more intense inflammation, but these two cases are the simplest and are empirically relevant, as discussed next.) Empirically, it appears that, once a sufficient quantity of asbestos (or similar carcinogens such as erionite) has been inhaled, it can cause MM decades later in a susceptible individual, even in the absence of further exposure (Carbone et al. 2012). Additional exposure does not appear to significantly increase future risk, or to reduce the latency period until MM develops (Carbone et al. 2012; Frost 2013). These observations favor the model in Fig. 5.9 over that in Fig. 5.8, so we use this "inflammation-mediated TSCE" (I-TSCE) model structure for MM as well as for LC, for which it was originally developed. To quantify an I-TSCE model for MM, we use the base (zero-dose) values from model T2 of Tan and Warren (2011), of $a1 = a2 = 2\text{E-}7$ for the initiation and progression rates (assuming 1E5 normal stem cells initially in the mesothelium) and $g = 0.042$

Table 5.5 Listing of TSCE model formulas and parameter values for LC model in Fig. 5.9

Initiated_cells(t) = Initiated_cells(t - dt) + (initiation + promotion - progression) * dt
INIT Initiated_cells = 0
INFLOWS:
initiation = Normal_cells*a1:_initiation_rate
promotion = Initiated_cells*g:_net_proliferation_rate_for_initiated_cells
OUTFLOWS:
progression = Initiated_cells*a2:_progression_rate
Malignant_cells(t) = Malignant_cells(t - dt) + (progression) * dt
INIT Malignant_cells = 0
INFLOWS:
progression = Initiated_cells*a2:_progression_rate
Normal_cells(t) = Normal_cells(t - dt) + (- initiation) * dt
INIT Normal_cells = 1e7
OUTFLOWS:
initiation = Normal_cells*a1:_initiation_rate
a1:_initiation_rate = 1.4e-7 {McCarthy et al. 2012, https://www.ncbi.nlm.nih.gov/pmc/articles/PMC3485693/}
a2:_progression_rate = 1.4e-7 {https://www.ncbi.nlm.nih.gov/pmc/articles/PMC3485693/}
E:_exposure_concentration = 0*10E6
F:_Threshold_fiber_burden_for_chronic_inflammation_in_lung = 100E6
g:_net_proliferation_rate_for_initiated_cells = IF(TIME < Time_of_onset_of_chronic_inflammation) THEN 0.075 else 0.09
h:_hazard_rate_x_100000 = progression*100000
P:_Probability_of_tumor_x_1000 = (1-exp(-Malignant_cells))*1000
Time_of_onset_of_chronic_inflammation = if (E:_exposure_concentration > 0) then F:_Threshold_fiber_burden_for_chronic_inflammation_in_lung/E:_exposure_concentration else 1E100

for the promotion rate if dose is zero. (For LC, the corresponding values are $a1 = a2 = 1.4\text{E-}7$ for the initiation and progression rates and $g = 0.075$ for the promotion rate.) Following onset of chronic inflammation, g increases relatively quickly; we approximate this as a step increase, of a size chosen to match the observation that sustained high exposure to asbestos starting from an early age can cause MM in up to 5% of individuals over a lifetime (e.g., by age 90) (Carbone et al. 2012). This corresponds to about a 3.5-fold increase in promotion rate g; this appears to be the main effect of asbestos exposure on carcinogenesis in TSCE models (Zeka et al. 2011, Table 5.2). Table 5.6 lists formulas and parameter values for the MM I-TSCE model. The two-stage clonal expansion (TSCE) modeling framework is undoubtedly only a relatively simple approximation to a more complex multistage clonal expansion (MSCE) process. However, it has provided useful empirical fits to several different data sets for asbestos-related cancer risks in past literature (e.g., Zeka et al. 2011; Tan and Warren 2011) and is adequate for the limited purposes of showing how inflammation-related changes in its transition rate parameters (Bogen 2019) can explain the main features of age-specific hazard rates for these data sets.

Table 5.6 Listing of I-TSCE model formulas and parameter values for MM model

Initiated_cells(t) = Initiated_cells(t - dt) + (initiation + promotion - progression) * dt
INIT Initiated_cells = 0
INFLOWS:
initiation = Normal_cells*a1:_initiation_rate
promotion = Initiated_cells*g:_net_proliferation_rate_for_initiated_cells
OUTFLOWS:
progression = Initiated_cells*a2:_progression_rate
Malignant_cells(t) = Malignant_cells(t - dt) + (progression) * dt
INIT Malignant_cells = 0
INFLOWS:
progression = Initiated_cells*a2:_progression_rate
Normal_cells(t) = Normal_cells(t - dt) + (- initiation) * dt
INIT Normal_cells = 1e5 {The value of 1e5 for the initial number of stem cells is from p. 25 of Tan and Warren 2011, http://www.hsc.gov.uk/research/rrpdf/rr876.pdf}
OUTFLOWS:
initiation = Normal_cells*a1:_initiation_rate
a1:_initiation_rate = 2e-7 {The value 2E-7 is from Tan and Warren (2011), http://www.hse.gov.uk/research/rrpdf/rr876.pdf, Table 5.6, p. 26}
a2:_progression_rate = a1:_initiation_rate {The constraint a2 = a1 is from p. 24 of Tan and Warren (2011), http://www.hse.gov.uk/research/rrpdf/rr876.pdf}
g:_net_proliferation_rate_for_initiated_cells = IF(TIME < Time_of_onset_of_chronic_inflammation) THEN 0.042 else 3.5*0.042 {The 0.042 background value is from Tan and Warren (2011), http://www.hse.gov.uk/research/rrpdf/rr876.pdf}
h:_hazard_rate_x_100000 = progression*100000
P:_Probability_of_tumor_x_1000 = (1-exp(-Malignant_cells))*1000
Time_of_onset_of_chronic_inflammation = 40 {years}
Time_of_onset_of_chronic_inflammation = if (E:_exposure_concentration > 0) then F:_Threshold_fiber_burden_for_chronic_inflammation_in_lung/E:_exposure_concentration else 1E100

Developing more realistic and detailed multistage clonal expansion models for asbestos-induced cancers appears to be a useful topic for future research. When such models become available, they can be used in place of the TSCE model in Table 5.4.

References

Abderrazak A, Syrovets T, Couchie D, El Hadri K, Friguet B, Simmet T, Rouis M. NLRP3 inflammasome: from a danger signal sensor to a regulatory node of oxidative stress and inflammatory diseases. Redox Biol. 2015;4:296–307. https://doi.org/10.1016/j.redox.2015.01.008.

Attanoos RL, Churg A, Galateau-Salle F, Gibbs AR, Roggli VL. Malignant mesothelioma and its non-asbestos causes. Arch Pathol Lab Med. 2018;142(6):753–60. https://doi.org/10.5858/arpa.2017-0365-RA.

Baas P, Burgers JA. Is one single exposure to asbestos life-threatening? Ned Tijdschr Geneeskd. 2014;158:A7653.
Baroja-Mazo A, Martın-Sanchez F, Gomez AI, et al. The NLRP3 inflammasome is released as a particulate danger signal that amplifies the inflammatory response. Nat Immunol. 2014;15(8):738–48. https://doi.org/10.1038/ni.2919.
Berman DW, Crump KS. Update of potency factors for asbestos-related lung cancer and mesothelioma. Crit Rev Toxicol. 2008;38(Suppl 1):1–47. https://doi.org/10.1080/10408440802276167.
Bernstein DM. The health risk of chrysotile asbestos. Curr Opin Pulm Med. 2014;20(4):366–70. https://doi.org/10.1097/MCP.0000000000000064.
Bittoni MA, Carbone DP, Harris RE. Ibuprofen and fatal lung cancer: a brief report of the prospective results from the Third National Health and Nutrition Examination Survey (NHANES III). Mol Clin Oncol. 2017;6(6):917–20. https://doi.org/10.3892/mco.2017.1239.
Bogen KT. Inflammation as a cancer co-initiator: new mechanistic model predicts low/negligible risk at noninflammatory carcinogen doses. Dose-Response. 2019;17(2):1559325819847834. https://doi.org/10.1177/1559325819847834.
Borm PJA, Schins RPF, Albrecht C. Inhaled particles and lung cancer, part b: paradigms and risk assessment. Int J Cancer. 2004;110:3–14.
Borm P, Cassee FR, Oberdörster G. Lung particle overload: old school -new insights? Part Fibre Toxicol. 2015;12:10. https://doi.org/10.1186/s12989-015-0086-4.
Browne K. Asbestos related malignancy and the Cairns hypothesis. Br J Ind Med. 1991;48(2):73–6.
Carbone M, Ly BH, Dodson RF, Pagano I, Morris PT, Dogan UA, Gazdar AF, Pass HI, Yang H. Malignant mesothelioma: facts, myths, and hypotheses. J Cell Physiol. 2012;227(1):44–58. https://doi.org/10.1002/jcp.22724.
Cho WC, Kwan CK, Yau S, So PP, Poon PC, Au JS. The role of inflammation in the pathogenesis of lung cancer. Expert Opin Ther Targets. 2011;15(9):1127–37. https://doi.org/10.1517/14728222.2011.599801.
Churg A. Fiber counting and analysis in the diagnosis of asbestos-related disease. Hum Pathol. 1982;13(4):381–92.
Comar M, Zanotta N, Zanconati F, Cortale M, Bonotti A, Cristaudo A, Bovenzi M. Chemokines involved in the early inflammatory response and in pro-tumoral activity in asbestos-exposed workers from an Italian coastal area with territorial clusters of pleural malignant mesothelioma. Lung Cancer. 2016;94:61–7.
Cox LA. Risk analysis implications of dose-response thresholds for NLRP3 inflammasome-mediated diseases: respirable crystalline silica and lung cancer as an example. Dose-Response. 2019;17(2):155932581983690. https://doi.org/10.1177/1559325819836900.
Cox LA Jr. Universality of J-shaped and U-shaped dose-response relations as emergent properties of stochastic transition systems. Dose-Response. 2006;3(3):353–68. https://doi.org/10.2203/dose-response.0003.03.006.
DeStefano A, Martin CF, Wallace DI. A dynamical model of the transport of asbestos fibres in the human body. J Biol Dyn. 2017;11(1):365–77. https://doi.org/10.1080/17513758.2017.1355489.
Donaldson K, Poland CA, Murphy FA, MacFarlane M, Chernova T, Schinwald A. Pulmonary toxicity of carbon nanotubes and asbestos - similarities and differences. Adv Drug Deliv Rev. 2013;65(15):2078–86. https://doi.org/10.1016/j.addr.2013.07.014.
Dostert C, Pétrilli V, Van Bruggen R, Steele C, Mossman BT, Tschopp J. Innate immune activation through Nalp3 inflammasome sensing of asbestos and silica. Science. 2008;320(5876):674–7.
Elliott EI, Sutterwala FS. Initiation and perpetuation of NLRP3 inflammasome activation and assembly. Immunol Rev. 2015;265(1):35–52.
Frost G. The latency period of mesothelioma among a cohort of British asbestos workers (1978-2005). Br J Cancer. 2013;109(7):1965–73. https://doi.org/10.1038/bjc.2013.514.
Galani V, Varouktsi A, Papadatos SS, Mitselou A, Sainis I, Constantopoulos S, Dalavanga Y. The role of apoptosis defects in malignant mesothelioma pathogenesis with an impact on progno-

sis and treatment. Cancer Chemother Pharmacol. 2019;84(2):241–53. https://doi.org/10.1007/s00280-019-03878-3.
Gamble JF, Gibbs GW. An evaluation of the risks of lung cancer and mesothelioma from exposure to amphibole cleavage fragments. Regul Toxicol Pharmacol. 2008 Oct;52(1 Suppl):S154–86. https://doi.org/10.1016/j.yrtph.2007.09.020. Epub 2007 Oct 22. PMID: 18396365.
Gilham C, Rake C, Burdett G, Nicholson AG, Davison L, Franchini A, Carpenter J, Hodgson J, Darnton A, Peto J. Pleural mesothelioma and lung cancer risks in relation to occupational history and asbestos lung burden. Occup Environ Med. 2016;73(5):290–9. https://doi.org/10.1136/oemed-2015-103074.
Gilroy D, Lawrence T. The resolution of acute inflammation: a 'tipping point' in the development of chronic inflammatory diseases. In: Rossi AG, Sawatzky DA, editors. The resolution of inflammation. Progress in inflammation research. Basel: Birkhäuser; 2008.
Gomes M, Teixeira AL, Coelho A, Araújo A, Medeiros R. The role of inflammation in lung cancer. Adv Exp Med Biol. 2014;816:1–23. https://doi.org/10.1007/978-3-0348-0837-8_1.
Grabiec AM, Hussell T. The role of airway macrophages in apoptotic cell clearance following acute and chronic lung inflammation. Semin Immunopathol. 2016;38(4):409–23. https://doi.org/10.1007/s00281-016-0555-3.
Hatna E, Benenson I. Combining segregation and integration: schelling model dynamics for heterogeneous population. J Artif Soc Soc Simul. 2015;18(4):1–15.
Hauenstein AV, Zhang L, Wu H. The hierarchical structural architecture of inflammasomes, supramolecular inflammatory machines. Curr Opin Struct Biol. 2015;31:75–83. https://doi.org/10.1016/j.sbi.2015.03.014.
He Q, Fu Y, Tian D, Yan W. The contrasting roles of inflammasomes in cancer. Am J Cancer Res. 2018;8(4):566–83.
Heid ME, Keyel PA, Kamga C, Shiva S, Watkins SC, Salter RD. Mitochondrial reactive oxygen species induces NLRP3-dependent lysosomal damage and inflammasome activation. J Immunol. 2013;191(10):5230–8. https://doi.org/10.4049/jimmunol.1301490.
Hillegass JM, Miller JM, MacPherson MB, Westbom CM, Sayan M, Thompson JK, Macura SL, Perkins TN, Beuschel SL, Alexeeva V, Pass HI, Steele C, Mossman BT, Shukla A. Asbestos and erionite prime and activate the NLRP3 inflammasome that stimulates autocrine cytokine release in human mesothelial cells. Part Fibre Toxicol. 2013;10:39. https://doi.org/10.1186/1743-8977-10-39.
Huang L, et al. NLRP3 deletion inhibits inflammation-driven mouse lung tumorigenesis induced by benzo(a)pyrene and lipopolysaccharide. Respir Res. 2019;20(1):4. https://doi.org/10.1186/s12931-019-0983-4.
Hutton HL, Ooi JD, Holdsworth SR, Kitching AR. The NLRP3 inflammasome in kidney disease and autoimmunity. Nephrology. 2016;21(9):736–44. https://doi.org/10.1111/nep.12785.
Ilgren EB. The Biology of Cleavage Fragments: A Brief Synthesis and Analysis of Current Knowledge. Indoor and Built Environment. 2004;13(5):343–356. https://doi.org/10.1177/1420326X04047563.
Jo EK, Kim JK, Shin DM, Sasakawa C. Molecular mechanisms regulating NLRP3 inflammasome activation. Cell Mol Immunol. 2016;13(2):148–59. https://doi.org/10.1038/cmi.2015.95.
Juliana C, Fernandes-Alnemri T, Kang S, Farias A, Qin F, Alnemri ES. Non-transcriptional priming and deubiquitination regulate NLRP3 inflammasome activation. J Biol Chem. 2012;287(43):36617–22. https://doi.org/10.1074/jbc.M112.407130.
Kadariya Y, Menges CW, Talarchek J, Cai KQ, Klein-Szanto AJ, Pietrofesa RA, Christofidou-Solomidou M, Cheung M, Mossman BT, Shukla A, Testa JR. Inflammation-related IL1β/IL1R signaling promotes the development of asbestos-induced malignant mesothelioma. Cancer Prev Res. 2016;9(5):406–14. https://doi.org/10.1158/1940-6207.CAPR-15-0347.
Kane AB, Hurt RH, Gao H. The asbestos-carbon nanotube analogy: an update. Toxicol Appl Pharmacol. 2018;361:68–80. https://doi.org/10.1016/j.taap.2018.06.027.
Karasawa T, Takahashi M. Role of NLRP3 inflammasomes in atherosclerosis. J Atheroscler Thromb. 2017;24(5):443–51. https://doi.org/10.5551/jat.RV17001.

Karki R, Man SM, Kanneganti TD. Inflammasomes and cancer. Cancer Immunol Res. 2017;5(2):94–9. https://doi.org/10.1158/2326-6066.CIR-16-0269.

Kaul H, Ventikos Y. Investigating biocomplexity through the agent-based paradigm. Brief Bioinform. 2015;16(1):137–52. https://doi.org/10.1093/bib/bbt077.

Kukkonen MK, Hämäläinen S, Kaleva S, Vehmas T, Huuskonen MS, Oksa P, Vainio H, Piirilä P, Hirvonen A. Genetic susceptibility to asbestos-related fibrotic pleuropulmonary changes. Eur Respir J. 2011;38(3):672–8. https://doi.org/10.1183/09031936.00049810.

Kukkonen MK, Vehmas T, Piirilä P, Hirvonen A. Genes involved in innate immunity associated with asbestos-related fibrotic changes. Occup Environ Med. 2014;71(1):48–54. https://doi.org/10.1136/oemed-2013-101555.

Kumagai-Takei N, Yamamoto S, Lee S, Maeda M, Masuzzaki H, Sada N, Yu M, Yoshitome K, Nishimura Y, Otsuki T. Inflammatory alteration of human T cells exposed continuously to asbestos. Int J Mol Sci. 2018;19(2):E504. https://doi.org/10.3390/ijms19020504.

Kuriakose T, Kanneganti TD. Regulation and functions of NLRP3 inflammasome during influenza virus infection. Mol Immunol. 2017;86:56–64. https://doi.org/10.1016/j.molimm.2017.01.023.

Lacourt A, Lévêque E, Guichard E, Gilg Soit Ilg A, Sylvestre MP, Leffondré K. Dose-time-response association between occupational asbestos exposure and pleural mesothelioma. Occup Environ Med. 2017;74(9):691–7. https://doi.org/10.1136/oemed-2016-104133.

Latz E, Xiao TS, Stutz A. Activation and regulation of the inflammasomes. Nat Rev Immunol. 2013;13(6):397–411. https://doi.org/10.1038/nri3452.

Lee DK, Jeon S, Han Y, Kim SH, Lee S, Yu IJ, Song KS, Kang A, Yun WS, Kang SM, Huh YS, Cho WS. Threshold rigidity values for the asbestos-like pathogenicity of high-aspect-ratio carbon nanotubes in a mouse pleural inflammation model. ACS Nano. 2018;12(11):10867–79. https://doi.org/10.1021/acsnano.8b03604.

Li M, Gunter ME, Fukagawa NK. Differential activation of the inflammasome in THP-1 cells exposed to chrysotile asbestos and Libby "six-mix" amphiboles and subsequent activation of BEAS-2B cells. Cytokine. 2012;60(3):718–30.

Liu G, Cheresh P, Kamp DW. Molecular basis of asbestos-induced lung disease. Annu Rev Pathol. 2013;8:161–87. https://doi.org/10.1146/annurev-pathol-020712-163942.

MacPherson M, Westbom C, Kogan H, Shukla A. Actin polymerization plays a significant role in asbestos-induced inflammasome activation in mesothelial cells in vitro. Histochem Cell Biol. 2017;147(5):595–604. https://doi.org/10.1007/s00418-016-1530-8.

Markowitz S. Asbestos-related lung cancer and malignant mesothelioma of the pleura: selected current issues. Semin Respir Crit Care Med. 2015;36(3):334–46. https://doi.org/10.1055/s-0035-1549449.

McCarthy WJ, Meza R, Jeon J, Moolgavkar SH. Chapter 6: lung cancer in never smokers: epidemiology and risk prediction models. Risk Anal. 2012;32(suppl 1):S69–84. https://doi.org/10.1111/j.1539-6924.2012.01768.x.

Miller JM, Thompson JK, MacPherson MB, Beuschel SL, Westbom CM, Sayan M, Shukla A. Curcumin: a double hit on malignant mesothelioma. Cancer Prev Res. 2014;7(3):330–40. https://doi.org/10.1158/1940-6207.CAPR-13-0259.

Moolgavkar SH, Anderson EL, Chang ET, Lau EC, Turnham P, Hoel DG. A review and critique of U.S. EPA's risk assessments for asbestos. Crit Rev Toxicol. 2014;44(6):499–522. https://doi.org/10.3109/10408444.2014.902423.

Moossavi M, Parsamanesh N, Bahrami A, Atkin SL, Sahebkar A. Role of the NLRP3 inflammasome in cancer. Mol Cancer. 2018;17(1):158. https://doi.org/10.1186/s12943-018-0900-3.

Motwani R, Raghavan P. Randomized algorithms. Cambridge: Cambridge University Press; 1995.

Offermans NS, Vermeulen R, Burdorf A, Goldbohm RA, Kauppinen T, Kromhout H, van den Brandt PA. Occupational asbestos exposure and risk of pleural mesothelioma, lung cancer, and laryngeal cancer in the prospective Netherlands cohort study. J Occup Environ Med. 2014;56(1):6–19. https://doi.org/10.1097/JOM.0000000000000060.

Oh JY, Ko JH, Lee HJ, Yu JM, Choi H, Kim MK, Wee WR, Prockop DJ. Mesenchymal stem/stromal cells inhibit the NLRP3 inflammasome by decreasing mitochondrial reactive oxygen species. Stem Cells. 2014;32(6):1553–63. https://doi.org/10.1002/stem.1608.

Page SE. The model thinker. New York: Basic Books; 2018.

Palomäki J, Välimäki E, Sund J, Vippola M, Clausen PA, Jensen KA, Savolainen K, Matikainen S, Alenius H. Long, needle-like carbon nanotubes and asbestos activate the NLRP3 inflammasome through a similar mechanism. ACS Nano. 2011;5(9):6861–70. https://doi.org/10.1021/nn200595c.

Paris C, Martin A, Letourneux M, Wild P. Modelling prevalence and incidence of fibrosis and pleural plaques in asbestos-exposed populations for screening and follow-up: a cross-sectional study. Environ Health. 2008;7:30. https://doi.org/10.1186/1476-069X-7-30.

Robb CT, Regan KH, Dorward DA, Rossi AG. Key mechanisms governing resolution of lung inflammation. Semin Immunopathol. 2016;38(4):425–48. https://doi.org/10.1007/s00281-016-0560-6.

Rödelsperger K, Mándi A, Tossavainen A, Brückel B, Barbisan P, Woitowitz HJ. Inorganic fibres in the lung tissue of Hungarian and German lung cancer patients. Int Arch Occup Environ Health. 2001;74(2):133–8.

Sayan M, Mossman BT. The NLRP3 inflammasome in pathogenic particle and fibre-associated lung inflammation and diseases. Part Fibre Toxicol. 2016;13(1):51. https://doi.org/10.1186/s12989-016-0162-4.

Schinwald A, Murphy FA, Prina-Mello A, Poland CA, Byrne F, Movia D, Glass JR, Dickerson JC, Schultz DA, Jeffree CE, Macnee W, Donaldson K. The threshold length for fiber-induced acute pleural inflammation: shedding light on the early events in asbestos-induced mesothelioma. Toxicol Sci. 2012;128(2):461–70. https://doi.org/10.1093/toxsci/kfs171.

Shao BZ, Wang SL, Pan P, Yao J, Wu K, Li ZS, Bai Y, Linghu EQ. Targeting NLRP3 inflammasome in inflammatory bowel disease: putting out the fire of inflammation. Inflammation. 2019;42(4):1147–59. https://doi.org/10.1007/s10753-019-01008-y.

Shukla A, Vacek P, Mossman BT. Dose-response relationships in expression of biomarkers of cell proliferation in in vitro assays and inhalation experiments. Nonlinear Biol Toxicol Med. 2004;2(2):117–28. https://doi.org/10.1080/15401420490464420.

Song N, Li T. Regulation of NLRP3 Inflammasome by Phosphorylation. Front Immunol. 2018;9:2305. https://doi.org/10.3389/fimmu.2018.02305.

Swanson KV, Deng M, Ting JP. The NLRP3 inflammasome: molecular activation and regulation to therapeutics. Nat Rev Immunol. 2019;19(8):477–89. https://doi.org/10.1038/s41577-019-0165-0.

Tan E, Warren N (2011) Mesothelioma mortality in Great Britain The revised risk and two-stage clonal expansion models Prepared by the Health and Safety Laboratory for the Health and Safety Executive. Harpur Hill Buxton Derbyshire SK17 9JN. www.hse.gov.uk/research/rrpdf/rr876.pdf

Tezcan G, Martynova EV, Gilazieva ZE, McIntyre A, Rizvanov AA, Khaiboullina SF. MicroRNA post-transcriptional regulation of the NLRP3 inflammasome in immunopathologies. Front Pharmacol. 2019;10:451. https://doi.org/10.3389/fphar.2019.00451.

Thompson JK, MacPherson MB, Beuschel SL, Shukla A. Asbestos-induced mesothelial to fibroblastic transition is modulated by the inflammasome. Am J Pathol. 2017;187(3):665–78. https://doi.org/10.1016/j.ajpath.2016.11.008.

Travis WD, Colby JV, Koss MN, Rosado-de-Christenson ML, Muller NL, King TE. Non-neoplastic disorders of the lower respiratory tract. In: Roasi J, editor. Atlas of nontumor pathology. Washington DC: American Registry of Pathologists, Armed Forces Institute Pathologists; 2002. p. 814–46.

Tuomi T, Segerberg-Konttinen M, Tammilehto L, Tossavainen A, Vanhala E. Mineral fiber concentration in lung tissue of mesothelioma patients in Finland. Am J Ind Med. 1989;16(3):247–54.

Vlachogianni T, Fiotakis K, Loridas S, Perdicaris S, Valavanidis A. Potential toxicity and safety evaluation of nanomaterials for the respiratory system and lung cancer. Lung Cancer. 2013;4:71–82. https://doi.org/10.2147/LCTT.S23216.

Warheit DB, Kreiling R, Levy LS. Relevance of the rat lung tumor response to particle overload for human risk assessment - update and interpretation of new data since ILSI 2000. Toxicology. 2016;374:42–59. https://doi.org/10.1016/j.tox.2016.11.013.

Woitowitz HJ, Baur X. Misleading new insights into the chrysotile debate. Pneumologie. 2018;72(7):507–13. https://doi.org/10.1055/s-0044-102169.

Yang Y, Wang H, Kouadir M, Song H, Shi F. Recent advances in the mechanisms of NLRP3 inflammasome activation and its inhibitors. Cell Death Dis. 2019;10(2):128. https://doi.org/10.1038/s41419-019-1413-8.

Zeka A, Gore R, Kriebel D. The two-stage clonal expansion model in occupational cancer epidemiology: results from three cohort studies. Occup Environ Med. 2011;68(8):618–24. https://doi.org/10.1136/oem.2009.053983.

Zhiyu W, Wang N, Wang Q, Peng C, Zhang J, Liu P, Ou A, Zhong S, Cordero MD, Lin Y. The inflammasome: an emerging therapeutic oncotarget for cancer prevention. Oncotarget. 2016; 7(31):50766–80. https://doi.org/10.18632/oncotarget.9391.

Zhou K, Shi L, Wang Y, Chen S, Zhang J. Recent advances of the NLRP3 inflammasome in central nervous system disorders. J Immunol Res. 2016;2016:9238290. https://doi.org/10.1155/2016/9238290.

Chapter 6
Nonlinear Dose-Time-Response Risk Models for Protecting Worker Health

Introduction

Why have occupational safety regulations in the United States not been more successful in protecting worker health from mesothelioma risks, while apparently succeeding relatively well in reducing silicosis risks? This chapter seeks to apply insights from the simulation models developed in Chaps. 4 and 5 to address this important practical question and to suggest possible directions for revising regulations of occupational exposures to make them more causally effective in protecting worker health. It is less technical than Chaps. 4 and 5, however. The main goal of this chapter is not to develop, but to apply, the pharmacokinetic (PBPK) models from Chaps. 4 and 5 to perform computational experiments illuminating how different time courses of exposure with the same time-weighted average (TWA) concentration affect internal doses in target tissues (lung for RCS and mesothelium for asbestos). Key conclusions are that (1) For RCS, but not asbestos, limiting average (TWA) exposure concentrations also tightly constrains internal doses and ability to trigger chronic inflammation and resulting increases in disease risks; (2) For asbestos, excursions (i.e., spikes in concentrations) and especially the times between them are crucial drivers of internal doses and time until chronic inflammation; and hence (3) These dynamic aspects of exposure, which are not addressed by current occupational safety regulations, should be constrained to better protect worker health. Adjusting permissible average exposure concentration limits (PELs) and daily excursion limits (ELs) is predicted to have little impact on reducing mesothelioma risks, but increasing the number of days between successive excursions is predicted to be relatively effective in reducing worker risks, even if it has little or no impact on TWA average concentrations.

To make this chapter relatively self-contained, we again briefly summarize from Chaps. 3, 4, and 5 the biological bases for thresholds and nonlinearities in exposure-response functions for respirable crystalline silica (RCS) and asbestos, based on

L. A. Cox Jr., *Quantitative Risk Analysis of Air Pollution Health Effects*,
International Series in Operations Research & Management Science 299,
https://doi.org/10.1007/978-3-030-57358-4_6

modeling the chronic inflammation mode of action discussed in Chap. 3, mediated by activation of the NLRP3 inflammasome, for both RCS and asbestos. We also refer to Appendices A–C from Chap. 5; throughout this chapter, references to the appendices are those for Chap. 5, since the same models are used in this chapter.

Science-Policy Background: Do Linear No Threshold (LNT) Assumptions Protect Worker Health?

For decades, low-dose linear no threshold (LNT) modeling assumptions implying that probabilities of adverse health responses are approximately proportional to cumulative exposure at low doses have facilitated regulatory risk assessments. Haber's rule, postulating that risks depends only on the product of exposure concentration and exposure duration (C × T) has been used to extrapolate risks across species and from observed exposure conditions to other exposure conditions of interest. These assumptions and exposure metrics have also been criticized—sometimes vehemently—as biologically unrealistic and descriptively incorrect for many chemical and health endpoints of practical concern, and alternatives such as peak exposures and time since peak exposure have been proposed as empirically better predictors of risk (Belkebir et al. 2011; Simmons et al. 2005; Kim et al. 2003). Noteworthy attempts to provide constructive alternatives to LNT models include physiologically based pharmacokinetic (PBPK) modeling of saturable kinetics, depletion of protective resources (e.g., detoxifying enzymes or antioxidant pools), and other sources of nonlinearity (Simmons et al. 2005); and two-stage or multistage clonal expansion (TSCE and MSCE) stochastic models of carcinogenesis that allow for co-initiation of carcinogenesis by inflammation (Bogen 2019; Li et al. 2019). Yet, simple, practical alternatives to LNT that are generally accepted as being more realistic are scarce. LNT and cumulative exposure are still widely used as default assumptions in regulatory risk assessments. They are often assumed to be health protective (especially for ionizing radiation), meaning that results of potentially more realistic models would be likely to vary in the direction of smaller risk estimates (Cardarelli and Ulsh 2018). When departures from cumulative and time-weighted average (TWA) exposure concentrations are considered in practice, they are usually addressed by excursion limits on the allowed frequency of exceeding a stated dose rate. For example, U.S. Occupational Safety and Health Administration (OSHA) standards for asbestos state that the "Permissible Exposure Limit (PEL) for asbestos is 0.1 fiber per cubic centimeter [f/cc] of air as an 8-h time-weighted average (TWA), with an excursion limit (EL) of 1.0 asbestos fibers per cubic centimeter over a 30-min period. The employer must ensure that no one is exposed above these limits." (www.osha.gov/Publications/OSHA3507.pdf).

Yet, establishing genuinely health-protective exposure standards for some important occupational exposures has remained frustratingly elusive. For example,

- *Asbestosis* deaths per year in the US increased dramatically (more than tenfold) between 1968 and 1999.
- *Mesothelioma*: A recent Centers for Disease Control and Prevention (CDC) report concluded that "However, although malignant mesothelioma deaths decreased in persons aged 35–64 years, the continuing occurrence of mesothelioma deaths among persons aged <55 years suggests ongoing occupational and environmental exposures to asbestos fibers and other causative EMPs, despite regulatory actions. …Despite regulatory actions and decline in asbestos use, the annual number of malignant mesothelioma deaths remains substantial. Contrary to past projections, the number of malignant mesothelioma deaths has been increasing. The continuing occurrence of mesothelioma deaths, particularly among younger populations, underscores the need for maintaining efforts to prevent exposure and for ongoing surveillance to monitor temporal trends (Mazurek et al. 2017).

Although long (e.g., multiple decade) latency periods between exposures and effects make it necessary to interpret such patterns cautiously, it appears plausible that exposure standards and protective measures for at least some mineral dusts and fibers have not been as effective as desired in reducing adverse health effects. There has been clear progress in related areas, however. For example, risks of silicosis mortality fell from 0.74 per million in 2001 to 0.39 per million in 2010 in the US population (Bang et al. 2015).

This history raises the following questions and challenges for occupational health risk analysis:

- Why have previous regulations and standards not been more effective in reducing risks? What are they missing that could explain continued risks despite reduced exposure concentrations?
- How might standards be revised to better protect workers against exposure-related risks?

This chapter proposes that *nonlinear* pharmacokinetics (PBPK) and pharmacodynamics (PD) may provide an important part of the answers because, in models with realistically nonlinear PBPK and PD responses to exposures, aspects of exposure other than TWA concentrations and excursions drive risk. In particular, the time until onset of chronic inflammation in target tissues, which may greatly amplify disease risks (Bogen 2019; Donaldson et al. 2010), depends on the *times between excursions* as well as on their sizes. Additional factors, such as imperfect monitoring and enforcement of existing standards, may also contribute to risk, but the focus of this chapter is on aspects of exposure-response dynamics that are not fully addressed by current and previous standards. Specifically, drawing on the PBPK and PD models for effects of respirable crystalline silica (RCS) and asbestos in Chaps. 4 and 5, respectively, we propose that *variability in exposure concentrations* over time is a critically important predictor for asbestos-related disease risk, including mesothelioma. It is less important for RCS than for asbestos, however, suggesting a possible explanation for why regulations and standards that control TWA exposure concentrations have been more effective for RCS than for asbestos. If this explanation is correct, then regulations might be made more health-protective by addressing variability as well as TWA concentrations and ELs. The remainder of this chapter focuses on RCS and asbestos, but similar considerations may apply to other substances with similar chronic inflammation-mediated modes of action.

Biological Background: NLRP3 Inflammasomes and Response Thresholds, Revisited

This section summarizes recent biological insights into exposure-related chronic inflammation. This material has already been discussed in more detail in Chaps. 3, 4, and 5; readers who have already those chapters may skip the rest of this section. In short, RCS and asbestos cause chronic inflammation in target tissues (lung, and pleural or peritoneal mesothelium, respectively). Chronic inflammation, in turn, appears to play an intrinsic role in the pathogenesis of malignant mesothelioma (MM), as well as increasing lung cancer risk and promoting tumor progression (Cho et al. 2011; Donaldson et al. 2010; Gomes et al. 2014; Kadariya et al. 2016; Sato et al. 2018; Thompson et al. 2017). Other diseases such as silicosis and asbestosis are also commonly preceded by chronic inflammation of the lung (Sato et al. 2018). Chronic inflammation in response to RCS and asbestos exposures is mediated by the NLRP3 inflammasome, a pro-inflammatory wheel-shaped signaling complex that self-assembles in the cytosol of alveolar macrophages (AMs) and some other cells in response to sustained danger signals (pathogen-associated molecular patterns (PAMPs) and damage-associated molecular patterns (DAMPs)) generated by a wide range of stimuli, from mineral dusts and fibers to bacterial pathogens to uric acid crystals (Kadariya et al. 2016; Sayan and Mossman 2016; Groslambert and Py 2018). Priming and activation of NLRP3 inflammasomes, the two steps in which its components are produced and self-assemble into a functioning inflammasome, are governed by several threshold mechanisms. These include two distinct thresholds (for NF-κB signaling and MAPK activation) acting as ultrasensitive switches for upregulating production of NLRP3 inflammasome proteins and inflammatory mediators (Gottschalk et al. 2016); cooperativity in the self-assembly of the inflammasome (via deubiquitination of NLRP3 protein filaments and oligomerization of these filaments and other components), creating a characteristic all-or-nothing response for self-assembly; and activation of the assembled inflammasome by a second signal (associated with potassium ion (K+) efflux, calcium ion (Ca2+) signaling, increased mitochondrial reactive oxygen species (mROS) and cytoplasmic ROS production, inhibition of antioxidant thioredoxins, mitochondrial damage, lysosome membrane permeabilization (LMP), and lysosomal rupture releasing cathepsins from the lysosome) (Groslambert and Py 2018). These thresholds prevent small and brief exposures from triggering an inflammatory response, while allowing sufficiently high and prolonged exposures to do so (Gottschalk et al. 2016). Once activated, the NLRP3 inflammasome recruits caspase-1, leading to generation of potent pro-inflammatory signals (cytokines IL-1β and IL-18), lysosome rupture and pyroptosis of cells, and rupture of the cell membrane, releasing pro-inflammatory contents (including the activated NLRP3 inflammasome) to the extracellular space, from which they can be taken up by, and inflame, nearby cells (Baroja-Mazo et al. 2014). This pro-inflammatory process competes with processes that ubiquitinate and degrade inflammasomes (via autophagosomes) to resolve inflammation (Groslambert and Py 2018). Mathematical modeling of this process (Chap. 4) predicts that the spread of inflammation in a population of cells has a threshold (somewhat analogous to the epidemic threshold for spread of contagious diseases in a population, or the percolation threshold for spread of a forest fire): if only a small fraction of cells is inflamed, then the inflammation is self-limiting and resolves itself when the inflammatory exposure ceases (acute inflammation); but above a critical threshold, the inflammation spreads and becomes self-sustaining even if exogenous exposure ceases (chronic inflammation). If chronic inflammation is triggered, the resulting high-ROS, pyroptosis environment increases cancer risk in the inflamed tissue (Bogen 2019). From this viewpoint, the key to protecting health is to keep exposures small and brief enough to avoid triggering chronic inflammation in target tissues.

For respirable crystalline silica (RCS), alveolar macrophages (AMs) constitute a key internal compartment. Sufficiently high concentrations of RCS is AMs trigger formation of the NLRP3 inflammasome, typically followed by frustrated phagocytosis (AMs cannot successfully degrade and remove some RCS particles, which may instead rupture the lysosome) (Nakayama 2018) and pyroptosis, as well as other changes in AM phenotype and performance. Sufficiently widespread activation of NLRP3 inflammasomes can induce chronic inflammation and elevated disease risks. To avoid these chronic inflammation-mediated increases in disease risks, therefore, exposures should be constrained to keep concentrations of RCS in AMs below the NLRP3 inflammasome activation threshold for a large enough fraction of AMs to avoid chronic inflammation, even in relatively susceptible workers. Similarly, for biopersistent asbestos fibers, accumulation of embedded fibers in the target tissue (e.g., at stomata of the parietal pleura, for pleural MM (Donaldson et al. 2010)) can trigger chronic inflammation and increased disease risk when the fiber load becomes sufficiently great. Appendices A-C of Chap. 5 describe some possible quantitative models for accumulation of fibers in the mesothelium over time, with sufficiently high loads leading to chronic inflammation and increased MM risk, by presenting PBPK, PD, and TSCE models for asbestos fibers and MM, respectively. Such models imply that sufficiently high cumulative exposure causes chronic inflammation and increased MM risk, and this qualitative conclusion is robust to many scientific uncertainties and variations in model details. Thus, a health-protective goal for avoiding MM risks is to keep cumulative load of asbestos fibers in the mesothelium small enough to avoid chronic inflammation.

Data and Methods: Physiologically-Based Pharmacokinetic (PBPK) Modeling

The following sections use previously published physiologically based pharmacokinetic (PBPK) models for RCS (Tran et al. 2001) and asbestos (DeStefano et al. 2017), discussed in Chaps. 4 and 5, respectively, to explore how predicted internal doses of RCS in AMs and of cumulative asbestos fibers lodged in the mesothelium, respectively, vary with how exposure is administered over time. They consider how these internal doses vary with both constant and time-varying exposure concentrations. The PBPK models used to perform these computational experiments are described in Appendix A of Chap. 5 and in Chap. 4 for amphibole asbestos fibers and RCS, respectively, and supporting data, methods, and validation experiments are detailed in the references.

Although PBPK modeling is a mature discipline, the specific models used here are not intended to be precisely accurate for any individual worker, nor to model interindividual variability in biochemistry and other phenotypic variations that plausibly occur in exposed worker populations. Rather, they are intended to illustrate the approximate distribution of mineral fibers or particles among compartments over time, recognizing that the precise values for individuals may vary several-fold around these simulated results (Chap. 4). What makes the results of these simulations potentially useful is the robust qualitative insights they provide, rather than the precise values of the simulated quantities. RCS and asbestos illustrate two different common patterns for pharmacokinetics: equilibration, in which intra-compartmental concentrations (here, of RCS in AMs) increase until the rate of removal (e.g., via phagocytosis and mucociliary clearance, translocation, and other pathways) equals the rate of addition; and accumulation, in which biopersistent fibers lodged in lung or mesothelial tissue accumulate with little or no degradation and no steady-state equilibrium is reached. For both patterns, it is useful to apply PBPK modeling to understand how the time to reach a threshold (e.g., the threshold value m_{crit} discussed in Chap. 4 for induction of inflammation by RCS in AMs, or the cumulative fiber load discussed in Appendices A–C of Chap. 5 for induction of chronic inflammation by fibers lodged in the mesothelium) depends on the time pattern of administered exposure concentrations.

The computational experiments reported in the following sections are designed as follows. Workers are assumed to be exposed to RCS or asbestos fibers over a number of consecutive time intervals (e.g., work weeks). Within each interval, the total amount (mass) of RCS or asbestos fibers is assumed to be the same, but it can be distributed over the interval in different ways, ranging from uniform concentration throughout the interval to a high concentration for the first part of the interval, followed by a zero-exposure recovery period for the rest of the interval. The PBPK models described in Appendix A of Chap. 5 and in Chap. 4 for asbestos and RCS, respectively, are used to calculate the resulting time courses of RCS in AMs and fibers lodged in the mesothelium. The asbestos PBPK model in Appendix A of Chap. 5, includes saturable kinetics for removal of asbestos fibers from blood by the kidney and for uptake of free fibers by alveolar macrophages and recruitment of alveolar macrophages to the lung, and the computational experiments reveal the effects of these nonlinearities on the predicted rate of accumulation of fibers in the mesothelium. By contrast, the RCS PBPK model from Chap. 4 is linear until inflammation begins. Since we take the time until RCS concentrations in AM become high enough to trigger inflammation as the output of principle interest, the RCS PBPK model is treated as linear throughout the simulations, recognizing that subsequent pharmacokinetics (including impaired phagocytosis by AMs) once the lung is inflamed are more complicated (Tran et al. 2001).

Results

Constant Exposure Concentrations

Figure 6.1, which was introduced in Chap. 4, shows the predicted time courses of RCS in AMs (in mg) for a worker described exposed to different constant concentrations of RCS in inhaled air. The 11 exposure scenarios shown cover a fairly narrow range, from 0.08 mg/m³ to 0.18 mg/m³ of RCS in inhaled air (bottom curve and top curve, respectively). In each case, the internal dose (concentration of RCS in AMs) increases following the start of exposure at time 0 and levels off at an equilibrium concentration that balances uptake and clearance of RCS. These curves are only intended to be descriptive of realistic responses while they are below the

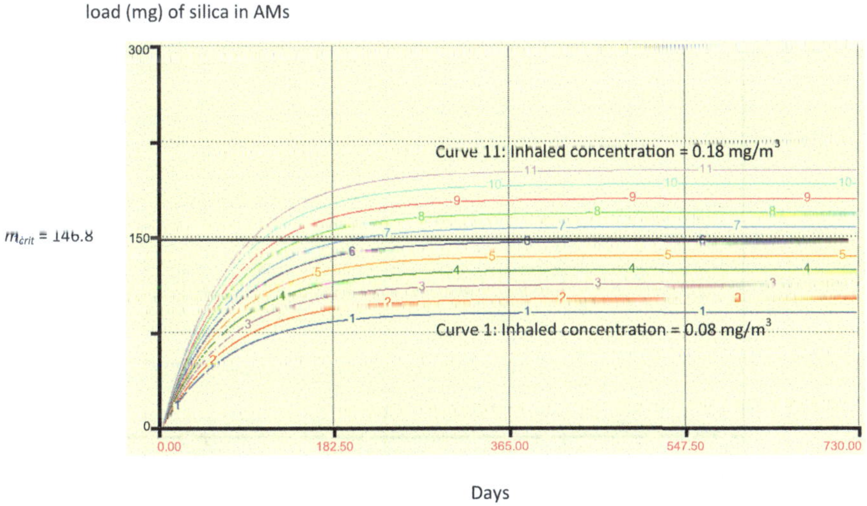

Fig. 6.1 Predicted time courses of silica load (mg) in AMs for 11 different exposure concentrations, 0.08 mg/m³ (curve 1) through 0.18 mg/m³ (curve 11), increments of 0.01 mg/m³. Predictions are made using the PBPK model in Chap. 4 (Tran et al. 2001)

horizontal black line (m_{crit} = 146.8 mg) at which inflammation begins; after the onset of inflammation, the parameters of the PBPK model change, e.g., to reflect reduced efficiency of clearance by AMs (Tran et al. 2001). Thus, the curves in Fig. 6.1 are of interest for showing how the predicted times until onset of inflammation (i.e., until the ascending curves reach the horizontal black line) depend on the administered concentration of RCS in inhaled air. (Again, inter-individual variability in model parameters is not modeled here, as the robust qualitative features of the curves are of principle interest, but see Tran et al. (2001) and Chap. 4 for further discussion of plausible variability). The key qualitative features of the answer are as follows:

a. Sufficiently low administered concentrations do not trigger inflammation (i.e., the internal concentration curves for low exposure concentrations level off below m_{crit}). For the specific parameter values in Chap. 4, the minimum exposure concentration needed to eventually cause inflammation is about 0.13 mg/m^3, although real-world values may vary several-fold based on differences in RCS potency and interindividual variability in parameter values (Cox 2019).
b. Most curves that do lead to inflammation do so within about 3–6 months (except for exposure concentrations very close to the threshold value of 0.13 mg/m^3).

As discussed in Chap. 3, similar qualitative features should be expected of any PBPK model that has approximately linear pharmacokinetics, for any internal dose metric that reflects the concentration or mass of RCS in some internal compartment that eventually reaches a long-run equilibrium with the concentration in inhaled air.

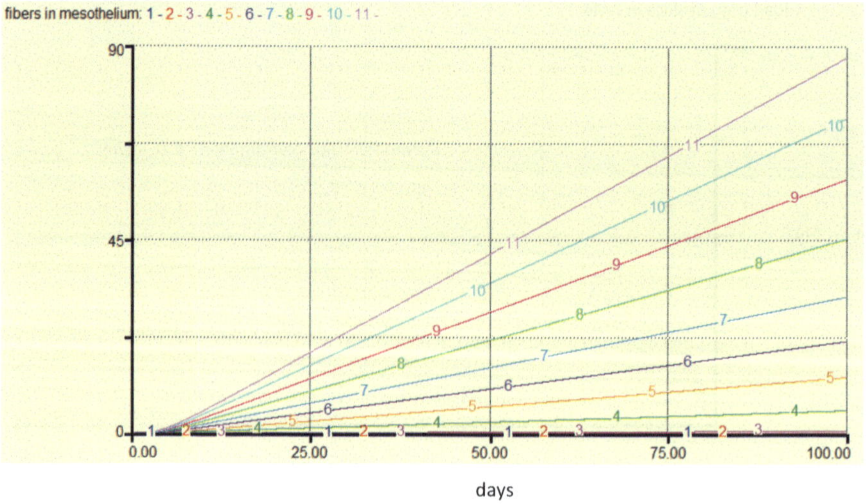

Fig. 6.2 Predicted time courses of asbestos fiber load for 11 different relative exposure concentrations, ranging from 0 (curve 1) through 5 (curve 11), in increments of 0.5. Predictions are made using the PBPK model in Appendix A of Chap. 5 (DeStefano et al. 2017)

Figure 6.2 presents analogous curves for asbestos, assuming for simplicity that all fibers are homogeneous, i.e., all of the same length, width, and stiffness. With this assumption, units of concentration are relative, with "1" (corresponding to curve 6) representing a baseline occupational exposure concentration for the hypothetical homogeneous fibers (DeStefano et al. 2017), and all other fiber count values scaled accordingly. (Thus, the units of fiber concentration on the vertical axis are notional units, e.g., each unit might correspond to tens or hundreds of thousands of physical fibers; if the concentration of asbestos in air is 0.1 fiber per cubic centimeter as an 8-h time-weighted average, and a worker inhales roughly 4000 l (4 million cc) of air per 8 h working day, then the expected number of inhaled fibers per day is 0.1 xs 4E6 = 400,000 fibers per day rather than "1," and all other fiber quantities are scaled accordingly.) The key qualitative features in this case are that

a. Accumulation of fibers in the mesothelium is approximately linear with respect to time; but
b. The rate of accumulation increases super-linearly (i.e., much more than proportionally) with exposure concentration.

To illustrate point (b) more clearly, Fig. 6.3 shows how the rate of predicted fiber accumulation in the mesothelium (proportional to relative units of fibers per

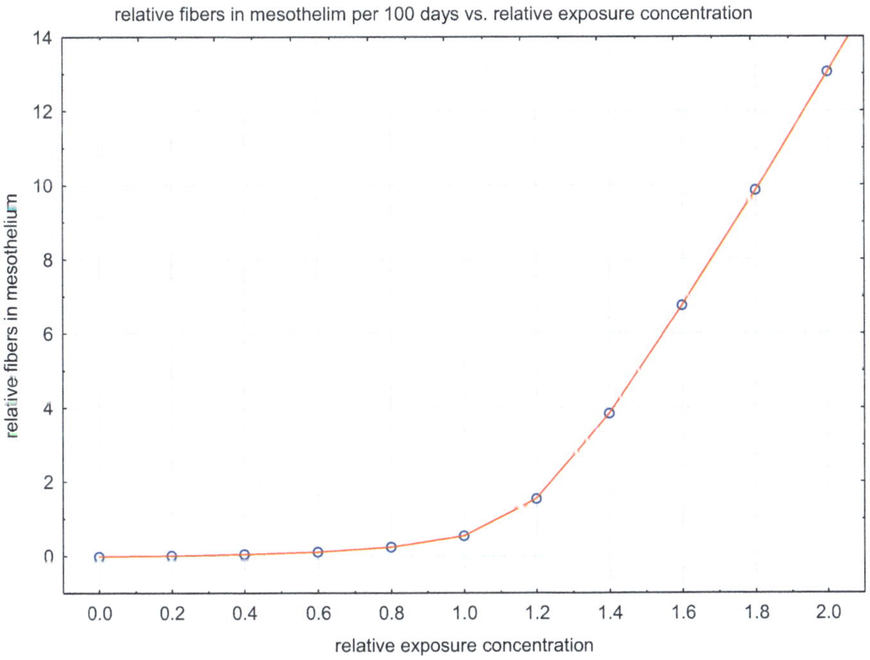

Fig. 6.3 Predicted relative accumulation rate of fibers in the mesothelium vs. relative exposure concentration of fibers in inhaled air. Doubling fiber concentration from 1 to 2 (on x axis) far more than doubles the rate of accumulation of fibers in the mesothelium (y axis)

100 days of constant exposure) varies as relative exposure concentration ranges from 0 to 2 (with 1 being the base case exposure scenario). Doubling exposure concentration from 1 to 2 increases the predicted rate of accumulation about 13-fold, illustrating the importance of nonlinear saturation processes in this PBPK model. Unlike Fig. 6.1 for RCS, there is no exposure concentration that will not eventually lead to accumulation of fibers in excess of any given threshold load for triggering chronic inflammation in the mesothelium.

However, a sufficiently small administered concentration keeps the rate of accumulation small enough so that the time until the inflammation threshold is reached is longer than a lifetime. The goal of protecting workers against increased risk from chronic inflammation can be achieved by keeping the time until enough asbestos accumulates to trigger chronic inflammation greater than a lifetime. For example, the curve in Fig. 6.3 has height 0.55 at a relative exposure concentration of 1 and height of only 0.02 at a relative exposure concentration of 0.2. Therefore, if the time until onset of chronic inflammation in the mesothelium for the baseline exposure scenario (constant exposure to "1" exposure concentration unit in Fig. 6.3) is at least 4 years, then the time until onset of inflammation when exposure concentration is 0.2 would be at least 4*(0.55/0.02) = 110 years. This might be regarded as a virtually safe level of exposure for chronic inflammation-mediated diseases.

Time-Varying Exposures

Which is more dangerous: 8 h/working day of exposure to 1 unit of concentration in inhaled air; 4 h/working day of exposure to 2 units of concentration, or 1 h of exposure per working day to 8 units of concentration? Or do they all create the same risk, since they entail the same cumulative exposure per day? As mentioned in Chap. 3, analogous questions can be asked for different time scales, e.g., is it less risky to have 2 months of daily exposure to a given concentration followed by 2 months of no exposure, repeated many times; or 1 month of exposure followed by 1 month of no exposure, repeated many times?

The curves in Fig. 6.4 suggest a partial answer for RCS by comparing predicted internal doses (mg of RCS in AMs) over time in response to different dose patterns in which the same total quantity of RCS per week is administered with different timing. For curve 1, equal amounts are inhaled on each of 5 consecutive days of a work week (with no exposures on weekends). For curves 2–5, concentrations that are 2–5 times higher, respectively, are administered for 1/2 to 1/5 as long (i.e., for the first 2.5 work days to 1 work day), respectively. Successive weeks repeat the same exposure pattern as the first week.

It is clear that the higher concentrations for shorter durations produce higher peak internal concentrations of RCS, even though all curves correspond to the same cumulative administered dose (AUC, in mg) per week. However, the effect is of limited size: the peak internal concentrations are less than 1.5 times higher for curve 5 than for curve 1. For asbestos, the analogous effects are much larger. Figure 6.5

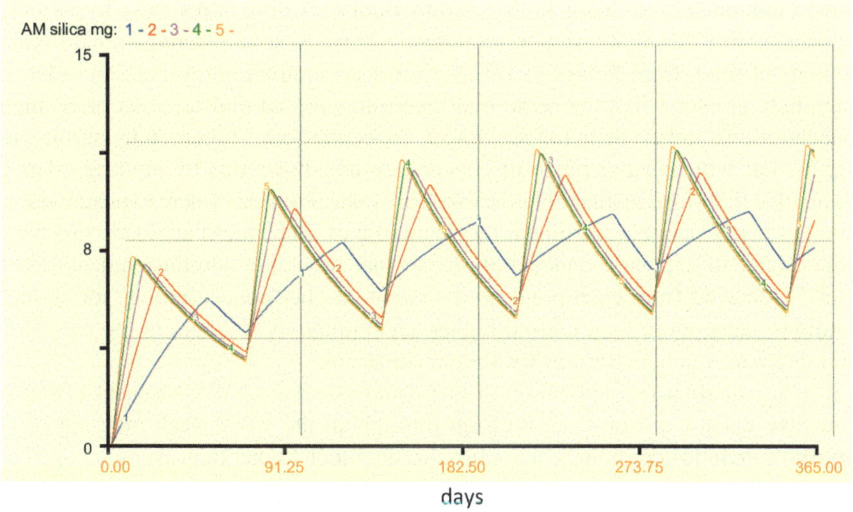

Fig. 6.4 Predicted time courses of silica load (mg) in AMs for 6 different exposure profiles giving the same cumulative exposure per unit time. Curve 1 is for 0.01 mg/m^3 of RCS for 5 days/week; curve 2 is for double the concentration for half the time (2.5 days) each week; and curve 5 is for 5 times the concentration for 1/5 the time (1 day) each week

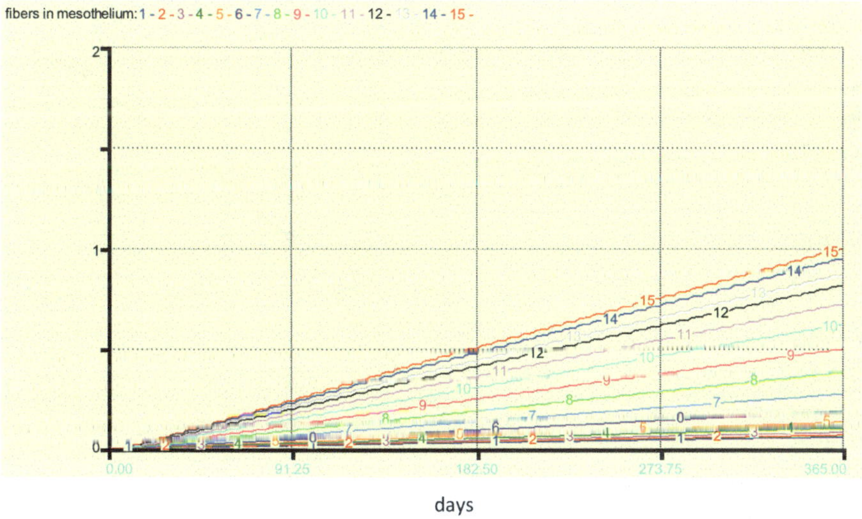

Fig. 6.5 Predicted time courses of cumulative asbestos fiber load lodged in the mesothelium for 15 different exposure profiles with the same total exposure per week. Relative exposure concentrations range from 0.2 for 5 days (curve 1) to 3 for 1/3 day (curve 15) in equal concentration increments of 0.2 (where 1 = baseline exposure concentration). Intervals of exposure are followed by intervals of no exposure for the rest of the week to achieve equal total exposures per week

shows cumulative fiber loads in the mesothelium over time in response to periodic exposures with the same average total dose per week administered over different amounts of time, from 5 days to 1/3 day. Fiber accumulation rates are an order of magnitude greater when the same total exposures are administered as brief, high concentrations rather than longer, lower concentrations. Figure 6.6 (similar to Fig. 6.3 but with a wider range of concentrations) shows this by plotting relative cumulative fiber load against relative exposure concentration, where exposure durations (in days per week, starting at the beginning of each week) are made inversely proportional to exposure concentrations so that the same administered dose per week is received from each point on the curve. Higher concentrations for shorter durations cause more than tenfold higher accumulations of fibers in the mesothelium than lower concentrations for shorter durations.

The key qualitative implication of this pattern is that a TWA-based PEL that is protective when exposures are uniform throughout the work week may not be if exposures include occasional excursions that are much higher than average, together with periods of low or zero exposure that, together with the excursions, achieve the required TWA target. For example, if a relative exposure concentration of 1 (i.e., baseline exposure concentration) received 5 days per week would lead to chronic

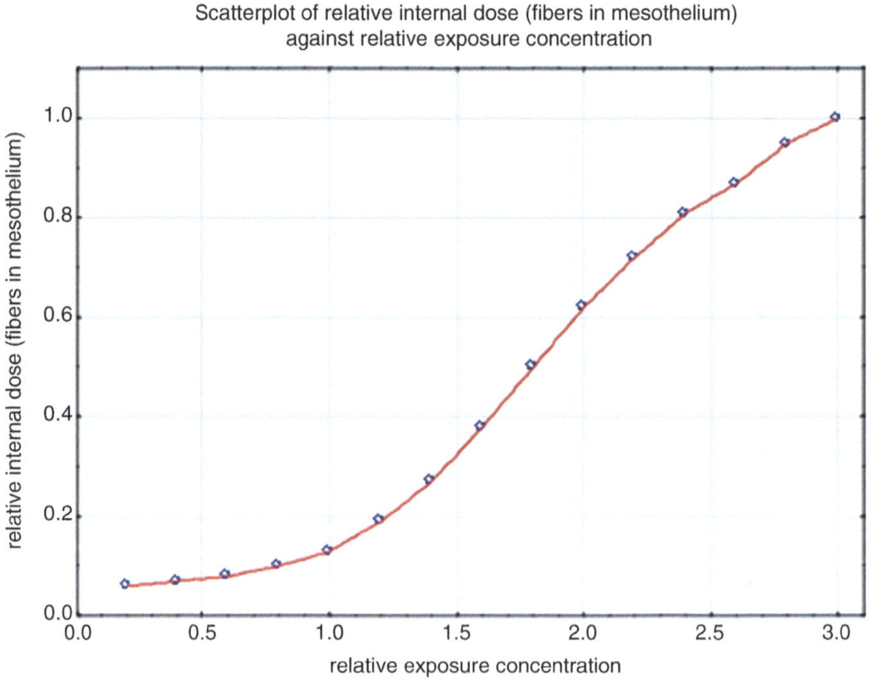

Fig. 6.6 Predicted relative accumulation rate of fibers in the mesothelium vs. relative exposure concentration of fibers in inhaled air. Doubling fiber concentration from 1 to 2 (on x axis) far more than doubles the rate of accumulation of fibers in the mesothelium (y axis). Fiber accumulation rates are normalized so that 1 on the y axis corresponds to 3 on the x axis

inflammation in 120 years, then an exposure concentration twice as great received 2.5 days/week (i.e., Monday, Tuesday, and half of Wednesday, followed by zero exposure until the following Monday) would lead to chronic inflammation in only 25 years. The first might well be deemed virtually safe (for chronic inflammation-mediated diseases) and the second not.

Exposures can also be spread over time by increasing the times between the starts of consecutive exposures. Additional simulations (not shown) reveal that doubling the baseline exposure concentration, but also doubling the time between starts of consecutive exposures from 1 week to 2 weeks to keep the average dose per week unchanged, reduces the predicted accumulation rate of fibers by about half compared to doubling the concentration but halving the duration received each week (from 5 to 2.5 days per week of exposure). Interpretively, spikes in concentration that are crowded together in time disproportionately increase accumulation of fibers in the mesothelium compared to fewer, more widely spaced spikes, even if they deliver the same average TWA exposure (e.g., average mg/week). These patterns also hold for exposure scenarios in which the sizes and times between exposure spikes are randomized. For example, Fig. 6.7 compares the accumulation of fibers in the mesothelium over 90 days for a deterministic exposure scenario (relative concentration of 1 for 5 days per week, shown as curve 1) and for 4 realizations of a randomized exposure scenario in which concentrations encountered throughout each day change every 18 min and are independently uniformly distributed between 0 and 2 (curves 2–5). The randomized scenarios tend to accumulate fibers more

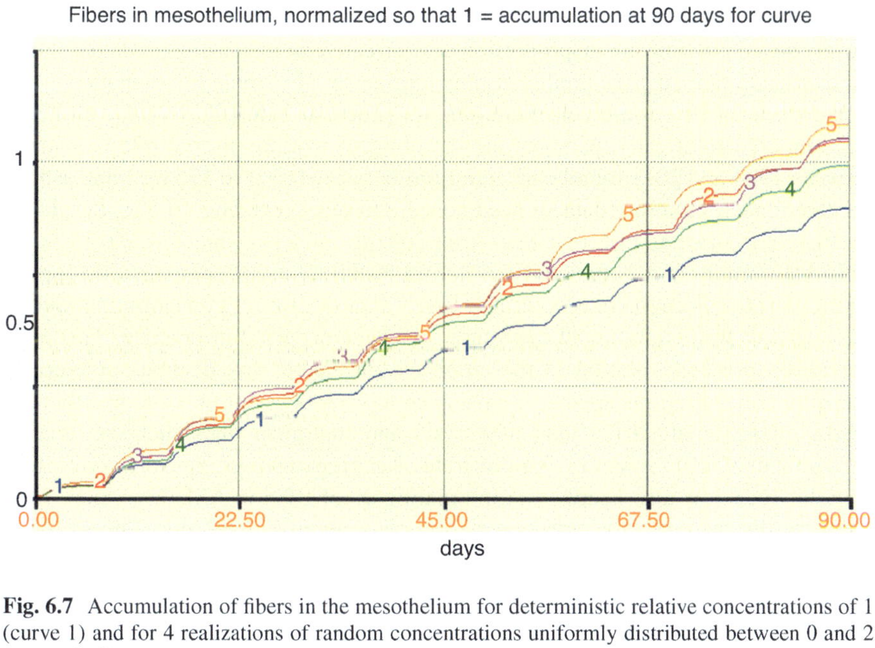

Fig. 6.7 Accumulation of fibers in the mesothelium for deterministic relative concentrations of 1 (curve 1) and for 4 realizations of random concentrations uniformly distributed between 0 and 2 (curves 2–5)

quickly than the deterministic scenario, even though their mean concentrations are identical (mean = 1), because, as shown in Fig. 6.6, random increases in concentration increase accumulation rates more than random decreases in concentration of the same size decrease accumulation rates.

Discussion

The simulation experiments presented here suggest that relatively brief excursions of relatively high concentrations, especially if they occur closely together in time (e.g., several times per week), contribute disproportionately to internal doses. For RCS, the effect of shorter durations of exposure to higher concentrations is relatively small (e.g., peak concentrations caused by different time patterns of administered concentration are within a factor of 1.5 of each other in Fig. 6.4). For asbestos, the effect is much larger, and can exceed a factor of 10 (Fig. 6.6). This amplification holds specifically for half-hour excursions that exceed average concentrations by tenfold, as in the current EL/PEL TWA limits: such excursions hasten the accumulation rate of fibers in the mesothelium and hence reduce the time until onset of chronic inflammation. The intuitive mechanistic explanation is that a spike of high concentration reduces elimination rates by partially clogging or saturating clearance pathways; this in turn increases the rate of accumulation of fibers in target tissues from exposures that follow. If another spike follows the first one while clearance remains impeded, it will be more efficient in depositing fibers in the mesothelium than it would be without the preceding spike. Thus, high-concentration excursions closely spaced in time are disproportionately effective in increasing predicted internal doses and reducing the predicted time until onset of chronic inflammation.

How relevant are these considerations for practical industrial hygiene and risk assessment? Empirically, personal air sampling of workers in multiple industries between 1984 and 2011 found concentrations between 0.001 to 175 f/cc, and asbestos compliance sampling data showed concentrations exceeding 10 f/cc, far above the PEL of 0.1 f/cc and EL of 1 f/cc, from 2001 to 2011 (Cowan et al. 2015). Such wide distributions of occupational exposure concentrations appear to provide opportunities for the phenomenon described here—that even brief concentration spikes increase risk from subsequent exposures—to have occurred in practice. This also provides a possible solution to the empirical puzzle of why declines in average exposure concentrations appear to have been more successful in reducing silicosis deaths (Bang et al. 2015) than asbestosis and malignant mesothelioma deaths (Mazurek et al. 2017). As Fig. 6.4 illustrates, keeping *average* exposure concentrations for RCS low also keeps *peak* internal doses of RCS in AMs low; more precisely, if the average concentration is less than about half of the concentration needed to eventually cause inflammation, then the peak internal dose remains below the inflammation threshold even if exposure is concentrated in a relatively short interval. By contrast, as illustrated in Figs. 6.5 and 6.6, average concentration of asbestos in air does not tightly constrain predicted internal doses, which may vary by over an order of magnitude for the same average TWA (and for the same EL), depending on how often spikes occur. In particular, an aspect of exposure that has a large impact on risk in these simulations for asbestos but that is not addressed by current PEL/EL-based regulations is that *high exposure concentrations on consecutive days* (or at least several times per week) are much more dangerous than the same high exposures if they are well separated in time. The first such exposure increases the internal dose formed by any subsequent exposure, and it takes time (e.g., several days) for this effect to wear off. Another high exposure concentration the following day would create a much higher accumulation of fibers in the mesothelium than if it does not occur until the effects of the first one have subsided. The calculations in Figs. 6.5 and 6.6 suggest that this effect can be at least an order of magnitude. Thus, an exposure history with occasional high concentrations bunched together in time might reduce the time until chronic inflammation from a virtually safe level (e.g., 120 years) to an unsafe level (e.g., 12 years) even if the same total dose administered as a constant exposure concentration would be virtually safe. Although we have not addressed variability and uncertainty in data and model parameters, additional simulations (not shown) confirm that these essential features of dynamic responses to different time patterns of exposure concentration—specifically, that closely spaced spikes are disproportionately hazardous—are robust to variations by factors of 2 in either direction around the specific numerical values of the PBPK parameters in these simulations.

As discussed in Chap. 3, the qualitative principle that even brief episodes of high concentrations closely spaced in time (e.g., separated by hours or days rather than weeks or months) are disproportionately effective in increasing disease risks is likely to apply to other substances that act via an NLRP3 inflammasome-mediated chronic inflammation mode of action (or other modes of actions with clear activation thresholds) if their pharmacokinetics are similar to those for asbestos in that high concentrations reduce clearance (or detoxification) of subsequent exposures for hours or days. It has recently become increasingly understood that inflammasome-mediated chronic inflammation plays a key role in the development of many types of cancer (Karki et al. 2017). It is tempting to speculate that mechanisms similar to those discussed here might help to explain longstanding observations in the literature showing experimental violations of Haber's rule, even for chemicals (such as volatile organic compounds) that are physically, chemically, and biochemically very different from mineral particles or fibers. For example, a 1996 study of isoprene exposures on cancer in mice (Cox Jr et al. 1996) concluded that "it appears that exposure concentration has a greater impact on tumor rates than weeks of exposure. To predict tumor probabilities, the time pattern of dose administration must be considered. The asymmetric effects of concentration, hours per-day, and number of days of exposure on tumor risks imply that complex dynamic risk models may be needed to accurately describe dose-time-response relations. The traditional concept of a dose-response relation as a static curve relating a numerical summary of dose to a numerical summary of response probability is probably not predictively useful." This conclusion from inhalation oncogenicity experiments in B6C3F1 mice also describes the results of the computational experiments in Figs. 6.5 and 6.6, with the PBPK model used to calculate them (and the I-TSCE model in Appendix C of Chap. 5 motivating the focus on time-to-inflammation as a proxy for more complex measures of risk) serving as the complex dynamic models needed to describe dose-time-response relations. The relevance of pharmacokinetics for RCS and asbestos risks seems clear based on results such as those in Figs. 6.4, 6.5, 6.6, and 6.7, but dose-response dynamics for isoprene may be more complicated, with multiple thresholds (or threshold-like nonlinearities) for tumors in different tissues (*ibid*). Although the calculations in this chapter focus specifically on RCS and asbestos, it may be worthwhile to explore the extent to which pharmacodynamic models of inflammation-mediated carcinogenesis (Bogen 2019) and pharmacokinetic models of the time required for administered exposure patterns to generate internal doses that trigger chronic inflammation can jointly explain observed dynamic exposure-response relationships for a wide range of chemicals and exposure-related diseases.

Some important limitations of the modeling and results presented here have already been briefly mentioned. For practical purposes, the main one is that the conclusions for asbestos rest on a single PBPK model for exposure to hypothetical idealized homogeneous fibers. In reality, fibers have a very wide range of shapes, lengths, stiffness coefficients, and chemical properties that can substantially affect their inflammatory and carcinogenic potencies, and exposed individuals have distributions of PBPK and PD parameters (e.g., NLRP3 inflammasome activation thresholds). The relevance of the models and insights in this chapter to real-world exposures is thus based largely on robust qualitative properties such as that high exposure concentrations make subsequent exposures more potent in forming internal doses, rather than on the specific numerical values in the simulations. In addition, we have focused on a single simple mode of action (accumulation of internal doses until NLRP3 inflammasome activation thresholds are exceeded, followed by spread of inflammation, chronic inflammation, and increased risk) and have ignored important and realistic complications and details such as the subsequent growth of tumors and the latency period between origination and clinical manifestation of tumors. The modeling reported here also does not address inter-individual variability or genetic contributions to susceptibility to asbestos exposures or in dose-response relationships. For example, Kukkonen et al. (2011) found that inherited genetic polymorphisms in enzymes involved in metabolizing oxidative agents were significantly associated with responses to asbestos exposures in Finnish workers, with the GSTT1 null genotype associated with a threefold increase in estimated risk of severe fibrotic changes and a more modestly elevated risk (OR 1.36, 95% CI 1.03–1.80) for thicker pleural plaques. They concluded that inherited detoxification capacity may affect nonmalignant pulmonary responses to asbestos exposures. Similarly, Hirvonen et al. (1996) reported that Finnish asbestos insulation workers with homozygous deletion of the GSTM1 gene and a NAT2 slow-acetylator genotype exposed had nearly fourfold higher risks of malignant mesothelioma that workers with a fast-acetylator genotype; but in an Italian population, the association was reversed, with the fast-acetylator workers having higher risks of malignant mesothelioma, and subsequent pooled analysis suggested no strong effect of NAT2 polymorphisms on risk of malignant mesothelioma (Betti et al. 2009). Such inconsistencies and challenges in interpreting and synthesizing results of genetic polymorphism studies are not unusual. Characterization of human inter-individual variability in health risk assessment remains far from an exact science. Nonetheless, the possibility of significant differences in individual-level dose-response relationships and in model parameters is an important omission from the current study, where future research and modeling may be valuable in better characterizing population risks.

Despite these omissions and simplifications, we hope that the insights from the simulation experiments presented can help to suggest how current regulations might be altered to better protect worker health, especially by addressing times between consecutive excursions, taking into account that narrowly spaced concentration spikes are predicted to be particularly dangerous.

Conclusions

For regulations to be effective in protecting the health of workers, they must constrain exposures in ways that avoid eliciting harmful biological responses. Chemically somewhat similar substances, such as mineral particles (RCS) and mineral fibers (asbestos), may have very different pharmacokinetics (Figs. 6.1 and 6.2) and pharmacodynamics, and may require different forms of regulatory rules to protect occupational health most effectively. In the examples of RCS exposures considered in this chapter, limiting mean (TWA) exposure concentrations also limits internal peak doses (Fig. 6.4) and resulting time until of chronic inflammation, which may be modeled very simply as occurring when internal concentrations of RCS in a sufficiently large fraction of alveolar macrophages exceed an inflammation activation threshold. By contrast, for asbestos, limiting mean (TWA) exposure concentrations does very little to limit risk (Fig. 6.6). Instead, excursions and the times between them dominate predicted risks. Current regulations do not address times between consecutive excursions, instead emphasizing 8-h TWA concentrations and number of excursions during the work day above a specified limit (such as 10 times the TWA). The modeling presented in this chapter and in Chaps. 3, 4, and 5 suggests that controlling time between excursions deserves more emphasis. It may be more effective for risk management than limiting concentration levels. If these results hold in practice, then increasing the time between successive excursions is likely to be more effective than further adjusting mean concentrations or excursion limits in reducing risks of inflammation-mediated diseases such as asbestosis and malignant mesothelioma.

References

Bang KM, Mazurek JM, Wood JM, White GE, Hendricks SA, Weston A. Silicosis mortality trends and new exposures to respirable crystalline silica - United States, 2001-2010. MMWR Morb Mortal Wkly Rep. 2015;64(5):117–20.

Baroja-Mazo A, Martín-Sánchez F, Gomez AI. The NLRP3 inflammasome is released as a particulate danger signal that amplifies the inflammatory response. Nat Immunol. 2014;15(8):738–48. https://doi.org/10.1038/ni.2919.

Belkebir E, Rousselle C, Duboudin C, Bodin L. Bonvallot N Haber's rule duration adjustments should not be used systematically for risk assessment in public health decision-making. Toxicol Lett. 2011;204(2-3):148–55. https://doi.org/10.1016/j.toxlet.2011.04.026.

Betti M, Neri M, Ferrante D, Landi S, Biava A, Gemignani F, Bertolotti M, Mirabelli D, Padoan M, Ugolini D, Botta M, Bonassi S, Magnani C, Dianzani I. Pooled analysis of NAT2 genotypes as risk factors for asbestos-related malignant mesothelioma. Int J Hyg Environ Health. 2009;212(3):322–9. https://doi.org/10.1016/j.ijheh.2008.08.001.

Bogen KT. Inflammation as a cancer co-initiator: new mechanistic model predicts low/negligible risk at noninflammatory carcinogen doses. Dose-Response. 2019;17(2):1559325819847834. https://doi.org/10.1177/1559325819847834.

Cardarelli JJ 2nd, Ulsh BA. It Is Time to Move Beyond the Linear No-Threshold Theory for Low-Dose Radiation Protection. Dose Response. 2018 Jul 1;16(3):1559325818779651. https://doi.org/10.1177/1559325818779651. PMID: 30013457; PMCID: PMC6043938.

Cho WC, Kwan CK, Yau S, So PP, Poon PC, Au JS. The role of inflammation in the pathogenesis of lung cancer. Expert Opin Ther Targets. 2011;15(9):1127–37. https://doi.org/10.1517/14728222.2011.599801.

Cowan DM, Cheng TJ, Ground M, Sahmel J, Varughese A, Madl AK. Analysis of workplace compliance measurements of asbestos by the U.S. Occupational Safety and Health Administration (1984-2011). Regul Toxicol Pharmacol. 2015;72(3):615–29. https://doi.org/10.1016/j.yrtph.2015.05.002.

Cox LA. Risk analysis implications of dose-response thresholds for NLRP3 inflammasome-mediated diseases: respirable crystalline silica and lung cancer as an example. Dose-Response. 2019;17(2):155932581983690. https://doi.org/10.1177/1559325819836900.

Cox LA Jr, Bird MG, Griffis L. Isoprene cancer risk and the time pattern of dose administration. Toxicology. 1996;113(1-3):263–72.

DeStefano A, Martin CF, Wallace DI. A dynamical model of the transport of asbestos fibres in the human body. J Biol Dyn. 2017;11(1):365–77. https://doi.org/10.1080/17513758.2017.1355489.

Donaldson K, Murphy FA, Duffin R, Poland CA. Asbestos, carbon nanotubes and the pleural mesothelium: a review of the hypothesis regarding the role of long fibre retention in the parietal pleura, inflammation and mesothelioma. Part Fibre Toxicol. 2010;7:5. https://doi.org/10.1186/1743-8977-7-5.

Doney BC, Blackley D, Hale JM, Halldin C, Kurth L, Syamlal G, Laney AS. Respirable coal mine dust in underground mines, United States, 1982-2017. Am J Ind Med. 2019;62(6):478–85. https://doi.org/10.1002/ajim.22974.

Gomes M, Teixeira AL, Coelho A, Araújo A, Medeiros R. The role of inflammation in lung cancer. Adv Exp Med Biol. 2014;816:1–23. https://doi.org/10.1007/978-3-0348-0837-8_1.

Gottschalk RA, Martins AJ, Angermann BR, Dutta B, Ng CE, Uderhardt S, Tsang JS, Fraser ID, Meier-Schellersheim M, Germain RN. Distinct NF-κB and MAPK activation thresholds uncouple steady-state microbe sensing from anti-pathogen inflammatory responses. Cell Syst. 2016;2(6):378–90. https://doi.org/10.1016/j.cels.2016.04.016.

Groslambert M, Py BF. Spotlight on the NLRP3 inflammasome pathway. J Inflamm Res. 2018;11:359–74. https://doi.org/10.2147/JIR.S141220.

Hall NB, Blackley DJ, Halldin CN, Laney AS. Current review of pneumoconiosis among US coal miners. Curr Environ Health Rep. 2019;6:137–47. https://doi.org/10.1007/s40572-019-00237-5.

Hirvonen A, Saarikoski ST, Linnainmaa K, Koskinen K, Husgafvel-Pursiainen K, Mattson K, Vainio H. Glutathione S-transferase and N-acetyltransferase genotypes and asbestos-associated pulmonary disorders. J Natl Cancer Inst. 1996;88(24):1853–6.

Kadariya Y, Menges CW, Talarchek J, Cai KQ, Klein-Szanto AJ, Pietrofesa RA, Christofidou-Solomidou M, Cheung M, Mossman BT, Shukla A, Testa JR. Inflammation-related IL1β/IL1R signaling promotes the development of asbestos-induced malignant mesothelioma. Cancer Prev Res. 2016;9(5):406–14. https://doi.org/10.1158/1940-6207.CAPR-15-0347.

Karki R, Man SM, Kanneganti TD. Inflammasomes and cancer. Cancer Immunol Res. 2017;5(2):94–9. https://doi.org/10.1158/2326-6066.CIR-16-0269.

Kim AH, Kohn MC, Nyska A, Walker NJ. Area under the curve as a dose metric for promotional responses following 2,3,7,8-tetrachlorodibenzo-p-dioxin exposure. Toxicol Appl Pharmacol. 2003;191(1):12–21.

Kukkonen MK, Hämäläinen S, Kaleva S, Vehmas T, Huuskonen MS, Oksa P, Vainio H, Piirilä P, Hirvonen A. Genetic susceptibility to asbestos-related fibrotic pleuropulmonary changes. Eur Respir J. 2011;38(3):672–8. https://doi.org/10.1183/09031936.00049810.

Kukkonen MK, Vehmas T, Piirilä P, Hirvonen A. Genes involved in innate immunity associated with asbestos-related fibrotic changes. Occup Environ Med. 2014;71(1):48–54. https://doi.org/10.1136/oemed-2013-101555.

Li L, Tian T, Zhang X. Stochastic modelling of multistage carcinogenesis and progression of human lung cancer. J Theor Biol. 2019;479:81–9. https://doi.org/10.1016/j.jtbi.2019.07.006.

Mazurek JM, Syamlal G, Wood JM, Hendricks SA, Weston A. Malignant mesothelioma mortality — United States, 1999–2015. MMWR Morb Mortal Wkly Rep. 2017;66:214–8.

Mazurek JM, Wood J, Blackley DJ, Weissman DN. Coal workers' pneumoconiosis-attributable years of potential life lost to life expectancy and potential life lost before age 65 years - United States, 1999-2016. MMWR Morb Mortal Wkly Rep. 2018;67(30):819–24. https://doi.org/10.15585/mmwr.mm6730a3.

Nakayama M. Macrophage recognition of crystals and nanoparticles. Front Immunol. 2018;9:103. https://doi.org/10.3389/fimmu.2018.00103.

Sato T, Shimosato T, Klinman DM. Silicosis and lung cancer: current perspectives. Lung Cancer. 2018;9:91–101. https://doi.org/10.2147/LCTT.S156376.

Sayan M, Mossman BT. The NLRP3 inflammasome in pathogenic particle and fibre-associated lung inflammation and diseases. Part Fibre Toxicol. 2016;13(1):51. https://doi.org/10.1186/s12989-016-0162-4.

Simmons JE, Evans MV, Boyes WK. Moving from external exposure concentration to internal dose: duration extrapolation based on physiologically based pharmacokinetic derived estimates of internal dose. J Toxicol Environ Health A. 2005;68(11-12):927–50.

Tan E, Warren N. (2011) Mesothelioma mortality in Great Britain The revised risk and two-stage clonal expansion models Prepared by the Health and Safety Laboratory for the Health and Safety Executive. Harpur Hill Buxton Derbyshire SK179JN. www.hse.gov.uk/research/rrpdf/rr876.pdf

Thompson JK, MacPherson MB, Beuschel SL, Shukla A. Asbestos-induced mesothelial to fibroblastic transition is modulated by the inflammasome. Am J Pathol. 2017;187(3):665–78. https://doi.org/10.1016/j.ajpath.2016.11.008.

Tran CL, Graham M, Buchanan D (2001) A biomathematical model for rodent and human lung describing exposure, dose, and response to inhaled silica. Institute of Occupational Medicine Technical Memorandum. 2001 TM/01/01. http://www.iom-world.org/pubs/IOM_TM0104.pdf

Part II
Statistics, Causality, and Machine Learning for Health Risk Assessment

Chapter 7
Why Not Replace Quantitative Risk Assessment Models with Regression Models?

Introduction: Use and Abuse of Regression Models in Public Health Risk Assessments

Chapter 2 suggested that dynamic simulation models, Bayesian networks, and causal analysis can add value to statistical regression modeling for understanding causal exposure concentration-response (C-R) relationships well enough to predict how changes in exposure would affect health risks—a task that typically requires causal insights that regression modeling alone cannot deliver (Pearl 2009). Chapters 3, 4, 5, and 6 have discussed dynamic simulation models. This part of the book turns to Bayesian networks and causal analysis (Chaps. 9, 10, and 11). First, however, this chapter and Chap. 8 examine some ways in which regression has been misapplied in public health risk analysis, both to motivate the need for other methods and to explain why regression alone is not an adequate substitute for quantitative risk assessment (QRA), with its explicit emphasis on preventable causes of disease and the quantitative causal relationships between reductions in exposures and resulting reductions in health risks. Part 3 will apply these lessons specifically to air pollution and public health, with greatest emphasis on National Ambient Air Quality Standards (NAAQS) for fine particulate matter (PM2.5). This chapter previews some of the issues developed in Part 3 by considering public health risks from a much more local form of air pollution: emissions from factory farms.

Do pollution emissions from livestock operations increase infant mortality rate (IMR)? In the past, some regression analyses of changes in IMR regressed against changes in aggregate "animal units" (a weighted sum of cattle, pig, and poultry numbers) over time, for counties throughout the United States, suggested the provocative conclusion that they do: "[A] doubling of production leads to a 7.4% increase in infant mortality" (Sneeringer 2009). This chapter observes that regressing IMR changes against changes in specific components of "animal units" (cattle, pigs, and broilers) at the state level reveals statistically significant *negative*

L. A. Cox Jr., *Quantitative Risk Analysis of Air Pollution Health Effects*, International Series in Operations Research & Management Science 299, https://doi.org/10.1007/978-3-030-57358-4_7

associations between changes in livestock production (especially, cattle production) and changes in IMR. It therefore appears that statistical associations between livestock variables and IMR variables are very sensitive to modeling choices (e.g., selection of explanatory variables, and use of specific animal types *vs.* aggregate "animal units). Chapters 1 and 2 emphasized that such model-dependent associations, whether positive or negative, do not warrant causal interpretation. Instead, standard methods of quantitative risk assessment (QRA), including emissions release (source) models, fate and transport modeling, exposure assessment, and causal dose-response modeling, are important—and, indeed, perhaps essential—for drawing valid causal inferences about health effects of exposures, to guide sound, well-informed public health risk management policy. Reduced-form regression models, which skip most or all of these steps, can only quantify *statistical* associations (which may be due to model specification, variable selection, residual confounding, or other non-causal factors). Sound risk management requires the extra work needed to identify and model valid *causal* relations (Pearl 2009; see also Chap. 9).

The striking proposition that air pollutants from cattle, swine, and poultry production operations measurably increase infant mortality rates (IMR) in the United States (Sneeringer 2009) was based on applying straightforward ordinary least squares regression statistical methodology to data and interpreting the coefficients as indicating causal relationships, in keeping with the WoE approach described in Chap. 1. This regression analysis did not seek to apply a standard quantitative risk assessment (QRA) approach, based on source modeling, fate-and-transport modeling, exposure assessment, dose-response characterization, and risk characterization. Yet, the main conclusions were of great potential interest for public health risk analysts and policy makers, as well as to various activist, agricultural, and public health communities. The conclusions were that (1) "This study uses geographic shifts in the industry to measure the impact of pollution on infant health." (2) "The article finds that a doubling of production leads to a 7.4% increase in infant mortality." (3) "*This finding is robust* with respect to inclusion of multiple fixed and time-varying controls." (4) "The mortality increases are driven by elevated levels of respiratory diseases, providing suggestive evidence of an air pollution mechanism." (Sneeringer 2009, emphases added) If true, these are startling and important risk assessment findings, suggesting that urgent action is justified.

But are they true? The key conclusions rest on subjective *causal interpretations* of ambiguous aggregate statistical associations (reduced-form regression model coefficients), rather than on a QRA decomposition into exposure-assessment and dose-response components that would demonstrate a valid empirical *causal* relation between exposure and risk. For example, the variables analyzed do not include any pollution measurements; thus, the claim that "This study uses geographic shifts in the industry to measure the impact of pollution on infant health" (emphasis added) is largely a matter of interpretation or conjecture. Linear regressions of changes in infant mortality against changes in aggregate animal units, and a selection of other covariates, constitute the heart of the analysis. This is a very different paradigm from standard QRA, which seeks to describe (and to document with relevant measurements and data) the links in the hypothesized causal chain leading from emissions of hazardous substances to adverse health effects.

Some important methodological points can be made about the differences between these statistical risk estimates, on the one hand, and the kinds of risk conclusions that QRA (based on exposure data and exposure-response relations) might support, on the other. To cast the relevant statistical issues into sharp relief, the following sections first use a simple thought experiment, and then examine available state-level data, to illustrate some pitfalls of assigning causal interpretations to purely statistical associations (reduced-form regression coefficients) between changes in livestock production and changes in IMR. (The original data set is partly proprietary, and is not readily available; hence we use similar data available at the state level for the years around 2005 to illustrate methodological points. County-level data for more years would certainly be preferable, but for the methodological purposes of this chapter, the less detailed state data suffice to illustrate key conceptual and methodological points.)

A Methodological Thought Experiment: Reduced-Form Model Regression Coefficients Are Not Causal Coefficients

In interpreting statistical regression models of exposure-response relations, a crucial methodological caveat is that reduced-form regression model coefficients do *not* necessarily have valid causal interpretations (Chaps. 1 and 2; Jacobs et al. 1979; Cox Jr. 2005). Even the sign, as well as the magnitude, of the reduced-form regression coefficient for a particular explanatory variable may be very sensitive to the modeler's selection of other explanatory variables to include in the model. Different, equally valid, selections may reverse the sign of the coefficient. The following simple conceptual example explains and illustrates these points.

Suppose that, unbeknownst to the risk assessor, the correct structural equations describing the causal relations affecting IMR are as follows:

$$\Delta IMR = Poverty - Education - \Delta Livestock + error \tag{7.1}$$

$$\Delta Livestock = 0.5 * Poverty \tag{7.2}$$

Here, *Poverty*, and *Education* are the fractions of a county's population recorded as below the poverty line, and having at least a Bachelor's degree, respectively. *ΔLivestock* is an appropriately scaled measure of animal units. Structural equation (7.1) is explicitly intended to be interpreted causally: *ceteris paribus*, an increase in *ΔLivestock* decreases *ΔIMR*. (This causal relation is specified only for purposes of argument.) Equation (7.2) indicates that *ΔLivestock* is proportional to *Poverty*.

In this simplified model, regressing *ΔIMR* against *ΔLivestock*, and *Education* would produce the following reduced-form regression model (as can be seen by substituting "*Poverty* = 2**ΔLivestock*", from Eq. (7.2), into (7.1):

$$\Delta IMR = \Delta Livestock - Education + error \quad (7.3)$$

Equation (7.3) provides a valid *statistical* description of the level of ΔIMR corresponding to any vector of explanatory variables (*ΔLivestock*, *Education*). But it is not a valid *causal* model for predicting how changes in *ΔLivestock* would change *ΔIMR*. Concluding from the reduced-form model (7.3) that increasing *ΔLivestock* would lead to an increase *ΔIMR* would be unwarranted. That the regression coefficient for *ΔLivestock* in model (7.3) is +1 has no valid causal interpretation. (As revealed by structural Eq. (7.1), increasing *ΔLivestock* would actually *decrease* *ΔIMR*, in this example.)

In the real-world application, interpreting a reduced-form regression model coefficient as implying that "[D]oubling of production leads to a 7.4% increase in infant mortality" is analogous to interpreting the +1 coefficient of *ΔLivestock* in model (7.3) as implying that increasing *ΔLivestock* would increase *ΔIMR*. It is not necessarily a valid causal interpretation. (Including multiple years of data and repeated observations, as in the original study, does not fix the problem, which does not arise from data, but from incorrect interpretation of the results.) In general, *reduced-form model coefficients are not causal, or structural, coefficients.*

A modeler's choices about which explanatory variables to include in a regression model can make even a negative causal relation look positive in a reduced-form model. For example, in our simple conceptual model, a modeler who includes *Poverty* as an explanatory variable, and hence fits model (7.1) instead of model (7.3) to the data, would obtain a regression coefficient of −1 for *ΔLivestock*, instead of a coefficient of +1. Conversely, a modeler who wanted to exhibit a positive regression coefficient for the relation between *ΔLivestock* and *ΔIMR* could easily do so, by using model (7.3) instead of model (7.1). Neither coefficient is incorrect as a statistical description: they simply rely on different subsets of variables. However, selecting a particular reduced-form regression model for *ΔIMR*, and then interpreting a positive coefficient for *ΔLivestock* as evidence of a causal "impact of pollution on infant health," overlooks the fact that equally valid regression models with a different choice of explanatory variables could have very different (even, oppositely signed) coefficients for *ΔLivestock*.

Statistical issues in assessing whether changes in IMR are robustly and positively associated with changes in livestock production can be illustrated with available state-level data for the years around 2005. Total annual production, in millions of pounds, of cattle and calves, broilers (for a subset of states), and pigs and swine, are available on-line from the Census bureau for 2005–2007, and infant mortality rate (IMR) data are available for 2000–2005 (Appendix) Annual rates of change in both IMR and livestock variables are relatively stable over time, so that it makes sense to ask whether states with increases in livestock variables tend to have significantly associated increases in IMR. As it turns out, the state-level data show that changes in cattle production are significantly *positively* correlated with changes in pig production, but *negatively* correlated with broiler production (and also with changes in IMR, as discussed below). Any summary measure of changes in aggregate "animal units" obscures these component-specific changes; thus, we consider further the relations between changes in IMR and changes in these specific components of livestock production.

Qualitatively, exploratory analysis of Census data shows a striking, although unexpected, pattern. States in which cattle production was *increasing* around 2005 (as measured by the difference in cattle production from 2005 to 2006, the last two years with data) tend overwhelmingly to be states in which human IMR was *decreasing* (as measured either by year-to-year changes, or by the longer-run change in IMR rates from 2000 to 2005, the first and last year for which IMR data are available.) This relation is highly statistically significant (but not necessarily causal): only 22% of states with increasing cattle production experienced increases in IMR rates, compared to 73% of states without increasing cattle production that experienced increases in IMR rates. Thus, increased cattle production is significantly associated with *reduced* risk of IMR increases: states with no increases in cattle production were more than 3 times more likely to have increased IMR than states with increases in cattle production (73% *vs.* 22%).

Quantitatively, Fig. 7.1a plots changes in IMR between 2000 and 2005 against changes in livestock (millions of pounds of cattle, pigs, and broilers) in 2005–2006, for states that have data on all three livestock variables. Figure 7.1b, c show analogous plots for ΔIMR *vs.* changes in cattle and pig production, respectively, for all 49 states other than Hawaii. (Hawaii is excluded from all analyses, as it has very different demographics than other states; including it does not change our conclusions, as can be readily verified using the data files in Appendix.) To make best use of the limited available data, we focus on 2005 as the reference year, with changes between 2004 and 2005, or between 2005 and 2006, assumed to have held for most or all of 2005. The simple linear regression lines shown have statistically significant *negative* slopes. (The plot for broilers lacks data for some states, and its regression line slope, though also negative, is not statistically significant.)

Covariates such as race, *Education*, and so forth are also available for each state, allowing multivariate regression models to be fit to state-level data. Table 7.1 shows such a multiple linear regression model, for IMR in 2005 *vs.* various explanatory variables, developed using automated (forward stepwise) variable selection from the list in Appendix. In this model, too, change in cattle production from 2005 to 2006 is significantly *negatively* associated with IMR levels. This negative association is also apparent in nonparametric multivariate analysis using classification trees (Chap. 1; Biggs et al. 1991); thus, it appears to be quite robust to changes in specific modeling methods. Table 7.2 shows a regression model with *change* in IMR between 2004 and 2005 as the dependent variable (rather than level of IMR in 2005, as in Table 7.1). The only two livestock-related predictors, change in broilers and change in cattle, both have *negative* coefficients, consistent with the results from exploratory data-plotting and simple (univariate) linear regression analyses.

To further examine multivariate dependencies, Table 7.3 shows statistically significant ($p < 0.05$) partial correlations between mortality rate variables (for adults as well as infants) and other variables, including *Poverty 2005*, while "controlling" (via multiple linear regression) for *Income 2005*, *Black 2005*, *Hispanic 2005*, *White*

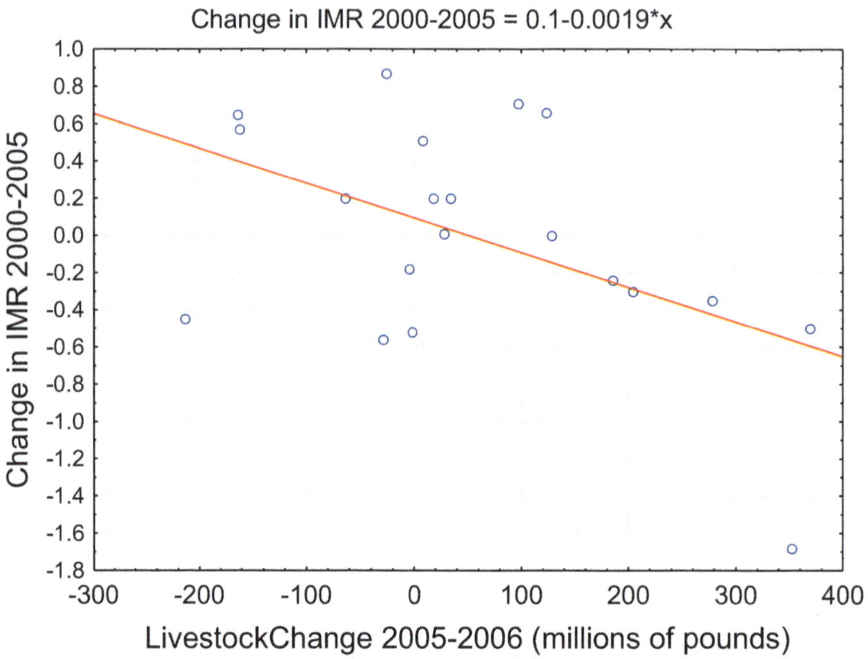

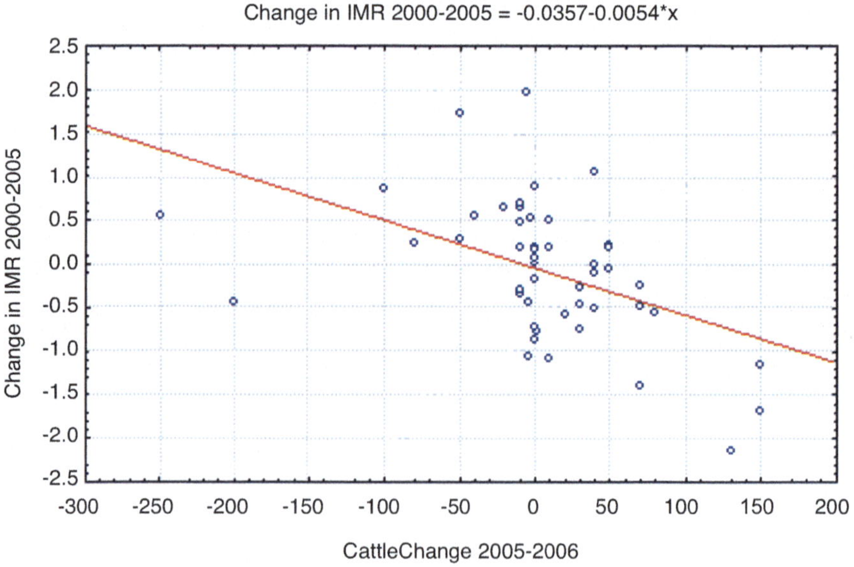

Fig. 7.1 Scatter plots for change in IMR vs. change in livestock production in the 49 Continental United States (i.e. Hawaii not included), *circa* 2005. (**a**) ΔIMR vs. $\Delta Livestock$ (millions of pounds) (*Data:* Appendix). States with missing data for poultry are not shown. (**b**) ΔIMR vs. $\Delta Cattle$ (millions of pounds) Data are from Appendix. The data are for the 49 continental states. (**c**) ΔIMR vs. $\Delta Pigs$ (millions of pounds). Data are from Appendix. The data are for the 49 continental states

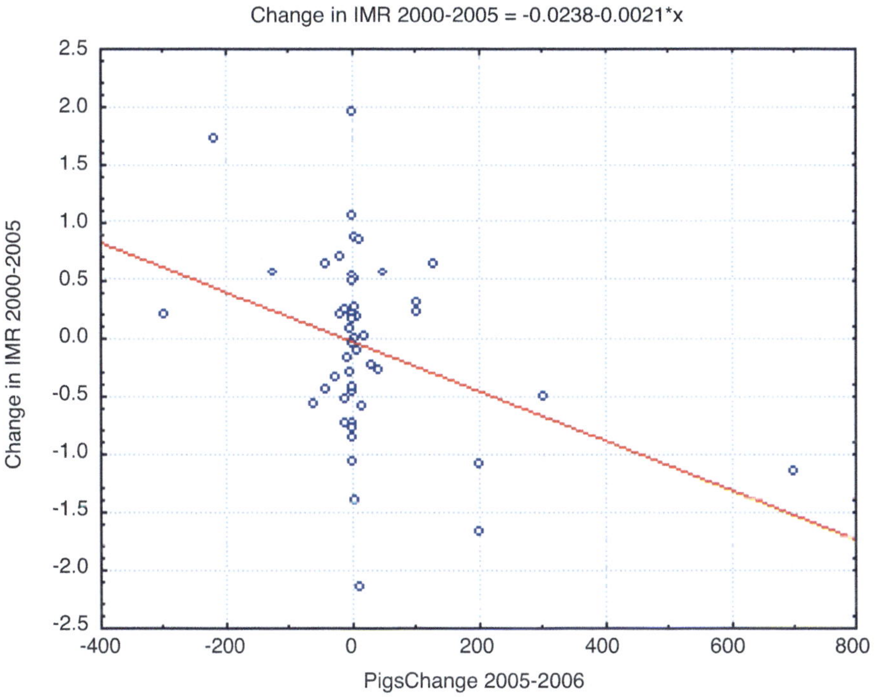

Fig. 7.1 (continued)

2005, and *Education 2005*. (*Poverty 2005* is a highly significant predictor of IMR, even after including *Income 2005*.) Although livestock production levels in 2005 are not themselves statistically significantly associated with mortality rates in 2005, *changes* in livestock production levels (bottom four rows of Table 7.3) are statistically significantly *negatively* correlated with IMR and with changes in IMR.

These state-level data for the years around 2005 are admittedly limited; they can and should be refined and extended. Doing so would permit a more direct comparison with the results of Sneeringer (2009). Yet, it seems clear that they do not indicate a robust positive association between livestock and infant mortality variables. To the contrary, interpreting these reduced-form coefficients causally (which we recommend against) would indicate a statistically significant *negative* (protective) relation between livestock production and infant mortality. While one could then speculate about the "health benefits" of farm air in this context, such speculation would be irresponsible: reduced-form regression coefficients—whether positive or negative—are *not* reliable indicators of true causal relations.

Discussion: Why the Difference?

The claim of a robust positive association between changes in animal production and changes in infant mortality rates at the county level (Sneeringer 2009) does not appear to be well supported by the robust, significantly negative ("protective") statistical relations between livestock production variables and infant mortality rate variables apparent in regression models for state-level data (Fig. 7.1 and Tables 7.1, 7.2 and 7.3). What explains the difference? Some *a priori* possible methodological explanations might include the following.

- *Simpson's Paradox and ecologic fallacy.* One possibility is that significant patterns at one level of data aggregation (county) are reversed at a different level of aggregation (state) (Simpson's Paradox). In this case, it might be possible to determine valid causal relations only by using individual-level data, which are not available. A substantial literature in risk assessment and epidemiology addresses the potential biases due to ecologic fallacy in studies based on proximity to exposure sources (Brodkin et al. 2006), census data for race and education (Kwok and Yankaskas 2001), and mortality effects associated with (but not necessarily caused by) air pollution (Selvin et al. 1984; Cui et al. 2003). Although a variety of promising proposals for using a combination of individual data and ecologic (e.g., county-level or state-level) data have been advanced recently (Wakefield 2009), neither studies based on purely ecologic data (Selvin et al. 1984), nor more recent multilevel study designs permit valid causal inferences in general (Greenland 2001).
- *Spurious regression.* Causally unrelated time series (including statistically independent random walks) that tend to drift upward or downward over considerable periods can give statistically significant regression coefficients when one is regressed against the other. Time acts much like a confounder in such cases: two series that are drifting over time can be used to help predict each other, since the passage of time is associated with changes in the levels of both. In such cases of "spurious regression" (Chen 2006), focusing on a narrower time window (e.g., years around 2005, for which state-level data are available) might reverse the sign of significant-looking regression coefficients based on a longer time window, simply because random drift can have different directions on different time scales.
- *Variable selection.* Omissions of some specific explanatory variables (such as *Black,* and *Poverty*, which, as previously noted, is not redundant with *Income*) in the original analysis might have unpredictably affected the regression coefficients for the remaining variables (Christenfeld et al. 2004). As illustrated in our simple thought experiment, different selections of other explanatory variables (covariates that the modeler seeks to "control" for) can completely change the regression coefficient(s) of the explanatory variable(s) of principle interest.
- *Residual confounding.* Attempting to "control" for confounders (such as income and poverty) using multiple linear regression, and then interpreting the regression coefficients for the main explanatory variables of interest (e.g., changes in aggre-

Table 7.1 A regression model for IMR 2005 vs. change in cattle production ("*CattleChange*"), controlling for *black*, *education*, and *growth in population*

	B	Std. err.	p-level
Intercept	9.614	0.64	0.000000
Black 2005	0.123	0.01	0.000000
Education 2005	−0.134	0.03	0.000139
CattleChange 2005–2006	−0.005	0.00	0.024912
Growth in population 2000–2006	−0.076	0.03	0.044170

Regression summary for infant mortality rate 2005
R = 0.93953789, R^2 = 0.88273145, Adjusted R^2 = 0.85145983

Table 7.2 A regression model for change in IMR between 2004 and 2005

	B	Std. err.	p-level
Intercept	−12.18	6.25	0.071
Poverty05	0.01	0.04	0.842
Black05	0.14	0.07	0.058
White05	0.13	0.07	0.085
BroilersChange 2005–2006	−0.0029	0.0012	0.029
CattleChange 2005–2006	−0.01	0.00	0.051

Regression summary for change in IMR 2004 to 2005
R = 0.73725, R^2 = 0.54353, Adjusted R^2= 0.38050709, casewise deletion of missing data

gate livestock production) as indicators of causal relations, can give false results. Residual confounding can explain many apparently statistically significant associations arrived at by this procedure (Fewell et al. 2007). As Christenfeld et al. (2004) warn:

- When experimental designs are premature, impractical, or impossible, researchers must rely on statistical methods to adjust for potentially confounding effects. Such procedures, however, are quite fallible. We examine several errors that often follow the use of statistical adjustment. The first is inferring a factor is causal because it predicts an outcome even after 'statistical control' for other factors. This inference is fallacious when (as usual) such control involves removing the linear contribution of imperfectly measured variables, or when some confounders remain unmeasured. …Statistical adjustment serves a useful function, but it cannot transform observational studies into natural experiments, and involves far more subjective judgment than many users realize.

  No state-level or county-level measurements of explanatory variables, such as income and race, fully controls for individual-level (within-county) variations in these variables. Residual confounding by incompletely controlled variables may create apparently significant reduced-form regression coefficients, even in the absence of any true causal relation (Fewell et al. 2007).
- *Differences in time periods and units of analysis.* As previously emphasized, our analysis has used state-level data rather than county-level data, and more recent data (for 2005/2006 instead of years prior to 2000).

Table 7.3 Partial correlations of mortality rates with other variables, controlling for *Income 2005*, *%Black 2005*, *%Hispanic 2005*, *%White 2005*, and *Education 2005*

	Crude death rate 2005	IMR 2004	IMR 2005	Change in IMR, 2004–2005
Infant mortality rate (IMR) 2005	**0.34**[a]	**0.55**[a]	1.00	**0.38**[a]
IMR 2003	0.17	**0.35**[a]	**0.60**[a]	0.21
IMR 2004	−0.08	1.00	**0.55**[a]	**−0.56**[a]
Chronic lower respiratory tract mortality rate 2005	0.72	0.13	**0.53**[a]	**0.39**[a]
Change in IMR 2000–2005	0.21	0.17	**0.49**[a]	**0.31**[a]
IMR 2000	0.10	**0.33**[a]	**0.40**[a]	0.03
Accident mortality rate 2005	0.29	**0.34**[a]	**0.40**[a]	0.02
Change in IMR, 2004–2005	**0.43**[a]	**−0.56**[a]	**0.38**[a]	1.00
Crude death rate 2005	1.00	−0.08	**0.34**[a]	**0.43**[a]
Car accident mortality rate 2005	0.11	**0.36**[a]	0.29	−0.11
Suicide mortality rate 2005	−0.19	**0.30**[a]	0.28	−0.06
Cancer mortality rate 2005	**0.94**[a]	−0.20	0.27	**0.49**[a]
Poverty 2005	**0.49**[a]	−0.14	0.26	**0.41**[a]
Firearms mortality rate 2005	−0.19	**0.34**	0.26	−0.12
Heart mortality rate 2005	**0.87**[a]	−0.21	0.18	**0.41**[a]
Influenza and pneumonia 2005	**0.58**[a]	−0.00	0.09	0.10
BroilersChange 2005–2006 (change in millions of pounds per year)	−0.45	0.44	−0.10	**−0.59**[a]
PigsChange 2005–2006	−0.03	**−0.31**	**−0.31**	0.02
CattleChange 2005–2006	−0.26	**−0.34**	**−0.49**	−0.12
LivestockChange 2005–2006 (sum of pigs, cattle, broilers)	−0.47	−0.22	**−0.55**	**−0.52**[a]

[a]Bold partial correlations are statistically significant at $p < 0.05$

Although these or other methodological factors mentioned in Chap. 2 (e.g., model specification errors) might explain the apparently contradictory findings, we prefer a simpler type of explanation: in an important sense, *there is nothing to explain!* Reduced-form regression models are not causal models, and—as illustrated in our simple thought experiment—the regression coefficient of an exposure variable may be positive in some model specifications and negative in others, without any implications for the causal relation (if any) between exposure and response. Thus, apparent "discrepancies" in the signs of regression coefficients for livestock variables as predictors of IMR (or other mortality) variables should not be distressing: since the coefficients are not causal (structural-equation) coefficients, there is no reason for them to have consistent signs in different reduced-form regression models.

Conclusions

This chapter has focused on methodological caveats about the dangers of applying statistical regression models, instead of traditional QRA models and methods (requiring exposure assessment and exposure-response models), to draw causal conclusions about human health risks from farm animal operations. Although informative causal modeling could be done, and valid causal inferences drawn, using appropriate techniques (such as path analysis and structural equation modeling, more general conditional independence tests and causal graph models (Chap. 9, Shipley 2000; Greenland and Brumback 2002; Ellis and Wong 2008), or risk simulation modeling (Shih et al. 2006)), doing so would require developing and validating a *structural* (or *causal*) model. Simply fitting a reduced-form regression model to data, and interpreting its coefficients causally, does not provide an adequate substitute for QRA.

Attempting to interpret reduced-form coefficients causally has led in this case to provocative policy conclusions, such as that "the new legislative efforts to regulate large-scale livestock farms under the CAA [Clear Air Act] are appropriate" (Sneeringer 2009). Yet, in general, effects of proposed regulatory interventions *cannot* be validly predicted from reduced-form models of historical associations (Lucas 1976; Jacobs et al. 1979). (As a simple analogy, observing a strong positive association between daily aspirin consumption and risk of heart attacks in historical data for elderly citizens would *not* imply that banning aspirin therapy would reduce heart attack risks!) Using regression model coefficients, rather than QRA, to drive policy conclusions, without developing or validating pollutant release models, fate and transport models, exposure models, or dose-response models that empirically causally link animal production operations or emissions to health impacts (e.g., Shih et al. 2006), is a recipe for misguided risk management—that is, for risk management based on an *assumed* causal relation, rather than on a *demonstrated* one. Indeed, as one anonymous reviewer of a draft of this chapter noted, "There is significant—even profound—over-interpretation of the findings by the author of the AJAE publication…. The combination of non-causal analysis and overinterpretation presents a frightening prospect of misdirected decision making and wasted resources." Part 3 of this book suggests that similarly frightening prospects apply at a national level when regulators misinterpret regression coefficients as causal and use them to set regulatory standards for air pollutants.

In summary, before interpreting regression results as suggesting that emissions from livestock operations cause any changes in IMR rates or trends—either adverse changes ("damage to the fetus") or positive changes, as suggested by Fig. 7.1 and Tables 7.1, 7.2 and 7.3—a more thorough examination of relevant emissions and mortality data appears to be well warranted. QRA provides one logical framework for assessing potential impacts of emissions from livestock operations on human health (e.g., Shih et al. 2006), but applying QRA requires collecting and using relevant data on emissions and exposure-response relations. Attempted short-cuts based on reduced-form regression modeling of historical and geographic data on livestock production numbers and IMR is less demanding, and therefore, may look tempting as an alternative to QRA. But such statistical shortcuts suffer from fundamental methodological limitations that prevent clear causal interpretation of their results, making them unsuitable as substitutes for QRA.

Appendix: Data Sources

Main data page	http://www.census.gov/compendia/statab/rankings.html
Violent crime per 100,000, 2004	http://www.census.gov/statab/ranks/rank21.html
Unemployment	http://www.census.gov/compendia/statab/ranks/rank25.html
Infant mortality 2003	http://www.census.gov/compendia/statab/ranks/rank17.xls
Infant mortality rates 2004, 2005	http://www.census.gov/compendia/statab/tables/09s0111.xls
Divorce	http://www.census.gov/compendia/statab/tables/09s0123.xls
Abortions	http://www.census.gov/compendia/statab/tables/09s0099.xls
Income 2005	http://www.census.gov/compendia/statab/ranks/rank29.xls
% Black, Hispanic, White	http://www.census.gov/compendia/statab/ranks/rank05.xls
Poverty 2005	http://www.census.gov/compendia/statab/ranks/rank34.xls
Education 2005	http://www.census.gov/compendia/statab/ranks/rank19.xls
Livestock ($M)	http://www.ors2.state.sc.us/abstract/chapter1/staterank12.php
Pigs&Hogs 2005–2007 (number, production)	http://www.census.gov/compendia/statab/tables/09s0829.xls
Cattle&Calves 2006–2008 (number, production)	http://www.census.gov/compendia/statab/tables/09s0830.xls
Broilers&Turkeys 2005–2007 (production)	http://www.census.gov/compendia/statab/tables/09s0835.xls
Farm acreage	http://www.census.gov/compendia/statab/tables/09s0797.xls
Death rates	http://www.census.gov/compendia/statab/tables/09s0108.xls
Causes of death	http://www.census.gov/compendia/statab/tables/09s0116.xls
Manufacturing employment % 2005	http://www.census.gov/compendia/statab/ranks/rank26.xls

Violent crime 2005	http://www.census.gov/compendia/statab/ranks/rank21.xls
Teacher salaries 2004	http://www.census.gov/compendia/statab/ranks/rank20.xls
Marriage rates	http://www.census.gov/compendia/statab/tables/09s0123.xls

References

Biggs D, de Ville B, Suen E. A method of choosing multiway partitions for classification and decision trees. J Appl Stat. 1991;18(1):49–62.

Brodkin CA, Balmes JR, Redlich CA, Cullen MR. Residential proximity to naturally occurring asbestos: health risk or ecologic fallacy? Am J Respir Crit Care Med. 2006;173(5):573.

Chen W-D. Testing for spurious regression in a panel data model with the individual number and time length growing. J Appl Stat. 2006;33(8):759–72.

Christenfeld NJ, Sloan RP, Carroll D, Greenland S. Risk factors, confounding, and the illusion of statistical control. Psychosom Med. 2004;66(6):868–75.

Cox LA Jr. Some limitations of a proposed linear model for antimicrobial risk management. Risk Anal. 2005;25(6):1327–32.

Cui Y, Zhang ZF, Froines J, Zhao J, Wang H, Yu SZ, Detels R. Air pollution and case fatality of SARS in the People's Republic of China: an ecologic study. Environ Health. 2003;2(1):15.

Ellis B, Wong WH. Learning causal Bayesian Network structures from experimental data. J Am Stat Assoc. 2008;103(482):778–89.

Fewell Z, Davey Smith G, Sterne JA. The impact of residual and unmeasured confounding in epidemiologic studies: a simulation study. Am J Epidemiol. 2007;166(6):646–55.

Greenland S. Ecologic versus individual-level sources of bias in ecologic estimates of contextual health effects. Int J Epidemiol. 2001;30(6):1343–50.

Greenland S, Brumback B. An overview of relations among causal modelling methods. Int J Epidemiol. 2002;31(5):1030–7.

Jacobs RL, Leamer EE, Ward MP. Difficulties with testing for causation. Econ Inq. 1979; 17(3):401–13.

Kwok RK, Yankaskas BC. The use of census data for determining race and education as SES indicators: a validation study. Ann Epidemiol. 2001;11(3):171–7.

Lucas R. Econometric policy evaluation: a critique. Carnegie-Rochester Conf Ser Public Policy. 1976;1:19–46.

Pearl J. Causal inference in statistics: an overview. Stat Surveys. 2009;3:96–146.

Selvin S, Merrill D, Wong L, Sacks ST. Ecologic regression analysis and the study of the influence of air quality on mortality. Environ Health Perspect. 1984;54:333–40.

Shih J-S, Burtraw D, Palmer K, Siikamäki J (2006) Air emissions of ammonia and methane from livestock operations: valuation and policy options. Resources for the future. Discussion paper. Washington, D.C. http://www.rff.org/Documents/RFF-DP-06-11.pdf

Shipley B. Cause and correlation in biology. a user's guide to path analysis, structural equations and causal inference. New York: Cambridge University Press; 2000.

Sneeringer S. Does animal feeding operation pollution hurt public health? A national longitudinal study of health externalities identified by geographic shifts in livestock production. Am J Agric Econ. 2009;91(1):124–37.

Wakefield J. Multi-level modelling, the ecologic fallacy, and hybrid study designs. Int J Epidemiol. 2009;38(2):330–6.

Chapter 8
Causal vs. Spurious Spatial Exposure-Response Associations in Health Risk Analysis

Introduction

This chapter extends to spatial statistics the main theme from Chap. 7: that positive exposure-response coefficients in regression models are not valid substitutes for quantitative risk assessment, because statistical coefficients do not usually reveal causal relationships. Many recent health risk assessments have noted that adverse health outcomes are significantly statistically associated with proximity to suspected sources of health hazard, such as manufacturing plants or point sources of air pollution. Using geographic proximity to sources as surrogates for exposure to (possibly unknown) releases, spatial ecological studies have identified potential adverse health effects based on significant regression coefficients between risk rates and distances from sources in multivariate statistical risk models. Although this procedure has been fruitful in identifying exposure-response associations, the resulting regression coefficients typically lack valid causal interpretations. Spurious spatial regression and other threats to valid causal inference discussed in this chapter undermine practical efforts to causally link health effects to geographic sources, even when there are clear statistical associations between them. This chapter demonstrates the methodological problems by examining statistical associations and regression coefficients between spatially distributed exposure and response variables in a realistic spatial data set. We find that distance from "nonsense" sources (such as arbitrary points or lines) are highly statistically significant predictors of cause-specific risks, such as traffic fatalities and incidence of Kaposi's Sarcoma. However, the signs of such associations typically depend on the distance scale chosen. This is consistent with theoretical analyses showing that random spatial trends (which tend to fluctuate in sign), rather than true causal relations, can create statistically significant regression coefficients: spatial location itself becomes a confounder for spatially distributed exposure and response variables. Hence, extreme caution, and careful application of spatial statistical methods are warranted before interpreting

L. A. Cox Jr., *Quantitative Risk Analysis of Air Pollution Health Effects*,
International Series in Operations Research & Management Science 299,
https://doi.org/10.1007/978-3-030-57358-4_8

proximity-based exposure-response relations as evidence of a possible or probable causal relation.

Spatial Exposure-Response Associations in Risk Assessment

Public health risk assessments often identify potential sources of adverse health effects by identifying a statistically significant positive association between exposure to the potential source of risk—the "hazard"—and prevalence or incidence rates of adverse health effects. Traditionally identified hazards include chemical pollutants, contaminants, and pathogens in air, food, and water; various forms of radiation (such as dental and medical X-rays, radioactivity from radon, and sunlight); and hazardous activities, such as driving a car or riding a bicycle. Recently, advances in spatial statistical modeling have inspired and enabled public health researchers to seek associations between various observed health effects and simply *living in proximity* to a specific facility, to facilities belonging to some group of industries, or to natural features containing hazardous substances (e.g., Goria et al. 2009; López-Abente et al. 2012). Proximity-based risk assessment has generated numerous papers attributing spatial variations in disease rates to (not necessarily specifically identified or observed) emissions from spatial sources. Table 8.1 lists some examples. In this literature, attributions of health effects to spatial sources are typically based simply on indicators of positive statistical associations between health effects and proximity to the source(s) of interest, such as statistically significant positive regression coefficients or relative risk (RR) ratios from logistic or other regression models fit to smoothed spatial data.

Statistically significant associations between exposures and health effects can result from many sources, as discussed in Chap. 2. For example, they might reflect genuine causation, if exposure increases the occurrence rates of adverse health effects. They might reflect false-positive correlations brought about by multiple testing bias (e.g., testing multiple industries, diseases, and sub-populations for associations, without reducing the *p*-levels used in each test); incompletely controlled confounders (e.g., confounders controlled by including categories for age, income, education, and other variables, but without controlling for confounding within each category); data and model selection biases (e.g., fitting a model that is smooth and linear at low doses, without evidence that thresholds or threshold-like nonlinearities do not exist); ignored model uncertainties (e.g., reporting significant results from a best-fitting model, without using Bayesian Model Averaging or other model ensemble methods to address how the results would change if the best-fitting model is incorrect); coincident trends in both exposure and response variables [e.g., caused by common dependence on a third factor or by historical trends in both]; or unmodeled errors in exposure estimates, and so forth (Chap. 2; Cox Jr. 2007).

Determining whether a statistical association between exposure to suspected sources and risk of adverse health effects is causal requires more than judging the

Table 8.1 Examples of spatial associations for various diseases

Source	Effect	Attribution method and reference
Municipal solid waste incinerators in four French administrative departments	Cancer morbidity	GAM and Bayesian hierarchical models of association between cancer risk and an index of estimated exposure derived from dispersion modeling assumptions. "GAM models were fitted. For all the localizations of cancer studied with the exception of lung cancer in men, a nearly linear association between the risk of cancer and the index of exposure to incinerators was observed" (Goria et al. 2009)
Spanish paper and board industries which report their emissions to the European Pollutant Emission Register (EPER)	Lung cancer mortality	This was an ecological study that modelled the Standardised Mortality Ratio (SMR) for lung cancer in 8073 Spanish towns over the period 1994–2003. Population exposure to industrial pollution was estimated on the basis of distance from town of residence to pollution source. An exploratory, near-versus-far analysis was conducted, using mixed Poisson regression models and an analysis of the effect of municipal proximity within a 50-km radius of each of the 18 installations. Results varied for the different facilities. In two instances there was an increasing mortality gradient with proximity to the installation, though this was exclusively observed among men (Monge-Corella et al. 2008)
Spanish industrial installations that are governed by the integrated pollution prevention and control (IPPC) directive and the European pollutant release and transfer Register Regulation	Pleural cancer mortality	Analysis showed statistically significant RRs in both sexes in the vicinity [defined as <2 km] of 7 of the 24 industrial groups studied… These results support that residing in the vicinity of IPPC-registered industries that release pollutants to the air constitutes a risk factor for pleural cancer (López-Abente et al. 2012)
Power plants in Spain	Lung, laryngeal, and bladder cancer	Ecological study designed to model sex-specific standardized mortality ratios… Population exposure to pollution was estimated on the basis of distance from town of residence to pollution source. Using mixed Poisson regression models… excess mortality… was detected in the vicinity [defined as within 5 km] of pre-1990 installations for lung cancer(1.066, 1.041–1.091 in the overall population; 1.084, 1.057–1.111 in men), and laryngeal cancer among men (1.067, 0.992–1.148) (García-Pérez et al. 2010)
Spanish metal industries included in the European Pollutant Emission Register	Leukemia-related mortality	Similar in design to previous paper by the same authors (García-Pérez et al. 2010)

(continued)

Table 8.1 (continued)

Source	Effect	Attribution method and reference
Ultramafic rock deposits	Mesothelioma	Logistic regression analysis from a subset of 1,133 mesothelioma cases and 890 control subjects with pancreatic cancer showed that the odds of mesothelioma decreased approximately 6.3% for every 10 km farther from the nearest ultramafic deposit (a putative source of asbestos), an odds ratio of 0.937 (95% confidence interval = 0.895–0.982), adjusted for age, sex, and occupational exposure to asbestos (Pan et al. 2005)
Spanish metal industries	Prostate cancer mortality	Extended standard Poisson regression model with a multiplicative non-linear function to model effect of distance from an industrial facility, assuming an elevated risk close to the source and not at large distance. Concluded there was "evidence of association between the spatial distribution of prostate cancer mortality aggregated by census tracts and proximity to metal industrial facilities located within the area, after adjusting for socio-demographic characteristics at municipality level" (Ramis et al. 2011)
Air pollution	Children's asthma	The distance from a point source to the subject's residence area represented a surrogate exposure measurement. Odds ratios for a 1-km decrease in distance were obtained by conditional logistic regression. Risk of asthma attack was associated with residing near a grain mill (odds ratio (OR) = 1.35), petroleum refinery (OR = 1.44), asphalt plant (OR = 1.23), or power plant (OR = 1.28) (all p's < 0.05). Residence near major air emissions sources (>100 tons/year) increased asthma attack risk by 108% (p < 0.05). These results showed that proximity to some air pollution sources is associated with increased risks of asthma attacks (Loyo-Berríos et al. 2007)
Unconventional gas development (UGD, "fracking" sites)	Preterm birth	We used logistic or linear regression with generalized estimating equations to assess associations between UGD-activity and preterm birth, small-for-gestational age (SGA), fetal death, or birthweight. … Our results are suggestive of an association between maternal residential proximity to UGD-activity and preterm birth and fetal death" (Whitworth et al. 2017)

strength, consistency, and other aspects of the association. Methodological steps used to test specifically for causality include the following.

- *Generate, test, and, if possible, refute plausible alternative (non-causal) explanations* for a positive association. Typical examples of alternate explanations (illustrated in the previous paragraph) are: multiple testing bias, model selection bias, variable-coding bias (e.g., from dichotomizing continuous variables), data selection bias, unmodeled errors in exposure estimates, unmodeled errors or uncertainties in model form, residual confounding by identified but incompletely controlled confounders, confounding by omitted variables, coincident historical trends, and regression to the mean. Appropriate technical methods for testing these alternative hypotheses using observational data have been well developed in statistics and epidemiology (e.g., Campbell and Stanley 1966; Maclure 1990; Cox Jr. 2007), although they are not always used.
- *Show that the association cannot be explained away by conditioning on other information.* The hypothesized effect of a genuine direct cause (such as exposure to a source) should not be statistically independent of its value, given information about other (e.g., potentially confounding) variables, such as socioeconomic and demographic variables. This is the basis for *conditional independence tests* widely used in causal graph models (Chap. 9, Freedman 2004, Friedman and Goldszmidt 1998). For example, if changes in daily mortality rates are statistically independent of changes in air pollutant concentrations, given the values for minimum daily temperature and month of the year (so that the changes in mortality rates are "conditionally independent" of changes in air pollutant levels, given this other information), then that would undermine the causal hypothesis that changes in air pollutant levels are causal drivers of changes in daily mortality rates.
- If possible, test whether *changes* in responses follow (and can be successfully predicted from) *changes* in individual exposures (Dash and Druzdzel 2008; Druzdzel and Simon 1993), e.g., whether people who move into or out of the vicinity of a suspected geographic source of health hazards subsequently experience corresponding increases or decreases, respectively, in risks of health effects, compared to similar people who did not move.

Table 8.2 summarizes some of the best-known methods for testing for possible causality in observational (non-experimental) studies. These methods are relatively objective, compared to less formal, intuitive causal interpretations of associations (e.g., based on narrative plausibility of a suggested causal explanation), which are notoriously unreliable guides to the truth about causal relations (Gardner 2009). They may help both to critically evaluate past study designs and causal conclusions based on them, and also to improve future studies and causal inferences. Chapter 9 discusses several of these ideas and methods further.

In non-spatial epidemiology, the importance of rigorous testing of causal hypotheses, rather than trusting informal personal or expert judgments and causal interpretations, has been emphasized by recent findings that many published and publicized scientific conclusions about health effects are mistaken. They are usu-

Table 8.2 Methods for testing causal hypotheses

Method and references	Basic idea	Appropriate study design
Conditional independence tests (Freedman 2004; Friedman and Goldszmidt 1998)	Is hypothesized effect statistically independent of other ("ancestor") variables, given values of hypothesized direct causes ("parents") in causal graph model? If so, this strengthens causal interpretation Example: Is dependence of disease risk on location explained by variations in measured exposures at different locations? Is hypothesized effect statistically independent of hypothesized cause, given values of other variables? If so, this undermines causal interpretation Example: Does effect of exposure on disease risk disappear after accounting for education and income?	Cross-sectional data Can also be applied to multi-period data (e.g., in dynamic Bayesian networks)
Panel data analysis (Angrist and Pischke 2009; Stebbings Jr. 1978)	Are changes in exposures followed by changes in the effects that they are hypothesized to help cause? If not, this undermines causal interpretation; if so, this strengthens causal interpretation Example: Are changes in air pollutant levels in different cities followed (but not preceded) by corresponding changes in respiratory mortality rates?	Panel data study: collect a sequence of observations on same subjects or units over time
Granger causality test (Eichler and Didelez 2010)	Does the history of the hypothesized cause improve ability to predict the future of the hypothesized effect? If so, this strengthens causal interpretation; otherwise, it undermines causal interpretation Example: Can daily mortality rates in different cities be predicted better from time series histories of daily pollutant levels and mortality rates than from the time series history of daily mortality rates alone?	Time series data on hypothesized causes and effects
Quasi-experimental design and analysis (Campbell and Stanley 1966)	Can control groups and other comparisons refute alternative (non-causal) explanations for observed associations between hypothesized causes and effects? For example, can coincident trends and regression to the mean be refuted as possible explanations? If so, this strengthens causal interpretation	Longitudinal observational data on subjects exposed and not exposed to interventions that change the hypothesized cause(s) of effects

(continued)

Table 8.2 (continued)

Method and references	Basic idea	Appropriate study design
Intervention analysis, change point analysis (Helfenstein 1991; Gilmour et al. 2006)	Does the best-fitting model of the observed data change significantly at or following the time of an intervention? If so, this strengthens causal interpretation Do the quantitative changes in hypothesized causes predict and explain the subsequently observed quantitative changes in hypothesized effects? If so, this strengthens causal interpretation Example: Do cardiovascular disease mortality rates fall significantly faster in cities that banned smoking in public areas than in cities that did not, with the rate of decrease jumping after the ban in each case?	Time series observations on hypothesized effects, and knowledge of timing of intervention(s) Quantitative time series data for hypothesized causes and effects
Counterfactual and potential outcome models (Robins et al. 2000; Moore et al. 2012)	Do exposed individuals have significantly different response probabilities than they would have had if they had not been exposed? Example: Do inhabitants of a city which bans coal-burning have lower mortality risk after the ban than they would have had without it?	Cross-sectional and/or longitudinal data, with selection biases and feedback among variables allowed
Causal network models of change propagation (Hack et al. 2010; Dash and Druzdzel 2008)	Do changes in exposures (or other causes) create a cascade of changes through a network of causal mechanisms (represented by equations), resulting in changes in the effect variables? Example: Do relatively large variations in daily levels of fine particulate matter (PM2.5) air pollution create corresponding variations in markers of oxidative stress in the lungs?	Observations of variables in a dynamic system out of equilibrium
Negative controls (for exposures or for effects) (Lipsitch et al. 2010)	Do exposures predict health effects better than they predict effects that cannot be caused by exposures? Example: Does being vaccinated against influenza have a stronger association with reduction in influenza hospitalizations than with reduction in trauma hospitalizations? Are hospitalizations reduced more during flu season than before or after flu season? If not, then the association may reflect biases or confounding (e.g., perhaps relatively low-risk patients are more likely to be vaccinated)	Observational studies

ally biased toward false-positive conclusions that significant associations and causation have been found, where more careful analysis would reveal that they have not (Sarewitz 2012; Lehrer 2012; Imberger et al. 2011; Ottenbacher 1998). The problem of false-positive conclusions has grown large enough, especially in the past two decades, to prompt some commentators to warn that science is failing us (largely due to mistakes about causation) (Lehrer 2012) and that most published research findings are wrong (Ioannidis 2005). Despite these growing concerns, tests such as those in Table 8.2 have seldom been applied to studies of risks identified via spatial associations, such as those in Table 8.1.

Substantial theoretical work has shown that spatial data, like temporal data, are subject to non-causal associations. An example is the phenomenon of *spurious regression*, i.e., statistically significant regression coefficients arising from spatial autocorrelation in the explanatory variables (e.g., spatial location), and, independently, in the dependent variable (e.g., illness rates) (Lauridsen and Kosfeld 2011). Just as two statistically independent random walks will tend to have values that are statistically significantly correlated with each other (and with time) over any interval, since such series typically trend either up or down by chance over any interval, so statistically independent (but spatially autocorrelated) spatial random variables will tend to have statistically significant correlations and regression coefficients when values of one (e.g., exposure) are used to predict values of others (e.g., disease rates). This suggests the importance of (re)assessing findings of statistically significant exposure-response relations in regression models applied to spatial data. It is important to determine the extent to which such findings can be explained by spurious (or other non-causal) associations arising from the fact that both the explanatory and the response variables have spatial distributions, even if neither causes the other.

To investigate the importance of non-causal spatial exposure-response associations in practical risk analysis settings, the remainder of this chapter examines associations in a real data set from California. The goal is to determine whether simple regression models exhibit significant regression coefficients even for pairs of exposure variables that are not plausibly causally related, when data are collected from spatial units such as counties or census tracts. This problem is addressed specifically for logistic regression models by Berman et al. (this issue) who also address statistical power and consider a spatial clustering mechanism to explain the observed findings. Here, we examine the effects of realistic spatial correlations, based on California data, without regard for specific mechanisms of clustering and using non-parametric as well as parametric (multiple linear regression) models to understand their implications for regression coefficients.

Data and Methods

To create a realistic spatial data set for exploring associations among spatially distributed variables, we combined census tract level demographic data in California (CSDC 2012) with cancer rate data from the California Cancer Registry (CCR 2012). The age-adjusted cancer incidence rates were available by county (with some counties grouped), year (1988–2009), ethnic group (White, Black, Hispanic, and Asian), and specific cancer type (we used Mesothelioma, Kaposi's Sarcoma, and Pancreatic cancer). Tract level disease incidence counts were computed by multiplying county level incidence rates by year 2000 tract populations for each ethnic group.

Next, we selected as possible sources, to which risks would be attributed via statistical risk models, the locations of deposits of ultramafic rocks (which contain very little quartz or crystalline silica, but often contain asbestos) available from the U.S. Geological Survey (Krevor et al. 2009). Although it is conceivable, if perhaps not very plausible, that such local geological features might have some bearing on some health effects (e.g., Pan et al. 2005), we selected causes of death, such as traffic accident fatalities (NHTSA 2012) and Kaposi's Sarcoma, which have no plausible direct causal relation to the distribution of rock deposits. We then tested whether fitting regression models to these data would show presence or absence of statistically significant regression coefficients for the relation between distance from ultramafic rock and various fatality rates. In additional analyses, we also selected arbitrary point locations (e.g., the centroid of all the census tracts) and tested for significant regression coefficients relating distance from them to health risks. We also applied non-parametric (distance-weighted least squares) regression models to fit smooth curves to effect-vs.-distance data, and used Spearman's Rank Correlation coefficients to quantify significant ordinal relations that are not necessarily linear.

We found it necessary to make adjustments to the cancer data from the California Cancer Registry to account for the fact that low case counts (less than 15) are suppressed in the public online database. We therefore aggregated data over each 5 year interval (with 1999–2004 selected for our final analysis). If values at the county level were still not available for a given county in that time interval, we multiplied the state average value for that time interval and ethnic group by the corresponding ratio of county to state values over all time periods (1988–2009). In the few cases where the latter ratio was not available, we used the state average value for that time interval (and ethnic group) for the county.

The data specifying the locations of ultramafic rocks (Krevor et al. 2009) was contained within ArcGIS format shapefiles, with California deposits defined by 1071 irregular polygon shapes (see Fig. 8.1). Polygon centroids were computed as averages of X,Y values in meters based on the Lambert Conformal Conic Projection. Boundaries and centroids for year 2000 census tracts and counties were obtained from the US Census Bureau (USCB 2012). They were converted from decimal latitude and longitude to the X,Y coordinate system of the ultramafic rocks using the M map MATLAB toolbox (Pawlowicz 2011).

Results

This section summarizes the results of analyses designed to explore whether non-causal regression and associations commonly occur in practice, using the spatial data from California as a case study.

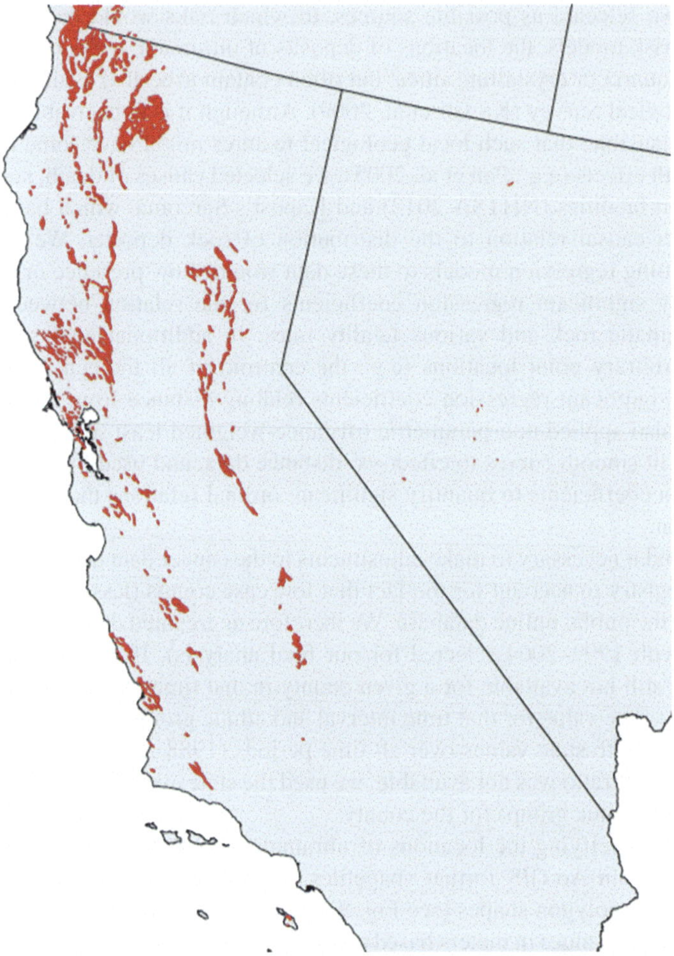

Fig. 8.1 Ultramafic rock deposits in California (Southern part, truncated in this figure, has no deposits)

Fatal Car Accidents and Proximity to Putative Asbestos Sources (Ultramafic Rock Deposits)

Table 8.3 shows the results of a multivariate linear regression model for traffic fatality rates (average annual traffic fatalities per 100,000 people per year (1999–2004)) as a function of distances from the five closest ultramafic rock deposits, UM1-UM5.

UM1, UM2, …, UM5 denote the distances between the centroids of census tracts for which data (e.g., on traffic fatality rates) are collected, and the centroids of the closest, second closest, …, fifth closest ultramafic rock deposits, respectively. The linear regression coefficients (determined by ordinary least squares) appear in the *b*

Table 8.3 Traffic fatality rates depend on distance from ultramafic rock

N = 7049	b*	Std. err. of b*	b	Std. err. of b	t(7043)	p-value
Intercept			92.01	1.54	59.6	0.000000
UM1	−0.04	0.07	−0.09	0.15	−0.6	0.562355
UM2	−1.71	0.21	−3.75	0.45	−8.4	0.000000
UM3	2.19	0.24	4.84	0.52	9.3	0.000000
UM4	0.34	0.13	0.76	0.29	2.7	0.008008
UM5	−0.69	0.02	−0.81	0.03	−29.6	0.000000

Regression summary for dependent variable: TRAFFICALL RATE (2000DataByTractExtended.sta) R = 0.42525247, R^2 = 0.18083966, Adjusted R^2 = 0.18025812 F(5.7043) = 310.97 $p < 0.0000$ std. error of estimate: 59.452

Table 8.4 Multivariate regression model for traffic fatalities within 50 km of ultramafic rock

N = 4942	b*	Std. err. of b*	b	Std. err. of b	t(4937)	p-value	Valid N
PopDensity	−0.43	0.01	−0.00	0.00	−33.1	0.000000	4948
UM1	−0.09	0.01	−0.37	0.05	−7.0	0.000000	4948
Pop65+Pov	0.08	0.01	0.14	0.02	6.4	0.000000	4948
pBlueCollar	0.37	0.01	135.94	4.75	28.6	0.000000	4948
Intercept			83.10	2.18	38.1	0.000000	

Regression summary for dependent variable: TRAFFICALL RATE, UM1 < 50, R = 0.51, R^2 = 0.26, Adjusted R^2 = 0.26

column, and their standardized values in the *b** column; in addition, standard errors are shown for both. Values of *t* tests for these coefficients to be statistically significantly different from zero, and *p* values for these hypothesis tests appear in the two right-most columns of the table, respectively.

Two striking features of Table 8.3 are that most of the distances from ultramafic rock deposits are highly statistically significant predictors of traffic fatalities (with the exception of UM1, the distance to the closest deposit); and the signs of the relations are both positive (for UM3 and UM4) and negative (for UM2 and UM5). Such highly statistically significant, yet inconsistently directed, associations are precisely what might be expected if the regression model is only revealing random trends in the spatial data, where the signs of the trends differ on different distance scales, e.g., due to random fluctuations in the spatially auto-correlated data.

In the regression models shown in Tables 8.3, 8.4, 8.5, 8.6, and 8.7, the usual regression assumption of statistically independent errors in the dependent values is violated because mortality rates for all tracts in the same county are equal to the county values, reducing the degrees of freedom. (This illustrates the importance of checking regression model diagnostics and testing for model specification errors, neither of which is usually reported in studies such as those in Table 8.1.) However, the key point that statistically significant-seeming regression coefficients result from applying standard regression procedures to spatial data, even if the assumptions underlying the procedures are not necessarily well justified, remains true, even if only counties, rather than tracts, are used. For example, U5 remains a statistically

Table 8.5 Multivariate regression model traffic fatalities within 10 km of ultramafic rock

N = 1616	b*	Std. err. of b*	b	Std. err. of b	t(1611)	p-value
Intercept			52.23	4.31	12.1	0.000000
UM1	0.05	0.02	1.26	0.54	2.3	0.019075
pBlueCollar	0.33	0.02	129.60	8.86	14.6	0.000000
Pop65+Pov	0.11	0.02	0.19	0.04	4.7	0.000002
PopDensity	−0.33	0.02	−0.00	0.00	−14.1	0.000000

Regression summary for dependent variable: TRAFFICALL RATE R = 0.48, R^2 = 0.23, Adjusted R^2= 0.23 include condition: UM1 < 10 km

Table 8.6 Relative risk of Kaposi's Sarcoma (relative to pancreatic cancer) decreases with distance from nearest ultramafic rock deposit (UM1), for UM1 <10 km

N = 1616	b*	Std. err. of b*	b	Std. err. of b	t(1611)	p-value
Intercept			0.14	0.01	23.1	0.000000
UM1	-0.27	0.02	-0.01	0.00	-14.4	0.000000
pBlueCollar	-0.20	0.02	-0.13	0.01	-10.5	0.000000
Pop65+Pov	0.10	0.02	0.00	0.00	5.2	0.000000
PopDensity	0.49	0.02	0.00	0.00	25.0	0.000000

Regression summary for dependent variable: KS:PANC ratio R = 0.69, R^2 = 0.47, Adjusted R^2 = 0.475 include condition: UM1 < 10 km

significant predictor of traffic mortality rates ($b^* = -1.45$, $p = 0.0016$) using only county-level data (with $n = 47$ counties). A companion paper (Berman et al. 2013) illustrates this point further for simulated data, without the constraint that all tracts in the same county have the same county-specific mortality rates. Recognizing that p levels are artificially small when county data are imputed to tracts and the tract data values are then treated as independent observations, we will nonetheless use such imputed data to illustrate key points about the limitations of regression modeling of the data. As noted later, key points can also be illustrated (albeit less vividly) using the county-level data.

Suppose now that, in the imputed tract-level data set, a researcher is convinced that proximity to ultramafic rock is a cause of car accidents, and that the closest source of ultramafic rock should exert a strong effect on car fatalities. How might he use the spatial data to derive apparently strong statistical support for these beliefs? Table 8.4 shows the result of fitting a multivariate linear regression model to all data from tracts with centroids within 50 km of an ultramafic rock deposit. Within this radius, a highly statistically significant *negative* regression coefficient relates UM1 to traffic fatalities, suggesting that, the closer a driver is to the nearest source of ultramafic rock, the higher is his risk of a car accident fatality, just as the researcher expected. The other variables in the model include a measure of poverty ("Pop65+Pov" refers to the fraction of the elderly (over 65) population below the poverty line), population density ("PopDensity," which helps to indicate urban-rural distinctions), and proportion of population with blue collar jobs (pBlueCollar).

Table 8.7 Relative risk of Kaposi's Sarcoma (relative to pancreatic cancer) increases with distance from nearest ultramafic rock deposit (UM1)

N = 7049	b*	Std. err. of b*	b	Std. err. of b	t(7044)	p-value
Intercept			0.08	0.00	43.1	0.000000
UM1	0.11	0.01	0.00	0.00	10.3	0.000000
pBlueCollar	−0.23	0.01	−0.08	0.00	−21.0	0.000000
Pop65+Pov	0.06	0.01	0.00	0.00	5.5	0.000000
PopDensity	0.51	0.01	0.00	0.00	47.3	0.000000

Regression summary for dependent variable: KS:PANC ratio $R = 0.51$, $R^2 = 0.26$, Adjusted $R^2 = 0.26$

These socioeconomic factors are also highly significant predictors of traffic fatality rates. The purpose of this model is not to develop in detail a model for predicting traffic fatality risks; if it were, many other variables, such as median household income, could be added to further improve model fit. Rather, it is to show that it is easy to find multivariate models in which the regression coefficient relating UM1 to traffic accident fatality risks is significantly negative. Table 8.4 illustrates one such model, and others can easily be generated by including other variables as predictors.

On the other hand, suppose that a different researcher feels equally passionately that proximity to ultramafic rock is beneficial, and that it significantly reduces the risk of car accidents. How might she use the same spatial data to support this opposite conclusion? Table 8.5 shows the result of fitting a different multivariate linear regression model (for the same variables as in Table 8.2) to all data from tracts with centroids within 10 km of an ultramafic rock deposit. Within this narrower radius, a statistically significant *positive* regression coefficient relates UM1 to traffic fatality rates, suggesting that, the closer a driver is to the nearest source of ultramafic rock, the *lower* is his risk of a car accident fatality. Again, the researcher's prior expectations appear to be well supported by the data.

Ability to support any desired conclusion (including contradictory ones) by choosing radii of different sizes for the spatial data to include in the analysis appears analogous to the ability, in time series analysis of autocorrelated processes with unit roots (e.g., random walks), to obtain either positive or negative trends, simply by adjusting the length of the interval analyzed (e.g., due to the arcsine law for random walks) (www.encyclopediaofmath.org/index.php/Arcsine_law).

Figure 8.2 fits a nonparametric smooth regression curve (using a distance-weighted least squares (local polynomial regression) smoothing algorithm, Statistica 2012) to the proximity-traffic fatality data. The result is a smooth, but clearly non-monotonic, exposure-response relation (using proximity to ultramafic rock as a surrogate for exposure). A parametric (e.g., linear or logistic regression model), which assumes a single coefficient relating UM1 to traffic fatality rates, can produce either a positive average slope or a negative one, depending on what part of the full range of data is included.

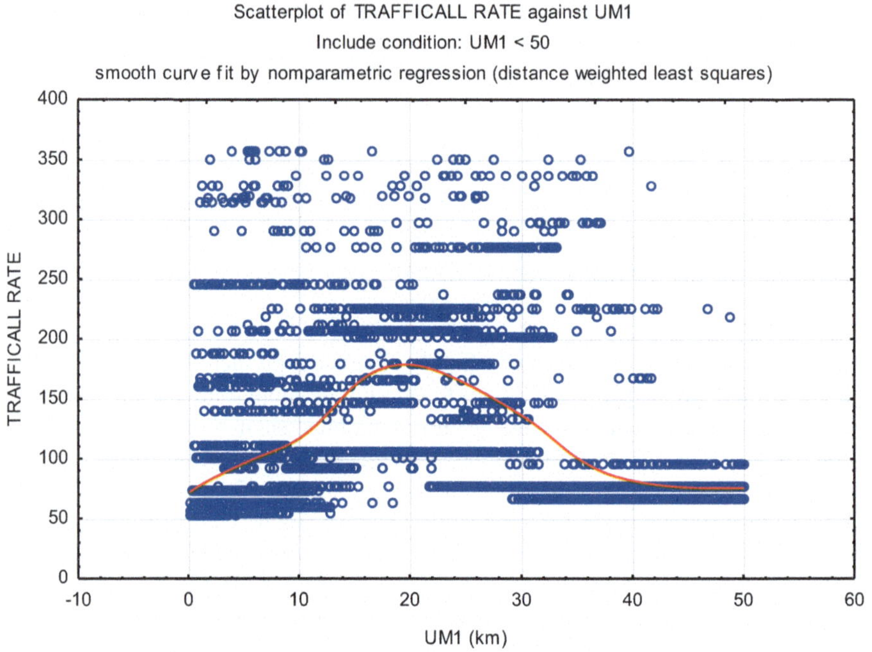

Fig. 8.2 Nonparametric regression of traffic fatality rates vs. UM1

Kaposi's Sarcoma and Proximity to Ultramafic Rock Deposits

What is true for car accidents is also true for other causes of death: *proximity to ultramafic rock is a strong predictor for cause-specific disease incidence.* For example, Table 8.6 shows the results of a multivariate linear regression model for incidence of Kaposi's Sarcoma, a tumor caused by human herpesvirus. Figure 8.3 shows a smooth exposure-response curve relating risk of Kaposi's Sarcoma to proximity to the closest ultramafic rock deposit (UM1). Tables 8.6 and 8.7, inspired by the analysis of Pan et al. (2005), treat the ratio of Kaposi's Sarcoma incidence counts to pancreatic cancer incidence counts as the dependent variable. Again, one can conclude that relative risk of Kaposi' Sarcoma is significantly positively or significantly negatively associated with distance to ultramafic rock (UM1), by selecting an appropriate radius (e.g., infinite or 10 km, respectively). (For county-level data, relative risk of Kaposi' Sarcoma is significantly negatively associated with UM1 if the radius is 30 km.) Again, the nonparametric regression relation (Figure 8.3) explains this by the fact that there are different average trends on different distance scales—all highly statistically significantly different from zero, for the imputed tract-level data, but with opposite signs at different distances. A curve very similar in shape to Fig. 8.3 describes the absolute risk of Kaposi's Sarcoma, as opposed to its ratio to pancreatic cancer (and absolute risk of Kaposi' Sarcoma is

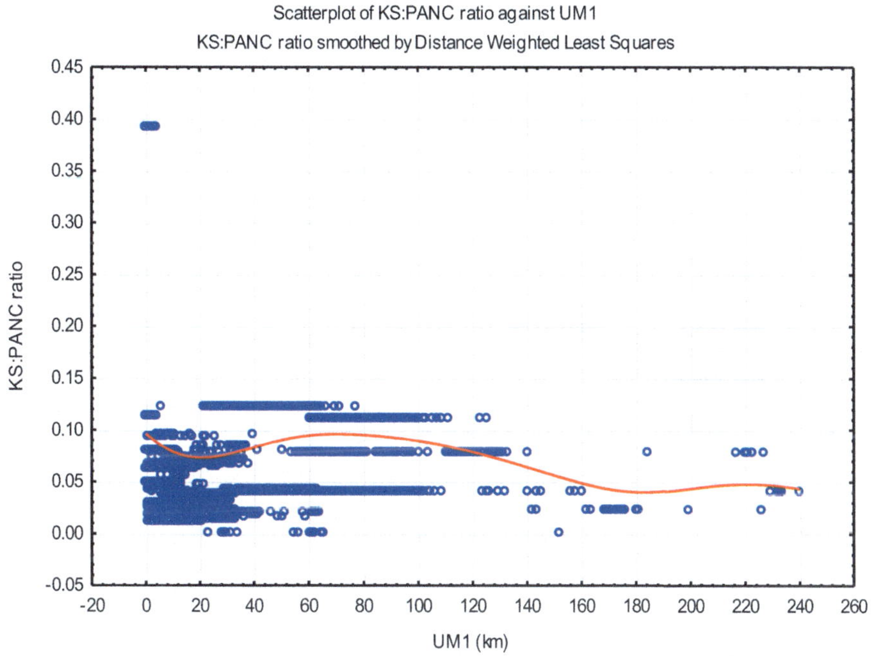

Fig. 8.3 Nonparametric regression of Kaposi's Sarcoma-to-pancreatic cancer incidence counts ratio vs. UM1 shows different associations at different distances

also significantly negatively associated with UM1 in county-level data if the radius is 30 km.).

Mesothelioma and Proximity to Ultramafic Rock Deposits

Figures 8.4 and 8.5 plot absolute and relative (to pancreatic cancer) mesothelioma incidence risks against distance from ultramafic rock deposits (UM1). Broadly consistent with the observations of Pan et al. (2005) (but without multivariate adjustments), both absolute and relative mesothelioma risks increase with proximity to ultramafic rock on some distance scales. However, the sign of the association reverses on larger distance scales, similar to associations for Kaposi's Sarcoma. Hence, fitting any model with a single regression coefficient to relate UM1 to risk would be misleading, as the sign of their association varies with distance. Spearman's rank correlation coefficients show that UM1 is significantly positively associated with mesothelioma rate (MESOALL), traffic fatality rate (TRAFICALL RATE) and relative risk of Kaposi's Sarcoma (relative to pancreatic cancer), with about equal strengths (Spearman's rank correlation between 0.27 and 0.28) for all three. However, Figs. 8.4 and 8.5 show that the signs of these associations simply reflect

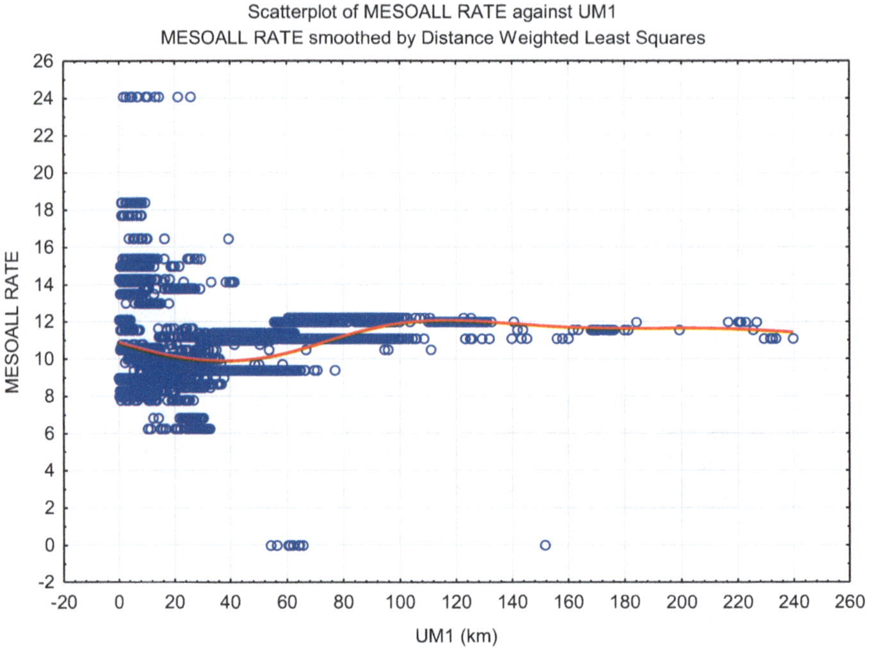

Fig. 8.4 Nonparametric regression of mesothelioma rate vs. UM1 shows different associations at different distances

where most people live (i.e., at distances with positive rather than negative associations), rather than any underlying consistent positive association. Together, Figs. 8.2, 8.3, 8.4, and 8.5 show that positive or negative statistical associations of health endpoints with distance from sites reflect where people live and the distance scales chosen for the analysis, rather than consistent underlying positive or negative effects.

Results for Point Sources

The preceding results use distance to ultramafic rock as an explanatory variable. In California, there are multiple deposits of ultramafic rock, running roughly in two main north-to-south stripes (see Fig. 8.2). Thus, the distance to UM can be thought of as (roughly) the distance to the nearer of two line sources on the scale of the state. But many of the studies in Table 8.1 are for point sources. It is therefore interesting to examine the dependence of other variables on distance from an arbitrary point source. Tables 8.8 and 8.9 choose the centroid of all the census tracts (median latitude and median longitude converted to X, Y) as such a point, and show that distance from this centroid is a statistically significant predictor of both traffic fatality

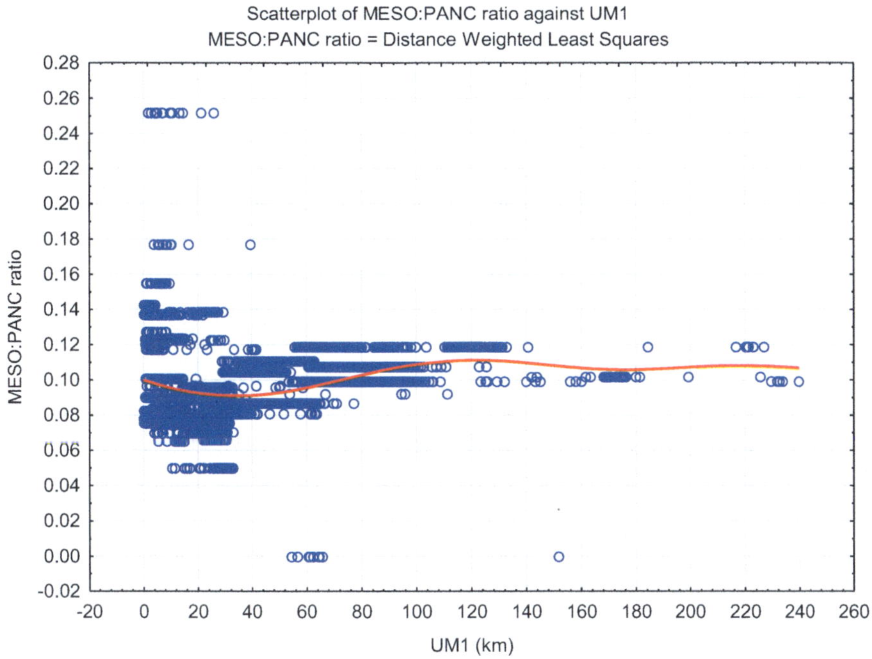

Fig. 8.5 Nonparametric regression of Mesothelioma-to-Pancreatic Cancer incidence count ratio vs. UM1 shows different associations at different distances

Table 8.8 Multivariate model for traffic fatalities vs. distance from centroid of data

N = 7049	b*	Std. err. of b*	b	Std. err. of b	t(7043)	p-value
Intercept			59.04	2.67	22.1	0.000000
UM1	0.08	0.01	0.18	0.02	8.1	0.000000
pBlueCollar	0.35	0.01	129.39	4.03	32.1	0.000000
Pop65+Pov	0.08	0.01	0.15	0.02	7.9	0.000000
PopDensity	−0.42	0.01	−0.00	0.00	−38.8	0.000000
distance from centroid	0.09	0.01	5.40	0.64	8.5	0.000000

Regression summary for dependent variable: TRAFFICALL RATE $R = 0.51$, $R^2 = 0.26$ Adjusted $R^2 = 0.26$

and Kaposi's Sarcoma risk, as well as mesothelioma risk, in the population (Table 8.10). The same is true for distance from other arbitrary points, selected at random; and for the ratio of mesothelioma to pancreatic cancer cases. Plotting risk vs. distance from a point again yields non-monotonic curves, with different average slopes for different distance scales, similar to what we have already seen for distance from ultramafic rock (e.g., Fig. 8.3).

These findings suggest that discovering a statistically significant regression coefficient linking a health risk variable to proximity to a source (or to multiple sources, as in the ultramafic rock example) should not, by itself, be taken as evidence that the source(s) probably cause or contribute to (or reduce) the health risk.

Table 8.9 Multivariate model for Kaposi's sarcoma vs. distance from centroid of data

N = 7049	b*	Std. err. of b*	b	Std. err. of b	t(7043)	p-value
Intercept			7.10	0.29	24.5	0.000000
UM1	0.10	0.01	0.02	0.00	10.0	0.000000
pBlueCollar	−0.22	0.01	−8.91	0.44	−20.4	0.000000
Pop65+Pov	0.06	0.01	0.01	0.00	5.5	0.000000
PopDensity	0.51	0.01	0.00	0.00	47.2	0.000000
distance from centroid	0.06	0.01	0.39	0.07	5.6	0.000000

Regression summary for dependent variable: KSALL RATE R = 0.51, R^2 = 0.26, Adjusted R^2 = 0.26

Table 8.10 Multivariate model for mesothelioma vs. distance from centroid of data

N = 7038	b*	Std. err. of b*	b	Std. err. of b	t(7032)	p-value
Intercept			8.85	0.09	103.2	0.000000
UM1	0.15	0.01	0.01	0.00	13.2	0.000000
pBlueCollar	0.04	0.01	0.46	0.13	3.5	0.000423
Pop65+Pov	−0.03	0.01	−0.00	0.00	−2.6	0.008805
PopDensity	−0.16	0.01	−0.00	0.00	−14.0	0.000000
distance from centroid	0.29	0.01	0.53	0.02	25.7	0.000000

Regression summary for dependent variable: MESOALL RATE R = 0.37, R^2 = 0.14, Adjusted R^2 = 0.14

Theoretical Interpretation: What Does a Spatial Regression Coefficient Show?

The preceding results raise the questions of what *does* explain observed distance-risk associations, and what the regression coefficients between spatially distributed variables *do* signify, if not causation. Figure 8.6 suggests a possible answer. Geographic contours of approximately equal risks for traffic fatalities, here interpolated by fitting a smooth (spline) surface to data on traffic fatalities in tracts with different latitudes and longitudes (for the tract centroids), show that risks tend to increase toward the northeast. (Only the risks contours that fall in the data cloud are empirically driven: the much higher risks to the extreme northeast are just extrapolations of the spatial trends from the limited data. Although more detailed spatial resolution of the data might give different surfaces and contours, our argument is general.) Given such a spatial distribution of risks, choosing *any* line that runs diagonally through the state from its northwest to its southeast—whether of ultramafic rock or something else—will create a line source with the property that populations to the left and below (southwest of) the line will tend to have smaller risks than populations above and to the right (northeast) of it. If the population on one side of the line is much bigger than the population on the other side of it, then average risk will change with proximity to the line, simply reflecting the fact that risk generally decreases from the northeast to the southwest, even if the line itself (or sources located along it) have no effect on risk. Similarly, if one selects a point at random,

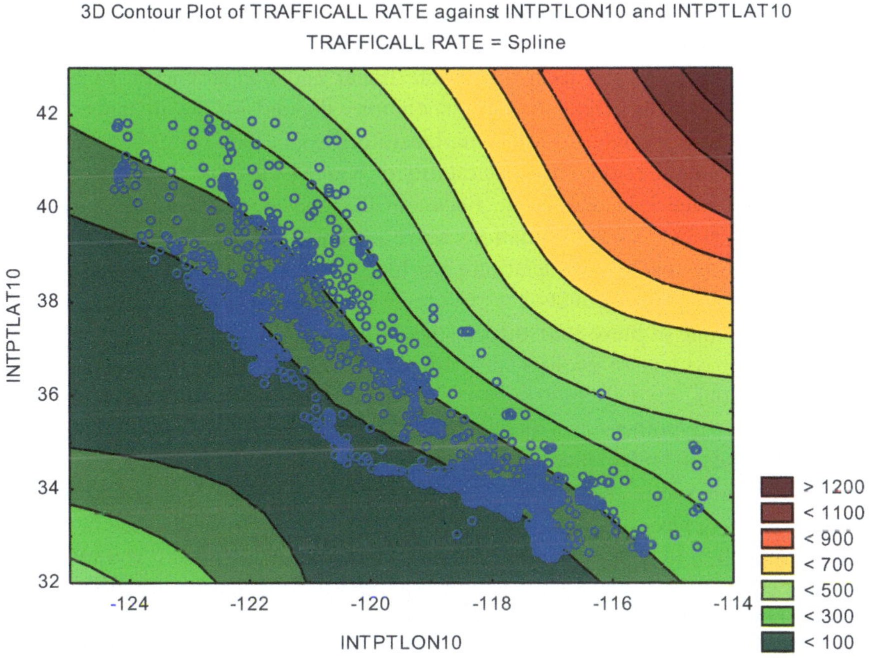

Fig. 8.6 Traffic fatality rates vs. latitude and longitude for 7049 California Census tracts (contour view)

and if population is not symmetrically distributed around it, then people living near the point may tend to be on a higher or lower contour (depending on the location of the point) than most of the population that lives far from it. (To see this, pick a point at random in Fig. 8.6, and confirm that near neighbors will, in most cases, live on different risk contours from those who live far away.) Therefore, what a regression coefficient of risk vs. distance indicates is not necessarily a causal effect of distance on risk, but rather a reflection of the spatial distribution of the population over risk contours. Thus, the results of studies such as that of Pan et al. (2005) and others in Table 8.1, which present regression coefficients without modeling the non-causal effects of such spatial distributions of populations, do not provide a reliable basis for drawing causal inferences.

Discussion and Conclusions

This chapter has examined whether non-causal spatial regression relations arise in realistic data sets, when proximity to ultramafic rock is used as an exposure surrogate, and various outcomes (traffic fatality rate, absolute risk of Kaposi's Sarcoma, and relative risk of Kaposi's Sarcoma and mesothelioma (both compared to pancreatic cancer)) are treated as response variables. The main finding is that statistically significant associations can readily be found in multivariate linear regression models (as well as in univariate linear and nonparametric regression analyses), although the signs of the associations depend on the distance scales considered. These include causally implausible associations (e.g., between distance from ultramafic rock and traffic fatality or Kaposi's sarcoma risks) as well as more causally plausible ones (e.g., between distance from ultramafic rock and mesothelioma risks). Such associations serve as negative controls (Lipsitch et al. 2010), indicating that the associations are likely to reflect bias or confounding, rather than purely causal effects.

Of course, none of these statistical associations should be interpreted as demonstrating that proximity to geological features either causes or protects against car accident fatalities, sexually transmitted diseases, or other health outcomes. A plausible alternative hypothesis is that *spatial location acts as a confounder* when there are spatial trends in both explanatory and dependent variables. Such trends can arise even in random spatial autoregressive processes (Lauridsen and Kosfeld 2011), although here we have focused on real-world trends arising from the specific population distribution in California. Such spatial trends in both exposure and response variables create highly statistically significant exposure-response regression coefficients, whether or not there is any causal relation between them.

For spatial ecological studies such as those in Table 8.1, traditional methodological concerns about ecological associations also apply. For example, multiple testing and confirmation biases (Sarewitz 2012) can produce excess false positives if investigators must make numerous modeling choices (e.g., of which confounders, covariates, and interactions to include in the model; what model forms to specify; how to treat exposure uncertainties, and so forth). Such concerns are addressed in a companion paper (Berman et al. 2013) via simulation modeling, where the underlying data-generating process is known exactly. Yet, the types of phenomena noted here (non-causal associations reflecting the underlying spatial distributions of variables) still arise. That even highly significant regression coefficients for health effects vs. proximity to sources may simply reflect the spatial distribution of a population, rather than an effect of proximity on health effects, adds to these traditional concerns. In addition, future experimental investigation might greatly clarify causality, e.g., by revealing whether measured concentrations in air of mineral products from ultramafic rock deposits are capable of causing changes in the lung environment (e.g., inflammation, increases in oxidative stress) of potential importance in the etiology of mesothelioma (e.g., Carbone and Yang 2012). This chapter has focused on causal interpretation of regression coefficients in observational (and ecological) data. Experimental and toxicological investigations could potentially provide a different basis for understanding whether causal interpretations of such associations is likely to be warranted. Meanwhile, our findings suggest that such additional insights would be valuable before causal interpretations of associations are acted on, since statistically significant associations commonly occur in regression analyses of observational data with spatially distributed variables, even when there are no known plausible causal relations between exposure and effect variables.

These findings suggest that extreme caution is needed in interpreting the conclusions from studies such as those in Table 8.1. Unless it is demonstrated that proximity-response associations are *not* explained by non-causal artifacts of spatial regression (e.g., spurious regression), using appropriate spatial statistical tests (e.g., Lauridsen and Kosfeld 2011) and also that they are not due to model-specification errors (e.g., attempting to fit regression models with constant coefficients to relations that are non-monotonic), the presence of positive regression coefficients should not be interpreted as providing any evidence for a true exposure-response relation. Our findings (as well as those in Berman et al. 2013) suggest that such regression relations may be ubiquitous in realistic spatial epidemiological data, even when no causal interpretation is plausible. Additional analyses, analogous to those in Table 8.2, are therefore likely to become recognized as essential preludes to valid causal interpretation of spatial exposure-response associations. Chapter 9 discusses principles for valid causal inference in greater detail.

References

Angrist JD, Pischke J-S. Mostly harmless econometrics: an empiricist's companion. Princeton, NJ: Princeton University Press; 2009.

Berman DW, Cox T, Popken D. A cautionary tale: the characteristics of two-dimensional distributions and their effects on epidemiological studies employing an ecological design. Crit Rev Toxicol. 2013,43(Suppl 1):1–25.

Campbell DT, Stanley JC. Experimental and quasi-experimental designs for research. Chicago: Rand McNally; 1966.

Carbone M, Yang H. Molecular pathways: targeting mechanisms of asbestos and erionite carcinogenesis in mesothelioma. Clin Cancer Res. 2012;18(3):598–604.

CCR (2012) Cancer incidence/mortality rates in California. http://www.cancer-rates.info/ca. Accessed 5 Jan 2012

Cox LA Jr. Regulatory false positives: true, false, or uncertain? Risk Anal. 2007;27(5):1083–6.

CSDC (2012) Selected census 1990 and 2000 population and housing estimates for tract 2000. California Dept. of Finance, Demographic Research Unit, California State Data Center. http://www.dof.ca.gov/research/demographic/state_census_data_center/products-services/. Accessed 5 Jan 2012

Dash D, Druzdzel MJ. A note on the correctness of the causal ordering algorithm. Artif Intell. 2008;172:1800–8.

Druzdzel MJ, Simon HA. Causality in Bayesian belief networks. In: Proceedings of the ninth annual conference on uncertainty in artificial intelligence (UAI-93). San Francisco: Morgan Kaufmann Publishers, Inc.; 1993. p. 3–11.

Eichler M, Didelez V. On Granger causality and the effect of interventions in time series. Lifetime Data Anal. 2010;16(1):3–32.

Freedman DA. Graphical models for causation, and the identification problem. Eval Rev. 2004;28(4):267–93.

Friedman N, Goldszmidt M. Learning Bayesian networks with local structure. In: Jordan MI, editor. Learning in graphical models. Cambridge: MIT Press; 1998. p. 421–59.

García-Pérez J, López-Cima MF, Boldo E, Fernández-Navarro P, Aragonés N, Pollán M, Pérez-Gómez B, López-Abente G. Leukemia-related mortality in towns lying in the vicinity of metal production andprocessing installations. Environ Int. 2010;36(7):746–53.

Gardner D. The science of fear: how the culture of fear manipulates your brain. New York: Penguin Group; 2009.

Gilmour S, Degenhardt L, Hall W, Day C. Using intervention time series analyses to assess the effects of imperfectly identifiable natural events: a general method and example. BMC Med Res Methodol. 2006;6:16.

Goria S, Daniau C, de Crouy-Chanel P, Empereur-Bissonnet P, Fabre P, Colonna M, Duboudin C, Viel JF, Richardson S. Risk of cancer in the vicinity of municipal solid waste incinerators: importance of using a flexible modelling strategy. Int J Health Geogr. 2009;8:31.

Hack CE, Haber LT, Maier A, Shulte P, Fowler B, Lotz WG, Savage RE Jr. A Bayesian network model for biomarker-based dose response. Risk Anal. 2010;30(7):1037–51.

Helfenstein U. The use of transfer function models, intervention analysis and related time series methods in epidemiology. Int J Epidemiol. 1991;20(3):808–15.

Imberger G, Vejlby AD, Hansen SB, Møller AM, Wetterslev J. Statistical multiplicity in systematic reviews of anaesthesia interventions: a quantification and comparison between cochrane and non-cochrane reviews. PLoS ONE. 2011;6(12):e28422. https://doi.org/10.1371/journal.pone.0028422.

Ioannidis JPA. Why most published research findings are false. PLoS Med. 2005;2(8):e124. https://doi.org/10.1371/journal.pmed.0020124.

Krevor SC, Graves CR, Van Gosen BS, McCafferty AE (2009) Mapping the mineral resource base for mineral carbon-dioxide sequestration in the conterminous United States: U.S. Geological Survey Digital Data Series, p. 414. http://pubs.usgs.gov/ds/414/

Lauridsen J, Kosfeld R. Spurious spatial regression and heteroscedasticity. J Spat Sci. 2011; 56(1):59–72.

Lehrer J (2012) Trials and errors: why science is failing us. Wired. http://www.wired.co.uk/magazine/archive/2012/02/features/trials-and-errors?page=all

Lipsitch M, Tchetgen Tchetgen E, Cohen T. Negative controls: a tool for detecting confounding and bias in observational studies. Epidemiology. 2010;21(3):383–8.

López-Abente G, Fernández-Navarro P, Boldo E, Ramis R, García-Pérez J. Industrial pollution and pleural cancer mortality in Spain. Sci Total Environ. 2012;424:57–62.

Loyo-Berríos NI, Irizarry R, Hennessey JG, Tao XG, Matanoski G. Air pollution sources and childhood asthma attacks in Catano, Puerto Rico. Am J Epidemiol. 2007;165(8):927–35.

Maclure M. Multivariate refutation of aetiological hypotheses in non-experimental epidemiology. Int J Epidemiol. 1990;19(4):782–7.

Monge-Corella S, García-Pérez J, Aragonés N, Pollán M, Pérez-Gómez B, López-Abente G. Lung cancer mortality in towns near paper, pulp and board industries in Spain: a point source pollution study. BMC Public Health. 2008;8:288.

Moore KL, Neugebauer R, van der Laan MJ, Tager IB. Causal inference in epidemiological studies with strong confounding. Stat Med. 2012;31(13):1380–404. https://doi.org/10.1002/sim.4469.

NHTSA (2012) Fatality analysis reporting system. http://www-fars.nhtsa.dot.gov/Main/index.aspx. Accessed 5 Jan 2012

Ottenbacher KJ. Quantitative evaluation of multiplicity in epidemiology and public health research. Am J Epidemiol. 1998;147:615–9.

Pan XL, Day HW, Wang W, Beckett LA, Schenker MB. Residential proximity to naturally occurring asbestos and mesothelioma risk in California. Am J Respir Crit Care Med. 2005; 172(8):1019–25.

Pawlowicz R (2011) M_Map: a Mapping Package for MATLAB. http://www2.ocgy.ubc.ca/~rich/map.html. Retrieved 5 Jan 2012

Ramis R, Diggle P, Cambra K, López-Abente G. Prostate cancer and industrial pollution: risk around putative focus in a multi-source scenario. Environ Int. 2011;37(3):577–85.

Robins JM, Hernán MA, Brumback B. Marginal structural models and causal inference in epidemiology. Epidemiology. 2000;11(5):550–60.

Sarewitz D. Beware the creeping cracks of bias. Nature. 2012;485:149.

Statistica (2012) Statistica 10 documentation. http://documentation.statsoft.com/STATISTICAHelp.aspx?path=Graphs/Graph/ModifyingGraphs/Notes/DistanceWeightedLeastSquaresFitting

Stebbings JH Jr. Panel studies of acute health effects of air pollution. II. A methodologic study of linear regression analysis of asthma panel data. Environ Res. 1978;17(1):10–32.

USCB (2012) Census 2000 TIGER/line files. http://www.census.gov/geo/www/tiger/tiger2k/tgr2000.html. Accessed 5 Jan 2012

Whitworth KW, Marshall AK, Symanski E. Maternal residential proximity to unconventional gas development and perinatal outcomes among a diverse urban population in Texas. PLoS One. 2017;12(7):e0180966.

Chapter 9
Methods of Causal Analysis for Health Risk Assessment with Observational Data

Introduction

Perhaps no other topic in risk analysis is more difficult, more controversial, or more important to risk management policy analysts and decision-makers than how to draw valid, correctly qualified conclusions from observational data. Chapters 1, 2, 7, and 8 have warned against the common practice of using statistical regression models in place of causal analysis (Pearl 2014), and have suggested some alternatives, including causal analysis and Bayesian network (BN) modeling. This chapter examines these methods in greater detail. It is a more technically dense chapter than others in this book, but the required effort pays substantial dividends in methods for more clearly assessing evidence of causality in exposure concentration-response (C-R) functions, as illustrated in the chapters that follow. Readers who are content to accept that "correlation is not causation" and who are satisfied with the high level descriptions of Bayesian networks and causal analysis given in Chaps. 1 and 2 can skip to Chap. 10.

The starting point for this chapter is that statistical models alone are not adequate to draw valid causal conclusions (Pearl 2014), as emphasized in Chaps. 7 and 8; but neither is it necessary to simply make untested assumptions to draw causal conclusions from data. Modern causal analysis and modeling methods provide a broad middle ground in which data are used to scrutinize the testable implications of causal claims, and causal models and predictions are constrained to be consistent with, and supported by, data that shows their predictive validity when tested. As emphasized in Chap. 1, this possibility of testing predictive generalizations against data not used in deriving them is a defining principle of sound science. Statistical methods can readily quantify associations between observed variables using measures such as relative risk (RR) ratios, odds ratios (OR), slope coefficients for exposure or treatment variables in regression models, and quantities derived from these measures. Textbooks of epidemiology explain how to calculate population

L. A. Cox Jr., *Quantitative Risk Analysis of Air Pollution Health Effects*, International Series in Operations Research & Management Science 299,
https://doi.org/10.1007/978-3-030-57358-4_9

attributable fractions, attributable risks, burden-of-disease estimates, and probabilities of causation from relative risk (RR) ratios. Despite their suggestive names, these association-based measures have no necessary connection to causation if the associations on which they are based arise from bias, confounding, p-hacking, coincident historical trends, or other non-causal sources. But policy analysts and decision makers need something more: trustworthy predictions—and, later, evaluations—of the changes in outcomes *caused* by changes in policy variables. This concept of *manipulative causation* (or *interventional* causation) differs from the more familiar concepts of associational and attributive causation most widely used in epidemiology. Drawing on modern literature on causal discovery and inference principles and algorithms for drawing limited but useful causal conclusions from observational data, this chapter proposes seven criteria for assessing consistency of data with a manipulative causal exposure-response relationship—mutual information, directed dependence, internal and external consistency, coherent causal explanation of biological plausibility, causal mediation confirmation, and refutation of no-causal explanations—and discuss to what extent it is now possible to automate discovery of manipulative causal dependencies and quantification of causal effects from observational data. We compare our proposed principles for causal discovery and inference to the traditional Bradford Hill considerations from 1965. Understanding how old and new principles are related can clarify and enrich both.

Science-Policy Setting: Bradford-Hill Considerations, False Causal Conclusions, and Reproducibility Challenges

Most health risk-related causal claims that garner headlines today, including frequent reports of adverse health effects being "linked" to various substances (see Chap. 19), reflect underlying subjective judgments and weight-of-evidence (WoE) determinations about causality based on statistical associations in selected study populations and analytic models. A recent study concluded that more than half of these associations turn out to be unsupported upon further investigation and meta-analyses, although such negative findings are seldom reported: "Whereas 234 newspaper articles covered the 35 initial studies that were later disconfirmed, only four press articles covered a subsequent null finding and mentioned the refutation of an initial claim" (Dumas-Mallet et al. 2017). Reported causal interpretations of associations are often justified as consensus judgments of selected scientific experts and authoritative bodies based on considerations such as whether observed exposure-response associations are judged to be strong, consistent, and biologically plausible. The left column of Table 9.1 provides a fuller list of these Bradford-Hill considerations. They are commonly used in WoE-based decision processes to structure and support such judgments about causality.

This approach sprang largely from an influential essay by Sir Austin Bradford Hill (1965). The considerations on the left side of Table 9.1 are often referred to as the "Hill criteria," although Hill referred to them as viewpoints, writing that "What

Table 9.1 Bradford-Hill and causal discovery principles

Bradford-Hill considerations	Modern causal discovery and inference principles
Strength of association: stronger associations are more likely to be causal	• *Mutual information principle*: causes are informative about their direct effects and help to predict their values • *Conditional independence version*: effects are not conditionally independent of their direct causes • Direct causes contain at least as much information about their effects as do more remote indirect causes
Consistency of findings across populations, study designs, times, locations, investigators, etc.	• *Internal consistency*: similar effects are estimated using different adjustment sets • *External consistency*: causal laws, expressed as conditional probability tables (CPTs), are invariant, homogeneous, and transportable across study settings
Temporality: causes precede their effects	• *Directed dependence principle*: information flows from causes to their effects over time • *Predictive causation principle*: changes in causes help to predict changes in their effects • Techniques: Granger causality tests, transfer entropy, directed information graphs, dynamic Bayesian networks • LiNGAM principle for linear non-Gaussian models: prediction error distributions vary simply and predictably with predicted values
Biological plausibility: exposure-response association has a plausible biological mechanism *Coherence*: agrees with knowledge of disease biology	• *Coherent structural causation*: effects are derived from their causes. Directed paths lead from causes (e.g., doses) to their effects (e.g., responses) in a causal graph • *Coherent causal explanation*: a valid causal graph explains the observed data • Exposures affect response probabilities via pathways (sequences of mechanisms in a causal biological graph)
Analogy: similar causes have similar effects	• Similar causes have similar effects if they act via the same mechanisms • Deep learning and automatic abstraction induce general rules from specific instances
Experiment: reducing exposure reduces effect	• Exogenous changes in causes produce predictable changes in the probability distributions of their effects that can be calculated via CPTs
Specificity of effects: a specific cause produces a specific effect	• Connectivity: one or more directed paths in a causal graph lead from causes to their effects • Direct effects of a cause are its children in a causal graph
Biological gradient: larger responses at higher exposures	• Variations in direct causes help to predict and explain variations in their (possibly joint) effects via a CPT

I do not believe—and this has been suggested—[is] that we can usefully lay down some hard-and-fast rules of evidence that must be obeyed before we can accept cause and effect. None of my nine viewpoints can bring indisputable evidence for or against the cause-and-effect hypothesis and none can be required as a *sine qua non*." Despite his modesty in making claims for the dispositive power of his viewpoints in determining causality, Hill's approach today informs numerous weight-of-evidence approaches for systematically documenting considerations and judgments about whether associations are likely to be causal (Fedak et al. 2015; Schünemann et al. 2011). It has been enormously influential in epidemiologic research, public health, and regulatory risk assessments over the past half century.

The performance of Hill's considerations outside their original historical context of helping people make up their minds that cigarette smoking causes lung cancer has long been questioned, however. A cogent evaluation by Ioannidis (2016) concluded that "of the nine criteria, experiment remains important and consistency (replication) is also very essential. Temporality also makes sense, but it is often difficult to document. Of the other six criteria, strength mostly does not work and may even have to be inversed: small and even tiny effects are more plausible than large effects; when large effects are seen, they are mostly transient and almost always represent biases and errors. There is little evidence for specificity in causation in nature. Biological gradient is often unclear how it should it modeled and thus difficult to prove. Coherence remains usually unclear how to operationalize. Finally, plausibility as well as analogy do not work well in most fields of investigation, and their invocation has been mostly detrimental." Likewise, Pearl and Mackenzie note that "Indeed, it is quite easy to find arguments against each of the [Hill] criteria… Consistency by itself proves nothing; if thirty studies each ignore the same confounder, all can easily be biased. Strength of association is vulnerable for the same reason… Specificity has always been a particularly controversial criterion," e.g., does the fact that smoking causes emphysema and cardiovascular disease really weaken the evidence that it causes cancer?

The following sections argue that the Hill considerations should be updated to include concepts, principles, and methods from modern causal discovery and inference algorithms to better serve the needs of scientists and policy makers and to overcome such limitations. Current ideas and methods can help to draw useful, though limited, valid causal conclusions from observational data. Although Hill did not explicitly define what he meant by "causation," his considerations support judgments about some types of causation (especially associational causation) much better than others that are more important to policy makers (especially manipulative causation), as discussed in the next section. The following sections therefore suggest ways to refine, extend, generalize, supplement, and operationalize Hill's considerations with ideas and principle from modern causal discovery algorithms to address questions of manipulative causation—how would changing exposures change probabilities of health outcomes?—which are not addressed by the original Hill considerations.

The intended audience for the remainder of this chapter includes not only scientists, policy analysts, and regulators interested in causality, causal exposure-response modeling, and evaluation of the effects caused by policies and interventions (i.e., "accountability analysis"), but also practitioners seeking to apply recent methods to their own data sets. The following sections and Appendixes 1 and 2 therefore mention relevant software packages for implementing these ideas and methods. Chapters 10, 11, 12, and 13 illustrate their practical applications to several epidemiological data sets.

Background on Modern Causal Discovery Concepts and Terminology

The Bradford Hill considerations on the left side of Table 9.1 are more familiar to many epidemiologists and risk analysts than the causal discovery principles and terminology on the right side. This section therefore briefly introduces key ideas and vocabulary of contemporary causal discovery algorithms to facilitate discussion of how they might be used to update the Hill considerations. Appendix 1 provides additional technical concepts and terminology on directed acyclic graph (DAG) methods and software and definitions of direct and total causal effects. Appendix 2 introduces causal graph learning ("causal discovery") methods and software.

Causal Graph Concepts and Terminology

The right side of Table 9.1 lists principles of modern causal discovery and inference that we propose should be used to update Hill's considerations. They draw on the theory of *causal graphs*—that is, networks of nodes, representing uncertain quantities or random variables, linked by arrows between some pairs of nodes. The arrows in a causal graph signify that changes in the value of a variable change the conditional probability distributions of variables into which it points (its "children," in graph-theoretic terminology). Quantitatively, each node with arrows pointing into it has a *conditional probability table* (CPT), or, more generally, a conditional probability model (such as a logistic regression model, classification tree, or other predictive analytics model) specifying conditional probabilities for its value, given the values of its parents. (The probability table for a node without parents, having only outward directed arrows, i.e., an "input node", represents its marginal distribution; we include these among CPTs.) A node's parents are interpreted as its *direct causes* with respect to the variables in the graph. Changing them is assumed to change its probability distribution. In risk analysis terms, this is a change in the *aleatory* probabilities for the node's possible values. By contrast, observing different values for the node's children can also change its inferred probability distribution, via Bayes' Rule, but this is an *epistemic* change: it allows different probabilistic inferences to be drawn about the value of the node, but does not change that value or its aleatory probability distribution.

Figure 9.1 shows an example of a *directed acyclic graph* (DAG) model—that is, one with no directed cycles (no node is its own parent or ancestor). It will be used to illustrate terms and ideas. The data sources and variables are detailed in Cox Jr. (2018), but the main idea of the DAG is fairly self-explanatory: the variable *HeartDiseaseEver* (with value 1 for respondents who report ever having been diagnosed with heart disease by a medical professional and 0 otherwise) has a CPT that depends on sex, age, income, and smoking status (coded as 1 for ever smokers, 0 for never smokers). Similar CPTs for the other variables describe the quantitative

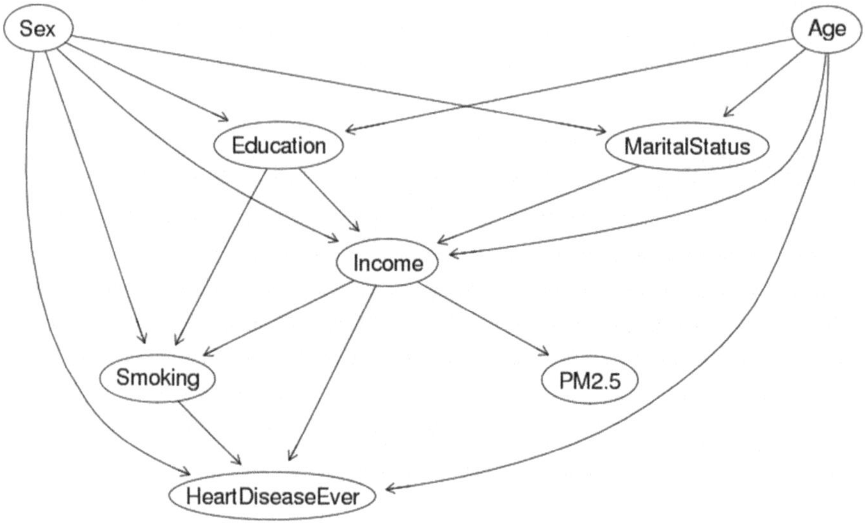

Fig. 9.1 A directed acyclic graph (DAG) model. An arrow between variables indicates that they are informative about each other. (*Source*: Cox Jr. 2018)

probabilistic dependencies among them. PM2.5, the average concentration of fine particulate matter to which respondents are exposed, is shown as depending on income; we shall discuss later what such a causal dependence of exposure on income might mean.

Figure 9.2 plots selected information from the CPT for heart disease (aggregated over smoking status to avoid complicating the graph. The full CPT gives this information separately for ever-smokers and never-smokers.) Figure 9.3 shows a nonparametric estimate of the *natural direct effect* of income on heart disease risk—that is, the effect of changes in income on changes in heart disease risk if only these two variables are allowed to change and all other variables are held fixed at the values they have in the data set (see Appendix 1 for details). This "partial dependence plot" (PDP) may be interpreted as follows: Given the joint frequencies of sex, age, and smoking in this data set, the probability of heart disease for a randomly selected respondent declines from about 0.10 to about 0.085 as income code increases from 1 to 8. The total-effect PDP curve is very similar to this natural direct effect curve, declining from about 0.10 to about 0.083 as income code increases from 1 to 8. This slightly greater effect reflects a decline in smoking as income increases to accurately describe the lower prevalence of smoking among higher-income respondents.

The tools of DAG models, conditional independence, CPTs, direct and indirect causes, and direct and total effects are discussed in more detail in Appendices A and B and by Pearl and Mackenzie (2018). They enable discussion of principles for causal discovery and inference, even before the intended meaning of "causation" is further defined. The following sections use them to discuss and explain further the principles referred to on the right side of Table 9.1.

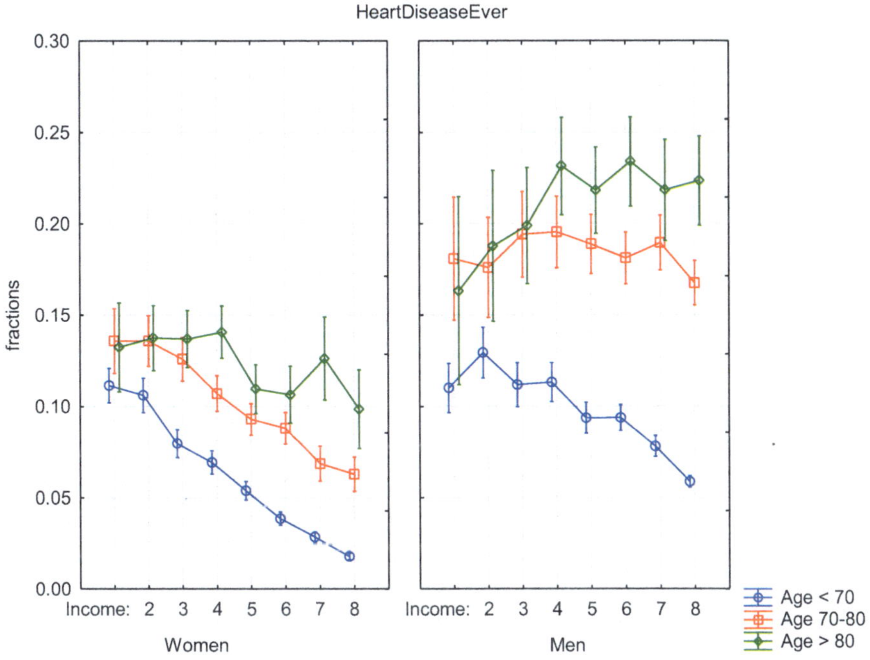

Fig. 9.2 Conditional probabilities for heart disease given sex, age, and income

Causal Concepts

Although Hill avoided defining causation, other scholars have elucidated several interrelated concepts of causality. Appendix 3 discusses nine different concepts of causation: probabilistic, associative, attributive, counterfactual-potential outcomes, structural, predictive, manipulative, explanatory-mechanistic, and but-for causation. Although each has a substantial specialized technical literature, Appendix 3 summarizes the main ideas, distinctions, and interrelationships among these concepts in a fairly informal and intuitive way, relying on the cited references for technical details. The most important distinction for our purposes is between *associative causation*, which follows in the tradition of Hill by seeking to assess causal impacts based on relative risks or similar statistical measures of association; and *manipulative causation*, which seeks to describe how the probability distribution of an effect changes in response to changes in its causes. *Attributive causation* is similar to associational causation but adds the idea of choosing one or more risk factors to blame for risk (e.g., exposure and smoking but perhaps not demographic factors such as age, sex, or income).

For risk managers and policy makers, *manipulative causation* is key, since the goal of decision-making is to choose acts that cause desired outcomes, in the sense of making them more probable. Manipulative causation is implied by mechanistic

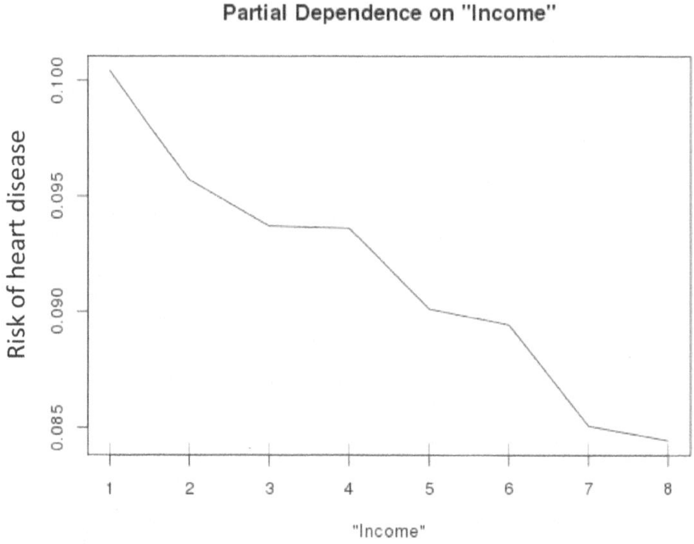

Fig. 9.3 A causal partial dependence plot (PDP) for the natural direct effect of income (horizontal axis) on absolute risk (i.e., probability) of self-reported heart disease (vertical axis), conditioning on variables in the minimal sufficient adjustment set {*Age*, *Sex*, *Smoking*}

causation—if there is a network of mechanisms by which acts change the probabilities of outcomes (mechanistic causation), then taking the acts will indeed change the probabilities of outcomes (manipulative causation). But neither one is implied by associational, attributive, counterfactual, or predictive concepts of causation (Pearl 2014; Pearl and Mackenzie 2018). Understanding and appropriately applying these distinctions among concepts of causation, and making sure that associational concepts are not misrepresented or misunderstood as manipulative causal ones in policy deliberations and supporting epidemiological analyses, provides a crucial first step toward improving current practice in epidemiology (Petitti 1991). At present, much of the epidemiological literature can be characterized as follows.

- *Associative causation* is used in the vast majority of papers that assert causal relations. Associative causation is often interpreted and applied using the Hill considerations or similar qualitative weight-of-evidence criteria.
- *Attributive causation* is used in a subset of papers that present "Burden of Disease" calculations (Murray and Lopez 2013; Fann et al. 2012).
- *Counterfactual and potential outcomes models* have started to be used much more frequently in the past few years, impelled in part by the ingenious work and enthusiasm of their developers. Zigler and Dominici et al. (2014) advocate the use of such models for air pollution epidemiology studies, but the results are often sensitive to untested assumptions (as well as to measurement errors and detailed modeling assumption). For example, Moore et al. (2012) find that a previously estimated significant causal effect of ozone on the proportion of asthma-related hospital discharges based on potential outcomes modeling "artificially

relies on untestable parametric modeling assumptions." The significant effect disappears when more realistic modeling is used that does not depend on these untestable modeling assumptions.

- *Predictive causation* is comparatively seldom used. There are a few exceptions. Chen et al. (2017) conclude from a predictive causality test (Granger causation) that fine particulate matter (PM2.5) on cold days is a predictive cause of influenza two days later in Taiwan. Cox Jr and Popken (2015) find clear exposure-response associations between concentrations of PM2.5 and ozone in air and all-cause mortality rates in 483 U.S. counties, but no evidence that either pollutant is a Granger cause of mortality. This illustrates association without predictive causation.
- *Structural, manipulative, mechanistic, and but-for causation* are comparatively neglected in most epidemiology publications, although a rapidly developing literature on causal mediation analysis addresses aspects of mechanistic causation. These concepts are closely interrelated (Druzdzel and Simon 1993), but not identical (Pearl and Mackenzie 2018). Our main focus in succeeding sections is on manipulative causation.

Why these Distinctions Matter in Practice: The CARET Trial as an Example

The practical importance of the distinction between associative and manipulative concepts of causation is well illustrated by the results of the CARET trial, a randomized, double-blind 12-year trial initiated in 1983 that sought to reduce risks of lung cancer by administering a combination of beta carotene and retinol to over 18,000 current and former smokers and asbestos-exposed workers (Omenn et al. 1996). This intervention was firmly based on epidemiological studies showing that relative risks of lung cancer were smaller among people with larger levels of beta carotene and retinol in diet or blood serum: the association was clear. However, unexpectedly, the effect of the intervention was to increase risk of lung cancer. In the words of the investigators, "The results of the trial are troubling. There was no support for a beneficial effect of beta carotene or vitamin A, in spite of the large advantages inferred from observational epidemiologic comparisons of extreme quintiles or quartiles of dietary intake or serum levels of beta carotene or vitamin A. With 73,135 person-years of follow up, the active-treatment group had a 28% higher incidence of lung cancer than the placebo group, and the overall mortality rate and the rate of death from cardiovascular causes were higher by 17% and 26%, respectively." That the intervention produced the opposite of its intended and expected effect is a valuable reminder that relative risks and other measures of association and epidemiological measures of causation and burden of disease derived from them do not necessarily predict how response probabilities will change if interventions are used to change exposures. In other words, associative exposure-response relationships do not necessarily predict corresponding manipulative causal exposure-response relationships.

The following sections propose criteria for assessing the consistency of observational exposure-response data with manipulative causal exposure-response relationships. The proposed criteria are inspired by Hill's considerations for assessing evidence of associative causation from observational data, but try to provide similar guidance for assessing evidence of manipulative causation, drawing as needed on recent literature on causal discovery algorithms and other causal concepts (especially, predictive, structural, and mechanistic causation).

Strength of Association and Mutual Information

This section reviews in more detail the concept of strength of association and suggests modifications based on information theory (Appendix 2) to develop an analogous criterion that is more useful for manipulative causation.

Some Limitations of Strength of Association

Hill wrote, "First upon my list I would put the strength of the association." Authorities including the International Agency for Research on Cancer (IARC) have interpreted this consideration to mean that "A strong association (e.g., a large relative risk) is more likely to indicate causality than a weak association" (IARC 2006). However, interpreted this way, the principle is not logically or empirically valid (Ioannidis 2016). On the one hand, a strong causal relation may have a zero statistical association between cause and effect in observational data. (As a simple example, velocity V and kinetic energy KE in the physical law $KE = 1/2MV^2$ for a particle of mass M have an average correlation of zero if the distribution of velocities is symmetric around the origin, even though V deterministically causes KE.) Likewise, if a cause has little or no variation in a data set, then there may be no association with the response that it causes. On the other hand, strong associations often exist without causation. A strong exposure-response association may—and perhaps usually does (Ioannidis 2016)—simply reflect strong sampling or selection biases, strong modeling errors and biases, strong coincidental historical trends, strong confounding, strong model specification errors, or other strong threats to internal validity (Campbell and Stanley 1963). For example, two statistically independent random walks usually have significantly correlated values over any selected time interval (since each tends to have a random drift with time), but such spurious correlations, no matter how strong, do not provide any evidence of causality (Yule 1926). Moreover, the strength and even the direction of association between exposure and response variables in a regression model often depend on modeling choices about which other variables to include in the model, how to code them, and what form of functional relation to assume for the regression model. For example, in the data set for Fig. 9.1, PM2.5 has a significant ($P < 0.002$) positive regression coefficient as a predictor of heart disease risk in a multiple linear regression model with age, sex, ever-smoker, and PM2.5 as predictors. If income is included as an additional predictor, however, then the association between PM2.5 and heart disease is no longer statistically significant ($P = 0.16$) (Cox Jr. 2018).

In some data sets, the sign of an association between a predictor and a response variable can be reversed, e.g., from a significant positive association to a significant negative association, depending on modeling choices about which other variables to include in a regression model or condition on in stratified analyses. This occurs for PM2.5 and response in the data set in Fig. 9.1 if heart attack rather than heart disease is used as the response variable (Cox Jr. 2017b). Simpson's Paradox, in which aggregating over levels of a discrete covariate reverses the direction of an association that holds within each level (i.e., conditioned on any level), illustrates the malleability of the concept of a positive association. Because the strengths and directions of exposure-response associations often depend on such details of modeling assumptions and choices, "There is a growing consensus in economics, political science, statistics, and other fields that the associational or regression approach to inferring causal relations—on the basis of adjustment with observable confounders—is unreliable in many settings" (Dominici et al. 2014).

The Mutual Information Criterion

To overcome these substantial challenges to the usefulness of association as an indicator of causality, many modern causal discovery algorithms replace *association* with the more general concept of *information* (Appendix 2 and Cover and Thomas 2006) and apply non-parametric methods to estimate statistical dependencies (i.e., information patterns) among variables. For practitioners, we recommend as our first criterion the following alternative to association-based considerations for screening for possible manipulative causal relations:

Mutual Information Criterion Use current graph-learning algorithms for causal discovery (Appendix 2) to identify detectable statistical dependence (i.e., mutual information) and conditional independence relations between variables. Only a variable's neighbors in such a graph are identified as its potential direct causes or direct effects. Only variables linked to it via one or more undirected paths are identified as potential indirect causes or effects.

To quantify and visualize the direct and total effects of one variable on another in a DAG, which may include multiple paths and confounding and selection effects, we recommend computing adjustment sets (Appendix 1) and creating partial dependence plots that condition on them, as in Fig. 9.3. The mutual information criterion implements the intuitive idea that *causes are informative about their effects.* A data set gives no evidence of predictive or manipulative causation between variables unless they are informative about each other, i.e., statistically dependent. Statistical dependence is indicated visually by arrows between variables in DAG models generated by modern causal discovery software packages (Scutari and Ness 2018) (or by undirected or bi-directed arcs between them in some more general graph models that allow for latent variables, e.g., the partial ancestral graphs used to display Markov equivalence classes of graphs in the presence of latent variables) (Heinze-Deml et al. 2018).

The mutual information criterion is only a first step toward establishing that data are consistent with a hypothesis of predictive causation, which is typically a prerequisite for manipulative causation. We recommend interpreting its results as follows: adjacency in a DAG or more general graph model provides evidence that one variable *might* be a predictive cause of another, in that knowing (i.e., conditioning on) the value of the first helps to predict the value of the second better than it could be predicted without this information (e.g., as measured by reductions in mean squared prediction error or mean conditional entropy). Such potential predictive causation is indicated by the arrows in DAG models learned from the data (Appendix 2). If two nodes are not adjacent in a graph, this does not prove that there is no causal relation between them, but only that no evidence of such a relation has been found. This in turn, can be used to put quantitative upper bounds on the plausible size of undetected effects. Conversely, finding that two variables *are* connected in a graph model invites further consideration of whether the statistical dependence between them is in fact due to causation instead of to other explanations such as reverse causation or unobserved (latent) confounders or selection bias or coincident trends. We next propose criteria for further testing detected statistical dependencies to determine whether they might arise from predictive or manipulative causation.

Criteria for Orienting Arrows in Causal Graphs: Temporality, Directed Information, Homoscedasticity, Exogeneity, Knowledge-Based Constraints, and Quasi-Experiments

The mutual information criterion does not use directional information about the arrows in a probabilistic graph model. This sharply constrains the causal inferences that can be drawn. The principle that causes are informative about their effects does not, by itself, distinguish among different causal graphs having the same dependence and conditional independence relations (i.e., in the same Markov equivalence class), e.g., between $X \rightarrow Y$ and $X \leftarrow Y$ (causation vs. reverse causation) or between $X \rightarrow Y \rightarrow Z$ and $X \leftarrow Y \rightarrow Z$ (causal chain vs. confounding). This deficiency can be remedied by applying additional criteria that address the orientation of arrows.

Temporality and Directed Information

If adequate longitudinal data are available, e.g., from panel studies, then arrows between variables can be oriented by applying the following refinement of Hill's temporality consideration: *Changes in causes precede and help to predict changes in their effects*. This adds to the traditional Hill temporality consideration (causes must precede their effects) the important proviso that variations in causes must also *help to predict* subsequent variations in their effects. As usual, by "help to predict" we mean: provide information that reduces average uncertainty about the predicted variable (e.g., as measured by its expected conditional entropy). By this criterion, if an exposure time series and a health effect time series are both declining over time, so that decreases in exposures are routinely followed by decreases in risk of an adverse health effect (and *vice versa*), this would *not* be considered evidence of a causal relation between them unless the changes in exposure can be used to predict future changes in health effects better than they could be predicted without this information, e.g., just from the secular trend—or, more generally, the past values—of the health effects time series itself. This proviso prevents the temporality criterion from being abused to support the *post hoc ergo propter hoc* fallacy.

A more operational version of this modified temporality criterion is that *future values of effects are not conditionally independent of the present and past values of their direct causes, even after conditioning on their own past values (and the past values of more remote ancestors, if any)*. This is typically an asymmetric condition, since information flows from causes to effects over time, but not *vice versa*. It can be implemented for longitudinal data sets containing time series of observations of exposures, effects, and covariates by applying the same machinery already discussed of DAG modeling, conditional independence tests, and mutual information to time-stamped values of variables to determine whether earlier values of hypothesized causes help to predict later values of their hypothesized effects. Appendix 4 mentions available software options for learning such dynamic *Bayesian networks* (DBNs) or closely related *directed information graphs* (DIGs) from data.

Directed Information Criterion for Longitudinal Data Use current DBN-learning software or DIG-learning algorithms (Appendix 4) to identify detectable directed statistical dependence (i.e., directed information) and conditional independence relations between past and future variables in longitudinal data sets. Only a variable's parents in such a directed graph are identified as its potential direct causes. Only its ancestors, linked to it via one or more directed paths, are identified as its potential indirect causes.

The following considerations for latent variables already discussed for the mutual information criterion also apply to the directed information criterion. Unobserved confounders or selectors (e.g., common ancestors not conditioned on, or common descendants conditioned on or used in selecting the data, respectively) can induce non-causal associations between observed variables, leading to false-positive arrows in DBN models for the observed variables. The effects of such latent variables can be detected and modeled based on correlations among estimated noise terms under certain conditions (Heinze-Deml et al. 2018). If the direction of information flow cannot be uniquely identified from available data, then using a bidirected arrow allows the directed information criterion to identify potential direct and indirect causes even in such ambiguous cases. If there might be unobserved confounders or selection conditions, then the criterion is best interpreted as a necessary but not sufficient condition for observational data to provide evidence of predictive causation. If it is known or assumed that all relevant variables have been observed, then directed information flow is both necessary and sufficient for predictive causation between variables except under unusual conditions, such as that the effects transmitted by different directed paths exactly cancel each other.

If longitudinal data are not available, then other techniques can be tried to establish the direction of information flow between variables under certain conditions. From this perspective, temporality can be viewed as just one of several ways to establish the direction of information flow among variables, thereby revealing the topology of a causal graph. Appendix 5 summarizes several alternatives for constraining the direction of information flow even in the absence of a clear temporal ordering between changes in variables. These include homoscedasticity of error terms if the effect is known to depend upon explanatory variables with an additive error term (the LiNGAM principle) (Shimizu et al. 2006; Tashiro et al. 2014); knowledge-based constraints; exogeneity-based constraints; and quasi-experimental studies and analyses. From the standpoint of orienting arrows in a causal graph to be consistent with a predictive or manipulative causal interpretation, these constraints play the same essential role as temporality: they specify or constrain the possible directions of information flows, and hence of causal influences, assuming that causes provide information about their effects. They can be used whether or not time series data are available.

The following general criterion capturing these principles is applicable to both cross-sectional and longitudinal data.

Directed Dependence Criterion (general version) Identify possible arrow directions between variables in a causal graph model of the data using applicable directed information (Appendix 4), error distribution, exogeneity, and knowledge-based constraints, including any constraints from experimental or quasi-experimental interventions as well as structural ordering constraints (Simon and Iwasaki 1988), if available (Appendix 5). Which of these constraints are applicable depends on study design (e.g., longitudinal vs. cross-sectional). A direction for an arrow is considered possible (relative to what is known) if it does not violate any of these identified constraints. Only a variable's possible parents (i.e., variables that could point into it without violating any of the constraints on arrow directions) are identified as its potential direct causes in such a constrained directed graph. Only its possible ancestors (i.e., variables that can be linked to it via one or more directed paths without violating any of the constraints on arrow directions), are identified as its potential indirect causes.

Like the mutual information criterion, this criterion is intended to constrain the set of plausible causal interpretations for observed dependencies among variables in a data set. To the extent that the directions of dependencies can be established by the data, based on temporality, homoscedasticity, exogeneity, or other criteria, they limit plausible causal interpretations to those that are consistent with these directions. Visually, arrows oriented via these constraints indicate the order in which causal information can flow among variables. If deliberate interventions are possible but expensive, as in animal experiments and some human studies with biomarkers, then a least-cost set of interventions can be designed that is sufficient to uniquely identify the directions of all arrows (Kocaoglu et al. 2017). This corresponds to Hill's "Experiment" consideration in Table 9.1.

The directed dependence criterion was used in Fig. 9.1 by specifying the knowledge-based constraints that *Age* and *Sex* were sources (i.e., with only outward-pointing arrows allowed) and *HeartDiseaseEver* a sink (only inward-pointing arrows allowed). With these constraints, the software package used (*bnlearn*, Scutari and Ness 2018) not only identified a plausible set of neighbors adjacent to each variable, but also identified plausible arrow directions for informational dependencies among them, as shown in Fig. 9.1. Caution is required in interpreting such DAGs, however, because their arrows represent a mixture of constraints on directions of information flow (causal ordering) and assignments of directions to remaining arrows that do not violate any known constraints but that do not necessarily represent the directions of information flows. This distinction is not apparent visually from the DAG model alone, so the constraints must be kept in mind in interpreting it. Thus, in Fig. 9.1, it is important to recognize that only the source and sink constraints have been specified for arrow directions. The arrow from *MaritalStatus* to *Income*, for example, does not have a corresponding constraint that justifies it, so causality might possibly run in the reverse direction, or in both directions, although a DAG model necessarily shows only one possible direction for each arrow. Whether high income increases probability of marriage, or marriage increases probability of high income, or both, cannot be determined from Fig. 9.1 and its underlying knowledge-based constraints. Likewise, the arrow from *Income* to *PM2.5* is not supported by a constraint, and hence it leaves unclear whether an increase in income would reduce PM2.5 exposure by enabling a respondent to move to a neighborhood with cleaner air, or whether an increase in air quality would cause people with higher incomes to move into the area, or both.

DAG models without directional constraints on their arrows do not necessarily reveal the order in which changes in one variable propagate to other variables, and hence they cannot answer such questions of causal interpretation except when directional constraints are known. (As previously noted, more general causal graphs show such ambiguities explicitly via undirected or bidirected arcs (Heinze-Deml et al. 2018).) In summary, for DAG models, it is necessary to keep track of the constraints on possible arrow directions. Arrow directions not implied by constraints simply indicate a way to factor the joint distribution of the variables, but have no further clear causal implications other than those that follow from the mutual information criterion.

To summarize the development to this point, a causal graph model is fully specified by three pieces of information: the pattern of adjacencies between its nodes (variables); the directions of arrows representing direct causality between adjacent nodes; and conditional probability tables (CPTs) at each node (including marginal probability tables for input nodes, as previously mentioned). CPTs quantify the dependence of each node on its direct causes, if any, including any interactions among them in affecting its conditional probability distribution. The mutual information criterion specifies that nodes are adjacent if and only if they are informative about each other. It constrains possible arrow directions to a Markov equivalence class via the rule that a variable is conditionally independent of its more remote ancestors and other non-descendents given the values of its parents. The directed dependence criterion uses constraints from data and user-supplied knowledge, including temporality, directed information flow, error distribution properties (e.g., homoscedasticity), exogeneity, and any available constraints from experiments, natural experiments, or quasi-experiments, to further constrain arrow directions. It identifies possible direct and indirect causal relations that are consistent with these constraints. The validity of such inferences rests on the validity of the constraints. Thus, if assumptions are used to justify some of these dependence constraints, as is typically the case for QE methods (Appendix 5), then the validity of the causal inferences rests on the validity of these supporting assumptions.

We turn next to principles of causal discovery that make use of the information in CPTs.

Consistency Checks: Internal and External Consistency and Generalization and Synthesis across Studies

Hill's first consideration was strength of association; his second was consistency, which he introduced as follows: "Next on my list of features to be specially considered I would place the *consistency* of the observed association. Has it been repeatedly observed by different persons, in different place circumstances, and times?" IARC (2006) adopted the following version of this principle for making judgments about causality based on exposure-response associations: "Associations that are replicated in several studies of the same design or that use different epidemiological approaches or under different circumstances of exposure are more likely to represent a causal relationship than isolated observations from single studies." An underlying intuition for this consideration is that *causal laws and mechanisms are universal*: they hold in different times and places and their operation can be observed and confirmed independently by different investigators. This intuition can be expressed in the language of causal graphs as the maxim that *causal CPTs are invariant*, meaning that a CPT that fully and correctly represents the causal relation between a variable's conditional probability distribution and the values of its direct causes remains the same independent of the context of other variables, values, causal networks, and interventions or manipulations in which they are embedded (Peters et al. 2016). More colloquially, "the same causes have the same effects" no matter where and when they occur. When this appears to be violated, it is usually because the causal model being used is incomplete, omitting some of the direct causes that, if included, would make it true: "If we consider all 'direct causes' of a target variable of interest, then the conditional distribution of the target given the direct causes will not change when we interfere experimentally with all other variables in the model except the target itself" Peters et al. (2016). This *invariant causal prediction* (ICP) principle, which holds whether variables are perturbed by chance or by deliberate manipulations, provides the basis for the ICP (Invariant Causal Prediction) algorithm for causal discovery in the *CompareCausalNetworks* package (Peters et al. 2016). A substantial philosophical literature addresses its foundations and validity for causal modeling (e.g., Cartwright 2002; Hausman and Woodward 1999).

Invariance of causal CPTs refines Hill's consistency consideration, clarifying that it is not the exposure-response relation (i.e., the conditional probability of response given exposure) itself that should be consistent (i.e., invariant) across studies, but rather the *conditional probability* of response (i.e, the CPT for the response variable) given exposure and the values of other direct causes of response. In Fig. 9.1, for example, the exposure-response relation between smoking and heart disease risk should not necessarily be expected to be the same across different study population except within subgroups ("strata") having the same sex, age, and income. Causal graphs illuminate exactly what should be consistent (CPTs) if causality holds and show how observed *inconsistencies*, or differences in exposure-response associations across different study populations, can provide empirical support for a causal model if they are predicted and explained by the response CPT and observed differences in values of other direct causes that are inputs to it (e.g., in the distributions of covariates such as sex and age) across the different populations being compared. Finally, causal graphs reveal conditions that can induce a strong, consistent exposure-response association without providing evidence that it is causal. These conditions include confounding due to failure to condition on a common cause; collider bias (also called stratification bias or selection bias) due to conditioning on a common effect (Cole et al. 2010); and reverse causation.

Causal graphs provide the following constructive approaches for testing various forms of consistency.

- *Internal consistency*: Are the estimated causal effects (e.g., causal PDP plots or other effects measures) calculated from different adjustment sets Appendix 1) significantly different? If not, this supports the internal consistency of the model, meaning that different ways of calculating the same causal effect do not lead to significantly different answers; otherwise, the DAG or causal graph model used to calculate the effects estimates is probably not valid (Oates et al. 2016). This approach provides a consistency check within a single data set ("internal" consistency), complementing the Hill approach which requires evaluating consistency across multiple data sets ("external" consistency).

- *External consistency*: Are estimates of the same causal CPT (i.e., estimates of the conditional probability distribution for the same dependent variable given the same values of its direct causes) in different data sets significantly different? If so, this violates the property of *invariant causal prediction* (ICP) (Peters et al. 2016) and the data sets jointly provide evidence that the CPT is not a valid description of a universal causal law. Otherwise, if the null hypothesis of identical CPTs in different data sets is not rejected, then the data sets are consistent with the ICP property and with the CPT describing a universal causal law.

These consistency criteria can be cast as the following recommended series of steps to be followed once a DAG model (or more general causal graph model)

Internal Consistency Criterion:

1. Specify a causal effect of interest (e.g., the natural direct effect or total effect of one variable on another, usually exposure on risk; see Appendix 1).
2. Calculate minimal sufficient adjustment sets for estimating the effect (assuming that one or more such sets exists; otherwise, this criterion cannot be used. The dagitty and CAT packages mentioned in Appendix 1 automate this computation and identify which effects can be estimated from data.)
3. Estimate a confidence interval or uncertainty interval for the specified causal effect of interest using each minimal sufficient adjustment set (e.g., Rodríguez-Entrena et al. 2018).
4. If these interval estimates are not significantly different (e.g., if they overlap), conclude, that the DAG model and causal effect estimates are consistent with the data by this internal consistency criterion; otherwise, conclude that the causal interpretation of the data shown in the DAG model is not consistent with the data based on this criterion.

If effects are estimated using PDPs, as in Fig. 9.3, then resampling techniques can be used to generate non-parametric confidence intervals around the PDP curves; this is done automatically by the CAT package in Appendix 1. Effects estimated from different minimal sufficient adjustment sets are significantly different if their confidence intervals do no overlap. The *p.adjust* package at https://stat.ethz.ch/R-manual/R-devel/library/stats/html/p.adjust.html can be used to correct significance levels for multiple comparisons bias when more than two adjustment sets are used.

External Consistency Criterion If multiple studies or data sets are available from which to estimate the CPT of a response variable of interest, then: (1) Test whether the CPTs estimated from the different data sets are significantly different. (2) If not, then conclude that the data pass this test for external consistency (i.e., they are consistent with the ICP property for the response variable of interest). Otherwise conclude that the data provide evidence that the ICP property (and hence external consistency) is violated for the response variable's CPT as currently defined and measured.

To implement this criterion, the CPT for a response variable can be estimated from a data set by standard BN learning algorithms provided by many commercial

BN packages as well as in free R packages (Appendix 2). (For large data sets, the conditional probabilities for the values of the response variables can be approximated by their observed empirical frequencies for different combinations of parent values, possibly with the help of a smoothing regression model. For small data sets, a Bayesian approach updates a Dirichlet prior for CPT probabilities with the observed frequencies (see e.g., Azzimonti et al. 2017). In both cases, responses must be observed for all combinations of parent values for which the model is to be used in order to avoid extrapolation beyond the range of the data.) Standard statistical tests for homogeneity of probability distributions in contingency tables (or of independence of the conditional distribution of the response from study ID), such as the chi square test, can be used to test the null hypothesis that the CPTs do not differ significantly across studies, treating study ID as a factor in the contingency table. Alternatively, a simple non-parametric approach is to merge the data from all studies and add a study ID field indicating which study each record (row) in the merged data set comes from. Each record in the combined data set then contains values for the selected response variable and its direct causes, as well as the study ID. Predictive analytics algorithms, including classification and regression trees and random forest ensembles, can then be used to quantify the conditional probability distribution of the response variable given the values of its predictors and to quantify the reduction in mean squared prediction error achieved by including each variable as a predictor. If the study ID variable is not identified as a useful predictor by such algorithms (i.e., if it does not help to reduce prediction errors or expected conditional entropy of the response variable's distribution), then the null hypothesis of invariant causal prediction (ICP) across studies is not rejected. Operationally, therefore, the test of whether estimated CPTs are significantly different in different data sets amounts to seeing whether the study ID enters is selected by these predictive algorithms as a predictor for the response variable.

Even if multiple data studies and data sets are not available to which to apply the external consistency criterion, it is often possible to create them from a single large data set by partitioning it into multiple sub data sets based on variables such as county or state that are not direct causes of the response variable. The external consistency criterion can then be applied to these sub data sets to test whether the ICP property is satisfied.

Learning from Apparent Inconsistencies

Although external consistency of exposure-response associations across studies—Hill's consistency consideration—can provide useful evidence supporting causality if the populations being compared are similar, it can also be misleading (Pearl and Mackenzie 2018), especially if assessed by subjective judgment. Like Hill's other principles of biological plausibility, coherence, and analogy, the consideration of consistency appeals strongly to intuition. However, psychologists have demonstrated convincingly that such intuitively appealing qualitative and holistic ("System

1") judgments are prey to potent psychological heuristics and biases such as motivated reasoning (finding what it pays us to find), groupthink (conforming to what others appear to think), and confirmation bias (finding what we expect to find and focusing on evidence that agrees with pre-existing hypotheses or beliefs rather than on potentially disconfirming or discordant evidence that could force us to correct initial misconceptions and to discover how the world works) (Kahneman 2011). Scientists and statisticians, like other people, are demonstrably subject to these and related heuristics and biases in their judgments under uncertainty (*ibid*, Chap. 11; Sacco et al. 2018). Consistency in estimates of effects and associations across multiple studies that arises from such biases and "p-hacking" (defined as varying modeling and data analysis methods, assumptions, and choices to obtain expected or desired results) does not provide evidence of causation (Fraser et al. 2018). "Causality implies consistency" does not necessarily justify the conclusion "Consistency implies causality." Conversely, consistent findings of similar exposure-response regression coefficients or effects estimates across different populations should invite the question of whether greater inconsistency should be expected accurately reflect differences in the study populations.

Apparent inconsistencies can be especially useful for learning causal graphs and correcting partly misspecified causal graph structures (Oates et al. 2017), as well as for detecting and modeling unobserved causes of an observed variable. A frequent reason for ICP violations in practice is that one or more direct causes that differ across data sets have been omitted from the DAG model and CPT for a response variable. Thus, data that do not pass the external consistency criterion may signal a need to search for additional explanatory variables or to model them via latent variables (see discussion at the end of Appendix 2). For example, if the CPT for heart disease in Fig. 9.1 were found to vary significantly across geographic regions or states, this might indicate a need to modify that section of the DAG to include currently omitted variables, such as high and low daily temperatures, that also affect risk of heart disease among elderly patients. (A second way to detect the presence of omitted causes is unexplained information dependencies across disjoint groups. For example, if days on which hospital admissions for heart disease are high among women under 70 also tend to have higher hospital admissions for heart disease among men over 70, this might again indicate a common environmental cause, e.g., daily temperature.)

In summary, *internal consistency*, meaning that similar estimates for specified total or direct causal effects are obtained using different adjustment sets for the same data, and *external consistency*, meaning that the same causal CPTs for the response variable are found in settings with very different joint distributions of the values of their direct parents, provide useful refinements of Hill's general consideration of "consistency." They can help to distinguish consistency that represents genuine predictive power (based on discovery of causal invariants that can be applied across multiple experimental and observational settings) from consistency arising from common errors (e.g., omitted confounders) across studies (Pearl and Mackenzie 2018) or arrived at by retrospective p-hacking to make results agree with expectations.

Causal Coherence, Biological Plausibility, and Valid Analogy: Explanations of Exposure-Response Dependencies via Directed Paths of Causal Mechanisms

Philosophers of science have articulated and studied various concepts of *coherence* carefully and critically since Hill's day. For example, Wheeler and Scheines (2013) show that coherence need not increase confirmation of a hypothesis when both are suitably formalized. They offer the following refinement: "We show that the causal structure among the evidence and hypothesis is sometimes enough to determine whether the coherence of the evidence boosts confirmation of the hypothesis, makes no difference to it, or even reduces it. We also show that, ceteris paribus, it is not the coherence of the evidence that boosts confirmation, but rather the ratio of the coherence of the evidence to the coherence of the evidence conditional on a hypothesis." Dragulinescu (2015) explains that, in epidemiology and other health sciences, it is often held that "[T]he health sciences make causal claims on the basis of evidence both of physical mechanisms, and of probabilistic dependencies. Consequently, an analysis of causality solely in terms of physical mechanisms or solely in terms of probabilistic relationships, does not do justice to the causal claims of these sciences… we need evidence of both production (which comes primarily from a study of mechanisms on a microstructural level and entails knowledge of entities, their spatiotemporal arrangement, etc.) and dependency or difference making (which comes primarily from a study of the probabilistic relationships on the level of population assessments), in order to formulate adequately medical causal claims." Our proposed criteria of mutual and directed information and internal and external consistency are based exclusively on probabilistic dependencies. Mechanisms have yet to be considered. Fortunately, graph algorithms have been widely developed and applied to describe molecular biological mechanisms, pathways (causal chains of mechanisms), and networks of multiple pathways at multiple levels of abstraction (e.g., Lähdesmäki et al. 2006; Lagani et al. 2016; Chang et al. 2015; Santra et al. 2013). We can apply these ideas to clarify concepts of coherence and mechanistic causation.

Causal inference without causal explanation, meaning understanding of the causal mechanisms that generate the observed data, is vulnerable to the notorious problem of induction: patterns that have held so far may cease to do so (especially after an intervention). A light bulb that until now has always turned on whenever a switch is flipped up, except during power outages, may abruptly cease to do so if and when its filament burns out or the bulb becomes loose in its socket. Nothing in the prior observations predicts this future change in behavior or the resulting loss of manipulative causation (as well as associational, predictive, and structural causation) between switch position and illumination of the bulb. What is missing in such cases is prediction of effects of manipulations based on understanding of causal mechanisms, e.g., that the illumination of a bulb is caused by heating of its filament due to its resistance to electric current flowing through it. Such an understanding makes clear that the direct causes of the bulb's illumination are just those conditions that allow or block the flow of electricity through the filament. With this understanding, it becomes possible to predict the effects of new conditions and manipulations (e.g., the filament burning out or the bulb being loosened in its socket) even in the absence of data on their effects. Without it, no amount of observation and statistical inference can necessarily predict effects of future actions.

Hill acknowledged the importance of considering causal mechanisms in his considerations of plausibility, coherence, and analogy, which he presented as follows:

- *Plausibility*: "It will be helpful if the causation we suspect is biologically plausible. But this is a feature I am convinced we cannot demand. What is biologically plausible depends on the biological knowledge of the day."
- *Coherence*: "On the other hand, the cause-and-effect interpretation of our data should not seriously conflict with the generally known facts of the natural history and biology of the disease."
- *Analogy*: "In some circumstances, it would be fair to judge by analogy. With the effects of thalidomide and rubella before us we would surely be ready to accept slighter but similar evidence with another drug or another viral disease in pregnancy."

Coherence is perhaps the central concept here. Although Hill defines it only as consistency (or at least not serious conflict) with generally known facts about the biology of a disease, we propose the following much stronger definition based on progress in systems biology and the data science of causal biological networks (Boué et al. 2015):

> An explanation for how exposure increases risk of disease is causally coherent if it is composed of one or more pathways (directed paths of mechanisms) leading through a causal biological network (CBN) from exposure to response (i.e., disease risk) and forming a unified whole. Such an explanation may also be called a coherent causal explanation.

The key concepts in this definition are as follows:

- A *causal biological network* is a directed graph in which the value of each variable (node) depends on the values of its parents and in which the variables represent biologically interpretable quantities, such as levels of proteins, gene expression, molecular signals, and so forth. For example, the Causal Biological Networks (CBN) database (Boué et al. 2015; http://causalbionet.com/) consists of over 120 such biological network models. In much of risk analysis, the major components in a coherent explanation of an exposure-disease relation are: (1) A front-end pharmacokinetics (PK) model, e.g., a physiologically-based pharmacokinetic (PBPK) model or a classical compartmental PK model that maps time courses of exposure to time courses of internal doses in target tissues or organs; (2) A pharmacodynamics (PD) model that explains how cells in the target tissue or organ respond, and how the aggregate responses of affected cell populations affect target organ homeostasis and functioning; and (3) A disease initiation and progression model (e.g., a multistage clonal expansion (MSCE) model of carcinogenesis) that maps changes in target tissues and organs to changes in probabilities of disease states. Despite these complexities, the concept of coherence as referring to directed paths of mechanisms that transmit changes in exposure to changes in risk is common to all of them.

- A *mechanism* determines the value of a variable (or its probability distribution) from the values of its parents in a causal biological network. A mechanism may be represented mathematically or computationally by a CPT, a formula, a structural equation or regression model, an ordinary or stochastic or partial differential equation with initial conditions, a simple set of logical rules (as in agent-based models), a state transition probability matrix, a stochastic simulation model, and so forth. These different formalism for describing biological mechanisms can all be viewed as ways to calculate child values (or conditional probabilities of child values) from parent values in a causal graph.
- An *explanation* for an exposure-response dependence is a sequence of mechanisms (a pathway) or a network of multiple pathways leading from exposure to response. It explains how changes in exposures can change the probabilities of response by changing the probability distributions of the parents (direct causes) of the response variable in the pathway or network.
- *Forming a unified whole* means that there are no explanatory gaps—exposure is a parent or ancestor of disease risk via one or more directed paths through the network—and changes at each node in the network are caused by changes in its parents (via the mechanism associated with the node) in a way that is consistent with known biology (Hill's coherence condition). Thus, the explanation gives a coherent end-to-end mechanistic description of how changes propagate through the network, from changes in exposure (an exogenous input) to changes in response probabilities or disease risks (outputs), with each change caused by those that precede it via identified mechanisms.

Armed with this concept of coherent causal explanation, *biological plausibility* of a proposed causal exposure-response relation can be interpreted as its having one or more coherent causal explanations, i.e., one or more identified pathways through a causal biological network that could explain how changes in exposure cause the corresponding changes in response probabilities. Each such explanation identifies sequences or networks of mechanisms via which changes in exposure might plausibly (i.e., consistently with currently available knowledge and data) propagate (i.e., change conditional probabilities of children at each successive node along a pathway) to produce changes in response probabilities. It answers the question "*How might changes in exposure cause changes in response probabilities?*" This strengthening of Hill's concept of coherence, from not being inconsistent with biological knowledge to being demonstrably consistent with it, requires sufficient substantive knowledge of relevant biology and toxicology to actually construct biologically plausible explanations. This might have been an unthinkably burdensome requirement in Hill's day, but seems more appropriate in the current age of bioinformatics, systems biology, and resources such as the Causal Biological Networks (CBN) repository (Boué et al. 2015; http://causalbionet.com/).

If no coherent causal explanation can be identified, then an observed dose-response association does not meet this proposed criterion for biological plausibility.

For example, if plausible PBPK models cannot generate sufficiently high internal doses in target cell populations to elicit the PD changes required to cause a disease condition attributed to an exposure, then explanations that posit that such high internal doses are produced and cause the required PD changes would not meet the criterion of biological plausibility. If benzene is found to cause changes in peripheral blood lymphocytes, this would not by itself constitute part of a biologically plausible explanation for how it might contribute to risk of myeloid-lineage diseases such as acute myeloid leukemia. *Analogy* suggests possible relevant similarities in exposure-response mechanisms, and hence in resulting exposure-response relations, for similar exposures and responses. What constitutes a relevant similarity is, of course, a crucial question on which rests the distinction between this proposed guide to scientific inference and the principle of sympathetic magic that "Like produces like." We propose that it can only be answered with confidence by an understanding of causal mechanisms: a valid analogy is one in which different exposure-response dependencies share some or all of the same coherent causal mechanisms and explanation(s). For example, if two different substances activate the same signaling pathway in different ways (e.g., via different receptors), then they are analogous in this respect.

The following criterion summarizes this proposed approach.

Causal Coherence and Biological Plausibility Criterion A qualitative claim that a specified exposure is a manipulative cause of a specified response, or a quantitative claim that a specified exposure-response relationship (e.g., a PDP plotting conditional probability of response against a metric of dose or exposure) describes such a manipulative causal relation, satisfies the **causal coherence criterion** if and only if has at least one identified coherent causal explanation. In this case, it may also be called **biologically plausible**. If no such biologically plausible explanation has been identified, then the proposed causal claim or relationship is not known to satisfy the criteria of causal coherence and biological plausibility.

Hill required only that a proposed dose-response or exposure-response association not violate currently accepted biological knowledge in order to consider it coherent. By contrast, the causal coherence criterion requires that at least one biologically plausible explanation for the proposed causal relation must be identified. The explanation need not be known to be correct, but it must not be known to be incorrect (i.e., violating biological knowledge and data constraints) if it is to be accepted as satisfying the requirement for biological plausibility. What makes a proposed causal exposure-response relation biologically plausible is that it has at least one identified causal explanation that *might* be correct, as far as we currently know: it fits all the facts that we currently have, even if its correctness has not been proved. This contrasts with some previous influential interpretations of biological plausibility that are not based on coherent causal explanations. For example, IARC (2006) considers it "biologically plausible" that a chemical that causes cancer in rats or mice will do so in people. Our version of causal coherence criterion requires more than this. Rodents have organs that people do not (e.g., Harderian and Zymbal glands) and develop cancers via mechanisms not found in other species (e.g., alpha

2 mu globulin protein drop nephropathy in male rats). Exposure-related adverse effects that occur only via these species-specific mechanisms and in species-specific organs that have no analogues in people are not necessarily relevant to human risk. They do not provide a coherent causal explanation of how exposure might cause cancer in people, and thus would not meet the causal coherence criterion for biological plausibility.

The following two additional criteria can strengthen conclusions based on causal coherence alone.

- **Causal mediation confirmation (CMC) criterion**: One or more variables ("mediators") on one or more pathways from exposure to response in a coherent causal explanation are observed to have different values for different exposures, as implied by the coherent causal explanation. The observed differences in values of these mediators are at least partly predicted and explained by the differences in exposures, and, in turn, they help to predict and explain observed differences in responses, consistent with the coherent causal explanation.
- **Refutation criterion** (Rothman and Greenland 2005): Common non-causal explanations for observed statistical dependencies between exposure and response are tested and rejected.

The left side of Table 9.2 lists some common alternative explanations for exposure-response dependencies. The refutation criterion exploits the process of elimination: eliminating hypothesized non-causal explanations raises the probability that a causal explanation for an observed dependency is correct. It can be implemented by using principles listed on the right side of Table 9.2 to test and refute or account for the non-causal explanations on the left side.

The CMC criterion strengthens coherence requirements from qualitative (does at least one biologically plausible causal pathway exist?) to quantitative (do variations along the pathway suffice to explain how variations in exposures affect variations in response probabilities or other observed effects?). Recall that, qualitatively, one variable helps to predict and explain a later one in a causal chain if they are not statistically independent of each other (Appendix 2). Quantitatively, if a coherent causal chain from exposure X to response Y contains an intermediate variable (a "causal mediator") Z, as in the DAG model $X \rightarrow Z \rightarrow Y$, then the exposure-response relation $P(y \mid x)$ should be consistent with the Chapman-Kolmogorov identity $P(y \mid x) = \Sigma_z P(z \mid x)P(y \mid z)$ (or, for conditional expected values, $E(y \mid x) = \Sigma_z P(z \mid x)E(y \mid z)$). This consistency check can be carried out by comparing observed conditional frequencies, corresponding to $P(y \mid x)$, to the model-predicted conditional frequencies, corresponding to $\Sigma_z P(z \mid x)P(y \mid z)$, e.g., using Pearson's chi-squared test of the null hypothesis that they are equal. Failure to reject this null hypothesis means that the CMC criterion is satisfied. (This is what it means for the causal mediator "to have different values for different exposures, as implied by the coherent causal explanation.") More generally, the CMC criterion checks that the effect of changes in one variable on changes in the conditional distribution of another is consistent with changes in the distributions of variables on the directed path(s) connecting them. For practical implementation, the *dagitty* package (Appendix 1) automatically lists

Table 9.2 Potential non-causal explanations for observed exposure-response dependencies, and methods to overcome them

Non-causal explanation	Methods for addressing non-causal associations
Unobserved (latent) confounders (Pearl and Mackenzie 2018)	These can be tested for and their effects modeled and controlled for using the *Tetrad, Invariant Causal Prediction,* and *BACKSHIFT* algorithms, among others
Spurious regression in time series or spatial observations with trends (Yule 1926)	Spurious regression arising from coincident trends can be detected and avoided by using conditional independence tests and predictive causation (e.g., Granger causality) instead of regression models
Collider bias; stratification or selection bias (Cole et al. 2010; Pearl and Mackenzie 2018)	A study that stratifies or matches individuals on certain variables, such as membership in an occupation, or an analysis that conditions on certain variables by including them on the right-hand side of a regression model, can induce exposure-response associations if the variables conditioned, matched, or stratified on are common descendents of the exposure and response variables. The association does not indicate causality between exposure and response, but that they provide alternative explanations of an observed value. Such biases can be avoided by using *dagitty* to compute adjustment sets and conditioning only on variables in an adjustment set
Other threats to internal validity (Campbell and Stanley 1963)	Threats to internal validity (e.g., regression to the mean, coincident historical trends, sample selection or attrition biases, reporting biases, etc.) were enumerated by Campbell and Stanley (1963), who also discuss ways to refute them as plausible explanations, when possible, using observational data
Model specification errors (Lenis et al. 2018; Linden 2017; Pirracchio et al. 2015)	Model specification errors arise when an analysis assumes a particular parametric modeling form that does not accurately describe the data-generating process. Assuming a linear regression model when there are nonlinear effects present is one example; omitting high-order interactions terms is another. Model specification errors can be avoided by using non-parametric model ensemble methods such as PDPs
P-hacking, i.e., adjusting modeling assumptions to produce an association (e.g., a statistically significantly positive regression coefficient) (Fraser et al. 2018)	Automated modeling using CAT or packages such as *randomForest* and *bnlearn* to automate modeling choices such as which predictors to select, how to code them (i.e., aggregate their values into ranges), and which high-order interactions to include can help to avoid p-hacking biases
Omitted errors in explanatory variables. (Rhomberg et al. 2011)	Using job exposure matrices, remote-sensing and satellite imagery for pollutant concentration estimation, or other error-prone techniques for estimating exposures, creates exposure estimates for individuals that can differ substantially from their true exposures. In simple regression models, omitting errors from the estimated values of explanatory variables tends to bias regression coefficients toward the null (i.e., 0), but the bias can be in either direction in multivariate models, and failing to carefully model errors in explanatory variables can create false-positive associations. These errors and biases can be avoided by modeling errors in explanatory variables

(continued)

Table 9.2 (continued)

Non-causal explanation	Methods for addressing non-causal associations
Omitted interdependencies among explanatory variables. (Pearl and Mackenzie 2018; Textor et al. 2016)	Regression models that ignore dependencies among right-hand side variables can create non-causal exposure-response associations. This can be avoided by using *dagitty* to compute adjustment sets for the causal effect of exposure on response and then conditioning on variables in an adjustment set to estimate that effect

testable conditional independence implications of DAG models (Textor et al. 2016). They can be used together with Chapman-Kolmogorov identities for coherence of probabilistic dependencies among variables in a DAG to generate tests of a coherent causal explanation presented as a causal DAG model. The CMC criterion can be applied provided that the DAG model describes multiple exchangeable instances for which data are available (e.g., different individuals with different exposures but with the same pathways through a biological causal network linking observed exposures to observed responses for each individual).

Intuitively, the CMC criterion uses a coherent causal explanation to make testable predictions about changes in observable variables in a postulated causal pathway or network in response to different exposures, and then tests the predictions by observing the values of the variables to see whether they vary with exposure as predicted. This strengthens the causal coherence criterion by moving from simply imagining a possible biologically plausible explanation for an exposure-response relationship to adducing evidence that its quantitative predictions are actually confirmed by observations. Thus, if a proposed biologically plausible explanation postulates that sufficiently high and prolonged exposures to a substance increase risk of lung inflammation by inhibiting phagocytosis and increasing pyroptosis in alveolar macrophages, then the causal mediation criterion could be satisfied by data showing that phagocytosis is inhibited and pyroptosis increased in alveolar macrophages of sufficiently highly exposed subjects compared to exchangeable unexposed or less exposed subjects.

Specificity and Biological Gradient: Special Cases of Testable Implications

Although Hill did not include confirmation of causal mediation among his considerations, he did consider two qualitative properties of exposure-response relations, *specificity* and *biological gradient*, that are testable implications of a particularly simple causal model: a single causal chain leading from exposure to response probability with each variable along the chain being an increasing function of its predecessor. The popular linear no-threshold (LNT) modeling assumptions imply a biological gradient, and an exposure-specific response (e.g., malignant mesothelioma in response to asbestos exposure) with a single causal chain certainly simplifies causal inference, but many epidemiologists no longer consider these two considerations to be very useful, as real-world causal exposure-response relationships often violate them (Rothman and Greenland 2005; Ioannidis 2016; Pearl and Mackenzie 2018). Many examples of U-shaped and J-shaped dose-response relations indicating hormesis are now known. The same exposure often causes multiple distinct effects (e.g., chronic inflammation, fibrosis, emphysema, lung cancer, cardiovascular disease, autoimmune responses, etc.)

The causal graph framework subsumes specificity and biological gradient as special cases: they are testable predictions made if the causal graph structure is a single chain and the CPTs are such that the conditional response probability increases monotonically with exposure, respectively. The causal mediation confirmation criterion generalizes such considerations by allowing for other testable predictions in more general causal network models. It can be satisfied not only by showing that a biological gradient occurs when one is predicted by the coherent causal explanation, but also by showing that J-shaped or U-shaped or n-shaped exposure-response relationships occur when they are predicted. Molecular epidemiological observations, e.g., comparing biomarker levels in exchangeable exposed and unexposed subjects, can provide powerful evidence that either supports a biologically plausible causal explanation by confirming its predictions, or refutes it by falsifying its predictions. The CMC criterion attempts to capture this aspect of the scientific process for establishing or refuting causal theories based on how well observations match their testable predictions.

Summary: Making Limited Progress in the Spirit of Hill

The main perspective offered in this chapter is that the following criteria or rules of evidence ought to be satisfied—or at least not known to be violated—before it is concluded that data indicate that exposure is a plausible manipulative cause of response:

1. *Mutual information*: Exposure should help to predict response; the two should be statistically dependent. (We propose this as a replacement for Hill's consideration of strength of association.)
2. *Directed dependence*: Information should flow from exposure to response over time, so that past changes in exposures help to predict future changes in responses. (This is proposed as a generalization and refinement of Hill's consideration of temporality. We also propose that, together with the CMC criterion, it can replace Hill's "Experiment" consideration, since information from experiments, natural experiments, quasi-experiments, and other sources can be used to help establish directions of dependence, as discussed in Appendix 5.)

3. *Internal consistency*: The predictive causal exposure-response relations calculated from different adjustment sets should be mutually consistent. (This is proposed as an addition to Hill's consideration of consistency.)
4. *External consistency*: Conditional probabilities for response given exposure and other direct causes should be the same in different populations and studies. (This invariant causal prediction (ICP) property is proposed as a refinement of Hill's consideration of consistency. It should hold if the response CPT correctly describes the causal dependence of response probability on its direct causes.)
5. *Causal coherence and biological plausibility*: There should be at least one identified biologically plausible explanation for the exposure-response dependency. (It can be represented by one or more directed paths through a causal biological network connecting exposure to response via a sequence of mechanisms. We propose this a replacement for Hill's much weaker considerations of coherence, biological plausibility and analogy.)
6. *Causal mediation confirmation criterion*: Quantitative testable implications of one or more biologically plausible explanations should be consistent with observations. More specifically, mediating variables should be related to both exposure and response as predicted by the biologically plausible explanations. (This is proposed as a replacement and generalization of Hill's considerations of specificity and biological gradient, which are testable predictions that only hold in special cases.)
7. *Refutation of non-causal explanations*: Non-causal explanations should not fully account for the observed exposure response dependency. (This is a proposed addition to Hill's considerations.)

Table 9.3 summarizes statistical tests for applying each criterion to data. We propose that clear violations of any of these criteria should be resolved before concluding that the evidence considered makes manipulative causation plausible or probable. Conversely, satisfying these criteria, while not necessarily proving that manipulative causation explains the observed exposure-response dependency, renders it plausible in light of existing evidence. We thus view criteria 1–7 as a useful screen for establishing whether data support testable implications of a manipulative causal exposure-response relation. Following Hill, we propose them not as logically necessary and sufficient conditions for establishing manipulative causality from observational data, but as constraints to help classify data as supporting, failing to support, or being ambiguous about whether exposure is a manipulative cause of a response. If none of them is satisfied, the data do not support this causal hypothesis. If all are satisfied, then the data are consistent with this hypothesis in nontrivial ways. If it is uncertain whether some of them are satisfied, then the evidence for manipulative causation is also intermediate between these relatively strong extremes.

Table 9.3 Tests and methods for implementing proposed criteria to assess consistency of data with a manipulative causal dependency of response *Y* on explanatory variable *X*

Criterion	Test	Methods
Mutual information	Reject null hypothesis that *Y* is conditionally independent of *X*	Reject null hypothesis if *X* and *Y* are linked in DAG models learned from data by causal discovery algorithms (e.g., those in the *bnlearn* package). Other tests for independence (e.g., chi-squared tests) can also be used
Directed dependence	For longitudinal data: reject null hypothesis that future values of *Y* are conditionally independent of past values of *X*, even after conditioning on past values of *Y* and other variables For cross sectional data: reject null hypothesis that direction of dependence is undetermined by data	Reject null hypothesis if *X* and *Y* are linked in DBNs or DIGs learned from data (e.g., via the *bnstruct* package). Granger tests can also be used for time series data Reject null hypothesis if constraints determining the direction of an arrow can be identified from data (e.g., using LiNGAM, BACKSHIFT, or Simon-Iwasaki causal ordering)
Internal consistency	Do not reject null hypothesis that effects estimated from different adjustment sets are the same	Reject null hypothesis if confidence bands for effects estimated from different adjustment sets do not overlap
External consistency	Do not reject null hypothesis that response CPTs estimated from different studies or data sets are the same	Reject null hypothesis if study ID is a parent of response in DAG models learned from data
Causal coherence and biological plausibility	Reject the null hypothesis that identified biologically plausible pathway(s) directed from *X* to *Y* cannot explain the observed dependence of *Y* on *X*	Reject the null hypothesis if one or more biologically plausible coherent causal explanations (pathways in a causal biological network) are identified that can explain the dependence of *Y* on *X*
Causal mediation confirmation	Do not reject the null hypothesis that variations in *Y* caused by variations in *X* are explained by resulting variations in mediating variables (e.g., as described by the Chapman-Kolmogorov identities implied by an explanation in the form of a probabilistic causal graph model)	Reject null hypothesis if variations in mediating variables do not explain variations in *Y* for different values of *X* (e.g., if a chi-squared test rejects the conditional independence and Chapman-Kolmogorov implications of the explanation)
Refutation of non-causal explanations	Reject the null hypothesis that the observed statistical dependence of *Y* on *X* has a non-causal explanation	Reject the null hypothesis if threats to validity are refuted (Campbell and Stanley 1963)

Discussion

A Dose of Humility: How Useful are Our Proposed Causal Criteria for Observational Data?

This completes our attempt to update Hill's considerations to assess consistency with manipulative causation. But how well do the proposed updated criteria work in enabling valid conclusions about manipulative causation to be drawn from observational data? Hill himself opined that "What I do not believe—and this has been suggested—is that we can usefully lay down some hard-and-fast rules of evidence that must be obeyed before we accept cause and effect." Half a century later, despite substantial progress and many useful applications of causal graph methods in epidemiology and other fields (e.g., Pearl and Mackenzie 2018), caution is still certainly warranted in using current data science principles and causal discovery algorithms to help discover causal graph models from observational data. As noted by Oates et al. (2016):

> There has long been (in our view justifiable) empirical skepticism towards *de novo* causal discovery… The issue is that it is difficult to empirically validate causal discovery on a given problem using data at hand, leaving the analyst unsure as to whether or not the output of a given procedure should be trusted. This goes a bit further than familiar issues of statistical uncertainty, since the underlying concern is of a potentially profound mismatch between critical assumptions and the real data-generating system.

The key issue here is not that causal discovery algorithms are unable to generate and quantify causal graph models from data, such as that in Fig. 9.1, but rather that it is not necessarily clear how credible they are and to what extent their predictions and effects estimates can be trusted. In Fig. 9.1, for example, how well justified would we be in concluding that increasing income would (probably) decrease heart disease risks? Conversely, can we confidently conclude that reducing fine particulate matter (PM2.5) concentrations would *not* detectably reduce heart attack risk in this study population? How can we know, and how sure can we be about the answers? Such questions must be addressed to create useful guidelines for inferring whether and to what extent data support conclusions about manipulative causation.

To help address them, the criteria we have discussed provide testable implications that usually hold when manipulative causation is present and not when it is absent, despite the logical possibility of devising counter-examples, e.g., with quantitative effects along different pathways exactly canceling each other. They attempt to approximate "some hard-and-fast rules of evidence that must be obeyed before we accept [manipulative] cause and effect." Specifically, if available data allow confident rejection of the null hypotheses of no mutual information (i.e., conditional independence) between exposure and response, no predictive causation or directed information flow from exposure to response, no biologically plausible pathway from exposure to response, and no refutation of alternative (non-causal) explanations or threats to validity of causal inference (Campbell and Stanley 1963), but do not allow confident rejection of the null hypotheses of internal consistency, external consistency (ICP), or causal mediation confirmation, then we would view these findings as providing substantial evidence that available data are consistent with a manipulative causal dependence of response on exposure. Satisfying the causal coherence/biological plausibility and CMC criteria makes it plausible that the variables in a causal graph mediate manipulative causation. However, additional variables not included in the data might be needed to give a more detailed causal explanation of how effects are transmitted. Moreover, the size, design, and diversity of available data (e.g., are responses observed for enough different combinations of explanatory variables so that their CPTs can be estimated?) limit the valid causal conclusions that can be drawn. Thus, while failure to detect a direct causal relation between exposure and response can be used to put an upper bound on the plausible size of a possible undetected relationship, it cannot be used to prove that no such relationship exists.

Being able to systematically generate and test implications of causal graph models and providing algorithms to test them with stated levels of statistical confidence are real and substantial accomplishments (Pearl and Mackenzie 2018). But they do not guarantee that a unique, well-validated causal model can be discovered from data, or that a unique, precise estimate of the causal impact of a change in exposure on the change in response probability can be inferred from data. Indeed, much of the theoretical and practical progress in causal discovery that has been made since Hill's day consists of clarifying when such "identifiability" of causal graph models and effects is possible and what other variables must be observed to identify specified causal effects from observational data—in other words, whether adjustment sets exist that allow them to be estimated (Textor et al. 2016). As discussed earlier, in epidemiology and risk analysis, it is often the case that the directions of some of the arrows in a causal graph are not uniquely determined by observational data alone: the direction of information flow between variables sometimes cannot be determined without outside knowledge or experimental interventions. This does *not* mean that a causal graph model learned from the data via automated algorithms is meaningless or wrong, even if its arrows cannot be interpreted as showing directions of causation. Rather, it means that only arrows with their directions determined and justified by explicit constraints (Appendix 5) can be interpreted as showing the correct direction of information flow. Remaining arrows only reveal mutual information, and their directions lack further causal interpretation. (They do show how the joint distribution of variables can be decomposed into CPTs, but this is a probability concept rather than a causal one.)

Such difficulties and nuances of interpretation have led to widespread misconceptions about what causal graph methods and their equivalent structural equation models (SEMs) can and cannot do. (SEMs use systems of equations and error terms instead of nodes, arrows, and CPTs to encode the dependencies among variables.

SEMs and Bayesian networks (BNs) provide two mathematically equivalent ways of describing the joint probability distribution of a systems of discrete random variables (Druzdzel and Simon 1993).) For example, Bollen and Pearl (2013) identify and address the following misconceptions: "(1) SEMs aim to establish causal relations from associations alone, (2) SEMs and regression are essentially equivalent, (3) no causation without manipulation, (4) SEMs are not equipped to handle nonlinear causal relationships, (5) a potential outcome framework is more principled than SEMs, (6) SEMs are not applicable to experiments with randomized treatments, (7) mediation analysis in SEMs is inherently noncausal, and (8) SEMs do not test any major part of the theory against the data." To these, we can add the following misconceptions that we have sometimes encountered:

- "The validity of predictions from causal graph models learned by causal discovery algorithms depends upon the validity of the model specification provided to the algorithms." This critique does not apply to causal discovery algorithms such as those in *bnlearn* (Appendix 2) for which no model specification is provided. These algorithms learn causal graphs from data using nonparametric methods without assuming any specification. However, it is a valid critique of causal discovery algorithms that assume linear models or additive errors, such as some LiNGAM algorithms.
- "The validity of predictions from causal graph models learned by causal discovery algorithms depends upon the validity of data used to fit it." This is incorrect insofar as many causal discovery methods, including ICP and LiNGAM (Appendix 5), are robust to data corrupted by measurement errors, missing values, omitted or latent variables, and possibly unknown transformations of variables.
- "Uncontrolled confounding can lead to false conclusions from a Bayesian network or causal graph model just as easily as from any other association method." This is incorrect because adjustment sets show precisely how to correct for confounding in DAG models. This solves the long-standing problem in regression analysis of selecting which variables to include on the right-hand side. (This objection is a variation of the "SEMs and regression are essentially equivalent" misconception discussed by Bollen and Pearl (2013).)
- "Cross-sectional data does not allow one to assess the direction of causality in a causal graph any better than by any other method." This is mistaken because causal graph methods often allow the directions of many (sometimes all) arrows to be determined from cross-sectional data, as discussed in detail for the directed dependence criterion. LiNGAM and Simon and Iwasaki (1988) structural causal ordering are examples of techniques that allow arrow directions to be constrained by cross-sectional data (Appendix 5). The true statement that often not *all* arrow direction can be uniquely determined from cross-sectional observational data should not be confused with the mistaken generalization that causal graph algorithms do not provide sound methods for determining *some* arrow directions from cross-sectional data.

In summary, we agree with Oates et al. (2016) that, very often, "it is difficult to empirically validate causal discovery on a given problem using data at hand." Even if misconceptions are avoided, it remains true that a unique causally interpretable DAG or other causal graph model may be underdetermined by observational data, leaving it unclear how well the output from a causal graph-learning program represents manipulative causal relationships rather than only mutual information relationships between variables. Yet, we also believe that the proposed list of testable criteria 1–7 can be very helpful in establishing the extent to which data are consistent with a hypothesized manipulative causal exposure-response relationship. They can serve this purpose whether or not a fully developed, uniquely specified, well validated causal model can be obtained from the data. Our relatively modest goal of testing the consistency of data with a manipulative causal claim does not require a fully validated causal model to be discovered, but only that at least some of a series of consistency tests can be carried out. The least demanding of these, mutual information, can be carried out with most data sets that have exposure, response, and covariate variables. More demanding tests require more data; the most demanding, the CMC criterion, can be carried out only when enough knowledge and data are available to allow specific hypotheses about mediating variables to be formulated and tested.

Returning to interpretation of the link between income and heart disease in Fig. 9.1 exposes some of the limitations and remaining challenges for our proposed criteria for manipulative causation. Even if all of our criteria 1–7 are satisfied, *how* income is changed might very well affect how, if at all, it affects heart disease risk. If income is increased by inflating the currency equally for everyone (so that nominal incomes increase but real incomes do not), the effects on intermediate variables such as stress hormones and blood pressure, as well as on heart disease risks, might be quite different than they would be if income were increased by a tax break or subsidy that increased real incomes, or by extending working hours and increasing overtime pay, or by instituting a minimum wage and employment guarantees. Nothing in Fig. 9.1 (which refers exclusively to nominal incomes) expresses such distinctions. The concepts of "direct cause," "manipulation," and "intervention" applied to income envision exogenously setting it to new levels and having effects propagate to its children through their CPTs without having to specify how the change in income is brought about. But for variables such as income that provide aggregate summaries of many details, the hidden details of how the exogenous change is accomplished may matter for quantifying its likely effects. For now, the art of modeling—which includes supplying sufficient detail in the set of variables considered to allow for realistic modeling of the effects of decisions and interventions on outcomes of interest—remains beyond the scope of mainstream causal discovery algorithms, and requires human knowledge. Ongoing research in deep learning for feature selection and abstraction is starting to yield ways to automatically form hierarchies of abstraction for describing time series to bring out the relations between them more clearly (LeCun et al. 2015), but automated approaches for forming and validating high-level variables that summarize lower-level data adequately for purposes of causal modeling are not yet well developed and validated.

In Fig. 9.1, the absence of an arrow between PM2.5 and heart disease risk indicates that this data set was not found to provide evidence that PM2.5 is a manipulative cause of heart disease risk (although they might well have a strong, consistent, coherent, graded, biologically plausible, temporal association as judged by the Hill considerations, e.g., due to strong confounding by income and perhaps other variables, such as sufficiently lagged daily temperatures, not included in Fig. 9.1). In Fig. 9.1, this is due to failure of the mutual information criterion: heart disease risk is conditionally independent of PM2.5 given income. But "PM2.5," like "Income," is a highly aggregate variable. The same numerical value might represent quite different mixes of detailed components in different data sets. Therefore, failure to establish manipulative causality in one does not necessarily imply that it does not hold in others. Such realistic limitations and complexities highlight the simplistic and restrictive assumptions of idealized criteria that search for invariant, generalizable relations between direct causes and their effects in real-world data where the same measured values of variables such as income or PM2.5 may describe importantly different configurations of the detailed variables that actually drive response probabilities. As long as such aggregate variables are used to summarize observations, resulting vagueness and ambiguity about exactly what their values and represent and how changes in their values are accomplished will limit the power of criteria such as external consistency, causal coherence, and causal mediation confirmation to detect and generalize reliable manipulative causal dependencies among them using only one or a few data sets. Techniques for causal inference with latent variables, mentioned throughout our technical discussion of the proposed criteria, will be important as long as important but unobserved details affect the manipulative causal dependencies among observed variables.

Criteria 1–7 are proposed to help improve the odds of finding useful manipulative causal exposure-response relations in observational data. Despite their acknowledged limitations for realistically imperfect, incomplete, and aggregate data, we believe that applying them may help to reduce both false positives and false negatives in detecting manipulative causation. At present, this is a hope rather than a finding: practical experience, and perhaps additional refinements and extensions, will be needed to assess their practical value.

The CARET Trial Reconsidered

How might our proposed criteria have fared with the CARET trial (Omenn et al. 1996), in which strong, consistent, coherent, biologically plausible, graded associations, as judged by the Hill considerations, between levels of beta carotene or vitamin A in diet and blood serum and reduced risk of lung cancer failed to predict that manipulations to increase these levels significantly increased risk of lung cancer? Arguably, the data available before the trial began showed that the criteria of mutual information, and perhaps to some extent refutation of non-causal explanations, were satisfied for observe dependencies between levels of beta carotene and vitamin A in diet and levels of lung cancer risk. The hypothesis that increasing levels of the anti-oxidants by administration of beta carotene and vitamin A would probably increase scavenging of free radicals, reduce oxidative damage, slow cell death and replacement, and reduce lung cancer risk was biologically plausible in the Hill sense that it did not contradict current biological knowledge. But increases in levels of beta carotene and vitamin A over time had not been observed to predict reductions in lung cancer risk over time (directed information), and no observations confirmed that increases in antioxidants reduced the net proliferation rate of premalignant (or malignant) cells, which is a presumed direct cause of lung cancer risk (coherent causal explanation, CMC). (Indeed, after the trial was stopped due to observed increases in lung cancer risks, some argued that the increased antioxidants scavenged reactive oxygen species that would have slowed or prevented the development of cancer.) Thus, our criteria for causal coherence and biological plausibility and causal mediation confirmation were not met. Nor were the criteria for internal and external consistency, since no causal DAG models, adjustments sets, or CPTs had been developed and no checks for the ICP property had been performed for response CPTs. (Hill consistency for associations held, but this did not establish ICP for response CPTs.) In short, at least 5 of our 7 proposed criteria for concluding that increased beta carotene and vitamin A would reduce lung cancer were not met. In this example, it appears that existing evidence gave good reason to conclude that the observed negative association between beta carotene and vitamin A in diet (and blood serum) and lung cancer risk was "causal" in Hill's sense (i.e., an appropriate label for associations based on his considerations of strength, consistency, coherence, and biological plausibility). However, the same evidence provided little or no reason to expect that increasing beta carotene and vitamin A would reduce lung cancer risk based on our proposed criteria for manipulative causation. Acknowledging hindsight bias, we nonetheless conclude that if our criteria had been used to guide the assembly of supporting evidence, a case for manipulative causation could not easily have been made, and a need for more research to seek a coherent causal explanation might have been identified.

Elevating the Goals for Causality Criteria: Returning to Manipulative Causation

Hill (1965) began with a question of great practical importance: "But with the aims of occupational, and almost synonymous preventive, medicine in mind the decisive question is where the frequency of the undesirable event B will be influenced by a change in the environmental feature A." This is a question about *manipulative* causation: how would changing exposure or environmental feature A affect the frequency or probability distribution of undesirable event B? However, rather than considering how to obtain valid scientific answers to this key question, Hill reformulated it as follows: "Disregarding then any such problem in semantics we have this situation. Our observations reveal an association between two variables, perfectly clear-cut and beyond what we would care to attribute to the play of chance. What aspects of that association should we especially consider before deciding that the most likely interpretation of it is causation?" This is a very different question. It is no longer about how to discover how changing one variable will change another. Rather, it is about what to consider before *deciding* that the *most likely* explanation or interpretation for an observed *association* between two variables is "causation" (without definition or explanation of what that label means, i.e., its semantics). This is a much less interesting question. Making a decision about how to label an association between two variables is less useful for effective decision-making than figuring out how changing one variable would change the other. The crucial idea of manipulative causation has disappeared. Moreover, the new question of what to consider before "deciding that the most likely interpretation of it [the observed association] is causation" imposes a false dichotomy: that an association is either causal or not, rather than some fraction of it is causal and the rest not.

Hill himself did not consider that his considerations solved the scientific challenge of discovering valid manipulative causal relationships from data. He offered them more as a psychological aid for helping people to make up their minds about what judgments to form, motivated largely by the specific historical context of deciding whether available evidence warranted a conclusion that smoking causes lung cancer: "None of my nine viewpoints can bring indisputable evidence for or against the cause-and-effect hypothesis and none can be required as a *sine qua non*. What they can do, with greater or less strength, is to help us to make up our minds on the fundamental question—is there any other way of explaining the set of facts before us, is there any other answer equally, or more, likely than cause and effect?" This is a very limited goal. It is often the case that even the most likely explanation is quite likely to be wrong if many other plausible explanations are available. Since Hill avoided specifying what he means by "cause and effect" in this context (instead "disregarding then any such problem in semantics"), his considerations must serve as an implicit definition: for Hill, "cause and effect" is a label that some people feel comfortable attaching to observed associations after reflecting on the considerations on the left side of Table 9.1. Deciding to label an association as "causal" based on the Hill considerations does not imply that a causal interpretation is more likely than not to be correct or that other non-causal explanations, such as those in Table 9.2, are unlikely to be correct. Indeed, by formulating the problem as "is there any other answer equally, or more, likely than cause and effect?" Hill allows labeling an association as causal even if it almost certainly isn't, as when there are many equally likely non-causal interpretations and one slightly more likely causal one. Deciding to label an exposure-response association as "causal" in this sense does not require or imply that it has any specific real-world properties, such as that changing exposure would change the probability of response. It carries no implications for consequences of manipulations or for decisions needed to achieve a desired change in outcome probabilities.

Given these limitations, Hill's considerations are usually not suitable for discovering or quantifying manipulative causal relationships. Hence, they are usually not suitable for supporting policy recommendations and decisions that require understanding how alternative actions change outcome probabilities. Associations can be useful for identifying non-random patterns that further investigation may explain. That associational studies are nonetheless widely interpreted as if they had direct manipulative causal implications for policy indicates an important need and opportunity to improve current practice.

Studies that explicitly acknowledge that statistical analyses of associations are useless for revealing how policy changes affect outcome probabilities are relatively rare. One exception is a National Research Council report on *Deterrence and the Death Penalty* that "assesses whether the available evidence provides a scientific basis for answering questions of if and how the death penalty affects homicide rates." This report "concludes that research to date on the effect of capital punishment on homicide rates is not useful in determining whether the death penalty increases, decreases, or has no effect on these rates" (National Research Council 2012). Such candid acknowledgements of the limitations of large bodies of existing research and discussion clear the way for more useful future research. In public health risk research, fully accepting and internalizing the familiar warnings that correlation is not causation, that associations are not effects (Petitti 1991), and that observations are not actions (Pearl 2014) may help to shift practice away from relying on associational considerations such as the Hill considerations on the left side of Table 9.1 toward fuller use of causal discovery principles and algorithms such as those on the right side of Table 9.1.

Doing so can potentially transform the theory and practice of public health research by giving more trustworthy answers to more useful causal questions such as how changing exposures would change health effects. This requires returning scientific focus to Hill's first question—"the decisive question is where the frequency of the undesirable event B will be influenced by a change in the environmental feature A"—rather than to the more limited question he ended up addressing, of how to help us make up our minds whether to attach a label of "causal," having no specific definition and no clear practical implications, to an association that might have multiple plausible explanations. The criteria we have proposed are intended to facilitate this reorientation.

Conclusions

This chapter how the classic Hill (1965) considerations for judging whether an observed exposure-response association should be labeled "causal" can be updated to take into account ideas, principles, and algorithms drawn from more recent work on causal graph models (Pearl and Mackenzie 2018). The main ideas for refining and augmenting Hill's considerations are as follows.

- *Focus specifically on manipulative causation.* The proposed criteria are intended to help decide whether available evidence is consistent with a conclusion of manipulative causation between exposure and response in an exposed population. This means that changing exposure will change response probabilities. Although our proposed criteria use a mix of concepts from predictive causation (mutual information and directed dependence criteria), structural causation (internal consistency criterion), mechanistic causation (causal coherence/ biological plausibility and causal mediation confirmation criteria), as well as other principles (external consistency, refutation of non-causal explanations), their common goal is to help determine whether and by how much changing exposure changes response probabilities.
- *Consider quantitative estimation of causal exposure-response functions rather than on qualitative assessment of whether causality is the best explanation for an observed exposure-response association.* We believe that the question that Hill ended up addressing—"What aspects of [an exposure-response] association should we especially consider before deciding that the most likely interpretation of it is causation?"—is not as useful as it could be insofar as it presupposes that a single interpretation (causal or not) is best. In reality, as causal graph modeling makes very clear, many (perhaps most) exposure-response associations have multiple contributing explanations. In Fig. 9.1, for example, the association between smoking and heart disease risk is explained in part by a direct effect of smoking (indicated by the arrow from *Smoking* to *HeartDiseaseEver*), which could well represent manipulative causation; but it is also explained in part by confounding due to the facts that men smoke more and, independently, have higher age-specific heart disease risks (Fig. 9.2); and that people with higher incomes are likely to smoke less and, independently, to have lower risk of heart disease. The machinery of adjustment sets (Appendix 1) and quantitative estimation of total and direct causal effects (Fig. 9.3) takes into account the multiple paths that can contribute to the total observed exposure-response association and corrects for non-causal sources of association (e.g., confounding or collider bias). This clear recognition that an exposure-response association may have multiple explanations, some causal and some not, and that possibly causal components (corresponding to effects along directed paths from exposure to response) can and should be clearly distinguished from non-causal components (corresponding to associations due to common ancestors or descendants) in quantify-

ing causal effects of exposure on response probability, marks a substantial conceptual advance over efforts to decide whether an observed association as a whole is "causal." The most usual correct answer is that some fraction is causal and the rest is not, and it is useful to quantify both.

- *When possible, use causal graphs to explicate causal hypotheses and theories connecting exposure to response via causal networks.* We propose that, when there is sufficient knowledge to do so, causal graphs or causal biological networks should be used to show hypothesized explanations for how changes in exposure might cause changes in response probabilities by changing one or more mediating variables on causal pathways between them. Creating such causal explanations is admittedly more burdensome than Hill's considerations of coherence and biological plausibility, which amount to not contradicting current biological knowledge. Although more demanding, constructing one or more biologically plausible explanations for how manipulative causation might work, consistent with current biological knowledge, can be very valuable in generating additional testable implications and improving understanding of risks suggested by epidemiological data.
- *Identify testable implications of the hypothesized manipulative causal exposure-response dependency.* We propose to model the causal dependency of response probabilities on exposure and other causes using causal graphs and CPTs, and then to generate testable implications via the mutual information, directed dependence, and external consistency criteria. If one or more biologically plausible (i.e., coherent causal) explanations of a manipulative causal exposure-response relationship is available in terms of proposed pathways of mechanisms leading from exposure to response, then its testable implications can also be used (internal consistency, causal coherence, and causal mediation confirmation criteria).
- *Use data to test the testable implications of causal hypotheses, theories, and explanations.* Our proposed criteria, other than refutation of non-causal explanations, suggest testable implications of the hypothesis of a manipulative causal dependency of response probability on exposure, as summarized in Table 9.3. Conceptually, each of the proposed criteria is tested by determining whether data allow confident rejection of the null hypothesis that it does not hold. The various specific statistical tests, algorithms, and software mentioned in the right column of Table 9.3 provide practical means for accomplishing this testing.
- *Refute non-causal explanations for observed exposure-response dependencies.* Manipulative causation becomes a more likely explanation for observed exposure-response dependencies when other explanations are ruled out, e.g., using the techniques on the right side of Table 9.2.

The following chapters undertake several case studies indicating how these criteria can be applied in practice to provide a useful complement and update to Hill's original considerations by providing a sharper focus on manipulative causation. They allow modern data science tools and causal graph methods to be used in formulating testable implications of hypothesized manipulative causal exposure-response dependencies and biologically plausible explanations for them; in testing these implications using available data; and in quantify causal effects when they exist. Doing so helps to pursue what Hill referred to as "the decisive question" of how the frequency of an undesirable event B will be influenced by a change in an environmental feature A.

Appendix 1: Directed Acyclic Graph (DAG) Concepts and Methods

Directed acyclic graph (DAG) models are widely used in current causal analytics. This appendix summarizes some key technical methods for causal graphs and DAG models; for details, see Pearl and Mackenzie (2018). A DAG model with CPTs for its nodes but with arrows that do not necessarily represent causality is called a *Bayesian network* (BN). In a BN, observing or assuming the values of some of the variables lets conditional probability distributions for the other variables be calculated based on this information; this is done by *BN inference algorithms*. If the BN is also a causal graph model, then it also predicts how changing some variables (e.g., smoking or income) will change others (e.g., risk of heart disease). Whether a given BN is also a causal graph model is determined by whether its arrows and CPTs describe how changes in the values of some variables change the conditional probability distributions of their children.

The arrows in a BN reveal *conditional independence* relations between variables, insofar as a variable is conditionally independent of its more remote ancestors and other non-descendants, given the values of its parents; in machine learning for BNs, this is commonly referred to as the *Markov condition*. It is often assumed to hold, but can be violated if unobserved latent variables induce statistical dependencies between a variable and its non-descendants, even after conditioning on its parents. The conditional independence relations implied by a DAG model can be systematically enumerated using graph-theoretic algorithms available in free software packages such as *dagitty* (Textor et al. 2016). They can be tested for consistency with available data by statistical independence testing algorithms included in statistical packages such as the R package *bnlearn* (Scutari and Ness 2018). For example, the three DAG models $X \rightarrow Y \rightarrow Z$, $X \leftarrow Y \rightarrow Z$, and $X \leftarrow Y \leftarrow Z$ all imply that X and Z are conditionally independent of each other given the value of Y; in the technical literature they are said to be in the same *Markov equivalence class*, meaning that they imply the same conditional independence relations. By contrast, the DAG $X \rightarrow Y \leftarrow Z$ implies that X and Z are unconditionally independent but dependent after conditioning on Y. Thus, in the model *Age* → *Heart_disease* ← *Sex*, data analysis should show that *Age* and Sex are independent (so that their joint frequency distribution is not significantly different from the product of their marginal distributions), but that they become dependent after conditioning on a value of *Heart_disease*.

(If, as in Fig. 9.2, heart disease risk increases with both age and sex (coded so that 0 = female, 1 = male) then among patients with heart disease, age and sex should be negatively associated, since a low value of either one implies that the other must have a high value to make heart disease a likely outcome. Thus conditioning on heart disease makes sex and age informative about each other even if they are statistically independent in the absence of such conditioning.) Since DAG models imply such testable conditional independence relations and dependence relations, the observed dependence and independence relations in a data set can be used to constrain the DAG models describing the joint probability distribution of its variables.

Adjustment Sets, Partial Dependence Plots, and Estimation of Predictive Causal Effects

Unbiased estimates of both the direct causal effect and the total causal effect of one variable on another in a DAG can be obtained by conditioning on an appropriate *adjustment set* of other variables (Greenland et al. 1999; Glymour and Greenland 2008; Knüppel and Stang 2010; Textor et al. 2016). The *direct effect* on *Y* of a unit change in *X* is defined as the change in the conditional mean (or, more generally, the conditional probability distribution) of *Y* in response to a unit change in *X* when only *X* and *Y* are allowed to change and all other variables are held fixed. By contrast, the *total effect* on *Y* of a unit change in *X* is the change in the conditional probability distribution (or its mean) for *Y* when *X* changes and other descendants of *X* are allowed to change in response. For example, in Fig. 9.1, the total effect of an increase in age on risk of heart disease includes the indirect effects of age-related changes in income (including any income-related changes in smoking behavior), as well as the direct effect of age itself on heart disease risk. An *adjustment set for* estimating the direct or total causal effect of *X* on *Y* in a causal DAG model is a set of observed variables to condition on to obtain an unbiased estimate of the effect (e.g., by including them on the right side of a regression model, random forest model, or other predictive analytics model with the dependent variable *Y* to be predicted shown on its left side, and explanatory variables (predictors) consisting of *X* and the members of the adjustment set on its right side). (The main technical idea for estimating total causal effects is that an adjustment set blocks all noncausal paths between *X* and *Y* by conditioning on appropriate variables (such as confounders) without blocking any causal path from *X* and *Y* by conditioning on variables in directed paths from *X* to *Y* and without creating selection, stratification, or "collider" biases by conditioning on common descendants of *X* and *Y*) (Elwert 2013).) A *minimal sufficient adjustment set* is one with no smaller subset that is also an adjustment set. Graph-theoretic algorithms for determining exactly which direct causal effects and total causal effects in a DAG model can be uniquely identified from data and for computing minimal sufficient adjustment sets for estimating them are now readily available; the *dagitty* package (Textor et al. 2016) carries out these calculations, among others. (It also determines which path coefficients can be estimated in a path analysis model; what instrumental variables can be used to estimate a given effect via regression modeling; and what testable conditional independence relations among variables are implied by a given DAG model.)

Given a cause such as income and an effect such as heart disease risk in a DAG model such as Fig. 9.1, adjustment sets can be automatically computed for estimating both the "natural direct effect" of the cause on the effect (i.e., how does the effect change as the cause changes, holding all other variables fixed at their current values?) and also the "total effect" (i.e., how does the effect change as the cause changes, allowing the effects on other variables to propagate via all directed paths from the cause to the effect?) In Fig. 9.1, {*Age*, *Sex*, *Smoking*} is an adjustment set for estimating the natural direct effect of income on heart disease risk, and {*Age*, *Sex*, *Education*} is an adjustment set for estimating the total effect of income on heart disease risk, part of which is mediated by the effect of income on smoking.

Given a causal DAG model, it is often useful to distinguish among several different measures of the "effect" of changes in one variable on changes in another. The *pure* or *natural direct effect* of one variable on another (e.g., of exposure on risk) is defined for a data set by holding all other variables fixed at the values they have in the data set, letting only the values of the two specified variables change; this is what a partial dependence plot shows. In Fig. 9.1, the natural direct effect of changes in income on changes in heart disease risk might be the effect of interest, in which case it can be estimated as in Fig. 9.3. Alternatively, the effect of interest could be the *total* effect of changes in income on changes in heart disease risk, calculated (via BN inference algorithms) by allowing changes in income to cause changes in the conditional probability distribution of smoking (as described by the CPT for smoking) to capture indirect, smoking-mediated effects of changes in income on changes in heart disease risk. In general, the effect of interest can be specified by selecting a target variable and one of its causes (parents or ancestors in the causal DAG model) and then specifying what should be assumed about the values of other variables in computing the desired effect (e.g., direct or total) of the latter on the former. This is relatively straightforward in linear causal models, where direct effects coincide with regression coefficients when appropriate adjustment sets are used (Elwert 2013). In general, however—and especially if the CPT for the response variable has strong nonlinear effects or interactions among its parents affecting its conditional probability distribution—it is necessary to specify the following additional details: (1) What are the initial and final values of the selected cause(s) of interest? The PDP in Fig. 9.3 shows the full range of incomes (codes 1–8) on the horizontal axis and corresponding conditional mean values for heart disease risk on the vertical axis. This allows the change in risk as income is changed from any initial level to any final level (leaving the values of other variables unchanged) to be ascertained. (2) When other variables are held fixed, at what levels are they held fixed? In quantifying the direct effect on heart disease risk of a change in income from one level to another, should the conditional probability of smoking be fixed at the level it had before the change, or at the level its CPT shows it will have after the change? The answer affects the precise interpretation of the estimated effect of the change in income on heart disease risk.

Such distinctions have been drawn in detail in epidemiology, yielding concepts of controlled direct effects (with other variables set to pre-defined levels), pure or natural direct effects (with other variables having their actual values in a data set and effects being averaged over members of the population), indirect and controlled mediated effects, total effects, and so forth (e.g., Tchetgen Tchetgen and Phiri 2014; VanderWeele 2011; VanderWeele and Vansteelandt 2009; McCandless and Somers 2017). The natural direct effect PDP in Fig. 9.3 shows the predicted conditional mean value of the selected dependent variable (here, heart disease risk) for each value of the selected explanatory variable (income), as estimated via a random forest algorithm holding other variable values fixed at their observed levels. Figure 9.3 was created (via the free Causal Analytics Toolkit (CAT) software, http://cloudcat.cox-associates.com:8899/) by applying the *randomForest* package in R to generate a partial dependence plot (PDP) (Greenwell 2017) for heart disease risk vs. income, conditioning on the adjustment set {*Age*, *Sex*, *Smoking*} (Cox Jr. 2018). DAG algorithms also allow other effects measures to be computed if they are identifiable from the data (VanderWeele 2011). In the presence of latent variables, they permit tight bounds and sensitivity analyses for various effects measures to be calculated (Tchetgen Tchetgen and Phiri 2014; McCandless and Somers 2017).

Appendix 2: Information Theory and Causal Graph Learning

Two random variables are *informative* about each other, or have *positive mutual information* (as measured in units such as bits, nats, or Hartleys), if and only if they are not statistically independent of each other (Cover and Thomas 2006). If two variables are mutually informative, then observing the value of one helps to predict the value of the other, in that conditioning on the value of one reduces the expected conditional entropy (roughly speaking, the "uncertainty") of the other. Two random variables can be mutually informative even if they have zero correlation, as in the $KE = 1/2MV^2$ example in the text; or they can be correlated and yet have zero mutual information, as in the case of the values of statistically independent random walks.

A key principle of many causal discovery algorithms is that *direct causes are informative about (i.e., help to predict) their effects*, even after conditioning on other information; thus, they have positive mutual information. Conversely, *effects are not conditionally independent of their direct causes*, even after conditioning on other information such as the values of more remote ancestors (except in trivial and statistically unlikely cases, such as that several variables are deterministically related, so that conditioning on one is equivalent to conditioning on all). Predictive analytics algorithms that select predictors for a specified dependent variable (such as random forest (https://cran.r-project.org/web/packages/randomForest/index.html)) automatically include its parents among the selected predictors, at least to the extent that the predictive analytics algorithm is successful in identifying useful predictors. If the Markov condition holds (Appendix 1), so that each variable is conditionally independent of its more remote ancestors after conditioning on its parents, then these algorithms will also exclude more remote ancestors that do not improve prediction once its parents' values are known. In this way, predictive analytics algorithms help identify potential causal DAG structures from data by identifying conditional independence constraints (represented by absence of arrows between variables in a DAG) and information dependency constraints (represented by arrows in a DAG).

Statistical software packages such as *bnlearn* (Scutari and Ness 2018), *CompareCausalNetworks* (Heinze-Deml and Meinshausen 2018), and *pcalg* (Kalisch et al. 2012) carry out conditional independence tests, assess mutual information between random variables, and identify DAG structures that are consistent with the constraints imposed by observed conditional independence relations in available data. The "structure-learning" algorithms in these packages complement such constraint-based methods with score-based algorithms that search for DAG models to maximize a scoring function reflecting the likelihood of the observed data if a model is correct (Scutari and Ness 2018). Many scoring functions also penalize for model complexity, so that the search process seeks relatively simple DAG models that explain the data relatively well, essentially searching for the simple explanations (DAG models) that cover the observed facts (data). These packages provide a mature data science technology for identifying DAG models from data, but they have several limitations, discussed next. If they were perfect, then Hill's question "Is X strongly associated with Y?" could be replaced with "Is X linked to Y in DAG models discovered from the data?" and the answer, which would be yes when X and Y are mutually informative and not otherwise, could be obtained by running these DAG-learning packages. In practice, current DAG-learning packages approximate this ideal in sufficiently large and diverse data sets (with all variables having adequate variability), but the approximation is imperfect, as discussed next.

Some Limitations of Graph-Learning Algorithms: Mutual Information Does not Necessarily Imply Causality

DAG structures clarify that a strong statistical association—or, more generally, a strong statistical dependency, or high mutual information—can arise between two variables X and Z not only if they are linked by a causal chain, as in the direct causal relation $X \rightarrow Y$ or the indirect causal relation in the chain $X \rightarrow Y \rightarrow Z$, but also because of confounding by another variable that is their common ancestor, such as

Y in $X \leftarrow Y \rightarrow Z$; or from reverse causation, as in $X \leftarrow Y \leftarrow Z$; or from selection bias created by stratification, selection, or conditioning on a common descendant, such as Y in $X \rightarrow Y \leftarrow Z$. To Hill's idea that a strong association suggests causality, DAG modeling adds the important refinement that strong mutual information suggests causality *if and only if it arises from one or more directed paths* between the cause and effect variables in a causal DAG model. An association or other statistical dependence, no matter how strong, does not suggest causality if it arises from other graph-theoretic relations such as confounding, selection bias, or reverse causation. Information theory implies also that direct causes are at least as informative about their effects as more remote indirect causes. (In the chain $X \rightarrow Y \rightarrow Z$, the mutual information between Y and Z is at least as large as the mutual information between X and Z (Cover and Thomas 2006).) Only variables that are linked to a node by an arrow are candidates to be its direct causes, assuming that direct causes are informative about their effects and that the DAG model faithfully displays these mutual information relations.

Non-parametric graph-learning algorithms in free software such as the *bnlearn*, *pcalg*,and *CompareCausalNetworks* packages in R can play substantially the role that Hill envisioned for strength of association as a guide to possible causation, but they do not depend on parametric modeling choices and assumptions; in this sense, their conclusions are more reliable, or less model-dependent than measures of association. DAGs are also more informative than associational methods such as regression in revealing whether an observed positive exposure-response association is explained by confounding, selection bias, reverse causation, one or more directed paths between them, or a combination of these association-inducing conditions.

Despite their considerable accomplishments, DAG-learning and more general graph-learning algorithms have the following limitations, which practitioners must understand to correctly interpret their results.

- First, graph-learning algorithms, like other statistical techniques, cannot detect effects (i.e., dependencies between variables) that are too small for the available sample size and noise or sampling variability in the data. However, they can be used constructively to estimate upper bounds on the sizes of hypothesized effects that might exist without being detected. For example, suppose it were hypothesized that PM2.5 increases heart disease risk by an amount *b*PM2.5*, where *b* is a potency factor, which the *bnlearn* algorithms used to construct Fig. 9.1 failed to detect. Then applying these same algorithms to a sequence of simulated data sets in which heart disease probabilities are first predicted from data using standard predictive analytics algorithms (e.g., BN inference algorithms or random forest or logistic regression models) and then increased by *b*PM2.5*, for an increasing sequence of *b* values, will reveal the largest value of *b* that does not always result in an arrow between PM2.5 and heart disease. This serves as a plausible upper bound on the size of the hypothesized undetected effect.
- Graph-learning algorithms are also vulnerable to false-positive findings if unmodeled latent variables create apparent statistical dependencies between

variables. However, various algorithms included in the *CompareCausalNetworks* package allow detection and modeling of latent variable effects using graphs with undirected or bidirected arcs as well as arrows, based on observed statistical dependencies between estimated noise terms for the observed variables (Heinze-Deml et al. 2018).

- Graph-learning algorithms, like other techniques, cannot resolve which (if any) among highly multicollinear variables is the true cause of an effect with which all are equally correlated (but see Jung and Park 2018 for a ridge regression approach to estimating path coefficients despite multicollinearity). More generally graph-learning algorithms may find that several different causal graphs are equally consistent with the data.
- If passive observations have insufficient variability in some variables, their effects on other variables may again be impossible to estimate uniquely from data. (For this reason the *experimental treatment assignment* (ETA) condition, that each individual has positive probability of receiving any level of exposure, regardless of the levels of other variables, is often assumed for convenience although it is often violated in practice (Petersen et al. 2012.)) If interventions are possible, however, then experiments can be designed to learn about dependencies among both observed and latent variables as efficiently as possible (Kocaoglu et al. 2017).

Arguably, these limitations due to limited statistical power, unobserved (latent) variables, and need to perturb systems to ensure adequate variability and unique identifiability, apply to *any* approach to learning about causality and other statistical dependencies among variables from data, from formal algorithms to holistic expert judgments. Using algorithms and packages such as *bnlearn, pcalg,* and *CompareCausalNetworks* to determine whether there are detected statistical dependencies (i.e., positive mutual information) between variables that persist even after conditioning on other variables simply makes explicit operational procedures for drawing reproducible conclusions from data despite such limitations.

Unobserved (latent) variables can play a key role in explaining dependencies among observed variables, but have only recently started to be included widely used causal analytics software packages. A simple light bulb example illuminates data analysis implications of omitting direct causes of a variable from a model. Suppose that a light bulb's illumination as a function of light switch position is well described in a limited data set by a deterministic CPT with *P*(*light on* | *switch up*) = *P*(*light off* | *switch down*) = 1. In a larger data set that includes data from periods with occasional tripped circuit breakers, burned-out fuses, or power outages, this CPT might have to be replaced by a new one with *P*(*light on* | *switch up*) = 0.99, *P*(*light off* | *switch down*) = 1. Scrutiny of the data over time would show strong autocorrelations, with this aggregate CPT being resolved into a mixture of two different shorter-term CPTs: one with *P*(*light on* | *switch up*) = 1 (when current is available to flow, which is most of the time) and the other with *P*(*light on* | *switch up*) = 0 (otherwise), with the former regimen holding 99% of the time. Such heterogeneity of CPTs over time or across study settings points to the possibility of omitted causes; these can be modeled using statistical techniques for unobserved causes of changes and heterogeneity in observed input-output behaviors, such as Hidden Markov Models (HMMs) and regime-switching models for time series data, or finite mixture distribution models for cross-sectional data. In population studies, the condition that the conditional expected value (or, more generally, the conditional distribution) of a variable for an individual unit is determined entirely by the values of its parents, so that any two units with the same values for their direct causes also have the same conditional probability distribution for their values, is referred to as *unit homogeneity* (Holland 1986; King et al. 1994; Waldner 2015). When it is violated, as revealed by tests for homogeneity, the causal graph model used to interpret and explain the data can often be improved by adding previously omitted variables (or latent variables, if they are not measured) or previously missing links (Oates et al. 2017).

From this perspective, inconsistency across study results in the form of unexplained heterogeneity in CPTs can reveal a need to expand the causal model to include more direct causes as inputs to the CPT to restore consistency with observations and homogeneity of CPTs. As discussed in the text, a more convincing demonstration of the explanatory power of a model and its consistency with data than mere agreement with previous findings (which is too often easily accomplished via p-hacking) is to find associations and effects estimates that *differ* across studies, and to show that these differences are successfully predicted and explained by applying invariant, homogeneous causal conditional probabilities to the relevant covariates (e.g., sex, age, income, health care, etc.) in each population. Modern *transport formulas* for applying causal CPTs learned in one setting to new settings allow such detailed prediction and explanation of empirically observed differences in effects in different populations (Heinze-Deml et al. 2017; Bareinboim and Pearl 2013; Lee and Honavar 2013). Results of multiple disparate studies can be combined, generalized, and applied to new settings using the invariance of causal CPTs, despite the variations of marginal and joint distributions of their inputs in different settings (Triantafillou and Tsamardinos 2015; Schwartz et al. 2011). Exciting as these possibilities appear to be for future approaches to learning, synthesizing, and refining causal models from multiple studies and diverse data sets, however, they are only now starting to enter the mainstream of causal analysis of data and to become available in well supported software packages (e.g., https://cran.r-project.org/web/packages/causaleffect/causaleffect.pdf). We view the full exploitation of consistency, invariance, and homogeneity principles in learning, validating, and improving causal graph models from multiple data sets as a very promising area of ongoing research, but not yet reduced to reliable, well-vetted, and widely available software packages that automate the process.

Appendix 3: Concepts of Causation

In public and occupational health risk analysis, the causal claim "Each extra unit of exposure to substance X increases rates of an adverse health effect (e.g., lung cancer, heart attack deaths, asthma attacks, etc.) among exposed people by R additional expected cases per person-year" has been interpreted in at least the following ways:

1. *Probabilistic causation* (Suppes 1970): The conditional probability of the health response or effect occurring in a given interval of time is greater among individuals with more exposure than in similar-seeming individuals with less exposure. In this sense, probability of response (or age-specific hazard rate for occurrence of response) increases with exposure. On average, there are R extra cases per person-year per unit of exposure. The main intuition is that causes (exposures) make their effects (responses) more likely to occur within a given time interval, or increase their occurrence rates. CPTs can represent probabilistic causation while allowing arbitrary interactions among the direct causes of an effect. This probabilistic formulation is more flexible than deterministic ideas of necessary cause, sufficient cause, or but-for cause (see concept 9) (Rothman and Greenland 2005; Pearl and Mackenzie 2018). All of the other concepts that we consider are developed within this probabilistic causation framework. However, most of them add conditions to overcome the major limitations that probabilistic causation does not imply that changing the cause would change the effect and that probabilistic causation lacks direction (i.e., $P(X \mid Y) > P(X)$ implies $P(Y \mid X) > P(Y)$, since $P(X \mid Y)P(Y) = P(Y \mid X)P(X)$ implies that $P(Y \mid X) = P(X \mid Y) P(Y)/P(X)$, which exceeds $P(Y)$ when $P(X \mid Y)/P(X) > 1$, or $P(X \mid Y) > P(X)$).
2. *Associational causation* (Hill 1965; IARC 2006): Higher levels of exposure have been observed in conjunction with higher risks, and this association is judged to be strong, consistent across multiple studies and locations, biologically plausible, and perhaps to meet other conditions such as those in the left column of Table 9.1. The slope of a regression line between these historical observations in the exposed population of interest is R extra cases per person-year per unit of exposure. The main intuition is that causes are associated with their effects.
3. *Attributive causation* (Murray and Lopez 2013): Authorities attribute R extra cases per person-year per unit of exposure to X; equivalently, they blame exposure to X for R extra cases per person-year per unit of exposure. In practice, such attributions are usually made based on measures of association such as the ratio or difference of estimated risks between populations with higher and lower levels of exposure, just as for associational causation, together with subjective decisions or judgments about which risk factor(s) will be assigned blame for increased risks that are associated with one or more of them. Differences in risks between the populations are typically attributed to their differences in the selected exposures or risk factors. The main idea is that if people with higher levels of the selected

variable(s), such as exposure, have higher risk for any reason, then the greater risk can be attributed to the difference in the selected variable(s). (If many risk factors differ between low-risk and high-risk groups, then the difference in risks can be attributed to each of them separately; there is no consistency constraint preventing multiples of the total difference in risks from being attributed to the various factors.) For example, if poverty is associated with higher stress, poorer nutrition, lower quality of health care, increased alcohol and drug abuse, greater prevalence of cigarette smoking, increased heart attack risk, residence in higher-crime neighborhoods, residence in neighborhoods with higher air pollution levels, higher unemployment, lower wages, fewer years of education, and more occupation in blue-collar jobs, then attributive causation could be applied to data on the prevalence of these ills in different populations to calculate a "population attributable fraction" (PAF) or "probability of causation" (PC) for the fraction of any one of them to be attributed to any of the others (Rothman and Greenland 2005). In the simplest case where all are treated as binary (0–1) variables and all are 1 for individuals in a low-income population and 0 for individuals in a high-income comparison group, the PAF and PC for heart attack risk attributed to residence in a higher air pollution neighborhood (or, symmetrically, to any of the other factors, such as blue collar employment) is 100%. This attribution can be made even if changing the factors to which risk is attributed would not affect risk. Relative risk (RR) ratios—the ratios of responses per person per year in exposed compared to unexposed populations—and quantities derived from RR, such as burden-of-disease metrics, population attributable fractions, probability of causation formulas, and closely related metrics, are widely used in epidemiology and public health to quantify both associational and attributive causation.

4. *Counterfactual and potential outcomes causation* (Höfler 2005; Glass et al. 2013; Lok 2017; Li et al. 2017; Galles and Pearl 1998): In a hypothetical world with 1 unit less of exposure to X, expected cases per person-year in the exposed population would also be less by R. Such counterfactual numbers are usually derived from modeling assumptions, and how or why the counterfactual reduction in exposure might occur is not usually explained in detail, even though such details might affect resulting changes in risk. The main intuition is that differences in causes make their effects different from what they otherwise would have been. To use this concept, what otherwise would have been (since it cannot be observed) must be guessed at or assumed, e.g., based on what happens in an unexposed comparison group believed to be relevantly similar to, or exchangeable with, the exposed population. Galles and Pearl (1998) discuss reformulation of counterfactual and potential outcomes models as causal graph models based on observable variables. The counterfactual framework is especially useful for addressing questions about causality that do not involve variables that can be manipulated, such as "How would my salary for this job differ if I had been born with a different race or sex or if I were 10 years older?" or "How likely is it that

this hurricane would have occurred had oil and gas not been used as energy sources in the twentieth century?" (Pearl and Mackenzie 2018).

5. *Predictive causation* (Wiener 1956; Granger 1969; Kleinberg and Hripcsak 2011; Papana et al. 2017): In the absence of interventions, time series data show that the observation that exposure has increased or decreased is predictably followed, perhaps after some delay, by the observation that average cases per person-year have also increased or decreased, respectively, by an average of *R* cases per unit of change in exposure. The main intuition is that causes help to predict their effects, and changes in causes help to predict changes in their effects. (The meaning of "help to predict" is explained later in the section on mutual information, but an informal summary is that effects can be predicted with less average uncertainty or error when past values of causes are included as predictors than when they are not.) Predictive causation still deals with observed associations, but now the associations involve changes over time.
6. *Structural causation* (Simon 1953; Simon and Iwasaki 1988; Hoover 2012): The average number of cases per person-year is derived at least in part from the value of exposure. Thus, the value of exposure must be determined before the value of expected cases per person-year can be determined; in econometric modeling jargon, exposure is exogenous to expected cases per person-year. Quantitatively, the derived value of cases per person-year decreases by *R* for each exogenously specified unit decrease in exposure. The main intuition is that effects are derived from the values of their causes. Druzdzel and Simon (1993) discuss the relations between Simon-Iwasaki causal ordering and manipulative and mechanistic causation.
7. *Manipulative causation* (Voortman et al. 2010; Spirtes 2010; Hoover 2012; Simon and Iwasaki 1988): Reducing exposure by one unit reduces expected cases per person-year by *R*. The main intuition is that changing causes changes their effects. How this change is brought about or produced need not be explained as part of manipulative causation, although it is crucial for scientific understanding, as discussed next. Manipulative causation is highly useful for decision analysis, which usually assumes that decision variables have values that can be set (i.e., manipulated) by the decision-maker. Bayesian networks with decision nodes and value nodes, called influence diagrams, have been extensively developed to support calculated manipulation of decision variables to maximize the expected utility of consequences (Howard and Matheson 2005, 2006; Howard and Abbas 2016).
8. *Explanatory/mechanistic causation* (Menzies 2012; Simon and Iwasaki 1988): Increasing exposure by one unit causes changes to propagate through a biological network of causal mechanisms. Propagation through networks of mechanisms is discussed further in the section on coherent causal explanations and biological plausibility. The main idea is simply that, following an exogenous change in an input variable such as exposure, each variable's conditional prob-

ability distribution is updated to reflect (i.e., condition on) the values of its parents, as specified by its CPT, and the value drawn from its conditional probability distribution is then used to update the conditional probability distributions of its children. When all changes have finished propagating, the new expected value for expected cases per person-year in the exposed population will be R more than before exposure was increased. The main intuition is that changes in causes propagate through a network of law-like causal mechanisms to produce changes in their effects. Causal mechanisms are usually represented mathematically by conditional probability tables (CPTs) or other conditional probability models (such as structural equation models, regression models, or non-parametric alternatives such as classification trees). These specify the conditional probability or probability density for a variable, given the values of its parents in a causal graph or network, and they are invariant across settings (Pearl 2014).

9. *But-for causation*: If the cause had not occurred (i.e., "but for" the cause), the effect would not have occurred. More generally, if the value of a causal variable had been different, the probability distributions for its effects would have been different. This concept of causation has long been a staple of tort litigation, where plaintiffs typically argue that, but for the defendant's (possibly tortious) action, harm to the plaintiff would not have occurred.

It is important to understand that these different concepts are not variations, extensions, or applications of a single underlying coherent concept of causation. A headline such as "Pollution kills 9 million people each year" (e.g., Washington Post 2017) cannot be interpreted as a statement of mechanistic or manipulative causation, since the number of people dying in any year is identically equal to the number born one lifetime ago, and pollution presumably does not retroactively increase that number. (It might affect life lengths and redistribute deaths among years, but it cannot increase the total number of deaths in each year.) But it is easily accommodated in the framework of attributive causality, where it simply refers to the number of "premature deaths" per year that some authorities (authors of a *Lancet* article, in this case, www.thelancet.com/commissions/pollution-and-health) attribute to air pollution using a "burden of disease" formula that is unconstrained by physical conservation laws. These are not variations on a single concept, but distinct concepts: the kind of causation implied by the headline (attributive) has no implications for the number of deaths that could be prevented by reducing pollution (manipulative).

Galles and Pearl (1998) discuss essential distinctions between structural and counterfactual/potential outcomes models of causation and special conditions under which they are equivalent. The non-equivalence of predictive and manipulative causation can be illustrated through standard examples such as nicotine-stained fingers being a predictive but not a manipulative cause of lung cancer (unless the only way to keep fingers unstained is not to smoke and the only want to stain them is to smoke; then nicotine-stained fingers would be a manipulative but not a mechanistic cause of lung cancer). In short, the various concepts of causation are distinct, although the distinctions among them are seldom drawn in headlines and studies that announce causal "links" between exposures and adverse health responses.

Appendix 4: Software for Dynamic Bayesian Networks and Directed Information

For practitioners, software packages with algorithms for learning dynamic Bayesian networks (DBNs) from time series data, automatically identifying their DAG structures (including dependencies among variables in different time slices, corresponding to lagged values of variables) and estimating their CPTs, are now readily available. The free R package *bnstruct* (Sambo and Franzin 2016) provides non-parametric algorithms for learning DBNs from multiple time series even if some of the data values are missing. (Detailed implementations must also address instantaneous causation, i.e., causal relations between variables in the same time slice, but this does not require any new principles, as it is just the usual case of a non-dynamic BN.) The *pgmpy* package for probabilistic graphical models in Python has a module for DBNs, and commercial packages such as GeNIe (free for academic research), Netica, BayesiaLab, and Bayes Server all support DBN learning and inference. Links to these and other BN and DBN software packages can be found at www.kdnuggets.com/software/bayesian.html.

Alternatively, within the context of traditional parametric time series modeling, Granger causality testing (performed using the *grangertest* function in R) provides quantitative statistical tests of the null hypothesis that the future of the effect is conditionally independent of the past of the cause, given its own past. Non-parametric generalizations developed and applied in physics, neuroscience, ecology, finance, and other fields include several closely related measures of transfer entropy and directed information flow between time series (Schreiber 2000; Weber et al. 2017). An elegant alternative to DBNs is *directed information graphs* (DIGs) (Quinn et al. 2015). *Directed information* generalizes mutual information by distinguishing between earlier and later values of variables (Costanzo and Dunstan 2014). It allows for realistic complexities such as feedback loops and provides a non-parametric generalization of earlier parametric statistical tests for predictive causality between pairs of time series (Wiener 1956; Granger 1969). Directed information graphs generalize these ideas from pairs to entire networks of time series variables (Amblard and Michel 2011). In such a graph, each node represents a time series variable (stochastic process). An arrow from one node to another indicates that directed information flows from the former to the latter over time. DIG-learning algorithms (Quinn et al. 2015) use nonparametric measures of directed information flows between variables over time to identify causal graph structures, assuming that information flows from causes to their effects. However, we are unaware of currently available mainstream statistical packages that automate estimation of DIGs from data (but see https://pypi.org/project/causalinfo/ for a start). Thus, we recommend using DBNs and packages such as *bnstruct* to identify the directions of arrows representing the temporal flows of information between variables, i.e., from earlier values of some variables to later values of others.

Appendix 5: Non-temporal Methods for Inferring Direction of Information Flow

This appendix describes several principles and methods used to infer directions of information flow (and hence possible causality) between variables when longitudinal data are not available. The last part of the appendix discusses quasi-experimental (QE) study designs and analysis methods.

Homoscedastic Errors and LiNGAM

If causes and effects are related by some unknown regression model of the form

$$effect = f\left(cause\right) + error$$

where *f* is a (possibly unknown) regression function and *error* is a random error term, possibly with an unknown distribution, then it may be possible to discern which variables are causes of specified effects by studying the distribution of the error term. In the special case where *f* is a linear function and *error* has a Gaussian (normal) distribution with zero mean, the linear regression model

$$effect = b_0 + b_1 * cause + error$$

can be rearranged as

$$cause = -\left(b_0 / b_1\right) + \left(1 / b_1\right) * \text{effect} + \left(1 / b_1\right) * error.$$

Because the normal distribution is symmetric around the origin and multiplying a normally distributed random variable by a constant gives another normally distributed random variable, this is again a linear function with a normally distributed error term. Thus, the two properties of (a) having a linear regression function relating one variable to another and (b) having a normally distributed error, do not reveal which variable is the cause and which is the effect, since these properties are preserved when either variable is solved for in terms of the other. But if the regression function is nonlinear, or if the error term is not normally distributed (i.e., non-Gaussian), then the following very simple test for the direction of information flow emerges: if $y = f(x) + error$, where error is a zero-mean random variable (not necessarily Gaussian), then a scatter plot of y values against x values has the same distribution of y values around their mean for all values of x (since this vertical scatter is produced by an *error* term whose distribution is independent of x). Typically, except in the case of linear regression, the scatter plot of x values vs. y values will then look quite different: $x = f^{-1}(y - error)$ will have a scatter around its mean values that depends on y. Visually, plotting *effect* vs. *cause* yields a scatter plot that is homoscedastic (has the same variance everywhere), while plotting *cause* vs. *effect* typically does not. This asymmetry reveals which variable is the cause and which is the effect under the assumption that *effect* = *f*(*cause*) + *error*. It is used in recent algorithms (e.g., the LiNGAM (linear non-Gaussian acyclic model) (Shimizu et al. 2006; Tashiro et al. 2014) and CAM (causal additive model) algorithms for causal discovery (Buhlmann et al. 2014), both of which are included in the *CompareCausalNetworks* package (Heinze-Deml et al. 2018)). Applications in epidemiology have shown encouraging performance (Rosenström et al. 2012).

Knowledge-Based and Exogeneity Constraints

A third way to orient the arrows in a causal graph is to use *knowledge-based constraints*, such as that cold weather might be a cause of increased elderly mortality rates, but increased elderly mortality rates are not a potential cause of cold weather. Similarly, sex and age are typically potential causes but not potential effects of other variables. Death is typically a potential effect but not a potential cause of covariates. Such relatively uncontroversial assumptions or constraints can be very useful for directing the arrows in causal graphs, for both longitudinal and cross-sectional data. They can be incorporated into causal discovery algorithms used in *bnlearn* and other packages using "white lists" and "black lists" (and, to save time, via "source" and "sink" designations in CAT) to specify required, allowed, and forbidden arrow directions. More generally *exogeneity constraints* specify that some variables have values derived from (i.e., endogenously determined by) the values of others; the values of variables that are not derived from others are determined from outside (i.e., exogenous to) the system being examined (Galles and Pearl 1998). Such exogeneity constraints have long been used in econometrics, social sciences, electrical engineering, and artificial intelligence to help determine possible causal orderings of variables in systems of equations based on directions of information flow from exogenous to endogenously determined variables (Simon 1953; Simon and Iwasaki 1988). These ordering have proved useful in modeling manipulative causation as well as structural and predictive causation (Druzdzel and Simon 1993; Voortman et al. 2010).

Quasi-Experiments (QEs) and Assumption-Based Constraints on Arrow Directions

A very useful type of exogeneity for inferring causality from observations arises from *actions* or *interventions* that deliberately set the values of some controllable variables. These exogenously controlled variables are often called *decision variables* in decision analysis, *factors* or *design factors* in design of experiments, preventable *risk factors* in epidemiology, or *controllable inputs* in systems engineering. As a paradigmatic example, suppose that flipping a switch at random times is immediately followed by a light turning on whenever the switch is flipped up and off whenever it is flipped down. After some experimentation, an investigator might conclude that there is substantial empirical evidence that flipping the switch up or down causes the light to be turned on or off, respectively, even without further understanding of how the change in switch position propagates through a causal network to change the flow of current through the light bulb. Hill's "experiment" consideration allows for such empirical evidence of causality when manipulation of risk factors is possible (Fedak et al. 2015). Several caveats on such inferences of causality from data are noted later; the challenging philosophical problem of justifying inductive inferences of causation from observations is by no means completely solved even by random experimentation unless it can be assumed that the (perhaps unknown) mechanisms tying interventions to consequence probabilities remain unchanged over time or successive trials, and will continue into the future.

When deliberate manipulation is impossible, or is sharply restricted (e.g., to avoid exposing individuals to potentially harmful conditions), as if often the case in practice, observational data can sometimes be used to achieve many of the benefits of experimentation if it reveals the responses of dependent variables to exogenous changes ("interventions") or differences in explanatory variables. This is the basic principle of *quasi-experiments* (QEs) (Campbell and Stanley 1963). These differ from true designed experiments in that they lack random assignment of individuals to treatment or control groups. To interpret their results causally, therefore, it is typically necessary to make strong assumptions, such as that the populations with different levels of the explanatory variables are otherwise exchangeable, with no latent confounders (Hernán and Robins 2006). A variety of methods have been devised for drawing causal inferences about the directions and magnitudes of effects from QE data with the help of such assumptions. For single, large interventions at known points in time, a long tradition in the social sciences and epidemiology of *intervention analysis*, also called *interrupted time series analysis* (ITSA), compares time series of response variables before and after the intervention to determine whether they have changed significantly (Box and Tiao 1975). If so, and if repeated observations in multiple settings make it unlikely that the changes occur by coincidence, then they can be used to estimate the effects of the observed interventions (or possibly of other unobserved changes that caused them). More recently "difference in differences" (DID) analyses have applied similar pre-post comparisons to estimate effects of interventions with the help of such simplifying assumption as that effects are additive and trends are linear both before and after the intervention (Bertrand et al. 2004). Again, the "interventions" studied by QE methods need not be deliberate or man-made: what matters is that they cause observable changes in responses that cannot plausibly be explained by other factors. Similarly, in *natural experiments*, exogenous changes in exposures or other conditions arise from nature or other sources that the investigator does not control (DiNardo 2008). Comparing the distributions of observed responses in exposed and unexposed populations, or more generally in populations with different levels of an exogenous variable (or the same population before and after the natural change in explanatory variables), reveals the effects of these differences on responses—at least if the populations of individuals being compared are otherwise exchangeable (or can be made so in the analysis by stratification and matching on observed variables). The famous analysis of data from the 1854 Broad Street cholera outbreak by John Snow Regression is an outstanding practical example of a natural experiment used to identify a cause. The BACKSHIFT algorithm in the *CompareCausalNetworks* package (Rothenhausler et al. 2015) exploits the fact that many variables in the real world are constantly bombarded by small random shocks and perturbations. Even in the absence of any deliberate interventions, these disturbances act as random interventions of possibly unknown sizes and locations in the causal network (sometimes called "fat hand interventions") that cause variables to fluctuate around their equilibrium values over time. The BACKSHIFT algorithm uses correlations in these observed fluctuations to learn linear causal network models (i.e., networks in which changes in causes transmit proportional changes in their direct effects), even if cycles and latent variables are present.

In econometrics and epidemiology, a QE technique called *regression discontinuity design* (RDD) has found growing popularity. RDD studies compare response distributions in populations on either side of a threshold that triggers an exogenous change in exposures or other conditions. Such thresholds include legal ages for smoking or drinking or driving; and geopolitical boundaries for implementation of different policies or programs (Thistlethwaite and Campbell 1960; Imbens and Lemieux 2008). The thresholds have the effect of creating interventions in the explanatory variables even without deliberate manipulation. Another QE technique, *instrumental variables* (IV) regression, is also now widely used in econometrics, social statistics, and epidemiology to estimate causal effects despite potential biases from reverse causation, omitted variables (e.g., unobserved confounders), and measurement errors in explanatory variables. IV studies collect observations on "instruments," exogenously changing variables that are assumed to affect the explanatory variables in a model but not to directly affect the dependent variable. Changes in the instrument change the explanatory variable(s) of interest, thereby mimicking the effects of interventions that set their values. This generates observations from which the effects of the changes in explanatory variables on the dependent variable(s) of interest can be estimated, provided that the instrument is valid, i.e., that changing it causes changes in the explanatory variable(s) but not in the dependent variable(s) of interest except via its effects on the explanatory variables (Imbens and Angrist 1994)..

In general, QE techniques rely on an assumption that observed changes in exogenous causes—whether random, deliberate, designed, or coincidental—propagate to changes in the distributions of the observed endogenous dependent variables. Whether a light switch is deliberately flipped on and off in a designed experiment, or is flipped at random by passersby or by acts of nature or by chance, does not greatly affect the task of causal inference, provided that the relation between observed changes in the switch's position and ensuing observed changes in the on-off status of the light bulb are not coincidental, are not caused by latent variables that affect both, and are not explained by other "threats to internal validity" for quasi-experiments (Campbell and Stanley 1963). If the effects of different settings of an explanatory variable on the mean level of a response variable in a population are estimated by the differences in its observed conditional mean values for different observed settings of the explanatory variable, then it is necessary to assume that the populations experiencing these different settings are comparable, or exchangeable—a crucial assumption, although often difficult or impossible to test, in interpreting QE results. Such conditions are necessary to justify any of these techniques for inferring that causal information flows from changes in explanatory variables to changes in dependent variables.

References

Amblard P-O, Michel OJJ. On directed information theory and Granger causality graphs. J Comput Neurosci. 2011;30:7–16. https://doi.org/10.1007/s10827-010-0231-x.

Azzimonti L, Corani G, Zaffalon M (2017) Hierarchical multinomial-dirichlet model for the estimation of conditional probability tables. https://arxiv.org/abs/1708.06935. Last accessed 21 Aug 18

Bareinboim E, Pearl J (2013) Causal transportability with limited experiments. In: Proceedings of the 27th AAAI conference on artificial intelligence, pp 95–101

Bertrand M, Duflo E, Mullainathan S. How much should we trust differences in-differences estimates? Quart J Econ. 2004;119(1):249–75.

Bollen KA, Pearl J. Eight myths about causality and structural equation models. In: Morgan SL, editor. Handbook of causal analysis for social research. Dordrecht: Springer; 2013. p. 301–28.

Boué S, Talikka M, Westra JW, et al. Causal biological network database: a comprehensive platform of causal biological network models focused on the pulmonary and vascular systems. Database. 2015;2015:bav030. https://doi.org/10.1093/database/bav030.

Box GEP, Tiao GC. Intervention analysis with applications to economic and environmental problems. J Am Stat Assoc. 1975;70:70–9.

Buhlmann P, Peters J, Ernest J. CAM: causal additive models, high-dimensional order search and penalized regression. Ann Stat. 2014;42(6):2526–56. https://doi.org/10.1214/14-AOS1260.

Campbell DT, Stanley JC. Experimental and quasi-experimental designs for research. Boston: Houghton Mifflin Company; 1963.

Cartwright N. Two theorems on invariance and causality. Philos Sci. 2002;70:203–24.

Chang R, Karr JR, Schadt EE. Causal inference in biology networks with integrated belief propagation. Pac Symp Biocomput. 2015;2015:359–70.

Chen CWS, Hsieh YH, Su HC, Wu JJ (2017) Causality test of ambient fine particles and human influenza in Taiwan: age group-specific disparity and geographic heterogeneity. Environ Int. https://doi.org/10.1016/j.envint.2017.10.011

Cole SR, Platt RW, Schisterman EF, Chu H, Westreich D, Richardson D, Poole C. Illustrating bias due to conditioning on a collider. Int J Epidemiol. 2010;39(2):417–20.

Costanzo JAWB, Dunstan J (2014) A survey of causality and directed information. https://pdfs.semanticscholar.org/e15c/84188c9fd75ed59b9f68fb2ca3ab34786478.pdf

Cover TM, Thomas JA. Elements of information theory. 2nd ed. Hoboken: Wiley; 2006.

Cox LA Jr, Popken DA. Has reducing fine particulate matter and ozone caused reduced mortality rates in the United States? Ann Epidemiol. 2015;25(3):162–73. https://doi.org/10.1016/j.annepidem.2014.11.006.

Cox LA Jr. Socioeconomic and air pollution correlates of adult asthma, heart attack, and stroke risks in the United States, 2010-2013. Environ Res. 2017b;155:92–107. https://doi.org/10.1016/j.envres.2017.01.003.

Cox LA Jr. Socioeconomic and particulate air pollution correlates of heart disease risk. Environ Res. 2018;167:386–92. https://doi.org/10.1016/j.envres.2018.07.023.

DiNardo J. Natural experiments and quasi-natural experiments. In: Durlauf SN, Blume LE, editors. The New Palgrave dictionary of economics. 2nd ed. London: Palgrave Macmillan; 2008. https://doi.org/10.1057/9780230226203.1162.

Ding P. A paradox from randomization-based causal inference. Stat Sci. 2017;32(3):331–45.

Dominici F, Greenstone M, Sunstein CR. Particulate matter matters. Science. 2014;344(6181):257–9. https://doi.org/10.1126/science.1247348.

Dragulinescu S. Mechanisms and difference-making. Acta Anal. 2015;2015:1–26.

Druzdzel MJ, Simon H. Causality in Bayesian belief networks. In: UAI '93 proceedings of the ninth international conference on uncertainty in artificial intelligence. San Francisco: Morgan Kaufmann Publishers Inc., 1993. p. 3–11.

Dumas-Mallet E, Smith A, Boraud T, Gonon F. Poor replication validity of biomedical association studies reported by newspapers. PLoS One. 2017;12(2):e0172650. https://doi.org/10.1371/journal.pone.0172650.

Elwert F (2013) Graphical causal models. In: Handbook of causal analysis for social research. pp 245–273. https://doi.org/10.1007/978-94-007-6094-3_13

Fann N, Lamson AD, Anenberg SC, Wesson K, Risley D, Hubbell BJ. Estimating the national public health burden associated with exposure to ambient PM2.5 and ozone. Risk Anal. 2012;32(1):81–95.

Fedak KM, Bernal A, Capshaw ZA, Gross S. Applying the Bradford Hill criteria in the 21st century: how data integration has changed causal inference in molecular epidemiology. Emerg Themes Epidemiol. 2015 Sep 30;12:14. https://doi.org/10.1186/s12982-015-0037-4. PMID: 26425136; PMCID: PMC4589117.

Fraser H, Parker T, Nakagawa S, Barnett A, Fidler F. Questionable research practices in ecology and evolution. PLoS One. 2018;13(7):e0200303. https://doi.org/10.1371/journal.pone.0200303.

Galles D, Pearl J. An axiomatic characterization of causal counterfactuals. Found Sci. 1998; 3:151–82.

Glass TA, Goodman SN, Hernán MA, Samet JM. Causal inference in public health. Annu Rev Public Health. 2013;34:61–75. https://doi.org/10.1146/annurev-publhealth-031811-124606.

Glymour MM, Greenland S. Causal diagrams. In: Rothman KJ, Greenland S, Lash TL, editors. Modern epidemiology. 3rd ed. Philadelphia: Lippincott Williams and Wilkins; 2008. p. 183–209.

Granger CWJ. Investigating causal relations by econometric models and cross-spectral methods. Econometrica. 1969;37(3):424–38.

Greenland S, Pearl J, Robins JM. Causal diagrams for epidemiologic research. Epidemiology. 1999;10:37–48.

Greenwell BM. pdp: an R package for constructing partial dependence plots. The R Journal. 2017;9(1):421–36.

Hausman DM, Woodward J. Independence, invariance, and the causal Markov condition. Br J Philos Sci. 1999;50(4):521–83. https://doi.org/10.1093/bjps/50.4.521.

Heinze-Deml C, Meinshausen N (2018) Package CompareCausalNetworks. https://cran.r-project.org/web/packages/CompareCausalNetworks/CompareCausalNetworks.pdf

Heinze-Deml C, Peters J, Meinshausen N (2017) Invariant causal prediction for nonlinear models. https://arxiv.org/pdf/1706.08576.pdf

Heinze-Deml C, Maathuis MH, Meinshausen N. Causal structure learning. Annu Rev Stat Appl. 2018;5:371–91.

Hernán MA, Robins JM. Estimating causal effects from epidemiological data. J Epidemiol Community Health. 2006;60(7):578–86. https://doi.org/10.1136/jech.2004.029496.

Hill AB. The environment and disease: association or causation? Proc R Soc Med. 1965;58:295–300.

Höfler M. The Bradford Hill considerations on causality: a counterfactual perspective. Emerg Themes Epidemiol. 2005;2:11.

Holland P. Statistics and causal inference. J Am Stat Assoc. 1986;81:945–60.

Hoover KD (2012) Causal structure and hierarchies of models. Stud Hist Philos Biol Biomed Sci. https://doi.org/10.1016/j.shpsc.2012.05.007

Howard RA, Abbas AE. Foundations of decision analysis. Upper Saddle River: Pearson; 2016.

Howard RA, Matheson JE. Influence diagrams. Decis Anal. 2005;2(3):127–43.

Howard RA, Matheson JE. Influence diagrams. Decis Anal. 2006;2(3):127–43.

IARC. IARC monographs on the evaluation of carcinogenic risk to humans: preamble. Lyons: International Agency for Research on Cancer (IARC); 2006.

Imbens G, Angrist J. Identification and estimation of local average treatment effects. Econometrica. 1994;62(2):467–76.

Imbens G, Lemieux T. Regression discontinuity designs: a guide to practice. J Econ. 2008;142(2):615–35. https://doi.org/10.1016/j.jeconom.2007.05.001.

Ioannidis JP. Exposure-wide epidemiology: revisiting Bradford Hill. Stat Med. 2016;35(11):1749–62. https://doi.org/10.1002/sim.6825.

Iserman R, Münchhof M. Identification of dynamic systems: an introduction with applications. New York: Springer; 2011.

Jung S, Park J. Consistent partial least squares path modeling via regularization. Front Psychol. 2018;9:174. https://doi.org/10.3389/fpsyg.2018.00174.

Kahneman D. Thinking fast and slow. New York: Farrar, Straus, and Giroux; 2011.

Kalisch M, Machler M, Colombo D, Maathuis MH, Buhlmann P. Causal inference using graphical models with the R Package pcalg. J Stat Softw. 2012;47(11):1–26.

King G, Keohane RO, Verba S. Designing social inquiry: scientific inference in qualitative research. Princeton: Princeton University Press; 1994.

Kleinberg S, Hripcsak G. A review of causal inference for biomedical informatics. J Biomed Inform. 2011;44(6):1102–12.

Knüppel S, Stang A. DAG program: identifying minimal sufficient adjustment sets. Epidemiology. 2010;21(1):159. https://doi.org/10.1097/EDE.0b013e3181c307ce.

Kocaoglu M, Shanmugam K, Bareinboim E (2017) Experimental design for learning causal graphs with latent variables. In: 31st Conference on neural information processing systems (NIPS 2017), Long Beach, CA, USA. https://papers.nips.cc/paper/7277-experimental-design-for-learning-causal-graphs-with-latent-variables.pdf

Lagani V, Triantafillou S, Ball G, Tegnér J, Tsamardinos I. Probabilistic computational causal discovery for systems biology: chapter 2. In: Geris L, Gomez-Cabrero D, editors. Uncertainty in biology: a computational modeling approach. Cham: Springer; 2016.

Lähdesmäki H, Hautaniemi S, Shmulevich I, Yli-Hari O. Relationships between probabilistic Boolean networks and dynamic Bayesian networks as models of gene regulatory networks. Signal Process. 2006;86(4):814–34. https://doi.org/10.1016/j.sigpro.2005.06.008.

LeCun Y, Bengio Y, Hinton G. Deep learning. Nature. 2015;521:436–44.

Lee S, Honavar V. (2013) m-Transportability: transportability of a causal effect from multiple environments. In: Proceedings of the twenty-seventh AAAI conference on artificial intelligence. www.aaai.org/ocs/index.php/AAAI/AAAI13/paper/viewFile/6303/7210

Lenis D, Ackerman B, Stuart EA. Measuring model misspecification: application to propensity score methods with complex survey data. Comput Stat Data Anal. 2018;128:48–57. https://doi.org/10.1016/j.csda.2018.05.003.

Li J, Ma S, Le T, Liu L, Liu J. Causal decision trees. IEEE Trans Knowl Data Eng. 2017;29(2):257–71.

Linden A. Improving causal inference with a doubly robust estimator that combines propensity score stratification and weighting. J Eval Clin Pract. 2017;23(4):697–702. https://doi.org/10.1111/jep.12714.

Lok JJ. Mimicking counterfactual outcomes to estimate causal effects. Ann Stat. 2017;45(2):461–99. https://doi.org/10.1214/15-AOS1433.

McCandless LC, Somers JM (2017) Bayesian sensitivity analysis for unmeasured confounding in causal mediation analysis. Stat Methods Med Res. https://doi.org/10.1177/0962280217729844

Menzies P. The causal structure of mechanisms. Stud Hist Phil Biol Biomed Sci. 2012;43(4):796–805. https://doi.org/10.1016/j.shpsc.2012.05.00.

Moore KL, Neugebauer R, van der Laan MJ, Tager IB. Causal inference in epidemiological studies with strong confounding. Stat Med. 2012;31(13):1380–404. https://doi.org/10.1002/sim.4469.

Murray CJ, Lopez AD. Measuring the global burden of disease. N Engl J Med. 2013;369(5):448–57. https://doi.org/10.1056/NEJMra1201534.

National Research Council. Deterrence and the death penalty. Washington, DC: The National Academies Press; 2012.

Oates CJ, Kasza J, Mukherjee A. Discussion of causal inference by using invariant prediction: identification and confidence intervals by Peters, Buhlmann and Meinshausen. J R Stat Soc Ser B. 2016;78:1003.

Oates CJ, Kasza J, Simpson JA, Forbes AB. Repair of partly misspecified causal diagrams. Epidemiology. 2017;28(4):548–52. https://doi.org/10.1097/EDE.0000000000000659.

Omenn GS, Goodman GE, Thornquist MD, Balmes J, Cullen MR, Glass A, Keogh JP, Meyskens FL, Valanis B, Williams JH, Barnhart S, Hammar S. Effects of a combination of beta carotene and vitamin A on lung cancer and cardiovascular disease. N Engl J Med. 1996;334(18):1150–5.

Papana A, Kyrtsou C, Kugiumtzis D, Diks C. Assessment of resampling methods for causality testing: a note on the US inflation behavior. PLoS One. 2017;12(7):e0180852. https://doi.org/10.1371/journal.pone.0180852.

Pearl J. Reply to commentary by Imai, Keele, Tingley, and Yamamo to Concerning causal mediation analysis. Psychol Methods. 2014;19(4):488–92.

Pearl J, Mackenzie D. The book of why: the new science of cause and effect. New York: Basic Books; 2018.

Peters J, Bühlmann P, Meinshausen N. Causal inference by using invariant prediction: identification and confidence intervals. J R Stat Soc Ser B. 2016;78(5):947–1012.

Petersen ML, Porter KE, Gruber S, Wang Y, van der Laan MJ. Diagnosing and responding to violations in the positivity assumption. Stat Methods Med Res. 2012;21(1):31–54. https://doi.org/10.1177/0962280210386207.

Petitti DB. Associations are not effects. Am J Epidemiol. 1991;133(2):101–2.

Pirracchio R, Petersen ML, van der Laan M. Improving propensity score estimators' robustness to model misspecification using super learner. Am J Epidemiol. 2015;181(2):108–19. https://doi.org/10.1093/aje/kwu253.

Quinn CJ, Kiyavash N, Coleman TP. Directed information graphs. IEEE Trans Inf Theory. 2015;61(12):6887–909.

Rhomberg LR, Chandalia JK, Long CM, Goodman JE. Measurement error in environmental epidemiology and the shape of exposure-response curves. Crit Rev Toxicol. 2011;41(8):651–71. https://doi.org/10.3109/10408444.2011.563420.

Rodríguez-Entrena M, Schuberth F, Gelhard C. Assessing statistical differences between parameters estimates in Partial Least Squares path modeling. Qual Quant. 2018;52(1):57–69. https://doi.org/10.1007/s11135-016-0400-8.

Rosenström T, Jokela M, Puttonen S, Hintsanen M, Pulkki-Råback L, Viikari JS, Raitakari OT, Keltikangas-Järvinen L. Pairwise measures of causal direction in the epidemiology of sleep problems and depression. PLoS One. 2012;7(11):e50841. https://doi.org/10.1371/journal.pone.0050841.

Rothenhausler D, Heinze C, Peters J, Meinschausen N (2015) BACKSHIFT: learning causal cyclic graphs from unknown shift interventions. arXiv pre-print https://arxiv.org/pdf/1506.02494.pdf

Rothman KJ, Greenland S. Causation and causal inference in epidemiology. Am J Public Health. 2005;95(Suppl 1):S144–50.

Sacco DF, Bruton SV, Brown M. Defense of the questionable: defining the basis of research scientists' engagement in questionable research practices. J Empir Res Hum Res Ethics. 2018;13(1):101–10. https://doi.org/10.1177/1556264617743834.

Sambo F, Franzin A (2016) bnstruct: an R package for Bayesian Network Structure Learning with missing data. https://cran.r-project.org/web/packages/bnstruct/vignettes/bnstruct.pdf

Santra T, Kolch W, Kholodenko BN. Integrating Bayesian variable selection with modular response analysis to infer biochemical network topology. BMC Syst Biol. 2013;7:57. https://doi.org/10.1186/1752-0509-7-57.

Schreiber T. Measuring information transfer. Phys Rev Lett. 2000;85(2):461–4. https://doi.org/10.1103/PhysRevLett.85.461.

Schünemann H, Hill S, Guyatt G, et al. The GRADE approach and Bradford Hill's criteria for causation. J Epidemiol Community Health. 2011;65:392–5.

Schwartz S, Gatto NM, Campbell UB. Transportabilty and causal generalization. Epidemiology. 2011;22(5):745–6.

Scutari M, Ness R (2018) Package 'bnlearn'. https://cran.r-project.org/web/packages/bnlearn/bnlearn.pdf

Shimizu S, Hoyer PO, Hyvärinen A, Kerminen A. A linear non-gaussian acyclic model for causal discovery. JMLR. 2006;7:2003–30.

Simon HA. Causal ordering and identifiability. In: Hood WC, Koopmans TC, editors. Studies in econometric method, Cowles commission for research in economics monograph No, vol. 14. New York: Wiley; 1953. p. 49–74.

Simon HA, Iwasaki Y. Causal ordering, comparative statics, and near decomposability. J Econ. 1988;39:149–73.

Spirtes P. Introduction to causal inference. J Mach Learn Res. 2010;11:1643–62.

Suppes P. A probabilistic theory of causality. Amsterdam: North-Holland Publishing Company; 1970.

Tashiro T, Shimizu S, Hyvärinen A, Washio T. ParceLiNGAM: a causal ordering method robust against latent confounders. Neural Comput. 2014;26(1):57–83. https://doi.org/10.1162/NECO_a_00533.

Tchetgen Tchetgen EJ, Phiri K. Bounds for pure direct effect. Epidemiology. 2014;25(5):775–6. https://doi.org/10.1097/EDE.0000000000000154.

Textor J, van der Zander B, Gilthorpe MS, Liskiewicz M, Ellison GT. Robust causal inference using directed acyclic graphs: the R package 'dagitty'. Int J Epidemiol. 2016;45(6):1887–94.

Thistlethwaite D, Campbell D. Regression-discontinuity analysis: an alternative to the ex post facto experiment. J Educ Psychol. 1960;51(6):309–17. https://doi.org/10.1037/h0044319.

Triantafillou S, Tsamardinos I. Constraint-based causal discovery from multiple interventions over overlapping variable sets. J Mach Learn Res. 2015;16:2147–205.

VanderWeele TJ. Controlled direct and mediated effects: definition, identification and bounds. Scand Stat Theory Appl. 2011;38(3):551–63.

VanderWeele TJ, Vansteelandt S. Conceptual issues concerning mediation, interventions and composition. Stat Interf. 2009;2:457–68.

Voortman M, Dash D, Druzdzel MJ. Learning causal models that make correct manipulation predictions with time series data. Proc Machine Learn Res. 2010;6:257–66.

Waldner D. Process tracing and qualitative causal inference. Secur Stud. 2015;24(2):239–50.

Washington Post (2017) Pollution kills 9 million people each year, new study finds. https://www.washingtonpost.com/news/energy-environment/wp/2017/10/19/pollution-kills-9-million-people-each-year-new-study-finds/?noredirect=on&utm_term=.8339ea9b914c

Weber I, Florin E, von Papen M, Timmermann L. The influence of filtering and downsampling on the estimation of transfer entropy. PLoS One. 2017;12(11):e0188210. https://doi.org/10.1371/journal.pone.0188210.

Wheeler G, Scheines R. Coherence and confirmation through causation. Mind. 2013;122(485):135–70.

Wiener N. The theory of prediction. In: Beckenbach EF, editor. Modern mathematics for engineers, vol. 1. New York: McGraw-Hill; 1956.

Yule GU. Why do we sometimes get nonsense-correlations between time-series? -- A study in sampling and the nature of time-series. J R Stat Soc. 1926;89(1):1–63. https://doi.org/10.2307/2341482.

Chapter 10
Clarifying Exposure-Response Regression Coefficients with Bayesian Networks: Blood Lead-Mortality Associations an Example

Introduction

This chapter examines how the Bayesian network (BN) learning and analysis methods discussed I Chap. 9 can help to meet several methodological challenges that arise in interpreting significant regression coefficients in exposure-response regression modeling. As a motivating example, consider the challenge of interpreting positive regression coefficients for blood lead level (BLL) as a predictor of mortality risk for nonsmoking men. Practices such as dichotomizing or categorizing continuous confounders (e.g., income), omitting potentially important socioeconomic confounders (e.g., education), and assuming specific parametric regression model forms leave unclear to what extent a positive regression coefficient reflects these modeling choices, rather than a direct dependence of mortality risk on exposure. Therefore, significant exposure-response coefficients in parametric regression models do not necessarily reveal the extent to which reducing exposure-related variables (e.g., BLL) alone, while leaving fixed other correlates of exposure and mortality risks (e.g., education, income etc.) would reduce adverse outcome risks (e.g., mortality risks). This chapter illustrates how BN structure-learning and inference algorithms and nonparametric estimation methods (partial dependence plots) from Chap. 9 can be used to clarity dependencies between variables, variable selection, confounding, and quantification of joint effects of multiple factors on risk, including possible high-order interactions and nonlinearities. It concludes that these details must be carefully modeled to determine whether a data set provides evidence that exposure itself directly affects risks, and that BN and nonparametric effect estimation and uncertainty quantification methods can complement regression modeling and help to improve the scientific basis for risk management decisions and policy-making by addressing these issues.

L. A. Cox Jr., *Quantitative Risk Analysis of Air Pollution Health Effects*,
International Series in Operations Research & Management Science 299,
https://doi.org/10.1007/978-3-030-57358-4_10

Science-Policy Challenge: Determining Whether Data Sets Provide Evidence that Risk Depends on Exposure

When does a data set provide evidence that one variable depends on another? In human health risk assessment, as discussed in Chaps. 1 and 2, a traditional answer regresses a dependent risk variable, such as an indicator of mortality or morbidity, against a measure of exposure and other covariates (e.g., potential confounders and modifiers), and concludes that the data provide evidence that risk depends directly on exposure if the regression coefficient for exposure is statistically significantly greater than zero (NIOSH 2020). This approach is simple and convenient. It has long been used successfully in health, safety, and environmental (HS&E) risk analysis to identify associations between exposures and adverse responses. However, as stressed in Chaps. 1 and 2 and illustrated in Chaps. 7 and 8, the regression approach is limited by the fact that its conclusions are often *model-dependent*, meaning that they are driven by modeling choices and assumptions, rather than necessarily reflecting truths about the world. Hence, conclusions from association-based analyses, including regression analyses, are increasingly seen as unreliable in many cases (Dominici et al. 2014).

The following sections examine the potential of BN learning to help meet the challenges of determining to what extent significant regression coefficients provide evidence of dependence between variables in a data set, especially between exposure and adverse health responses. BN techniques can potentially help to improve the reliability of conclusions about whether and how much risk depends on exposure by using nonparametric tests to identify significant statistical dependencies between variables that remain after adjusting for values of other variables. This approach helps to avoid dependence on parametric modeling assumptions. Testing whether the null hypothesis of no relation (i.e., statistical independence) between variables is rejected after conditioning on other relevant variables—known in the BN learning literature as *conditional independence* testing (see Chap. 9)—helps to identify and avoid spurious correlations created by failing to condition on common ancestors (e.g., confounders) or by conditioning on common descendants (e.g., selection biases). The following sections examine the value of BN learning as a heuristic for identifying and helping to correct deficiencies in regression modeling due to model specification errors (e.g., omitted interaction terms), omitted confounding, and residual confounding.

For practical motivation, we consider how to determine whether a data set provides evidence that mortality risk in nonsmoking men depends on blood lead level (BLL). For concreteness, we assemble a simple illustrative data set to analyze. We then analyze it using both simple (main-effects) logistic regression models and Bayesian network methods. The data set is not intended to be comprehensive, and both the data and the logistic regression modeling are deliberately kept simple; specialists can and should consider additional variables and models. However, they are intended to illustrate methodological points clearly that are important in addressing the following key question: *Given* a data set—limitations and all—what methods can be used to determine whether it provides evidence that risk depends directly on exposure? More precisely, what data-analyses can be used to determine whether a given data set provides reason to believe that changing the level of an exposure variable would change the probability distribution of a response variable, even if other variables (e.g., confounders) are held fixed? If it does so, then we say that the data set provides evidence that the response variable *depends on* the exposure variable; otherwise, it does not, even if the exposure and response variables are strongly and significantly associated in a regression model.

Example Data Set

Data for this study were assembled from the NHANES III data set, which has been used in many studies that examine exposure-response associations, including BLL-mortality (e.g., Lanphear et al. 2018). The NHANES III study sampled households in 81 counties across the United States between 1988–1994, completing extensive interviews and medical examinations for nearly 40,000 people (CDC 1994). To construct an example data set for BLL and mortality that controls for potential confounding by sex and smoking (since women tend to have lower BLL and age-specific mortality rates than men, and smokers tend to have higher BLL and age-specific rates than never-smokers), we selected male nonsmokers with measured values for BLL (2903 cases). Variables were given short, mnemonic names for purposes of display, as detailed in Appendix. The variables considered are as follows.

- *Mortality* is a 0–1 variable with a value of 0 for people who had not yet died as of the end of the follow-up period (averaging nearly 20 years) and a value of 1 otherwise. This chapter treats mortality within the follow-up period as a binary dependent variable to provide simple illustrations of parametric and nonparametric tests for dependence on BLL and other factors. Although more detailed survival data analysis, with time until death or right-censoring as a dependent variable, would also be worthwhile, a simple binary outcome suffices to address the methodological challenge of deciding whether the data set provides evidence that mortality depends on BLL. The answer might differ in other data sets with other variables.
- *Heartdeath* is a 0–1 variable indicating whether mortality occurred before the end of the follow-up period, with the cause of death identified as heart-related (1 = yes, 0 = no). This variable, primarily reflecting CVD mortality, was derived from the NHANES III variable UCOD for underlying cause of death. It is defined as *heartdeath* = 1 if UCOD = 1, the code for "diseases of heart," and *heartdeath* = 0 otherwise.
- *Lead* (also called *BLL*) is the measured blood lead level (BLL), expressed in concentration units of μg/dL (micrograms per deciliter).
- Other variables are as follows: *age* in years, *family* size, highest *grade* of education attained (with a median value of 12, corresponding to high school graduates),

income expressed as a multiple of the poverty level, *marriage* (0 for never married, else 1); 0–1 indicators for *Hispanic* and *Black* (1 for people with each attribute, else 0); and 0–1 indicators for residence in a *small_metropolitan* area or in the *West* or *South* regions. To illustrate the importance of coding choices—specifically, dichotomizing continuous variables, as in Lanphear et al. (2018)—we also derived a binary *poverty* indicator from the *income* variable, with values of 1 for households below the poverty line and 0 otherwise.

Other health-related variables, such as comorbidities and blood concentrations of substances other than lead, were not included, both to prevent possible collider bias arising from conditioning on potential common consequences of lead exposure and medical conditions, and to keep the example simple.

These variables are intended to provide a simple but realistic example of epidemiological data for purposes of illustrating important methodological points and comparing different methods for quantifying dependence relations among variables. Although the NHANES III data set is carefully designed for use with weights to allow extrapolation to the US population, we do not use these weights or make such extrapolations here, but use the collected data (which deliberately oversampled Black and Hispanic subpopulations relative to the US population) to quantify relations among the variables. In addition, we only use records with values recorded for lead (BLL) (including the imputed value supplied in NHANES III for lead exposures below the detection limit, namely, the detection limit divided by the square root of 2). NHANES III handles missing data for many variables by providing 5 possible imputed values for each missing value, but the accuracy and bias of the imputed values is uncertain. Using measured values only, and dropping cases with missing BLL values, suffices for our illustrative purposes and avoids model uncertainties due to imputation. Since data are not necessarily missing at random, we again do not extrapolate beyond the specific data set analyzed.

Statistical Analysis Methods

A research question of central interest in quantitative risk assessment (QRA) modeling is how the probability distribution of a response variable such as mortality varies with an indicator of exposure, such as BLL, when values of other variable are held fixed. The concept "held fixed" is a causal, rather than a statistical, construct (Pearl 2009). We compare the following methods to examine this question for the example data set.

- *Logistic regression modeling*. Logistic regression was selected as a common parametric modeling method to allow comparison of results from parametric and non-parametric analyses. Analyses using more or different variables and other parametric or semiparametric (e.g., proportional hazards) models might also be useful, but the logistic regression model suffices to illustrate methodological points.

- *Bayesian network learning* to explore and visualize significant dependencies between variables were performed using Bayesian network (BN) learning (using the *bnlearn* package in R).

Both logistic regression and BN learning were carried out in R via using the Causal Analytics Toolkit (CAT) software (http://cloudcat.cox-associates.com:8899/), which provides online documentation for all R packages used. For readers who are not already familiar with BN modeling, additional explanation of the main ideas and interpretations of the outputs are provided in the section "Results".

Results

Results from Logistic Regression Modeling

Table 10.1 shows the results of a logistic regression model for all-cause mortality with *lead*, *age*, and *poverty* as predictors. Following Lanphear et al. (2018), Table 10.1 treats *age* as continuous and *income* as dichotomous (via the *poverty* indicator). (*Sex* is not relevant, as we only consider men.)

This model shows a highly statistically significant ($p < 0.004$) positive regression coefficient for *lead*. By contrast, Table 10.2 shows the model modified to treat *income* as a continuous variable instead of dichotomizing it. This changes the significance level for the *lead* regression coefficient to just over 0.05, suggesting the importance of not dichotomizing continuous predictors if residual confounding is to be avoided. Table 10.3 shows the model modified to include *grade* (highest grade completed) as an additional predictor. This more than doubles the *p* value for lead, to over 0.11, suggesting that omitting grade might fail to control for an important source of confounding (perhaps because people in the same birth cohort who completed fewer years of school are more likely to have higher BLL levels and higher mortality rates later in life). Finally, Table 10.4 adds *family* size as a predictor, which increases the significance (i.e., lowers the *p* value) of the regression coefficient for BLL again, from about 0.11 to just under 0.06. Including *family* as a predictor

Table 10.1 A main-effects logistic regression model for *mortality* with dichotomized *income* (*poverty*)

	Estimate	Std. Error	z value	Pr(>|z|)	Signif	OR	2.50%	97.50%
(Intercept)	−5.966	0.202	−29.564	<0.001	***	0	0	0
lead	0.051	0.018	2.887	0.00389	**	1.05	1.02	1.09
age	0.099	0.004	27.924	<0.001	***	1.1	1.1	1.11
poverty	0.705	0.153	4.598	<0.001	***	2.02	1.5	2.73

Signif. codes: 0 '***' 0.001 '**' 0.01 '*' 0.05 '.' 0.1 ' ' 1. "Estimate" = estimated logistic regression coefficient; OR = odds ratio; 2.50% = lower 95% confidence limit; 97.5% = upper 95% confidence limit

Table 10.2 A main-effects logistic regression model for *mortality* with continuous *income* variable

	Estimate	Std. error	z value	Pr(>|z|)	Signif	OR	2.50%	97.50%
(Intercept)	−5.281	0.198	−26.704	<0.001	***	0.01	0	0.01
Lead	0.035	0.018	1.948	0.05137	.	1.04	1	1.07
Age	0.102	0.004	27.871	<0.001	***	1.11	1.1	1.11
Income	−0.223	0.034	−6.588	<0.001	***	0.8	0.75	0.85

Signif. codes: 0 '***' 0.001 '**' 0.01 '*' 0.05 '.' 0.1 ' ' 1. "Estimate" = estimated logistic regression coefficient; OR = odds ratio; 2.50% = lower 95% confidence limit; 97.5% = upper 95% confidence limit

Table 10.3 A main-effects logistic regression model for *mortality* with continuous *income* variable and with *grade* as an additional predictor

	Estimate	Std. error	z value	Pr(>|z|)	Signif	OR	2.50%	97.50%
(Intercept)	−4.782	0.265	−18.026	<0.001	***	0.01	0	0.01
Lead	0.029	0.018	1.586	0.11278		1.03	0.99	1.07
Age	0.099	0.004	27.036	<0.001	***	1.1	1.1	1.11
Income	−0.173	0.038	−4.582	<0.001	***	0.84	0.78	0.91
Grade	−0.044	0.016	−2.713	0.00668	**	0.96	0.93	0.99

Signif. codes: 0 '***' 0.001 '**' 0.01 '*' 0.05 '.' 0.1 ' ' 1. "Estimate" = estimated logistic regression coefficient; OR = odds ratio; 2.50% = lower 95% confidence limit; 97.5% = upper 95% confidence limit

makes *lead* a more significant predictor of *mortality* and increases its regression coefficient. This can happen because of model specification error, when the implicit assumption that a single regression coefficient holds throughout the range of a predictor is violated for each of two predictors. Then, including both in a (misspecified) regression model can reduce total prediction error, compared to including just one of them, if the model specification errors partly offset each other. The value of the regression coefficient for BLL is changed by the model variations in Tables 10.1, 10.2, 10.3, and 10.4, from just over 0.05 in Table 10.1 to slightly under 0.03 in Table 10.3. In other examples of regression modeling for different pollutants, both significant positive coefficients and significant negative coefficients for the effect of exposure have been produced by varying the choice of other variables to include as predictors (Dominici et al. 2014).

Tables 10.1, 10.2, 10.3, and 10.4 illustrate that standard regression outputs, such as regression coefficients and their *p* values and confidence intervals, do not show how much of the value of a significant positive regression coefficient for an exposure-related variable (e.g., *lead*) is explained by omitted confounders (e.g., *grade* in Table 10.1); by residual confounding, if continuous confounders have been dichotomized or categorized (e.g., *income* in Table 10.1); or by model specification errors, e.g., from ignored interactions or nonlinearities. These are points of more than theoretical interest; they arise in practical work. For example, Lanphear et al. (2018) used 2 or 3 categories to dichotomize or categorize continuous potential confounders, including income, smoking history, alcohol consumption, exercise,

Table 10.4 A main-effects logistic regression model for *mortality* with continuous *income* variable and with *grade* and *family* size as additional predictors

	Estimate	Std. error	z value	Pr(>|z|)	Signif	OR	2.50%	97.50%
(Intercept)	−3.93	0.312	−12.604	<0.001	***	0.02	0.01	0.04
lead	0.035	0.019	1.893	0.05837	.	1.04	1	1.07
age	0.096	0.004	25.739	<0.001	***	1.1	1.09	1.11
Income	−0.206	0.038	−5.355	<0.001	***	0.81	0.75	0.88
grade	−0.053	0.016	−3.214	0.00131	**	0.95	0.92	0.98
family	−0.158	0.033	−4.791	<0.001	***	0.85	0.8	0.91

Signif. codes: 0 '***' 0.001 '**' 0.01 '*' 0.05 '.' 0.1 ' ' 1. "Estimate" = estimated logistic regression coefficient; OR = odds ratio; 2.50% = lower 95% confidence limit; 97.5% = upper 95% confidence limit

and healthy eating index. Other confounders, such as *grade*, were omitted (although the authors did note that, descriptively, "Participants who had the highest concentrations of lead in blood were older, less educated, and more likely to be male, to smoke cigarettes, to consume larger amounts of alcohol, and to have less healthy diets.") Our experience with *income* in Tables 10.1 and 10.2, as well as decades of methodological warnings (Streiner 2002; Naggara et al. 2011; Groenwold et al. 2013) indicate that use of a few discrete categories to represent continuous confounders can substantially increase the apparent significance of exposure (e.g., *lead* in this example) as a predictor of risk (e.g., *mortality*). Residual confounding, as well as omitted confounders, can make the regression coefficient for exposure positive and significant even if mortality does not depend on exposure, or depends on it less strongly than the model with residual confounding and omitted confounders suggests. Thus, although the simple regression models in Tables 10.1, 10.2, 10.3, and 10.4 can probably be improved in many ways (e.g., by using regression diagnostics to improve model specification, e.g., by modeling interactions among variables), they suffice to raise and illustrate the key methodological questions of (a) How to decide which variables to include in a regression model; and (b) How to obtain estimates of the effects of exposure on risk that have a clearer interpretation than a regression coefficient does when its significance, magnitude, or sign (positive or negative) depend on modeling choices about the functional form assumed (model specification) and the variables included on the right side of the regression equation. We next examine how Bayesian networks can help to address these questions

Results for Bayesian Network (BN) Modeling

Bayesian network (BN) methods for data analysis in quantitative risk assessment have three main parts, ad discussed further in Chap. 9: (1) Algorithms for determining *BN structure*, showing which other variables (if any) each variable is found to depend on; (2) Quantification of *conditional probability tables* (CPTs) or models showing how the conditional probability distribution of each variable changes as the values of the variables that it depends on (its direct "parents," if it has any) in the BN are changed; and (3) *Estimation* of the effects of changing exposure on risk, controlling for (by holding fixed at specified levels) the values of other variables on which risk also depends (Chap. 9; Nagarajan et al. 2013). In the terminology of causal mediation analysis, the *direct effect* of exposure on risk shows how risk changes as exposure changes, holding all other variables fixed; while the *total effect* of exposure on risk show how risk changes as exposure changes and other variables that depend on exposure, and that mediate its effects on risk, are allowed to change in response to changes in exposure. In the BLL-mortality, the relevant exposure variable is BLL, and we do not consider mediating variables, so the only effect of interest is the direct effect of BLL on mortality risk, showing how changing BLL changes the age-specific probability of mortality, given the values of age and other variables on which mortality risk depends. The full methodology of BN learning is quite detailed, and we defer to the references for details (Cox Jr. 2018; Nagarajan et al. 2013). However, the main principles and outputs are relatively easy to understand, and we focus on these here.

Figure 10.1 shows the structure of a BN learned from our example data set. BN structure-learning algorithms do not distinguish one dependent variable of interest, as in regression models, but rather address the more general question of how each variable depends on the rest.

The interpretation of such a network is that two variables are linked by an arrow if and only if they are found to be statistically dependent even after conditioning on the levels of other variables (e.g., via a conditional independence test). For example, an arrow between *Hispanic* and *West* signifies that the probability that a respondent lives in the *West* region differs significantly for people who are Hispanic compared to people who are not, even after conditioning on the values of other variables.

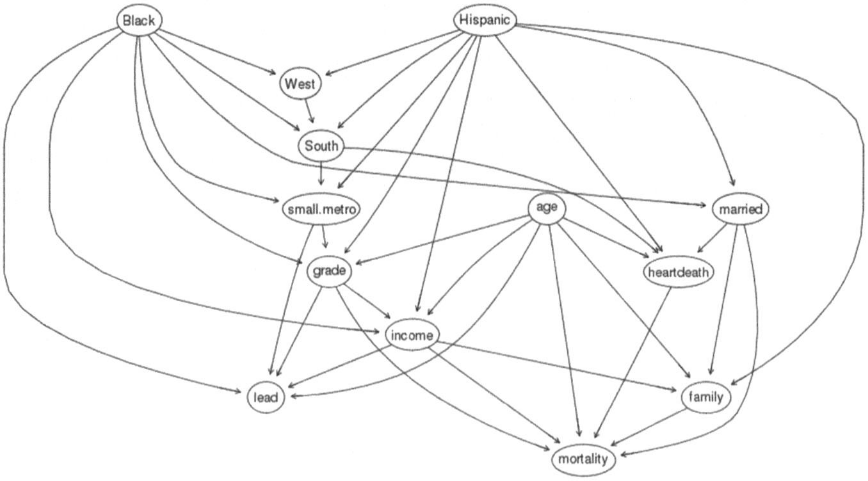

Fig. 10.1 Structure of a BN for all variables, assuming that *Black*, *Hispanic*, and *Age* are sources or inputs (no incoming arrows) and that *mortality* is a sink or output (no outgoing arrows)

Equivalently, any two variables are joined by an arrow if and only if they are found to be mutually informative about each other, in the sense of information theory (Chap. 9). Conversely, absence of an arrow between two variables indicates that they are not found to be mutually informative about each other, given the values of other variables: the null hypothesis of conditional independence is not rejected. For example, the absence of an arrow between *income* and *West* indicates that the probability that a respondent lives in the *West* region is independent of the respondent's income, given the values of the other variables (*Hispanic*, *Black*, and *South*) linked to *West*. (The absence of an arrow between *Black* and *Hispanic*, however, reflects a constraint that both are considered inputs, i.e, potential causes but not potential effects of other variables; this is discussed later.) Conditional independence does not usually imply independence. For example, in Fig. 10.1, *family* and *grade* both depend on *age* and on *Hispanic*, so it is not surprising that *family* and *grade* are significantly associated with each other (e.g., their linear (Pearson's) correlation of −0.16 is statistically significant at $p < 0.01$, and a multiple linear regression model for *grade* regressed against other variables shows a significant ($p < 0.001$) negative coefficient for *family*, presumably reflecting model specification error), even though they are shown as conditionally independent in Fig. 10.1. In a BN, the quantitative dependence of a variable (a node in the network) on the variables that point into (its parents) is represented by a conditional probability table (CPT). Conceptually, the CPT lists the conditional probability distribution of the variable for each setting of its parents. In practice, to avoid combinatorial explosion, more efficient data structures (such as regression models, trees, or nonparametric function approximators) may be used to quantify the dependence of each node's probability distribution on the values of its parents, but in any BN, the dependence of one variable on another (e.g., of risk on exposure) is modeled via CPT tables or models that explicitly take into account the levels of other variables (Nagarajan et al. 2013).

By Bayes' Rule, statistical dependence (or mutual information) is a symmetric relation. (If the conditional probability distribution of Y differs significantly for different values of X, signified by $X \rightarrow Y$ in BN notation, then the converse is also true, since the joint distribution of X and Y can be factored equally well as $P(X = x$ and $Y = y) = P(X = x \mid Y = y)*P(Y = y)$ or as $P(X = x$ and $Y = y) = P(Y = y \mid X = x)*\ P(X = x)$. Expressed more briefly, $P(x, y) = P(x \mid y)P(y) = P(y \mid x)P(x)$ In either case, if $P(x, y)$ differs from $P(x)P(y)$, it must be because $P(x \mid y)$ differs from $P(x)$ and also, symmetrically, because $P(y \mid x)$ differs from $P(y)$.) Thus, the same joint probability distribution of variables can be decomposed into marginal distributions and joint distributions in different ways, corresponding to different orientations of the arrows (e.g., $X \rightarrow Y$ corresponds to $P(y) = \Sigma_x P(y \mid x)P(x)$ and $Y \rightarrow X$ corresponds to $P(y) = \Sigma_y P(x \mid y)P(y)$; both are fully determined by the joint distribution $P(x, y)$.) Therefore, the arrows in a BN cannot necessarily be given a unique causal orientation and interpretation based only on the joint distribution of observations. Indeed, for some pairs of variables, such as *South* and *West*, it is not clear what it would mean to say that either causes the other, although it is perfectly clear that they must be negatively associated, insofar as they are mutually exclusive.

However, for many other pairs of variables, orientation assists causal interpretation. BN learning software such as the *bnlearn* R package, allows knowledge-based constraints on allowed orientations of arrows (using lists of allowed and forbidden arrow directions, where known and applicable). For example, the BN structure in Fig. 10.1 was learned from the example data set with the constraints that *Hispanic*, *Black*, and *age* are potential causes, but not potential effects, of other variables (and hence can only have outward-pointing arrows if the arrows are to be interpreted causally); while *mortality* is a potential effect, but not a potential cause of other variables (and so can have only inward-pointing arrows). Arrows with causal interpretations imply that changing the values of the variables that point into a node (i.e., into a variable, since BN network nodes represent variables) changes its conditional probability distribution, as specified by its CPT. Whether or not arrows are given causal interpretations, a BN represents a decomposition of the joint distribution of variables into marginal distributions for input nodes (those with only outward-pointing arrows) and conditional distributions (CPTs) for other nodes (those with inward-pointing arrows). If is assumed that part of the intended definition of "cause" is that direct effects are not conditionally independent of their causes (or, conversely, that causes provide unique information about their direct effects, implying that their mutual information is positive), then causes and their direct effects must be joined by arrows in a BN. This is true even if not all of the BN's arrows have causal interpretations or unique directions, as in the *West* → *South* example. In this sense, a BN provides a simple screen for a necessary condition for evidence of causation: *if two variables are not directly linked by an arrow in a BN, then the BN does not provide evidence that either variable is a direct cause of the other.* Conversely, if two variables are linked by an arrow (in either direction), then the BN provides evidence that one might be a direct cause of the other.

There are many algorithms for identifying BN structures from data, and they do not always give identical answers. Figure 10.2 summarizes the results of applying four different BN structure-learning algorithms to identify what *mortality* depends on in the example BLL data set (excluding the obvious dependence on *heartdeath*, which is a tautology): two algorithms essentially seek BNs to maximize a score reflecting the likelihood of the data, and two seek BNs satisfying the conditional independence and other constraints found in the data. (The four algorithms are the hill-climbing (HC) algorithm, which is the default choice in *bnlearn*, and the one used to generate the BN in Fig. 10.1; Tabu Search (Tabu), which helps to avoid local optima; the Grow-Shrink (GS) algorithm; and the Incremental Association Markov Blanket (IAMB) algorithm. Each has a large technical literature and is supported in software packages such as *bnlearn* and *CompareCausalNetworks*; see Nagarajan et al. 2013 for details.) Figure 10.2 uses heavier (thicker) lines for arrows that are identified by more of these algorithms, with the heaviest arrows being supported by all four algorithms and the lightest being supported by only one. Thus all four algorithms identify an arrow from *age* to *mortality*, and an arrow from *age* to *lead*; but only two algorithms (HC and Tabu) identify arrows from *income* and *grade* to *lead*. Non-existent arrows between two variables (e.g., between *West* and *income*) indicate that none of the algorithms found a direct dependence between them. Thus,

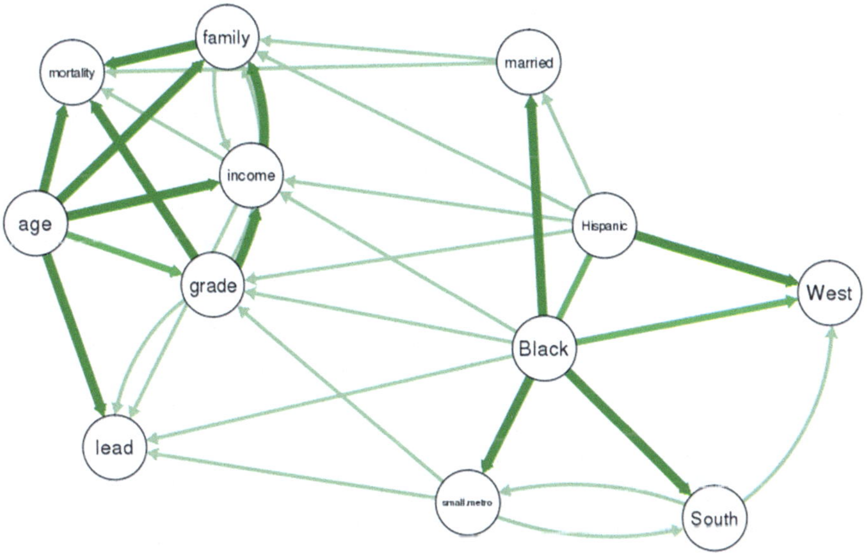

Fig. 10.2 Summary of BN structure learned by four algorithms. Heavier lines and shorter distances between nodes show dependence relations supported by more of the algorithms (lightest = supported by only 1; heaviest = supported by all 4). Algorithms GS and IAMB may generate undirected arcs: these arcs are counted once in each direction (e.g., between *South* and *small_metro*)

Fig. 10.2 helps to distinguish between relatively firm conclusions (e.g., that *age* is a confounder of the *lead-mortality* association, but there is no arrow directly between *lead* and *mortality*) and those that are supported by only a subset of the algorithms (e.g., that *grade* and *income* are additional confounders of the *lead-mortality* association, even after controlling for direct confounding by *age*).

The structure of a BN determines which variables must be conditioned on, and which must not be conditioned on, to obtain effects estimates that are not biased by the selection of variables (Arnold et al. 2020). The answers constitute *adjustment sets*, i.e., subsets of variables that, logically, must be "adjusted" for (by conditioning on their values) to isolate the effect of one variable on another. Adjustment sets generalize the idea that confounders (common parents) must be conditioned on, but that colliders (common children) must not be conditioned on, to isolate the effect of one variable on another. Graph-theoretic algorithms can be applied to a BN's structure to automatically identify minimal sufficient adjustment sets for estimating both (a) The *direct effect* of one variable on another, holding the levels of all other variables fixed at specified levels; and (b) The *total effect* of one variable on another, allowing the probability distributions of mediating variables along the directed paths leading from the first to the second to change in response to exogenous changes in the first, as described by their CPTs (Textor et al. 2016). For example, to estimate the effect of *income* on *mortality* in Fig. 10.1, any of the three minimal sufficient adjustment sets {*age, family, grade, married, heartdeath*}, {*age, family, grade, married, Black, Hispanic*}, or {*age, family, grade, married, Hispanic, South*} may be used. By contrast, to estimate the total effect of *income* on *mortality*, any of the three adjustment sets {*Hispanic, age, grade, heartdeath, married*}, {*Hispanic, South, age, grade, married*}, or {*Black, Hispanic, age, grade*} may be used, but *family* must be allowed to adjust to reflect the effects of different values of *income*. (These adjustment sets are calculated in CAT by applying the *dagitty* package (Textor et al. 2016) to the BN structure learned by the *bnlearn* package.) Such adjustment sets provide a principled answer to the question of which variables to include as predictors in a regression model. The selection of variables should be based on the purpose of the regression, i.e., on whether the direct or the total effect of a specified explanatory variable on the dependent variable is to be estimated. The answer, together with the structure of the BN, then determines which other predictors should be included: those in an adjustment set for the desired (direct or total) effect. For purposes of risk analysis, adjustment sets specify what to include in a regression model to avoid biasing the estimate of the desired effect (direct or total) of exposure on risk by failing to condition appropriately on other variables

When there are multiple adjustment sets, they can be used to help to indicate variability in estimates and to check the internal validity of the BN model: the effects estimates should not be significantly different if the BN model is valid, but their variability reveals how imprecise the estimates are. If they are significantly different, then there may be omitted variables or the structure of the BN may be wrong. If the structure is in doubt (as in Fig. 10.2, which reveals that some of the arrows in Fig. 10.1 are more robustly supported than others), then different structures and their implied adjustment sets can be used to calculate the causal effect of interest, and the set of possible answers reflects uncertainty about the correct model structure. As an example, Fig. 10.3 shows the estimated direct effect of *grade* on *mortality* using the BN structure in Fig. 10.1. Instead of a single numerical coefficients, as in parametric regression models, the estimated effect is a curve, called a partial dependence plot (PDP). The PDP shows how predicted *mortality* probability (vertical axis) varies with *grade* (horizontal axis) if variables in an adjustment set are held fixed at their current values for each individual. To minimize the potential model misspecification errors, an ensemble of 500 nonparametric models (created by the *randomForest* package in R (Liaw and Wiener 2002)) is used to generate a distribution of predicted values of *mortality* for each individual, given the recorded values for that individual for each variable in the adjustment set, and for each possible value of *grade*. (For details, see the *randomForest* documentation (*ibid*) and Cox Jr. 2018.) Figure 10.3 shows the means of these distributions for four different adjustment sets. The effects estimates (PDP curves) are similar but not identical.

(Note that the vertical scales in Fig. 10.3 run from about 0.30 to about 0.33, so the magnitude of the effect of *grade* is only about 10% of the average value of about 0.3 of this population dying during the follow-up period.) To better characterize uncertainty and variability, the left side of Fig. 10.4 shows the 95% uncertainty bands (expressed as differences from the mean value) for the PDP in the upper left of Fig. 10.3. For each *grade* level (0–17), the upper and lower 95% bands enclose 95% of the population distribution of predicted *mortality* values for that *grade* level.

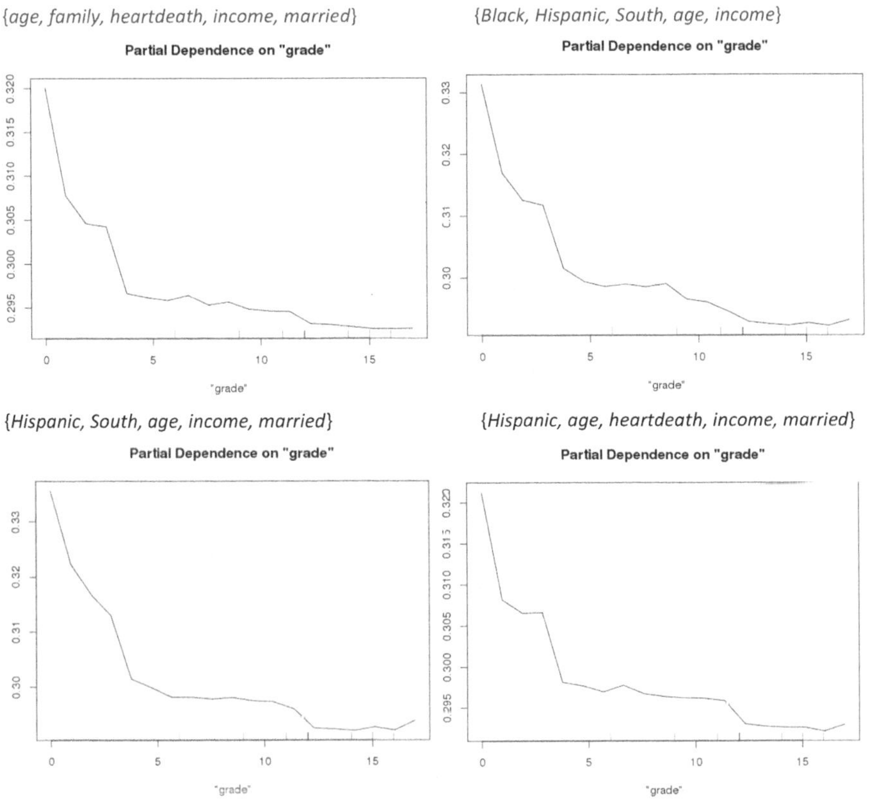

Fig. 10.3 Estimated direct effect of *grade* on *mortality* using four different minimal sufficient adjustment sets. The adjustment sets are listed above each causal partial dependence plot

The right panel shows the means of three clusters of individual-level effect curves, called individual conditional expectation (ICE) plots (generated by the *ICEbox* package in R) (Goldstein et al. 2015). (Three clusters are used to illustrate the key concepts, but the number of clusters can be varied to explore heterogeneity in more or less detail.) These curves show that a subset of individuals (about 31%), represented by the ICE curve that starts lowest on the left and end highest on the right end, are relatively insensitive to the effect of *grade* on *mortality*, meaning that their mean effect (the arithmetic average of their individual conditional expectation (ICE) curves for this effect) is relatively flat compared to the other curves. The differences in sensitivities of *mortality* to *grade* between different individuals are explained by differences in the remaining factors (*age, family, heartdeath, income, married*) that interact with *grade* to affect *mortality* probabilities, as described by the CPT for *mortality*.

Taken together, the two graphs in Fig. 10.4 illustrate that (a) The "effect" of one variable on another (here, of *grade* on *mortality*), whether direct or total, is better characterized by a PDP curve than by a single regression coefficient; (b) The PDP

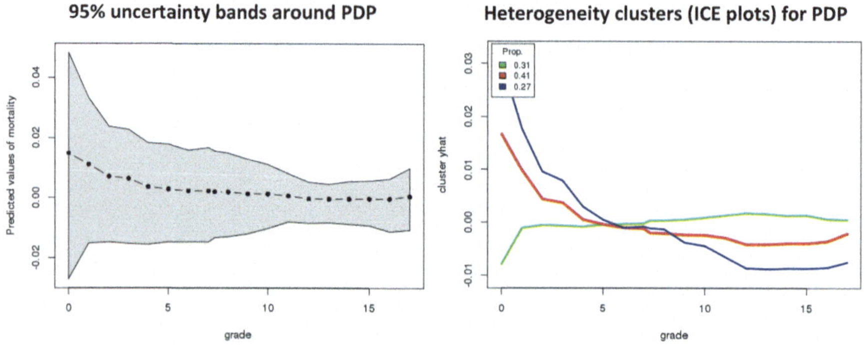

Fig. 10.4 Uncertainty and variability characterization. Left panel: upper and lower 95% uncertainty bands for deviations around the PDP mean values in Fig. 10.3 for the adjustment set {*age, family, heartdeath, income, married*}. Right panel: three individual conditional expectation (ICE) clusters illustrating variability in individual-level effects of *grade* on *mortality*

curve shows the mean values of distributions of the dependent variable for different values of the explanatory variable (e.g., *grade*). The widths of these distributions often vary with the value of the explanatory variable, as shown by uncertainty bands. These distributions cannot necessarily be accurately represented by simple monotonic (e.g., additive or multiplicative) error terms that could be rendered approximately normal and homoscedastic by appropriate transformations (e.g., taking the logarithms) of one or more variables. (c) The overall PDP curve describing the effect of changing one variable on the conditional expectation (and distribution) of another variable in a population of individuals can be disaggregated to show how it varies in different subpopulations. The distinction between direct and total effects, combined with calculations of appropriate adjustment sets for calculating each; the use of PDPs and uncertainty bands to quantify effects and uncertainties about them; and ICE plots to show interindividual heterogeneity, unpack and make explicit different concepts that are usually conflated in regression modeling of exposure-response relationships in risk analysis (Arnold et al. 2020).

BNs and regression can be used to together to reinforce each other in several ways. On the one hand, as already emphasized, BNs can be used to identify the variables to include in a regression model in order to quantify the desired effect (direct or total) of one variable on another, holding fixed the values of other causally relevant variables in an adjustment set for the effect of interest (Pearl 2009; Arnold et al. 2020). On the other hand, regression models can be used in place of discrete conditional probability tables (CPTs) when variables are continuous (or mixed continuous and discrete), to avoid residual confounding arising from discretization (Nagarajan et al. 2013). Parametric regression models risk model specification errors and resulting biases, but flexible nonparametric methods such as PDPs (Fig. 10.3) avoid this source of error. PDPs can also be used to identify model specification errors in parametric regression models. For example, Fig. 10.3 shows that a parametric regression model that uses a single regression coefficient to adjust for the effects of *grade*, when estimating effects of other variables on *mortality*, would not fully control for the effect of *grade*, since its effect is different (steeper) below grade 5 than above it. Similarly, PDPs reveal that only adjusting for *age* and age^2 in regression modeling of the effects of other variables on mortality risk, as is sometimes done in practice (e.g., Lanphear et al. 2018), would not fully account for the empirical dependence of mortality risk on ages above 60 years old, which is S-shaped. More generally, the use of flexible nonparametric techniques can help to identify departures from parametric modeling assumptions. This information can be used to improve the specification of the parametric models, e.g., by introducing breakpoints at which coefficients change.

Discussion

Limitations of Bayesian Network Modeling

Although BNs can be useful in hazard identification by identifying significant direct and indirect dependencies among variables, they are not infallible guides for determining whether a data set provides evidence that risk depends directly on exposure, for several reasons. A sufficiently weak dependence may not be detected by BN-learning algorithms, leading to false negatives. Simulation can be used to determine the smallest hypothesized effects that can be reliably detected in a given data set (e.g., by simulating how exposure would increase risk for different hypothetical potency values, and using the BN-learning software to identify the smallest potencies that generate arrows between exposure and risk in the BN) (Cox Jr. 2018). More simply, although less precisely, the sizes of any undetected direct effects of measured variables on an outcome such as mortality are probably smaller than those that are successfully identified by all algorithms (e.g., the difference between about 0.33 for low grade levels and about 0.29 for high grade levels for the effect of *grade* on *mortality* risk in Fig. 10.3). Conversely, omitted variables, as well as residual confounding if continuous variables are categorized, as is often done to obtain discrete CPTs when both continuous and discrete variables are present, can create false positives and obscure the causal interpretation of arrows in a BN. Figure 10.2 shows that different BN structure-learning algorithms yield somewhat different results, implying that a single algorithm might yield some false-positive or false-negative results. However, all algorithms agree on key findings, such as the arrow from *age* to *mortality*, implying a relatively confident conclusion that *mortality* risk depends on *age*.

The causal interpretation of dependencies is often not precisely clear, especially when key variables are not measured. For example, in Fig. 10.1, the arrow between *age* and *income* suggests that *age* might be a direct cause of *income*, and yet this connection is presumably explained by other variables not included in the data set (e.g., older people may have had more promotions at work, higher savings and investments, etc.) Similarly, the arrow between *grade* and *income* presumably does not imply that bestowing a high school diploma on all respondents would necessarily change their incomes (or the probability distribution of their incomes): the effect of *grade* on *income* in the data set reflects historical conditions that generated the data, and future interventions that change the distribution of *grade* might also change those conditions, so that the future effect of *grade* on *income* may differ from that in the data set (Lucas 1976). To help meet this challenge, recent causal analysis software checks whether CPTs learned from multiple studies remain invariant across the different settings and interventions in the different studies (Chap. 9). If abundant data from multiple diverse studies are available, this can be accomplished straightforwardly by testing whether a CPT (e.g., for risk) is conditionally independent of the study providing the data, given the values of the variables in the CPT (e.g., the parents of risk). If this "invariant causal prediction" (ICP) property (Heinze-Deml et al. 2018) has not been tested and verified, however, then it cannot necessarily be assumed, since effects of omitted (latent) variables may create different CPTs for the same observed variables in different studies. If only a single data set is being assessed, as in our example, then the ICP property cannot be verified by comparing CPTs across studies. In this sense, causality—more precisely, the ICP property for the CPT relating risk to its direct parents in a BN—cannot be established based on a single study in isolation. It is necessary to use data from multiple studies to verify that the estimated CPTs satisfy ICP under a range of conditions, and hence that using it to predict effects of interventions in new studies or situations that fall within this range is supported by relevant empirical experience.

In light of the limitations of BN methods, we regard them as potentially valuable heuristics that can help an investigator identify and quantify dependencies among variables. They are most useful when used in conjunction with common-sense causal knowledge, such as that death is a potential effect but not a potential cause of other variables; or that sex and ethnicity are possible causes but not possible effects of other variables. Such knowledge-based constraints help to interpret the network of empirical statistical dependencies among variables.

Contributions of Bayesian Network Modeling to Improving Risk Assessment

Despite their limitations, Bayesian networks can complement regression modeling approaches to determining whether a data set provides evidence that risk depends directly on exposure in at least the following ways. First, by distinguishing between direct effects and total effects, they help to increase clarity about precisely what is being asked in hazard identification questions such as "Does an increase in exposure cause an increase in risk?" The answer might be yes for a total effect but no for a direct effect (e.g., if increased levels of pollutant *A* produce increased levels of pollutant *B*, which directly increases risk of harm, even though *A* itself has no direct effect). Likewise, it might be no for a total effect but yes for a direct effect (e.g., if increasing *A* reduces *C*, which reduces risk of harm, and this outweighs a direct effects on risk of increasing *A*). In general no single regression coefficient can represent both a direct and a total effect, since the two can be quite different, and it is often unclear whether or to what extent a reported positive regression coefficient represents either one (or neither, as when it arises from residual confounding or from model specification errors) (Arnold et al. 2020). By contrast, BNs make it easy to distinguish between them: a direct effect is represented by an arrow from exposure to risk, while total effects are represented by all directed paths from exposure to risk.

Second, by providing multiple methods and underlying principles for identifying evidence of dependence between each pair of observed variables, BN structure-learning algorithms can be used both to discover dependencies (perhaps including unanticipated ones) and also to determine the degree to which they are supported by different algorithms (Fig. 10.2). Thus, they are useful not only for examining whether a data set provides evidence that risk depends significantly on exposure—the usual question of hazard identification in risk assessment—but also for discovering what else risk might depend on, and how firmly these dependencies are supported by the different considerations reflected in different algorithms (such as conditional independence, mutual information, maximizing scores reflecting measures of explanatory power (e.g., likelihood of the data), invariant causal prediction (ICP), and others (Cox Jr. 2018)). The dependencies uncovered by these algorithms might include some not previously suspected, or they might fail to include some that had previously been assumed. For example, for the BLL example data set, none of the four BN structure-learning algorithms found a direct effect of *lead* on *mortality*, but all identified *age* as a confounder (as expected), and all detected an effect of *grade* on *mortality* (Fig. 10.2). *Grade* is omitted from some previous regression analyses of BLL-mortality associations (e.g., Lanphear et al. 2018), although there is significant previous literature on education-mortality associations (Hummer and Hernandez 2013). The BN approach identifies such dependencies empirically, whether or not they have previously been considered in a particular stream of investigations.

Third, BN methods allow adjustment sets of variables to be conditioned on in calculating a desired specific effect (i.e., the direct or total effect of one variable on another, such as of exposure on risk) to be automatically calculated from the structure of a BN, thereby providing a constructive answer to the question of which other variables to consider in estimating the effect of interest (Textor et al. 2016). When the effect is then estimated by flexible nonparametric methods such as partial dependence plots (Fig. 10.3), nonlinearities and interactions are automatically included while avoiding the biases that can arise from misspecified parametric regression modeling. It is often possible to estimate the same effect in multiple ways using different adjustment sets, thus allowing the internal consistency of the BN model and of the effects estimates based on it to be examined (Fig. 10.3). The CPTs of a BN provide a flexible nonparametric way to model high-order and nonlinear interactions among variables by specifying the conditional probability distribution for a variable for each combination of values of its parents (assuming that continuous variables have been discretized via a preprocessing step (Nagarajan et al. 2013)). This is important for variables that have many parents, such as *mortality* in Figs. 10.1 and 10.2. Modeling interaction effects among 3 or more variables in regression analysis is frequently cumbersome or impossible due to combinatorial explosion. BNs exploit the sparsity of empirical dependencies to limit the combinations considered. Variables such as *mortality* depend directly on only a subset of other variables, namely its parents (comprising no more than 5 of the 12 variables in Fig. 10.2), and not all logically possible combinations of parent values lead to significantly different conditional probability distributions in a CPT. This makes consideration of relevant high-order interactions possible even with relatively small data sets (<3000 records in this example), for which modeling all possible interaction terms is not possible.

In the best case, BN analysis can help investigators identify and quantify relatively robust and reliable dependencies among variables based on properties (such as whether variable are found to be informative about each other, apart from any information obtained from other variables in an appropriate adjustment set) using multiple algorithms implementing diverse principles for detecting dependencies (Fig. 10.2). Such properties are less restrictive than typical parametric modeling assumptions (e.g., a generalized linear modeling assumption that a function of risk depends only on a linear combination of predictors), insofar as they specify qualitative properties that can be tested by nonparametric methods (Arnold et al. 2020). Even when BN methods give only partial insights and answers, as Fig. 10.2 where the existence or directions of some direct dependencies (arrows) are decided differently by different algorithms, they may be useful to investigators by suggesting hypotheses worth exploring further, such as that *grade* is a confounder of the *lead-mortality* association.

Practical Implications

Our illustrative analyses have highlighted the importance, as well as the possibility, of distinguishing between association and direct dependence or causation in interpreting statistical associations and regression coefficients in health risk assessments. BN analyses of our example data set suggest that various factors associated with mortality risks are interdependent in complex ways (Figs. 10.1 and 10.2). These complexities make traditional statistical modeling and causal interpretation of regression model coefficients and related quantities (odds ratios, hazard ratios, burden of disease estimates, population attributable fractions, and so forth) challenging, as they require accounting for the high-order and nonlinear interactions of multiple variables simultaneously. The BN in Fig. 10.1 suggests some of this complexity: *lead* has 5 direct parents, and *mortality* has 6, and these variables share numerous common ancestors. For the effects of blood lead level (BLL) and other variables on mortality risk, the main practical conclusion from Figs. 10.1 and 10.2 is that BN analyses of this data set by different structure-learning algorithms does not find evidence of a direct dependence of *mortality* risk on *lead* this data set, but does find evidence that *grade* and *income*, as well as *age*, are possible confounders of the BLL-mortality associations found in some regression models (e.g., Table 10.1). A resulting practical recommendation is that these variables should be fully controlled for in regression modeling by including both *grade* and *income* as predictors. Regression analyses that omit *grade* and dichotomize *income* and then find significant regression coefficients indicating an association between *lead* and *mortality*, as in the illustrative example in Table 10.1 and in some past influential real-world studies (e.g., Lanphear et al. 2018), leave open the question of how much of the association is due to confounding by *grade* and to residual confounding by *income*. Including *grade* and including *income* as a continuous variable would close that gap. Although perhaps no BN analysis was needed to make these points—they may seem obvious in retrospect—the BN analyses in Figs. 10.1, 10.2, 10.3, and 10.4 illustrate a systematic approach to data analysis using BNs that automatically produces such recommendations when needed by identifying adjustment sets for estimating specified effects of interest.

Regression models seldom consider more that two-way interactions between explanatory variables. Model specification errors arising from oversimplified model forms may contribute to apparent statistically significant exposure-response risk associations in parametric regression models, even if these associations occur only because of residual confounding or omitted confounders or because including exposure helps to reduce error due to model specification errors. Such model-dependent associations do not necessarily describe how or whether risks would change if exposure were reduced, or provide a stable, reliable basis for informing risk management, as illustrated by the variations in Tables 10.1, 10.2, 10.3, and 10.4. The BN methods we have illustrated can help to address this epistemic uncertainty about the extent to which implications of models are well supported by data by displaying the ranges of dependence relations between variables (i.e., BN structures) that are inferred from the data by different methods (Fig. 10.2), as well as by showing the range in effects estimates calculated from different adjustment sets for each structure (Fig. 10.2), and the uncertainty and variability of individual-level estimates for each adjustment set (Fig. 10.3). This analysis of uncertainty and variability is more thorough and informative than the point estimates, confidence intervals, and *p* values typically reported for regression coefficients.

Conclusions

Many studies in public health and regulatory risk assessment use regression models to estimate associations between exposure-related variables and elevated rates of adverse health effects in large observational data sets such as NHANES, concluding that significant observed associations (exposure-response regression coefficients) imply that reducing exposure could protect public health (NIOSH 2020). We have presented illustrative analyses of the multivariate associations among variables in an example data set that suggest that this causal inference is far from straightforward and is not necessarily justified in the absence of careful analysis of omitted and residual confounding, high-order interactions, and model specification errors. For the specific example of BLL-mortality associations examined as a case study in this chapter, it is already well known from past studies and regression analyses, and confirmed in our examples based on the NHANES III data, that elevated BLL is significantly associated with mortality in regression models (e.g., Table 10.1), as well as with many other factors, such as low educational attainment, that are also correlated with mortality risks (Figs. 10.1 and 10.2) (Lanphear et al. 2018). A simple mental model in which *mortality* depends directly on *lead*, so that elevated BLL directly causes increased mortality risk, is not supported by our BN analyses of this example data set (Figs. 10.1 and 10.2). Such a negative finding does not imply that an effect too small for the BN algorithms to detect does not exist in this data set, or that an effect large enough to detect by these methods (e.g., comparable to the effects of *grade* on *mortality* in Fig. 10.3) would not be found in other data sets (since the ICP property of invariance of direct effects across data sets has not been established by this single example).

Despite these limitations, the analysis does indicate that both *grade* and *income* should be fully controlled for in modeling effects of *lead* on *mortality*, without dichotomizing them or using only a few categories to represent them (Streiner 2002; Naggara et al. 2011; Groenwold et al. 2013). To the extent that this advice has not already been followed, existing data sets might be reexamined using BN methods to help clarify the fractions of previously noted BLL-mortality associations that are explained by direct dependence vs. by omitted and residual confounding or model specification errors (e.g., neglect of interaction terms). More generally, we have illustrated methods to help clarify whether and to what extent exposure-response associations found in via regression models represent evidence that response probabilities depend directly on exposure variables, rather than simply being associated with them.

Appendix: Variables and Data

The following table lists the short, mnemonic names given to variables used throughout the paper, in its left column; the corresponding NHANES III variable names in the middle column; and descriptions of the variables in its right column. Some of the variables were reoriented in going from the original variables to those in Table 10.1 (e.g., "Hispanic" is the reverse of "Not Hispanic" with a value of 1 if Not Hispanic = 0 and a value of 0 if Not Hispanic = 1; and "marriage" is the reverse of "Never Married.")

Name	Corresponding variable in NHANES IIII and linked files	Description
Mortality	MORTSTAT	Mortality
Heartdeath	UCOD = 1	Underlying cause of death (UCOD) is heart disease
Lead	PBPMI2	Blood lead level in μg/dL
Age	HSAGEIR	Age
Marriage	HFA12_7	Ever married (reverse of never married)
Family	HSFSIZER	Family size
Grade	HFA7	Highest grade of school
Income	DMPPIRMI3	Poverty income ratio
Hispanic	DMAETHNR_3	Hispanic (reverse of not hispanic)
Black	DMARACER_2	Race black
Small metro	DMPMETRO_2	< 1M+ metro area
West	DMPCREGN_4	Census region = West
South	DMPCREGN_3	Census region = South

References

Arnold KF, Davies V, de Kamps M, Tennant PWG, Mbotwa J, Gilthorpe MS. Reflections on modern methods: generalized linear models for prognosis and intervention—theory, practice and implications for machine learning. Int J Epidemiol. 2020;7:49. https://doi.org/10.1093/ije/dyaa049.

CDC. Plan and operation of the third national health and nutrition examination survey 1988-94 series 1: programs and No. 32. U.S. Department of Health and Human Services, public health service/Centers for Disease Control and Prevention. Hyattsville: National Center for Health Statistics; 1994.

Cox LA Jr. Modernizing the Bradford Hill criteria for assessing causal relationships in observational data. Crit Rev Toxicol. 2018;48(8):682–712. https://doi.org/10.1080/10408444.2018.1518404.

Dominici F, Greenstone M, Sunstein CR. Science and regulation. Particulate matter matters. Science. 2014;344(6181):257–9. https://doi.org/10.1126/science.1247348

Goldstein A, Kapelner A, Bleich J, Pitkin E. Peeking inside the black box: visualizing statistical learning with plots of individual conditional expectation. J Comput Graph Stat. 2015;24(1):44–65.

Groenwold RH, Klungel OH, Altman DG, van der Graaf Y, Hoes AW, Moons KG. Adjustment for continuous confounders: an example of how to prevent residual confounding. CMAJ. 2013;185(5):401–6. https://doi.org/10.1503/cmaj.120592.

Heinze-Deml C, Peters J, Meinshausen N. Invariant causal prediction for nonlinear models. J Causal Inf. 2018;6(2):20170016. https://doi.org/10.1515/jci-2017-0016.

Hummer RA, Hernandez EM. The effect of educational attainment on adult mortality in the United States. Popul Bull. 2013;68(1):1–16.

Lanphear BP, Rauch S, Auinger P, Allen RW, Hornung RW. Low-level lead exposure and mortality in US adults: a population-based cohort study. Lancet Public Health. 2018;3(4):e177–84. https://doi.org/10.1016/S2468-2667(18)30025-2.

Liaw A, Wiener M. Classification and regression by random forest. R News. 2002;2(3):18–22.

Lucas R. Econometric policy evaluation: a critique. In: Brunner K, Meltzer A, editors. The Phillips curve and labor markets. Carnegie-Rochester conference series on public policy, vol. 1. New York: Elsevier; 1976. p. 19–46.

Nagarajan R, Scutari M, Lebre S. Bayesian Networks in R with applications in systems biology. New York: Springer; 2013.

Naggara O, Raymond J, Guilbert F, Roy D, Weill A, Altman DG. Analysis by categorizing or dichotomizing continuous variables is inadvisable: an example from the natural history of unruptured aneurysms. AJNR Am J Neuroradiol. 2011;32(3):437–40. https://doi.org/10.3174/ajnr.A2425.

NIOSH (2020). Current intelligence bulletin 69: NIOSH practices in occupational risk assessment. By Daniels RD, Gilbert SJ, Kuppusamy SP, Kuempel ED, Park RM, Pandalai SP, Smith RJ, Wheeler MW, Whittaker C, Schulte PA. Cincinnati, OH: U.S. Department of Health and Human Services, Centers for Disease Control and Prevention, National Institute for Occupational Safety and Health. DHHS (NIOSH) Publication No. 2020-106 (revised 03/2020), https://doi.org/10.26616/NIOSHPUB2020106revised032020.

Pearl J. Causal inference in statistics: an overview. Stat Surv. 2009;3:96–146.

Streiner DL. Breaking up is hard to do: the heartbreak of dichotomizing continuous data. Can J Psychiatr. 2002;47(3):262–6.

Textor J, van der Zander B, Gilthorpe MS, Liskiewicz M, Ellison GT. Robust causal inference using directed acyclic graphs: the R package 'dagitty'. Int J Epidemiol. 2016;45(6):1887–94.

Chapter 11
Case Study: Does Molybdenum Decrease Testosterone?

Introduction

This chapter applies Bayesian network (BN) learning methods and other techniques from Chap. 9 to the following health risk question: Does exposure to the metal molybdenum (*Mo*), as indicated by *Mo* concentrations in urine, reduce total testosterone (*T*) concentrations in blood serum of men, posing a potential risk to male fertility? Some recent papers have suggested that it might. For example, Lewis and Meeker (2015) state that "In adjusted analyses where metals were modeled as a continuous variable, we found significant inverse associations between urinary molybdenum and serum copper and serum testosterone. …These findings add to the limited human evidence that exposure to molybdenum and other metals is associated with altered testosterone in men, which may have important implications for male health" (Lewis and Meeker 2015). The adjusted analyses referred to consist of multiple linear regression models of data for 484 men aged 18–55 years old with data on *Mo* and other metals, serum *T*, body mass index (BMI), poverty income ratio (PIR), race, serum cotinine, and urinary creatinine who were enrolled in the National Health and Nutrition Examination Survey (NHANES 2011–2012). Similarly, Wang et al. (2016) concluded that, among men attending an infertility clinic, environmental exposure to tin, nickel, zinc and molybdenum may be associated with decreased total testosterone, although further prospective research is needed to confirm these associations in the general population. Conversely, however, some animal studies suggest that *Mo* exposure may actually improve male fertility. For example, Liu et al. (2019) found that *Mo* could repair sterility in male mice under experimental conditions, and proposed that it might be a new candidate medicine for treatment of male infertility. For humans, Rotter et al. (2016) found positive correlations between *Mo* and *T*. Finally, some investigators have found no relation between *Mo* and *T*. For example, using multiple linear regression, Zeng et al. (2013) found significant negative associations between urinary Mn and Zn and

L. A. Cox Jr., *Quantitative Risk Analysis of Air Pollution Health Effects*, International Series in Operations Research & Management Science 299,
https://doi.org/10.1007/978-3-030-57358-4_11

serum levels of T among men attending an infertility clinic, but no relationship between urinary *Mo* and *T*. Thus, the literature appears to be mixed and inconclusive on whether *Mo* is associated with reduced (or increased or unchanged) levels of *T* in serum in multiple regression models.

This chapter reexamines the question of whether there is evidence that *Mo* reduces *T* in the NHANES data set using, techniques from Chap. 9 to identify and control for potential confounders. Similar to the analysis of the blood lead data in Chap. 10, this chapter applies BN learning and nonparametric comparisons of conditional distributions of *T* as *Mo* varies, holding other variables fixed, to complement the insights from multiple linear regression. As discussed in Chaps. 9 and 10, in contrast to multiple linear regression, Bayesian networks do not select a single dependent variable, but estimate the probabilistic dependencies among all variables simultaneously, i.e., how the conditional probability distribution of each variable changes as the values of variables on which it depends vary. Nonparametric statistical analyses do not presuppose a linear relationship between variables (i.e., that the change in one is proportional to the change in another), but allow nonlinear relationships and interactions among variables. The main purpose of the investigation in this chapter is to assess the extent to which these methods reveal aspects of the data, and specifically of the dependence of serum T on urinary Mo, that are not clear from multiple linear regression alone and that help to assess whether there is evidence that *Mo* reduces *T* after accounting for values of other covariates (including potential confounders and modifiers of the *Mo-T* relationship). Although NHANES contains hundreds of variables, including blood levels of multiple metals, demographic and lifestyle variables, and smoking and diet variables, we focus on the *Mo-T* association and on the methodological questions of whether and to what extent multiple linear regression provides reliable evidence about whether reducing *Mo* would increase T, and the extent to which other techniques can provide additional insights into this question beyond those from multiple linear regression.

Data

From the NHANES data collected over the three cycles 2011–2012, 2013–2014, and 2015–2016, we selected all men aged 18–55 with measured data on both *Mo* and *T* (2271 cases) for initial analyses. Subsets of this data set that also contain data on other variables were used for additional analyses, as described for specific analyses in the results section. This data set is substantially larger than the 484 cases for Lewis and Meeker (2015), which used data from 2011–2012 only. Additional details of the NHANES data set are available from the Centers for Disease Control and Prevention (CDC) web site (https://www.cdc.gov/nchs/nhanes/index.htm). For our purposes, this data set is used only to provide a realistic example of epidemiological data for illustrating methodological points and different methods for quantifying conditional probability relations among variables. Although the NHANES data set is carefully designed for use with weights to allow extrapolation to the US population, we do not use these weights or make such extrapolations here, but use the collected data (which deliberately oversampled Black and Hispanic subpopulations relative to the US population) to quantify conditional probability relations among the variables. Using only cases with measured values of *T* and *Mo* and dropping cases with missing values from analyses suffices for our illustrative purposes and avoids uncertainties due to imputation of missing values, but since data are not necessarily missing at random, we again do not extrapolate beyond the specific data set analyzed. Thus, we interpret the data as adequate for addressing the methodological questions of whether new statistical analyses can complement the insights from multiple linear regression with additional understanding of the *Mo-T* association, but we do not attempt here to generalize to the US population of men, or beyond the confines of the sample of men for whom NHANES provides data on the variables analyzed.

Statistical Analysis Methods

To study how the conditional probability distribution of a response variable—here, *T*—varies with an indicator of exposure (here, *Mo*) when values of other variable are held fixed, we apply the following statistical and data science methods to study the dependence of *T* on *Mo* for varying levels of other variables.

- Standard descriptive analyses (means, medians, etc.) and interaction plots (showing how the conditional mean value of T varies with the level of *Mo*, conditioning on levels of one or more other variables) were carried out with the *Statistica 12* commercial statistical software. This package was also used to generate various tables (Tables 11.1, 11.2, and 11.3) and plots (Figs. 11.1 and 11.4).
- Exploratory analyses and visualizations to detect and visualize significant dependencies between conditional probabilities of mortality and values or distributions of other variables were performed using Bayesian network (BN) learning (using the *bnlearn* package in R) and classification and regression tree (CART) analyses (using the *party* package in R). The needed R packages and supporting documentation were accessed via the Causal Analytics Toolkit (CAT) software package (http://cloudcat.cox-associates.com:8899/) and used for all other analyses and figures.
- Quantitative modeling was conducted using random forest analysis (with the *randomForest* package in R) to generate partial dependence plots (PDPs). These show the predicted conditional mean values of a dependent variable (e.g., *T*) as an explanatory variable (e.g., *Mo*) varies, holding the values of other variables for each individual subject fixed at their actual levels in the data set. Predicted values are obtained by averaging the predictions from an ensemble of 500 CART trees for the dependent variable, for each value of the predictor. This provides a non-parametric model ensemble prediction of the natural direct effect of the predictor on the expected value of the response holding other values fixed.

Table 11.1 Linear correlations between *T* (middle column and top row), *Mo* (right column), and other variables (left column)

	Marked correlations (*) are significant at p < 0.05	
Variable	T (testosterone, LBXTST) (ng/dL)	Mo (molybdenum, URXUMO) (ug/L)
T (testosterone, LBXTST) (ng/dL)	1.00	−0.06
Sex hormone-binding globulin (LBXSHBG) (nmol/L)	0.65*	−0.06*
Estradiol (LBXEST) (pg/mL)	0.40*	−0.05
Fasting time (hours)	0.20*	−0.07*
Iron, refrigerated serum (LBXSIR) (ug/dL)	0.14*	−0.06*
Cotinine (LBXCOT) (ng/mL)	0.13*	−0.08*
Cadmium (LBXBCD) (ug/L)	0.12*	−0.03
Household size (DMDHHSIZ)	−0.06*	0.05*
Mo (molybdenum, URXUMO) (ug/L)	−0.06*	1.00
Cesium (URXUCS) (ug/dL)	−0.09*	0.33*
Income (INDFMPIR)	−0.09*	−0.02
Copper (LBXSCU)	−0.10*	0.00
Age (RIDAGEYR)	−0.13*	−0.06*
Time of day (PHDSESN)	−0.18*	0.08*
Uric acid in blood (LBXSUA) (mg/dL)	−0.26*	−0.02
Weight (BMXWT) (kg)	−0.39*	0.03
BMI (BMXBMI)	−0.42*	0.03

Before turning to these new analyses, we begin by applying linear correlation and multiple linear regression modeling to check whether we can replicate past findings of significant negative *Mo-T* associations in multiple linear regression models for men in the NHANES sample. Lewis and Meeker (2015) state that "The associations between metal and *T* concentrations were first assessed using simple linear regression in unadjusted models. These associations were then examined using multiple linear regression in models that were adjusted for the following potential confounding variables: age, BMI [body mass index], PIR [ratio of household income to poverty level], and serum cotinine. …Urinary creatinine concentration was also added to adjusted models for metals in urine to adjust for variability in urinary output. In unadjusted and adjusted models, *T* was log-transformed because it did not follow a normal distribution. Serum cotinine and age were modeled untransformed, whereas all metal biomarkers were log-transformed to improve model fit." They then discuss how they categorized BMI and other continuous variables in seeking potentially significant associations. For comparability, we also develop linear regression models with log-transformed values of *Mo* and *T*, denoted by *log(Mo)* and *log(T)*, respectively; and with untransformed values of age and serum cotinine, BMI, and PIR. However, we do not categorize these continuous

Table 11.2 Summary of descriptive statistics (mean, median, minimum and maximum values, quartiles and 10% and 90% percentiles, standard deviation) for variables

Variable	Descriptive statistics									
	Valid N	Mean	Median	Minimum	Maximum	Lower quartile	Upper quartile	Percentile 10.00000	Percentile 90.00000	Std. dev.
Age (RIDAGEYR)	2271	36.3	36.0	18.0	55.0	27.0	46.0	21.0	52.0	11.1
Household size (DMDHHSIZ)	2271	3.7	4.0	1.0	7.0	2.0	5.0	2.0	6.0	1.7
Fasting time (hours)	2271	7.4	8.0	0.0	36.0	2.0	12.0	1.0	14.0	5.5
Time of day (PHDSESN)	2271	0.7	1.0	0.0	2.0	0.0	1.0	0.0	2.0	0.8
log(T)	2271	6.0	6.0	2.8	7.6	5.7	6.3	5.4	6.5	0.4
T (testosterone, LBXTST) (ng/dL)	2271	436.3	407.6	16.3	2000.0	308.0	534.0	230.0	673.4	186.7
Urinary cesium (URXUCS) (ug/L)	2271	5.1	4.5	0.1	110.9	2.7	6.6	1.6	9.0	3.8
Mo (molybdenum, URXUMO) (ug/L)	2271	55.3	43.5	1.3	541.0	21.9	74.7	11.5	108.7	48.8
Cotinine (LBXCOT) (ng/mL)	2269	123.4	18.3	0.0	1030.0	0.0	226.0	0.0	368.0	162.4
Uric acid in blood (LBXSUA) (mg/dL)	2269	6.0	5.9	0.4	11.1	5.1	6.8	4.5	7.6	1.3
Height (BMXHT) (cm)	2260	175.0	175.0	136.5	204.5	170.0	180.1	165.7	184.4	7.5
Iron, refrigerated serum (LBXSIR) (ug/dL)	2259	92.2	87.0	10.0	557.0	68.0	111.0	53.0	138.0	36.2
Weight (BMXWT) (kg)	2258	87.0	83.1	39.2	207.2	71.8	97.7	63.5	115.1	22.0
BMI (BMXBMI)	2258	28.3	27.4	15.4	74.1	23.7	31.5	21.2	36.6	6.5
Income (INDFMPIR)	2076	2.3	1.8	0.0	5.0	1.0	3.7	0.5	5.0	1.6
Cadmium (LBXBCD) (ug/L)	1961	0.6	0.3	0.1	9.3	0.2	0.7	0.1	1.4	0.8
Copper (LBXSCU)	1645	103.5	101.7	24.7	203.0	89.7	114.6	81.2	128.4	19.7
estradiol (LBXEST) (pg/mL)	1495	24.6	23.5	2.1	88.7	18.3	29.3	14.6	35.5	9.4
Sex hormone-binding globulin (LBXSHBG) (nmol/L)	1382	37.8	33.4	6.9	290.7	24.3	46.0	18.0	62.3	21.1
Urinary creatinine (URXUCR) (mg/dL)	1335	143.5	[illegible]	4.0	800.0	75.0	191.0	44.0	268.0	89.8

"Valid N" indicates the number of cases with data for each variable; it is less than 2271 if some cases have missing data for a variable

Table 11.3 Multiple linear regression model for *log(T)* vs. *log(Mo)* and four potential confounders identified by Lewis and Meeker (2015)

	Regression summary for dependent variable: *log(T)* R = 0.456, R^2 = 0.21, adjusted R^2 = 0.21						
N = 2068	b*	Std. err. of b*	b	Std. err. of b	t(2062)	p-value	Valid N
Intercept			7.023	0.059	118.4	0.000000	
log(Mo)	−0.05	0.020	−0.025	0.010	−2.6	0.010287	2271
Age (RIDAGEYR)	−0.10	0.020	−0.004	0.001	−4.8	0.000002	2271
Income (INDFMPIR)	−0.04	0.020	−0.010	0.005	−1.9	0.059674	2076
BMI (BMXBMI)	−0.41	0.020	−0.028	0.001	−20.7	0.000000	2258
Cotinine (LBXCOT) (ng/mL)	0.05	0.021	0.000	0.000	2.4	0.018258	2269

The b* coefficients are standardized regression coefficients; the b coefficients are unstandardized. There is a significant negative linear association between *log(Mo)* and *log(T)* (p = 0.01)

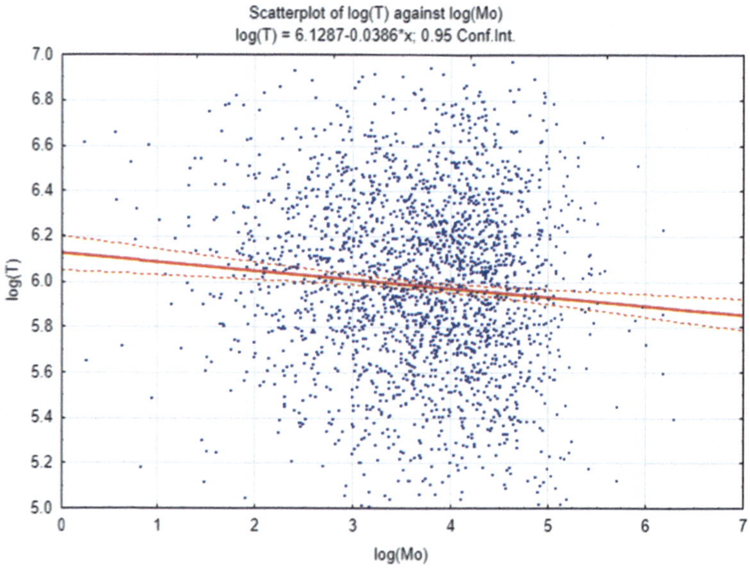

Fig. 11.1 Scatterplot and simple linear regression model of *T* (y axis, ng/dL) vs. *Mo* (x axis, µg/L)

variables, as doing so loses some information and multiple linear regression can be applied to the continuous values. We also seek other significant correlates of *T* of *Mo* that might confound the association between them and identify some, as discussed in the following section, that have not been highlighted in previous research.

Results

The following sections present a series of results from various statistical analyses of the relationships among *Mo*, *T*, and covariates such as age, height, and fasting time before blood was drawn from study participants. In addition to tables of key results, various graphs are used to visualize the main findings. The statistical techniques used to generate these results and graphs are briefly identified as the results that use them are presented.

Linear Correlation and Descriptive Statistics Results

Table 11.1 show linear (Pearson's product-moment) correlations between *T* (middle column), *Mo* (right column), and other variables (left column) that are significantly correlated with *T*. Cases with missing values for either variable in a correlation were skipped, i.e., we used pairwise deletion of missing data. Table 11.2 summarizes descriptive statistics for the variables and indicates the number of missing data values for each variable (via values of "Valid N" less than the total number of men in the sample, 2271). In both tables, the names of the variables as given in NHANES (e.g., LBXTST for blood serum testosterone concentration, URXUMO for urinary molybdenum concentration) are shown in parentheses, along with more descriptive names that we have assigned.

The linear correlations in Table 11.1 suggest that, in addition to the potential confounders of age, income-to-poverty ratio, body mass index (BMI), and serum cotinine considered by Lewis and Meeker (2015) (corresponding to the variables age, income, BMI, and cotinine in Tables 11.1 and 11.2), two time variable should also be further assessed as potential confounders in multiple linear regression modeling. These are fasting time (measured in hours, derived from the NHANES variables PHAFSTHR and PHAFSTMN variables for hours and minutes of fasting since the last mean) and time of day at which blood was sampled. Urinary cesium, iron in refrigerated blood serum, sex hormone-binding globulin concentration in blood, and number of people living in the same house ("household size") are also identified in Table 11.1 as significant correlates of both *Mo* and *T*.

Scatterplot and Simple Linear Regression Results

Figure 11.1 shows a scatterplot of *log(T)* on the y axis (units of nanograms per decaliter) against *log(Mo)* on the x axis (units of micrograms per liter) for the 2271 men aged 18–55 with measured values of both variables. A simple linear regression line with 95% confidence intervals is superimposed on the plot to provide a visual indication of the significant negative association between *Mo* and *T*. As noted by Lewis, there is a negative linear association between them. (The vertical scale is cropped to show most of the data points while allowing the negative slope of the ordinary least squares regression line to be clearly visible.) A nonparametric smoothing regression curve (not shown) lies close to the linear regression line, suggesting that a linear model specification is not unreasonable for these data.

Figure 11.2 provides visualizations of histograms for several variables with relatively little missing data (shown on the main diagonal) and scatterplots for pairs of variables with nonparametric (smoothing regression) curves overlaid on them (below the diagonal with each cell showing the scatterplot for the pair of variables pictured on the main diagonal for the cell's row and column). (The variables are ordered so that potential causes, such as age and BMI, precede potential effects, such as *log(T)*.) Corresponding linear correlation coefficients are shown in the upper right of this scatterplot matrix. This visualization shows that BMI is strongly

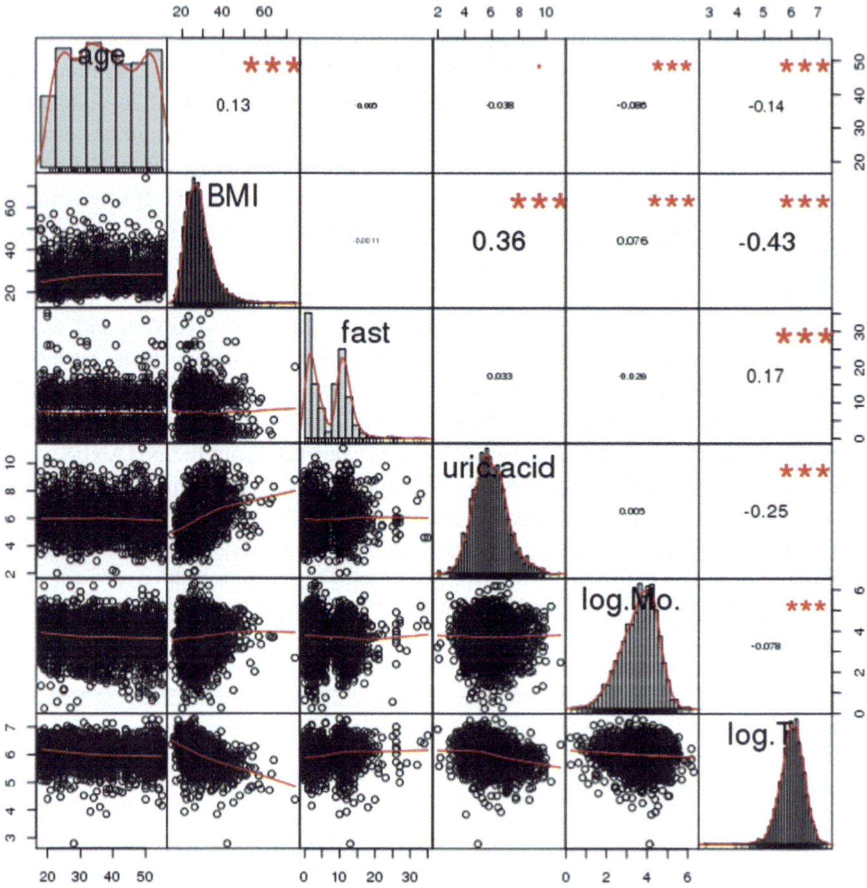

Fig. 11.2 Scatterplot matrix, with histograms, kernel density overlays, absolute correlations, and significance asterisks (* = 0.05, ** = 0.01, *** = 0.001)

positively associated with uric acid and strongly negatively associated with *log(T)* compared to other associations in the matrix, as indicated by the relatively steep slopes of the regression curves. Some of the regression curves are noticeably non-linear. For example, BMI increases with age for younger ages, but levels off at about age 35. (Horizontal scales are shown alternately at the bottoms and tops of successive columns, and vertical scales are shown alternately to the left and right of the rows, to avoid crowding the visual display.) The dependent variable of greatest interest, *log(T)*, depends nonlinearly on several variables, including uric acid, in these plots. Although the nonparametric regression curves are monotonic, they do not have constant slopes. This exploratory visual analysis suggests that, in addition to multiple linear regression (discussed next), it is worth applying nonparametric modeling that allows for departures from linearity.

Multiple Linear Regression Results

To check whether we could reproduce the main finding of a significant negative linear association between *log(Mo)* and *log(T)* in a multiple linear regression model for our expanded NHANES data set, we first fit a model with *log(T)* as the dependent variable and *log(Mo)*, age, income, BMI, and cotinine as predictors (and pair-wise deletion of missing data). Table 11.3 summarizes the regression model. The regression coefficient for *log(Mo)* is indeed significantly negative, consistent with previous findings (Lewis and Meeker 2015).

This finding is expected: adding two more cycles (4 more years) of data did not change the key finding from the first cycle in 2011–2012. However, the meaning and robustness of the linear regression model in Table 11.3 become less clear when additional correlates of *T* in Table 11.1 are included as predictors.

Table 11.4 shows a larger multiple linear regression model, with additional correlates of T (and *log(T)*) from Table 11.1 included as predictors. (Henceforth, abbreviated names of variables, as shown in the left column of Table 11.4, are used where convenient; e.g., "fast" is "fasting time (hours)"; "household" is "household size"; "income" is the household income to poverty ratio; and so forth.) Unsurprisingly, the larger model provides a better fit, as measured by the adjusted R^2 for the regression model. What is perhaps less expected is that cotinine and *log(Mo)* are no longer significant predictors of *log(T)* when these other variables are included ($p = 0.16$). Backward variable selection (which admittedly should not be over-interpreted) identifies age, BMI, household size, fasting time, time of day, and uric acid as independent predictors of *log(T)*. (If uric acid is excluded, no other variables enter; thus, the potential explanation that perhaps *Mo* affects *T* only through uric acid is not supported.) This extended analysis suggests that *log(Mo)* is a significant predictor of *log(T)* in Table 11.3 only because some of the other predictors in Table 11.4 were not included in the model. Thus, the dependence of *log(T)* on *log(Mo)* in the multiple linear regression in Table 11.3 is a *model-dependent* finding, meaning that it is driven by the choice of predictors to include in the regression model: other multiple

Table 11.4 Multiple linear regression model for *log(T)* vs. *log(Mo)* with additional predictors

	Regression summary for dependent variable: *log(T)* R = 0.501, R^2 = 0.251, Adjusted R^2 = 0.247					
N = 2064	b*	Std. err. of b*	b	Std. err. of b	t(2053)	p-value
Intercept			7.210	0.074	98.1	0.000000
log(Mo)	−0.030	0.021	−0.015	0.010	−1.4	0.161299
Age	−0.103	0.020	−0.004	0.001	−5.2	0.000000
Income	−0.043	0.020	−0.012	0.006	−2.1	0.036377
Household	−0.079	0.020	−0.021	0.005	−4.0	0.000064
Fast	0.120	0.024	0.010	0.002	5.0	0.000001
Time of day	−0.083	0.024	−0.049	0.014	−3.5	0.000521
BMI	−0.360	0.021	−0.024	0.001	−17.1	0.000000
Cotinine	0.032	0.020	0.000	0.000	1.6	0.115477
Cesium	−0.041	0.021	−0.005	0.002	−1.9	0.053871
Uric acid	−0.120	0.021	−0.042	0.007	−5.8	0.000000

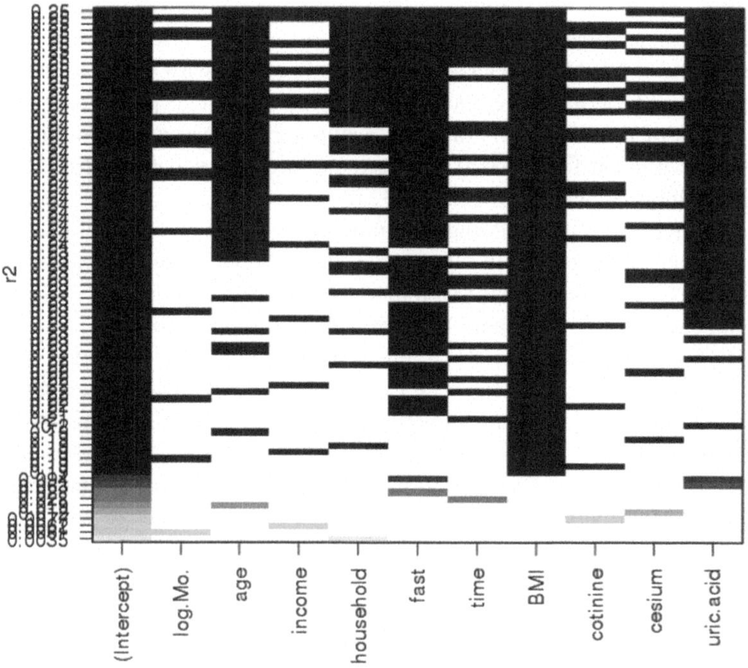

Fig. 11.3 Visualization of all-subsets regression results for dependent variable *log(T)*. This visualization indicates robust predictors (long black vertical strips) and less robust predictors. Each row of the figure represents a multiple linear regression model. Columns represent variables (predictors) and an intercept term. In each row, the presence of each variable is indicated visually (black = yes, white = no). Rows are ordered from poorest fit (lowest R^2) at the bottom to best fit at the top

linear regression models that fit the data better using different predictors lead to different conclusions, that *log(T)* does not depend significantly on *log(Mo)*.

To examine model-dependence more systematically, Fig. 11.3 visualizes the results from all-subsets regression modeling. This figure consists of 10 columns, one for each predictor, and an 11th column, on the left side, for the intercept term. It has a large number of rows, one for each regression model. In each cell (row and column), black indicates that the model in the cell's row includes the variable in the cell's column as a predictor of the dependent variable, *log(T)*. (In principle, there are $2^{10} = 1024$ possible models, since each of the 10 predictors can be included or not. For purposes of visualization, many of the worst-fitting models are not included individually, but are averaged, as indicated by the grey tones near the bottom of the figure.) The models are sorted, from those with the least explanatory power or poorest linear fit (lowest R^2) at the bottom to those with the highest R^2 at the top. (Although these R^2numbers, show on the vertical axis, are crowded because of the large number of models, they run from 0.0035 at the bottom to 0.25 at the top, i.e., the best-fitting multiple linear regression models explain 25% of the variance in *log(T)*.) From this visualization, it is clear that BMI must be included in any model with relatively high explanatory power, as indicated by the long black vertical strip ("icicle") for BMI. Likewise, uric acid, age, and fasting time are robust predictors, included in all of the best-fitting models. By contrast, *log(Mo)*, income, cotinine, and cesium are not robust predictors: they can be included or excluded from models without significantly affecting explanatory power, as measured by R^2. The specific regression models in Tables 11.3 and 11.4 (corresponding to two different rows in Fig. 11.3) illustrate this more general pattern.

Nonparametric Modeling Results: Splines, CART Trees and Random Forest Partial Dependence Plots

Multiple linear regression is limited by its assumption of linear relationships between its explanatory variables and the dependent variable (as well as by other assumptions such as normally distributed error terms). Exploratory visualizations indicate substantial departures from linearity, however. For example, Fig. 11.4 shows contour plots (produced by fitting nonparametric splines) for *log(T)* as a function of selected pairs of predictors. If the linearity assumption were correct, these contours should approximate parallel straight lines. That they are not suggests that nonparametric models allowing for nonlinear interactions might reveal insights not clear from multiple linear regression modeling.

A nonparametric alternative to multiple linear regression is classification and regression tree (CART) analysis, which seeks to partition records into subsets with significantly different values of the dependent variable, *log(T)*, by conditioning on values of predictors. (We use "CART" for the class of recursive partitioning algorithms that produces classification and regression trees; there are many variations.

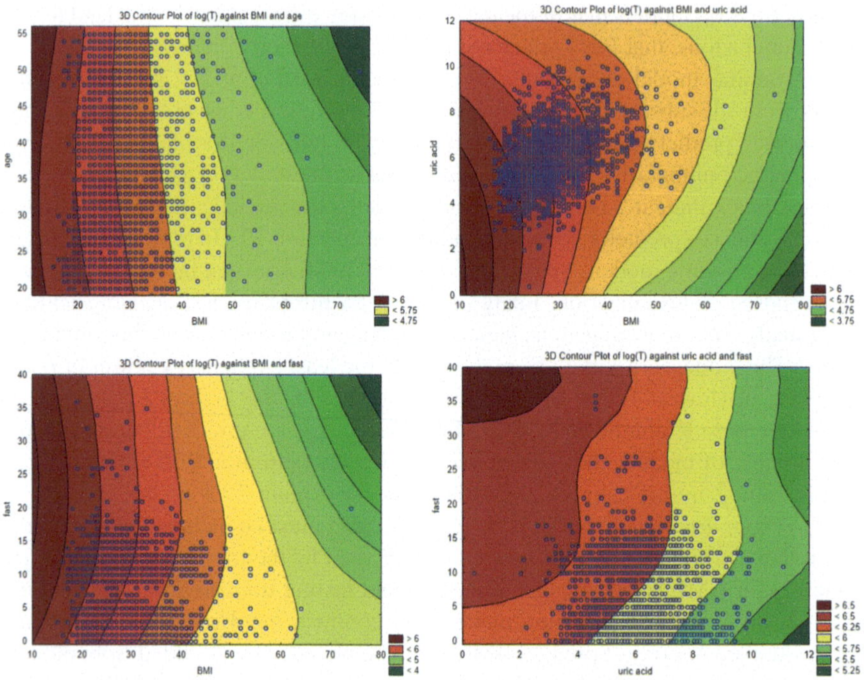

Fig. 11.4 Spline contour plots for *log(T)* vs. selected pairs of predictors (BMI and age, uric acid, or fasting time; and uric acid and fasting time). Deviations of the contours from parallel straight lines indicate that interactions and nonlinear effects should be considered

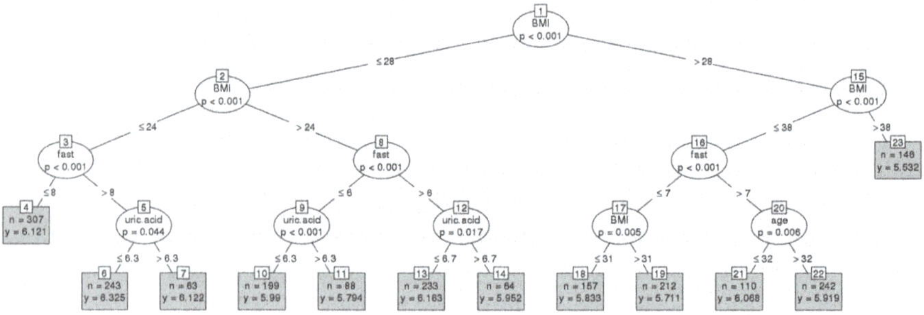

Fig. 11.5 CART tree for dependent variable *log(T)*. BMI, age, fasting time, and uric acid are automatically identified as significant predictors of *log(T)*. The conditional mean values of *log(T)* are shown in the shaded nodes ("leaf nodes") at the bottom of the tree (y values). Each value is the average for the subjects (numbers of subjects indicated by the n numbers) satisfying the conditions leading from the top of the tree to the leaf node

Details on the specific CART algorithm used here, as with other analytics techniques applied here, can be found in the online documentation in the CAT package used for these analyses, and in the appendices and references in Chap. 9). Figure 11.5 shows the results of applying CART analysis to the variables in Fig. 11.3. The

CART algorithm (as implemented in the R recursive partitioning package *partytree*, used in the CAT software) finds that conditioning on ranges of values (shown by inequalities on the branches) for BMI, fasting time, uric acid, and age predicts significantly different conditional distributions for *log*(*T*). These correspond to the most robust predictors identified in Fig. 11.3 (those with the longest black vertical rectangles). Other variables not shown in the tree are not identified as leading to significantly different distributions of *log(T)*; intuitively, they were not found to provide significant additional predictive power after conditioning on these major drivers. Age plays only a minor role (one split out of 11 in the tree). This is consistent with the contour plot in the upper left of Fig. 11.4, which shows nearly vertical contours over much of the range of data. (Narrow vertical contours would correspond to strong effects of BMI but no effects of age.) Thus, these results from CART non-parametric analysis are consistent with the results obtained by other methods that assume linear (Fig. 11.3) or smooth (Fig. 11.4) relationships between the predictors and *log(T)*.

CART algorithms have the advantage of making no specific parametric assumptions, such as linearity, about the form of the function relating predictors to conditional mean values of the dependent variable. They can automatically handle nonlinearities and high-order interactions among predictors. However, they are often not robust, in the sense that trees fit to different random subsets of the data may be quite different. To overcome this limitation, it is common practice in machine learning to average predicted values of the dependent variable (here, *log(T)*), for different values of an explanatory variable (here, *log(Mo)*), over an ensemble of hundreds of CART trees fit to different samples of data. This creates a partial dependence plot (PDP), such as the ones in Fig. 11.6. These plots show how predicted values of *log(T)* vary as *log(Mo)* varies (left panel); and as BMI varies (right panel), holding all other variables fixed at their observed values in the NHANES data set. There is a small but clear negative dependence of *log(T)* on BMI, i.e., the PDP

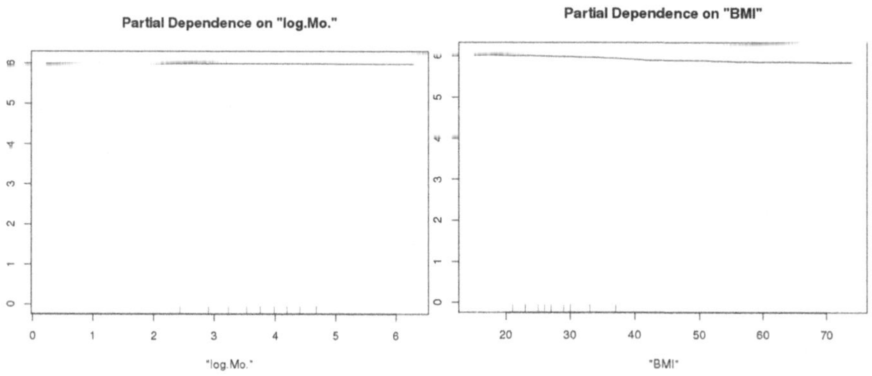

Fig. 11.6 Partial dependence plots for *log(T)* vs. *log(Mo)* (left panel) and for *log(T)* vs. BMI (right panel)

slopes down for BMI. By contrast, the PDP for *log(Mo)* is relatively flat, indicating no strong dependence of *log(T)* on *log(Mo)* when other variables are held fixed.

Bayesian Network Analysis Results

PDPs are limited by their construction, which holds fixed the values of all variables other than those being plotted against each other. They are also limited by the fact that each tree in the underlying ensemble assumes a constant value of the dependent variable for each leaf node in the tree; thus, slight associations due to residual confounding (from within-leaf dependences, which are not modeled) are not completely eliminated. In reality, of course, changing a single explanatory variable, such as BMI, might cause the values of multiple other variables to change, making it unrealistic to hold their values fixed in assessing total effects on the dependent variable. In addition, the dependent variable may (and probably does) change smoothly with the values of variables on which it depend. Bayesian network (BN) learning algorithms overcome these limitations. Figure 11.7 shows the structure of a BN learned from the NHANES data (using the default settings, including the hill-climbing (HC) algorithm in the *bnlearn* R package. The full methodology for BN learning is quite detailed; we refer the reader to Nagarajan et al. (2013) for details of the *bnlearn* package used to generate Fig. 11.7). The interpretation of such a BN is that an arrow between variables indicates a statistical dependence between them: knowing the value of one helps to predict the value of the other. (More formally, the null hypothesis of conditional independence between them, given the values of other variables, is rejected in favor of the alternative hypothesis that there is positive mutual information between them. By Bayes's Rule, this is a symmetric rela-

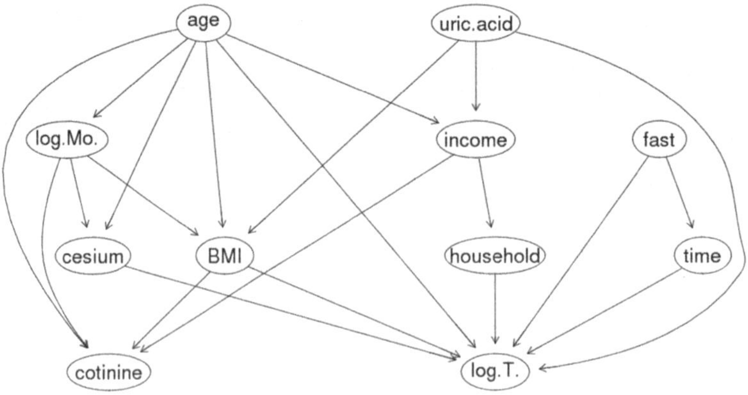

Fig. 11.7 Bayesian network (BN) structure learned from the NHANES data, with age specified as a source (outward-pointing arrows) and *log(T)* specified as a sink (inward-pointing arrows) to facilitate causal intepretation. An arrow between variables indicates a statistical dependence between them

tionship, so arrow directions are not important in revealing (mutual) statistical dependence between variables, but the arrow directions do indicate a way to factor the joint distribution of random variables into products of marginal and conditional distributions. Further details on interpretation of BNs are given in Chap. 9 and Nagarajan et al. 2013.) To facilitate a partial causal orientation of the arrows, we impose the constraints that age has only outward-pointing arrows (it can be a cause but not effect of other variables), while *log(T)* has only inward-pointing arrows (it can be an effect but not a cause of the variables that point into it). These constraints do not suffice to give unique, sensible causal interpretations to all of the arrows. For example, it is not clear how the arrow between household (i.e., number of people in a household) and income (household income) should be directed; and the arrow from fast (hours since eating) to time (time of day when blood was drawn) is reversed for a causal interpretation, if time of day determines time since eating. Other arrows, such as those from uric acid or *log(Mo)* to BMI, are also reversed if uric acid and *Mo* in fact are determined in part by BMI, rather than the reverse. Some of the arrows, such as that between uric acid and income, do not have obvious causal explanations (although it is tempting to speculate about diet as a mediating factor). Nonetheless, even disregarding arrow directions, the fact that *log(Mo)* is not directly linked to *log(T)* is consistent with the null hypothesis that T is conditionally independent of Mo, given the values of the other variables. We interpret not as proving that *Mo* is not a direct cause of *T*, but rather as indicating no clear evidence that *T* depends directly on Mo. The possibility of an indirect dependency mediated by cesium, or perhaps by other variables not included in this analysis, cannot be ruled out, and appears to be worth exploring further.

Insights from Comparing Linear and Nonparametric Nonlinear Regression Models

As a final step, we consider why *log(Mo)* might appears as a significant predictor of *log(T)* in some multiple regression models (Fig. 11.3), even if there is no clear dependence between them in others, or in nonparametric analyses such as those in Figs. 11.5, 11.6, and 11.7 (CART tree, random forest PDP, and BN analyses). In Fig. 11.7, both age and BMI are directly connected to both *log(Mo)* and *log(T)*. Thus, BMI and age are identified as potential confounders of the *Mo-T* (or *log(Mo)*-*log(T)*) association. To control for such potential confounders, multiple linear regression modeling includes them as predictors on the right side of the regression modeling. The estimated effect (or regression coefficient) of the predictor *log(Mo)* on the dependent variable *log(T)* is then said to have controlled for the potential confounders; it does so by conditioning on their values when estimating the association between values of *log(Mo)* and *log(T)*. However, this control is imperfect if, contrary to the assumptions of the multiple linear regression modeling framework, the association between the dependent variable and the potential confounders is not

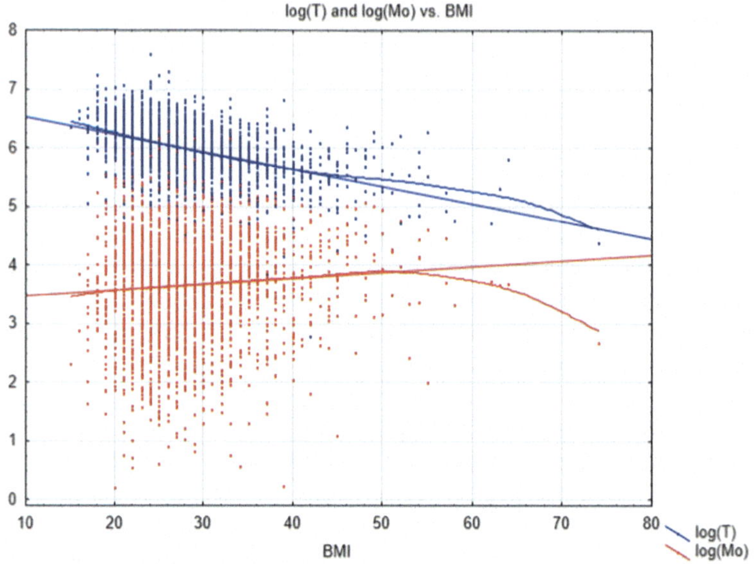

Fig. 11.8 Linear and nonlinear (smoothing) regressions for *log(T)* and *log(Mo)* as functions of BMI

linear. Figure 11.8 shows an example with BMI as the confounding variable (x axis) and with *log(T)* as a dependent variable (y axis, top curves and data points); in addition *log(Mo)* is plotted against BMI (lower curves and data points). For both *log(T)* and *log(Mo)*, simple linear regression lines for their values as functions of BMI are superimposed on the data points, along with a nonparametric (smoothing) regression curve, so that departures from linearity can be seen. Although the nonparametric regression curves for both *log(T)* and *log(Mo)* nearly coincide with the linear regression lines for BMI between 25 and 35, there are deviations from linearity at high and low values of BMI. The fact that these deviations occur over overlapping ranges of BMI values for both *log(Mo)* and *log(T)* (albeit with opposite signs, i.e., the regression line is below the nonparametric regression curve for *log(T)*, but below it for *log(Mo)*) implies that the mean squared error (MSE) from using a regression line to predict *log(T)* from BMI can be reduced by including *log(Mo)* as a predictor, since subtracting a multiple of *log(Mo)* from the linearly predicted value of *log(T)* can help to reduce the prediction error caused by the departures from linearity. In other words, including *log(Mo)*, in addition to BMI, as a predictor in a multiple linear regression model for *log(T)* reduces prediction error compared to using BMI as the only predictor, because it helps to correct for model misspecification error. This explains why *log(Mo)* might be a significant predictor for *log(T)* in a linear regression model, even if it adds no explanatory power to a nonparametric regression model: the incorrect assumption of linearity creates prediction errors that can be reduced by including *log(Mo)* because both *log(Mo)* and *log(T)* contain systematic departures from linearity over similar ranges of BMI values.

The methodological point that a significant regression coefficient in a multiple linear regression model may simply signify that the corresponding predictor can help reduce model specification error is somewhat subtle, but potentially of great practical importance in interpreting whether regression coefficients provide evidence that a variable poses a health risk. To clarify the logic with a simple arithmetic example (similar to those discussed in Chaps. 1 and 2), suppose that a confounder Z completely determines the values of both a dependent variable Y, indicating an adverse health effect; and also another variable, X. Moreover, suppose that the causal relationships determining the values of X and Y from the value of Z are nonlinear, described by the structural equations $Y = Z^2$ and $X = Z^{0.5}$. Then a multiple linear regression model for predicting Y from both X and Z has lower prediction error (MSE) than one for predicting Y from Z alone, even though, by construction, only Z and not X determines the value of Y. (If the values of Z are uniformly randomly distributed between 0 and 2, for example, then the best-fitting multiple linear regression model is $E(Y \mid Z, X) = 0.87 + 4.27Z - 4.05X$, with an adjusted R^2 of 0.99; by contrast, the best-fitting regression model with only Z as a predictor is $E(Y \mid Z) = -0.68 + 2Z$, with an adjusted R^2 of 0.94.) Departures of X ($=Z^{0.5}$) from what a linear regression of X against Z would predict can be used to help correct for departures of Y ($=Z^2$) from what a linear regression of Y against Z would predict, making X a significant predictor of Y in multiple linear regression model even though it contains no non-redundant information in addition to Z. By contrast, in a flexible nonparametric regression model, only Z, and not X, is a significant predictor of Y, since the regression curve $E(Y \mid Z) = Z^2$ has no prediction error. Analogously, for the NHANES data in Fig. 11.8, restricting the range of BMI values considered to the interval from 25 to 35, where the linear and nonparametric regression models nearly coincide, makes *log(Mo)* no longer a significant predictor of *log(T)*: BMI alone is a sufficient predictor over this range. Extending the range of BMI values makes *log(Mo)* a significant predictor of *log(T)*, but a potential noncausal interpretation is that this is because the departures of *log(Mo)* from what a linear regression against BMI would predict can be used to help correct for departures of *log(T)* from what a linear regression against BMI would predict.

Discussion and Conclusions

The main findings from our analysis are as follows.

1. We confirmed that urinary *Mo* is significantly negatively associated with serum testosterone in several multiple linear regression models (Table 11.3 and Fig. 11.3), consistent with previous findings in the literature (e.g., Lewis and Meeker 2015; Wang et al. 2016).
2. However, we found that urinary *Mo* is not a significant predictor of T in other multiple linear regression models (Table 11.4). Whether it is a significant predictor (at a 5% conventional significance level or at other significance levels)

depends on which other variables are selected to include in the model as predictors (Fig. 11.3).

3. Previous literature has not emphasized some relevant predictors of *T*, such as uric acid, fasting time, and time of day when blood was sampled (Fig. 11.7).
4. Including urinary *Mo* as a predictor in multiple linear regression modeling does not significantly improve ability to predict or explain testosterone when other relevant variables are also included as predictors. It does not significantly reduce mean squared prediction error or increase the proportion of variance in testosterone explained (R^2). *Mo* is automatically dropped in backward stepwise variable selection because it is not a useful predictor of testosterone unless other relevant variables with which it is correlated are excluded from the model.
5. Urinary *Mo* does not appear to be an informative predictor of serum testosterone in non-parametric analyses that are not limited by the assumptions of the multiple linear regression modeling framework (Figs. 11.5, 11.6, and 11.7).
6. A general methodological insight is that *log(Mo)* might be a significant predictor of *log(T)* in some multiple linear regression models because the linearity assumption is not entirely correct, and including *log(Mo)* can help to compensate for violations of the linearity assumption (Fig. 11.8). The phenomenon of a multiple linear regression model form (or other specific parametric modeling form) inducing significantly non-zero regression coefficients for predictors that do not help to predict the dependent variable in more flexible nonparametric models may have many other applications in epidemiology

In short, concerns about statistically significant negative associations (regression coefficients) between urinary *Mo* levels and serum testosterone levels in multiple linear regression models of NHANES data are not supported by our analyses, insofar as these associations disappear when more variables are included as predictors and when more flexible and general non-parametric models are applied.

We do not interpret our findings as showing that such concerns cannot be justified. In particular, Fig. 11.7 leaves open the possibility that some mechanisms of pathways that also affect urinary cesium might link *Mo* and *T*. Rather, we interpret our findings as illustrating a general methodological caveat that may also be important in other applications: a regression coefficient that is significantly different from zero does not necessarily provide evidence that an exposure variable causally affects a dependent variable that it helps to predict, even if potential confounders have been controlled for by including them in the regression model. The significant regression coefficient may arise from model specification errors. If this explanation cannot be ruled out, then the coefficient should not be interpreted as providing evidence that changing exposure could change risk (the dependent variable), or as a cause for alarm. Instead, further analysis of causality is needed (perhaps using causal Bayesian networks for observational data, or randomized experiments if possible) when the regression findings *per se* do not necessarily have a causal interpretation and are found to be model-dependent.

References

Cox LA Jr. Modernizing the Bradford Hill criteria for assessing causal relationships in observational data. Crit Rev Toxicol. 2018;48(8):682–712.

Lewis RC, Meeker JD. Biomarkers of exposure to molybdenum and other metals in relation to testosterone among men from the United States National Health and Nutrition Examination Survey 2011-2012. Fertil Steril. 2015;103(1):172–8.

Liu FJ, Dong WY, Zhao H, Shi XH, Zhang YL. Effect of molybdenum on reproductive function of male mice treated with busulfan. Theriogenology. 2019;126:49–54. https://doi.org/10.1016/j.theriogenology.2018.12.002.

Nagarajan R, Scutari M, Lebre S. Bayesian networks in R with applications in systems biology. New York: Springer; 2013.

Rotter I, Kosik-Bogacka DI, Dołęgowska B, Safranow K, Kuczyńska M, Laszczyńska M. Analysis of the relationship between the blood concentration of several metals, macro- and micronutrients and endocrine disorders associated with male aging. Environ Geochem Health. 2016;38(3):749–61. https://doi.org/10.1007/s10653-015-9758 0.

Wang YX, Sun Y, Huang Z, Wang P, Feng W, Li J, Yang P, Wang M, Sun L, Chen YJ, Liu C, Yue J, Gu LJ, Zeng Q, Lu WQ. Associations of urinary metal levels with serum hormones, spermatozoa apoptosis and sperm DNA damage in a Chinese population. Environ Int. 2016;94:177–88. https://doi.org/10.1016/j.envint.2016.05.022.

Zeng Q, Zhou B, Feng W, Wang YX, Liu AL, Yue J, Li YF, Lu WQ. Associations of urinary metal concentrations and circulating testosterone in Chinese men. Reprod Toxicol. 2013;41:109–14. https://doi.org/10.1016/j.reprotox.2013.06.062.

Chapter 12
Case Study: Are Low Concentrations of Benzene Disproportionately Dangerous?

Introduction: Are Benzene Dose-Response Functions Supralinear at Low Concentrations?

It has sometimes been proposed that supralinear dose-response functions—that is, dose-response functions in which low doses are disproportionately potent in causing harm—describe the risks from many well-studied public and occupational hazards including asbestos, benzene, lead, particulate air pollution, and ionizing radiation; and that this implies that exposure concentrations must be driven to zero, or close to it, to adequately protect human health (e.g., Hornung and Lanphear 2014; Lanphear 2017). Some of these concerns have been challenged by other investigators as statistically flawed or as biologically unrealistic (e.g., Waddell 2006). Fors example, studies that identify supralinear dose-response relationships can be questioned if they fail to control fully for important potential confounders, such as misclassified smoking (Hamling et al. 2019). Residual confounding by smoking can create significant associations between some common exposure metrics (e.g., blood lead levels) and adverse health effects (e.g., cardiovascular mortality and morbidity) whether or not the former affect the latter. Misattributing risk caused by a confounder to exposure creates the appearance of a supra-linear dose-response function since risk fails to decline as exposure decreases, in effect spreading the same risk over fewer units of exposure.

As discussed in Chaps. 3, 4, 5, and 6, for many other agents, including asbestos, mineral dusts and particulate matter, and ionizing radiation, it has been discovered that major exposure-associated adverse health effects are mediated by chronic inflammation and activation of the NLRP3 inflammasome signaling complex. This has an activation threshold, such that low exposure concentrations do not trigger inflammasome assembly and activation and resulting increases in disease risks (Bogen 2019; Du et al. 2019; Wei et al. 2019). However for other chemicals, the causal mechanisms involved in adverse outcome pathways and modes of action are

L. A. Cox Jr., *Quantitative Risk Analysis of Air Pollution Health Effects*, International Series in Operations Research & Management Science 299,
https://doi.org/10.1007/978-3-030-57358-4_12

less clear. For these chemicals, careful review is needed to assess the biological plausibility of supralinear dose-response function for low exposure concentrations. This chapter provides such an assessment for benzene, focusing specifically on whether production of putative harmful metabolites of benzene is supralinear at low concentrations.

The suggestion that low-dose metabolism of benzene is supralinear has previously been advanced using data from factory workers in Tianjin, China (Rappaport et al. 2010). Applying nonlinear parametric regression models that take a ratio Y/X as the dependent variable and X as the independent variable, where Y = measured quantity of a benzene metabolite and X = estimated concentration of benzene in inhaled air to data for factory workers in Tianjin revealed that the ratio Y/X increases as X decreases, for various metabolites Y (Rappaport et al. 2010). This observation was interpreted as strong evidence for a supralinear dose-response explained by an unknown enzyme that becomes saturated at concentrations below the detection limit for benzene, in preference to an alternative explanation that the apparent supralinearity reflects unrealistic assumptions about unmeasured background levels of benzene exposure (since the supralinearity finding disappears if "background" concentrations are defined as those for workers without occupational exposures) (Rappaport et al. 2013). More recent work applying nonparametric statistical and machine learning techniques to the Tianjin data to avoid potential biases from parametric regression modeling found no departures from linear metabolism at low exposure concentrations, but clarified that a ratio Y/X can increase as X approaches zero even if metabolism is linear, if estimated values are measured with error variance, as in the Tianjin data (McNally et al. 2017; Cox Jr. 2019).

To further advance understanding of the shapes of benzene metabolism and dose-response relationships at low exposure concentrations in air, the following sections address the following five related research questions:

1. Which metabolites and markers best predict human health risks from low levels of benzene exposure (and why)?
2. How do levels of metabolites and markers change with levels of benzene in inhaled air, and with levels of other factors (e.g., diet, alcohol consumption, cigarette smoking)?
3. How much do inter-individual heterogeneity and variability affect benzene metabolism?
4. How much intra-individual variability is there in benzene metabolism?
5. At what levels do different metabolites and relevant markers predict increased risk of adverse health effects from benzene exposure?

Questions 1 and 5 are addressed based on existing literature. To obtain new insights into questions 2–4, we also apply techniques from Chap. 9 to identify and quantify potential causal relationships between inhaled benzene and levels of relevant benzene metabolites in the Tianjin factory worker data.

Benzene Markers and Metabolites Relevant for Human Health

Benzene Metabolism and Pharmacokinetics

Human metabolism of benzene has been intensively studied and elucidated. Figure 12.1 shows major benzene metabolic pathways, both as sequences of molecular structures (left) and as a pathway schematic (right) with metabolites (blue) and metabolizing enzymes (black); these can be explored via the interactive WikiPathway (Slenter et al. 2017) from which the right side of Fig. 12.1 is taken. Physiologically based pharmacokinetic (PBPK) models have been developed and validated for benzene metabolism in humans, mice, rats, and other species (Knutsen et al. 2013). Metabolism and distribution of benzene in liver, bone marrow, and other tissue groups have been compared across species and their interindividual and intraindividual variability in human populations have been studied and modeled quantitatively in population PBPK models (Yokley et al. 2006). In discussing benzene metabolism, different authors focus on trans,trans-muconaldehyde (ttM), E,E-muconaldehyde, or trans,trans-muconic acid (t,t-MA or simply MA) (i.e., the diene dialdehyde and its dicarboxylic acid metabolite, respectively) as indicators of open-ring products formed by cytochrome P4502E1 (CYP) metabolism. We follow this mixed usage in summarizing relevant results from the literature.

Figure 12.2 shows the compartmental structure of a recent benzene PBPK model for humans (Knutsen et al. 2013). Figure 12.3 compares model predictions to observations of metabolites in data not used in building the model. (The scales in Fig. 12.3

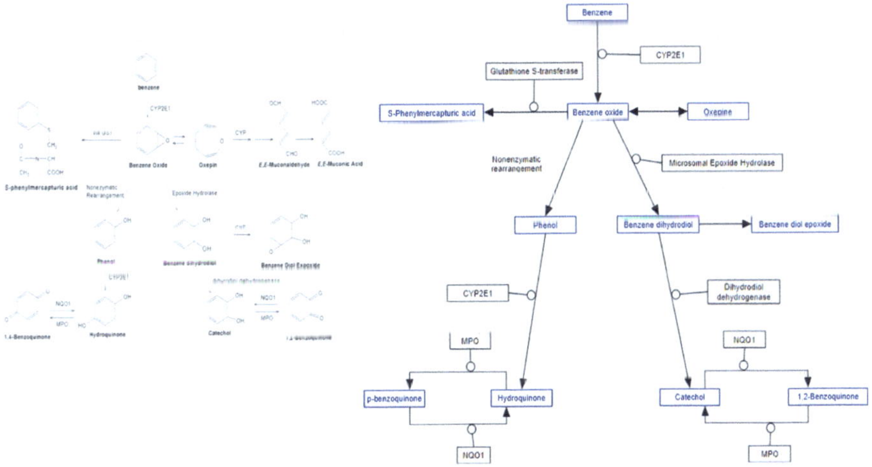

Fig. 12.1 Human metabolism of benzene. (left) Chemical structures of metabolites. (right) Pathway schematic showing major metabolites (blue) and metabolizing enzymes (black) (muconic acid (MA) not shown). Sources: Left side: Rappaport et al. (2010). Right side: side pathway schematic is from: https://www.wikipathways.org/index.php/Pathway:WP3891#nogo2

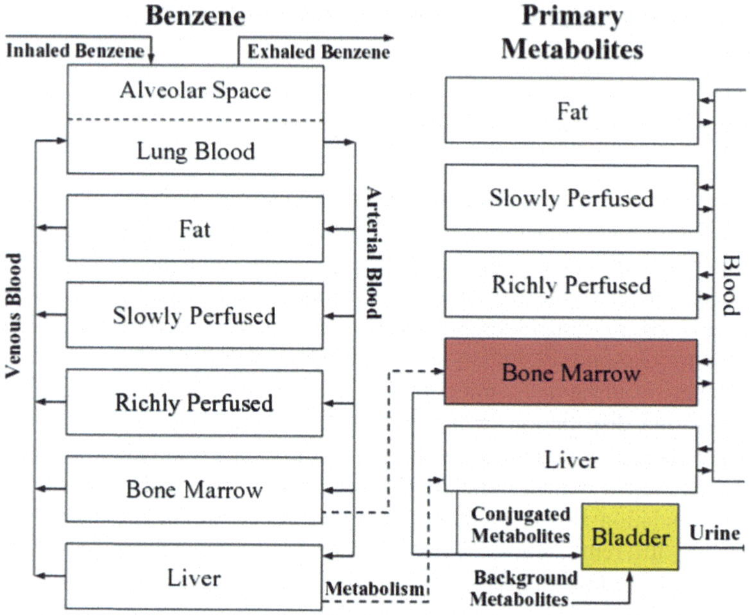

Fig. 12.2 Compartmental structure of a recent PBPK model for human metabolism of benzene (Source: Knutsen et al. 2013)

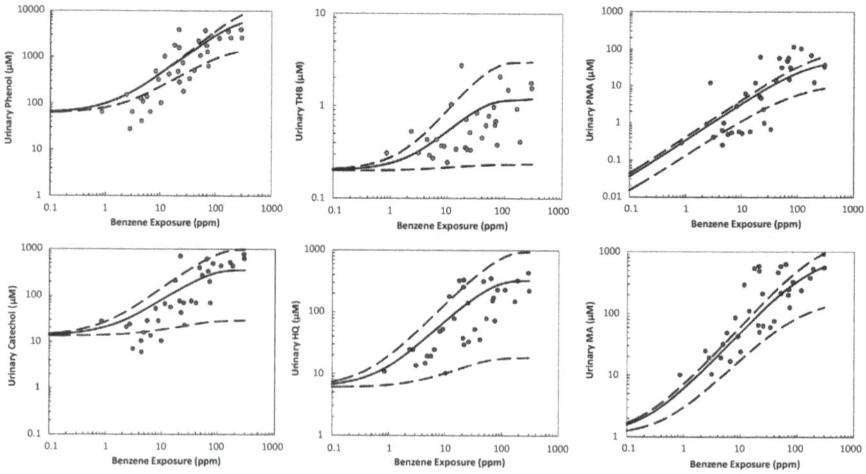

Fig. 12.3 Comparison of benzene PBPK model predictions (solid lines) and uncertainty bands (dashed lines, representing ±2 standard deviations) to data (points) not used in building the model. Abbreviations: *MA* muconic acid, *PMA* phenylmercapturic acid, *PH* phenol, *CAT* catechol, *HQ* hydroquinone, *BT* benzenetriol. Source: Knutsen et al. 2013

are logarithmic, so predictions and data are compared across estimated exposure concentrations that differ by several orders of magnitude.) Estimated rates of human conversion of benzene to muconic acid (MA) and phenylmercapturic acid (PMA), phenol (PH), catechol (CA), hydroquinone (HQ), and benzenetriol (BT) in the model were validated using urinary benzene metabolite data from Chinese benzene workers. Although there is substantial interindividual variability, as reflected in the large vertical scatter in Fig. 12.3, the PBPK model predictions pass through the observed data clouds, and in this sense roughly describe the data. Other PBPK models for benzene have been developed over the past 30 years and at least partially validated by comparing their predictions to observations in data sets not used in developing them, although most do not show bone marrow as a separate compartment. Similar to Fig. 12.3, these PBPK models predict linear or sublinear (upward-curving) production of metabolites as a function of air benzene concentrations, for air benzene concentrations below about 10 ppm. Saturation of metabolite production begins at air benzene concentrations between about 10 ppm and about 100 ppm.

Benzene Health Effects and Hypothesized Causal Biological Mechanisms

The major health effects of concern for benzene-exposed workers are acute myeloid leukemia (AML) and myelodysplastic syndrome (MDS), although high concentrations (especially >100 ppm) can also damage bone marrow and kill hematopoietic progenitor cells, causing bone marrow failure and aplastic anemia (Das et al. 2012; Rushton et al. 2014; Schnatter et al. 2012; Smith 1996). Some epidemiological data sets and statistical models have also identified associations between benzene exposure and other illnesses, especially non-Hodgkin's lymphomas, although the evidence is mixed and causality is uncertain (Galbraith et al. 2010; Teras et al. 2019).

A central challenge for dose-response modeling of benzene health effects at relatively low concentrations of air benzene is that the specific metabolites and biological mechanisms involved in causing MDS and AML and other effects have not yet been definitively identified, although several theories and supporting evidence have been advanced (Arnold et al. 2013). Some papers interpret the fact that benzene and its reactive metabolites react with a variety of proteins and DNA in a variety of test systems as mechanistic evidence that occupational levels of benzene cause AML. For example, Grigoryan et al. (2018) state that "The fact that these diverse human serum albumin modifications differed between benzene-exposed and control workers suggests that benzene can increase leukemia risks via multiple pathways involving a constellation of reactive molecules." However, most of the changes discussed (e.g., formation of albumin adducts in blood serum in this example, or induction of micronuclei in peripheral blood lymphocytes in other papers) have no clear causal relevance for health risks, as they are not validated markers for increased risk of MDS, AML or other adverse health effects caused by benzene (Hack et al. 2010). We focus on health-relevant markers, i.e., observable changes that act through identified causal pathways or modes of action to increase MDS and AML risks.

Most current research accepts that benzene acts by forming myelotoxic metabolites that concentrate in the bone marrow, where they damage early hematopoietic stem cells (HSCs), myeloid progenitor cells, and the stromal microenvironment, increasing the rate of cell death and replacement. However, there are diverse theories about how these events increase risk of MDS and AML. Among them are the following.

- *Two-stage clonal expansion (TSCE) mode of action for benzene carcinogenesis.* A proposed traditional mode of action (MoA) for benzene-induced AML begins with enzyme-dependent metabolism, mainly via cytochrome P450-2E1 (Fig. 12.1). The resulting toxic metabolites interact with HSCs or other target cells in the bone marrow, creating mutated ("initiated") target cells. Proliferation of these initiated cells ("promotion") (perhaps stimulated by cell death due to poisoning by toxic benzene metabolites) then increases the risk of transformation of at least one of them to AML ("progression") (Meek and Klaunig 2010). This MoA is consistent with the general two-stage clonal expansion (TSCE) models of carcinogenesis, and quantitative TSCE modeling suggests that benzene acts as a promoter, increasing the risk of AML by increasing the rate of proliferation of initiated cells (Richardson 2009).
- *HSC damage, genomic instability, and reduced immune surveillance.* Alternatively, it has been proposed that key events in development of benzene-induced leukemia include exposure-related increases in intracellular oxidative stress in HSCs or hematopoietic progenitor cells, dysregulation of the aryl hydrocarbon receptor (AhR), and reduced immune surveillance of HSCs in the bone marrow (Hirabayashi and Inoue 2010; McHale et al. 2012; Wang et al. 2012; Snyder 2012; Kerzic and Irons 2017). McHale et al. (2012) further postulate that benzene metabolites interact with critical genes and pathways in the HSC, leading to genotoxic, chromosomal or epigenetic abnormalities and genomic instability; stromal cell dysregulation; increased rate of death (via apoptosis) of affected HSCs and stromal cells; and proliferation of altered HSCs with unrepaired damage and chromosomal aberrations, culminating in creation of leukemic stem cells and subsequent leukemia. Others have proposed that benzene metabolites (especially 1,4-benzoquinone, i.e., p-benzoquinone) cause MDS and AML by increasing chromosomal instability and unrepaired DNA damage and AML-causing chromosomal rearrangements in HSCs (e.g., via interference with topoisomerase-mediated DNA ligation and nicking) (Son et al. 2016). However, Kerzic and Irons (2017) note that such hypotheses, which predict that cytogenetic damage in benzene-induced leukemogenesis should be similar to that in chemotherapy-induced leukemogenesis, are not consistent with cytogenetic data from benzene-exposed workers with hematopoietic diseases showing that benzene-induced AML is much more similar to *de novo* AML than to therapy-related AML.

- *Chronic inflammation MoA.* In Chinese worker data, chronic inflammation followed by an immune-mediated inflammatory response, rather than cytogenetic abnormalities *per se*, appears to drive initiation and progression of benzene-induced MDS and AML (Gross and Paustenbach 2018). This may help to explain earlier observations that rheumatoid arthritis and other autoimmune conditions (Anderson et al. 2009), as well as respiratory infections and other common infections (Titmarsh et al. 2014), are significantly associated with increased MDS and AML risk. The inflammatory MoA is also consistent with often-noted associations between increased reactive oxygen species (ROS) and oxidative stress in affected cells and increased risk of MDS and AML. It has recently been discovered that inflammation mediated by a specific intracellular signaling platform (the NLRP3 inflammasome) that self-assembles in the cytosol of immune cells in response to exposures above an activation threshold (Cox Jr. 2019) drives the pathogenesis of MDS (Sallman and List 2019). Recent findings that benzene induces inflammatory programmed cell death (pyroptosis) (Guo et al. 2019; Sallman et al. 2016) and autophagy (Qian et al. 2019) underscore the importance of inflammatory responses in benzene-induced pathogenesis. Although inflammation increases ROS and oxidative stress and cell turnover, thus potentially increasing formation and proliferation of HSCs with unrepaired cytogenetic damage and chromosomal aberrations, the inflammatory MoA suggests that it is the underlying inflammation that drives health risks.

The TSCE and the inflammatory MoAs raise the possibility of exposure concentration thresholds below which inhaled benzene does not increase disease risks. Promotion is typically a threshold phenomenon (Neumann 2009), e.g. because concentrations must be high enough to cause toxic damage and resulting compensating proliferation of damaged HSCs. Inflammatory responses have activation thresholds or threshold-like steep nonlinearities (e.g., ultrasensitive switches) in their dose-response functions, so that tissues are not continually being inflamed by low-level exposures; the mechanisms of these activation thresholds have been substantially elucidated (Cox Jr. 2019). (Inflammation might also play a role in clonal expansion of damaged stem cells due to continuing hyperplasia acting as a co-initiator of carcinogenesis, but this also implies a threshold (or threshold-like nonlinearity) in the dose-response function (Bogen 2019).) Thus, MoA theories for benzene-induced MDS and AML in which benzene metabolites cause chronic inflammation and resulting proliferation of damaged HSCs suggest highly sub-linear (threshold or threshold-like) dose-response functions, rather than supra-linear ones. If this is correct, then benzene exposures increase risks of MDS and AML only if concentrations and durations of exposure raise concentrations of toxic metabolites in bone marrow microenvironments sufficiently high and/or long to activate inflammasomes and cause pyroptosis and other inflammatory responses. This shifts attention back to PBPK considerations.

Relevance and Use of Biomarkers in Predicting Health Effects and Benzene Exposures: BN Models

Biomarkers are potentially useful for estimating unobserved benzene exposures (or improving estimates of concentrations measured with error) and for predicting health risks caused by exposures. The predictive value of a biomarker depends on how well it can be measured and on by how much conditioning on its measured value reduces uncertainty (e.g., the average entropy of the conditional probability distribution) of the dependent variable being predicted. For example, Fig. 12.4 shows the structure of a Bayesian network (BN) model for quantifying conditional probabilities of some variables (e.g., various metabolites, markers and AML (the *Leukemia* node at the lower right)), given observed or assumed values of others. The conditional probability distribution of each variable (node in the network) depends on the values of the variables that point into it. This BN, from Hack et al. (2010), was constructed manually based on a detailed literature review of candidate markers and outcomes and verifiable, usefully accurate predictive relationships among them, given limitations in measurement techniques.

Other potentially causally relevant variables (e.g., p-benzoquinone, NLRP3 inflammasome activation, age of patient, co-exposures and co-morbidities, and so forth) are "marginalized out" of the BN model in Fig. 12.4, meaning that they can still implicitly affect the probability of leukemia, but are not explicitly shown or conditioned on. Other investigators (or automated machine-learning programs for learning BNs directly from data) might select different markers and develop other

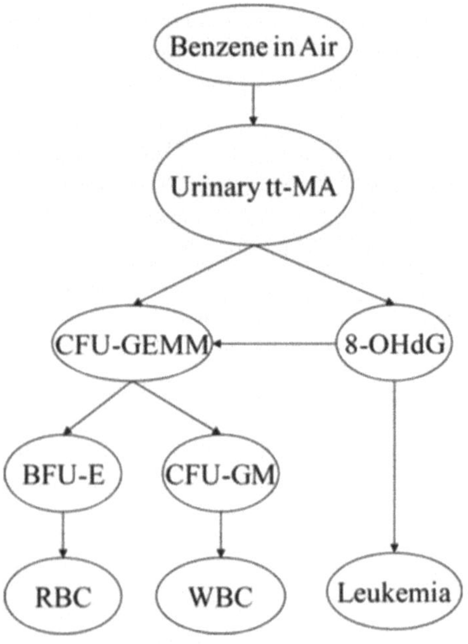

Fig. 12.4 Proposed Bayesian network (BN) model structure. Abbreviations: *8-OHdG* 8-hydroxyguanosine (a biomarker of oxidative stress), *CFU-GEMM* colony-forming unit-granulocyte, erythrocyte, monocyte, megakaryocyte (a precursor to RBCs and WBCs), *BFU-E* burst-forming unit-erythroid (a RBC precursor cell type), *CFU-GM* colony forming unit—granulocyte-macrophage (a WBC precursor), *RBC* red blood cell count, *ttMA* trans,trans muconic acid, *WBC* white blood cell count (Source: Hack et al. 2010)

BN models and perhaps get narrower uncertainty bands. For example, previous studies have identified urinary concentrations of SPMA (S-phenylmercapturic acid) and benzene itself as reliable indicators of recent benzene exposures, even at relatively low concentrations (e.g., 0.1 ppm for SPMA), whereas urinary tt-MA, used in Fig. 12.4, has inadequate specificity for benzene exposures below about 1 ppm when dietary sources of sorbic acid are also present (Arnold et al. 2013; Campagna et al. 2012; Fustinoni et al. 2011; Qu et al. 2003). It might therefore be advantageous to develop a BN with SPMA and urinary benzene (UB) included as variables that can be used (i.e., conditioned on) to help predict the conditional probability distributions for other variables.

Unlike benzene PBPK modeling, which has been developed, tested and refined for decades, benzene BN modeling of benzene dose-response relationships is relatively new. Future benzene BN models may well improve on the seminal work of Hack et al. summarized in Fig. 12.4, both by including more (and perhaps different) variables and by distinguishing between estimated and true values of benzene concentrations (and other variables measured with substantial error). The most predictively useful combinations of measurements for predicting MDS as well as AML and air benzene exposure levels have yet to be identified for such multi-variable BN models. Additional markers of chronic inflammation may be needed to complement the information already captured by the BN in Fig. 12.4 (e.g., 8-OHdG). However, despite these limitations compared to possible future BN models, it is clear in model curves shown in Hack et al. (2010) that dose-response results implied by the BN model in Fig. 12.4 do not suggest supralinearity. Rather, they are consistent with a threshold or very sublinear (J-shaped) dose-response function.

Dependence of Metabolites and Markers on Inhaled Air Benzene (AB) and Other Factors

Simple (Univariate) Regression Models

A well-developed literature documents increases in various markers of benzene exposure and activity with increasing air benzene (AB) concentrations, even at low doses. For example:

- Urinary benzene (UB) levels have been observed to be linearly related to work shift AB concentrations over a wide range of exposures, down to <0.2 ppm, providing a useful biomarker of benzene exposure over approximately the previous day (Weisel 2010).
- Urinary S-phenylmercapturic acid (SPMA) levels increase approximately linearly with AB concentrations between 0 and 80 ppm (Lin et al. 2008, Fig. 12.2; see also Knutsen et al. 2013).

- tt-MA is a less specific marker for low-level benzene exposure. The problem of dietary sources for tt-MA at low exposure concentrations has already been mentioned.
- S-phenylcysteine (SPC) in albumin in human workers increased linearly with average occupational benzene exposures ranging from 0 to 23 ppm (for 8 h/day, 5 day/week) (Bechtold and Henderson 1993).

Turning from literature review to data analysis, Fig. 12.5 shows how average metabolite quantities (curves) increase with air benzene (AB) concentrations among Tianjin, China factory workers (Rappaport et al. 2010; Price et al. 2012; Cox et al. 2017). Figure 12.5 shows non-parametric (lowess smoothing) regression curves fit to scatter plots of metabolite levels (normalized by dividing by their mean values, so that 1 on the vertical axis represents the mean value, for each metabolite) against AB concentrations from 1–3 ppm on the horizontal axis. (The data points shown are for UB vs. AB for workers with exposure concentrations in this range, with the vertical axis truncated at 3 to avoid scale compression due to outliers.) All curves are approximately linear for AB concentrations below 1 ppm.

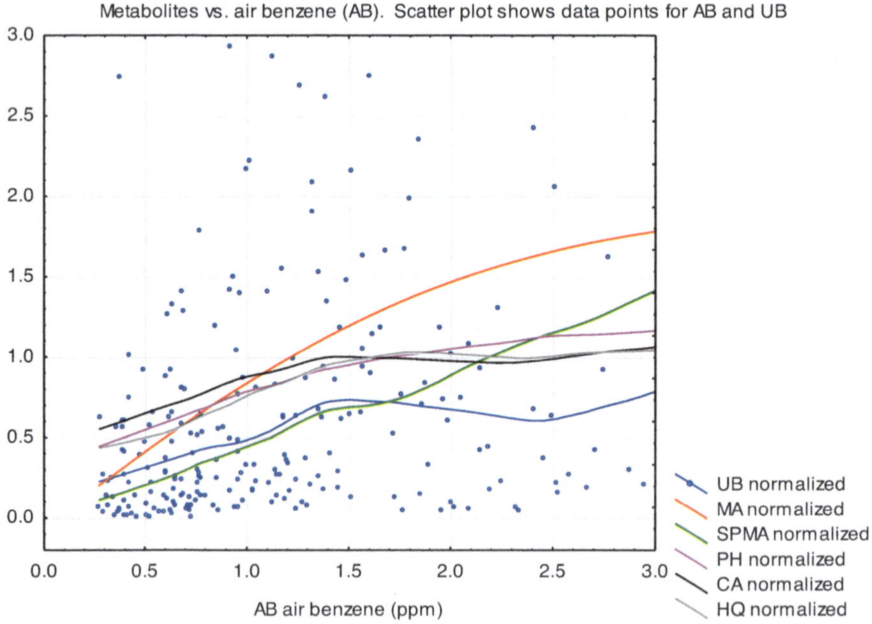

Fig. 12.5 Metabolite levels vs. air benzene (AB) at AB concentrations below 3 ppm. (Metabolites are scaled so that 1 = mean value of each metabolite in the worker population.) Curves are non-parametric (smoothing) regression curves. Data points are for urinary benzene (UB). Abbreviations: *UB* urinary benzene, *MA* E,E muconic acid, *SPMA* S-phenylmercapturic acid, *PH* urinary phenol, *CA* urinary catechol, *HQ* urinary hydroquinone. Data are from Price et al. (2012)

Exploratory Multivariate Modeling: Correlations

Univariate regression curves, with AB as the sole explanatory variable, do not explicitly account for the effects of other variables, such as age, sex, or co-exposures to toluene, in quantifying how metabolite amounts vary with AB. To better understand the interrelationships among variables, we apply the Causal Analytics Toolkit (CAT) software and methods described in Cox et al. (2017) to the Tianjin data for 177 workers with AB exposure concentrations of no more than 5 ppm and without missing data on AB or metabolites. Figure 12.6 shows traditional linear (Pearson's product-moment) correlations between pairs of variables, along with frequency histograms (main diagonal) and scatter plots. These are not ideal for binary variables

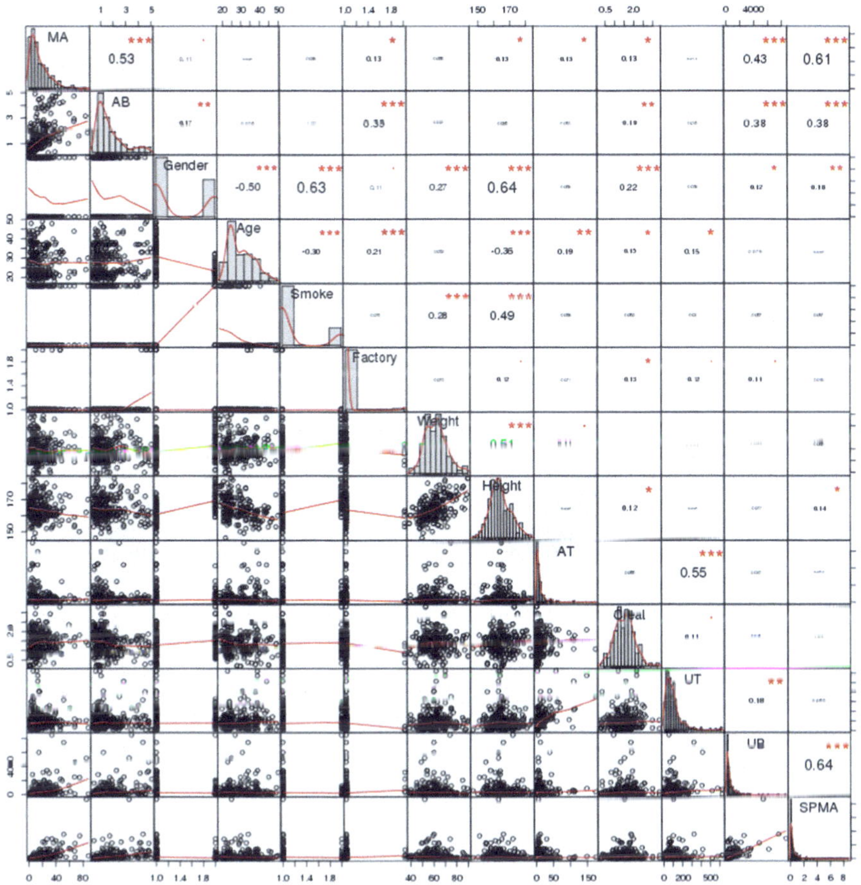

Fig. 12.6 Frequency histograms (main diagonal), scatter plots with kernel smoothing regression curves (below diagonal) and linear (Pearson's product moment) correlations and significance asterisks (above diagonal) (* = 0.05, ** = 0.01, *** = 0.001). *AT* air toluene, *UT* urinary toluene, *Creat* creatinine

or for highly nonlinear relationships, but allow for quick visual exploration of the association between pairs of variables. Likewise, Fig. 12.7 visualizes the linear correlations in Fig. 12.6 as a corrgram (heat map diagram at the upper left) and as a network (lower left) with stronger correlations indicated by closer distances and thicker arcs (green = positive correlation, red = negative) between variables. The upper right panel of Fig. 12.7 provides analogous information for partial linear correlations, in which the association between each pair of variables is adjusted for levels of other variables via multiple linear regression; and the lower right presents network visualization when polychloric correlations are used instead of linear correlations for the binary variables (factory, sex, and smoking). (Only data from the 2 factories with occupational benzene exposures are included.) Again, correlations are not ideal measures for binary variables, but provide for a quick exploration.

We do not base any final conclusions on the pairwise correlations in Figs. 12.5, 12.6, and 12.7, but note that age, gender, height, weight, and smoking cluster together, and that air benzene (AB) and benzene metabolites often proposed as markers of benzene exposure (MA, UB, and SPMA) also cluster together (and apart from the demographic variables and the toluene variables, AT and UT). At these low

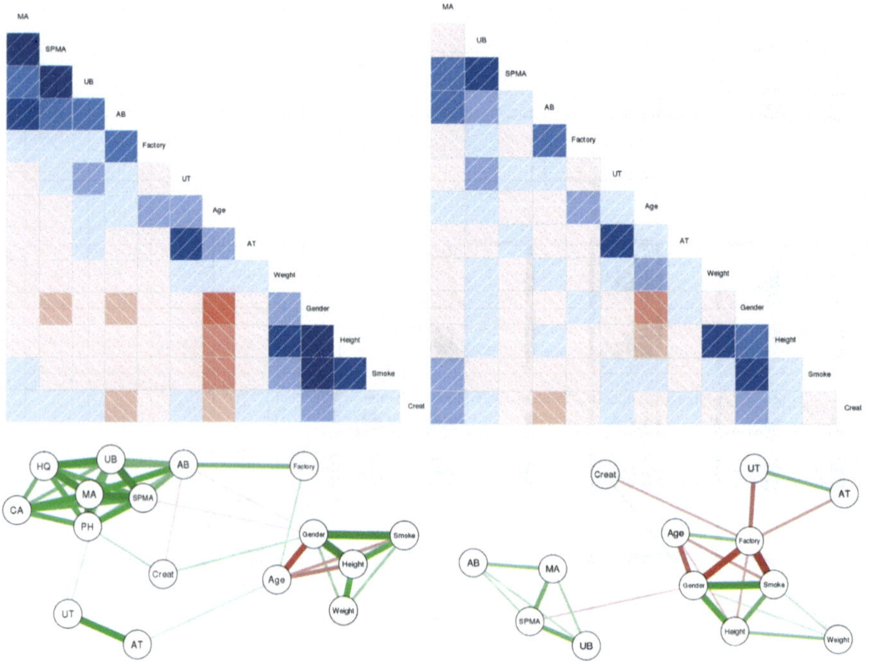

Fig. 12.7 Visualizations of correlations. (Upper left) Linear (Pearson's) correlation corrgram (blue = positive correlation, red = negative, stronger correlations are darker). (Upper right) Partial linear correlation corrgram. (Lower left) Network visualization of linear correlations. Green arcs represent positive correlations, red = negative, and thicker arcs and closer distances indicate stronger correlations. (Lower right) Network visualization with polychloric correlations for binary variables

concentrations, AT and UT have no significant correlations with AB and UB. This pattern is consistent with previous studies indicating that MA, UB, and SPMA are useful predictors of AB.

Partial Dependence Plots (PDPs)

A more thorough analysis examines how concentrations of metabolites increase with increasing AB, similar to Fig. 12.5, while controlling for (i.e., conditioning on) the values of all other variables. Such an analysis can be performed using ensembles of non-parametric models to minimize dependency on modeling assumptions (using the *randomForest* package in R). The top row of Fig. 12.8 shows the resulting partial dependency plots (PDPs) for dependency of SPMA and UB, the two metabolites identified as the best markers for low-level benzene exposures, on air benzene concentration (AB) in ppm, holding other variables (gender, age, smoke, factory,

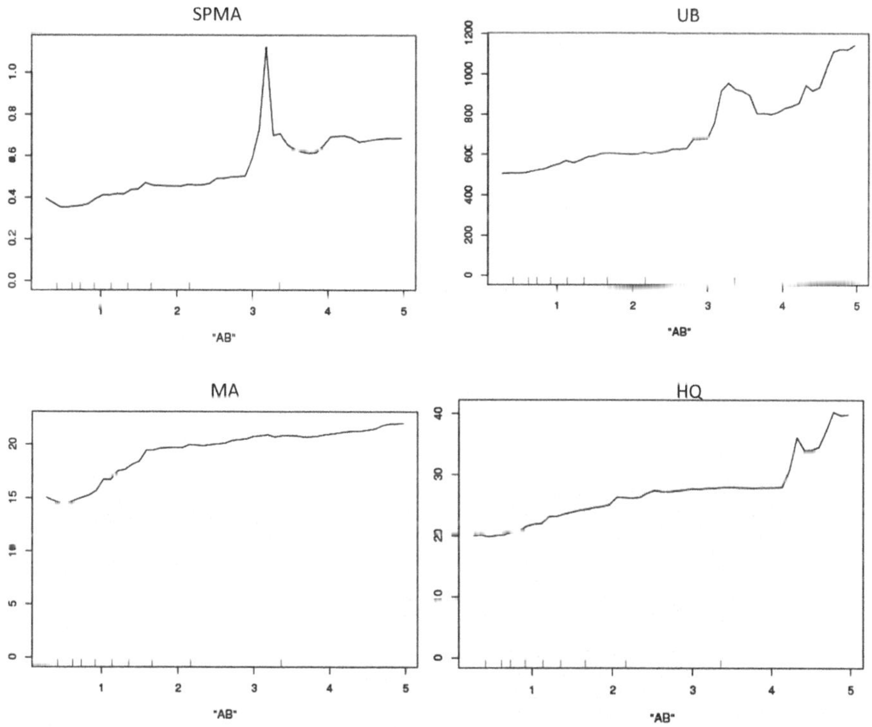

Fig. 12.8 Partial dependence plots (PDPs) for urinary metabolites vs. air benzene (AB) concentration (ppm), controlling for gender, age, smoke, factory, weight, height, AT, creatinine. PDPs were generated by the CAT software (http://cloudcat.cox-associates.com:8899/), which uses the *randomForest* package in R. (https://cran.r-project.org/web/packages/randomForest/randomForest.pdf)

weight, height, AT, creat) fixed at their recorded levels for each worker. The bottom row shows PDPs for MA and HQ. (Spikes at around 3.25 ppm in the PDPs for SPMA and UB reflect outliers, i.e., exceptionally high levels of these metabolites recorded for workers with that estimated exposure concentration.) All four PDPs show significant ordinal associations between AB and levels of each metabolite in urine ($p < 0.0001$ in Spearman's rank correlation test). UB concentrations appear to increase approximately linearly with AB for AB values below 3 ppm. None of these PDPs reveals a supralinear relationship between AB levels and levels of these metabolites at low AB concentrations.

Different metabolites are best for predicting different outcomes. For example, SPMA is the best single predictor for UB, but MA is the best single predictor for AB (as measured by increase in mean squared prediction error in *randomForest* if each variable is omitted). Unsurprisingly, better predictions than for any single predictor can usually be achieved by using sets of several other variables to predict a dependent variable of interest. For example, for the specific task of estimating AB concentration from urinary metabolites, the pair of metabolites MA and UB is much more informative than the pair MA and SPMA, as illustrated in Fig. 12.9.

The left side of Fig. 12.9 shows contours of equal AB (increasing from lower left to upper right) for different pairs of MA and SPMA values, and the right side shows a similar contour plot for AB given values of MA and UB. (For ease of comparison, we use the normalized values of SPMA, MA, and UB, i.e., axes are scaled so that 1 is the mean value of the corresponding variable.) The color keys show that the right-side diagram (MA and UB) has double the resolution of the left-side diagram (MA and SPMA), with contours ranging from 0.5 to 4 instead of 0 to 2 for AB concentrations. However, these diagrams show that, at low MA values, UB has little predictive power (the AB contours are nearly vertical, indicating AB independent of UB), whereas SPMA has better predictive power at lower MA values.

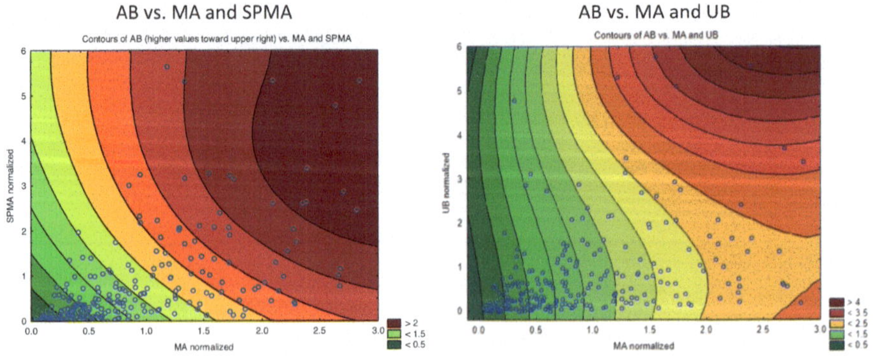

Fig. 12.9 Air benzene contours show that air benzene can be predicted better from MA and UB (right side) than from MA and SPMA (left side)

Bayesian Network (BN) Analysis

To obtain the best possible prediction of a variable, of course, it is necessary to condition on information from multiple variables that are informative about it. This can be done conveniently using Bayesian network (BN) methods. Figure 12.10 shows the structure of a BN model learned from the Tianjin data (with the knowledge-based constraint that AB should be an input). In this BN, two variables are informative about each other (i.e., not conditionally independent of each other, given the values of other variables, at least as far as the BN-learning can determine via statistical tests for conditional independence) if and only if they are joined by an arrow. The conditional probability distribution of each variable is determined by the values of the variables to which it is directly connected, and, given their values, is conditionally independent of the values of all other variables.

Unlike the exploratory correlation networks in Fig. 12.7, the BN is fully non-parametric (with no assumptions of linear or ordered relations or normally distributed error terms) and multivariate (the conditional probability distribution of each variable can be conditioned on the observed or assumed values of any number of other variables). Although Gender, Height, Age, and Weight are still clustered in the BN, as are the metabolites, the BN summarizes the joint distribution of all the variables, and thus contains far more information than the corresponding pairwise correlations. It reveals patterns not obvious from correlations alone, such as a slight positive relation between UT and UB, holding AB and other variables fixed. It can be used to automatically compute the subsets of variables ("adjustment sets") that should be conditioned on to estimate direct causal impacts of changes in one variable, such as AB, on changes in another, such as SPMA, MA, or UB (Cox Jr. 2018). For example, to estimate the causal PDP for the direct effect of AB on SPMA, any

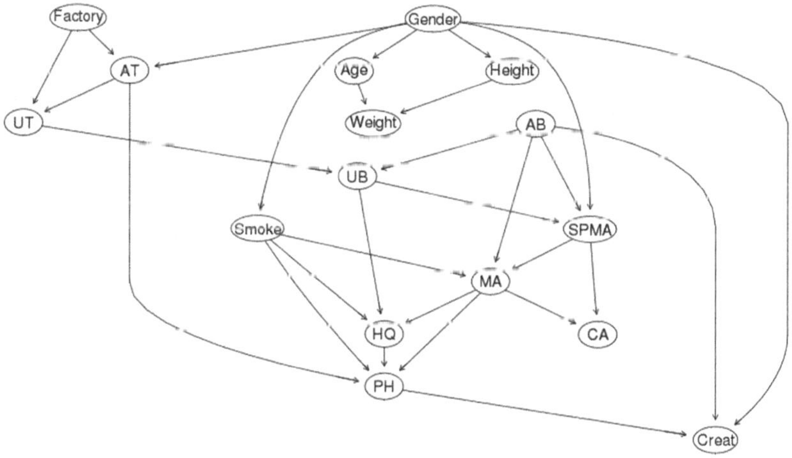

Fig. 12.10 A Bayesian Network (BN) model learned from the Tianjin data. Arrows connect pairs of variables that are identified as being informative about (i.e., help to predict) each other

of the three adjustment sets {Gender, UB}, {AT, Factory, UB}, or {UB, UT} can be conditioned on to remove potential confounding. Any of these choices produces an estimated causal PDP very similar to the one in the upper left of Fig. 12.8. For each of the metabolites SPMA, UB, MA, and HQ, the causal PDPs calculated from the BN are very similar to the PDPs in Fig. 12.8, with no indication of supralinearity at low doses.

Inter-individual Heterogeneity in Benzene Metabolism

Figure 12.1 shows that metabolism of benzene depends on several enzymes. Inter-individual differences in enzyme activity levels—possibly due to differences in genetic polymorphisms, lifestyle factors (e.g., alcohol consumption), or other causes—can lead to substantial differences in benzene metabolism across individuals, as reflected in the very wide variability of metabolite production across workers exposed to the same levels of air benzene (AB) (Fig. 12.5). The potential roles of gene polymorphisms (especially those affecting NQO1 (NADPH quinone oxidoreductase), CYP2E1, MPO, GSTM1, and GSTT1 genes and activity levels) have been much discussed in literatures on inter-individual variability in susceptibility to benzene exposures in humans and rodents. Although polymorphism data are often notoriously difficult to interpret reliably, due in part to multiple testing biases and latent (unobserved) covariates, both animal and human data have been interpreted as suggesting increased production of toxic metabolites associated with specific polymorphisms (e.g., Carrieri et al. 2018; Carbonari et al. 2016). For example, mouse data have led to the conclusion that NQO1 deficiency results in sex-specific increases in benzene toxicity, as assessed by induction of micronuclei in peripheral blood cells (Bauer et al. 2003). For human data, SPMA levels of men working in a chemical plant who had GSTT1-positive genotype were reported to be significantly higher than those of subjects with GSTT1-deficient genotype ($P = 0.041$) (Lin et al. 2008); and petrochemical plant employees with null GSTT1 or both null STT1 and GSTM1 genotypes had significantly reduced platelets, higher leukocyte (white blood cell) counts, and higher risks of hematological disorders than other employees (Nourozi et al. 2018).

Our data set for the Tianjin factory workers lacks polymorphism information, precluding a direct examination of how they might affect benzene metabolism. Yet, the extent of inter-individual heterogeneity in benzene metabolism among Tianjin factory workers exposed to low concentrations (<5 ppm) can be illuminated by statistical analyses. Clustering can be used to identify distinct “types” (clusters) of workers, defined by different patterns of metabolite production in response to AB, even though we cannot identify specific underlying polymorphisms or other unmeasured factors that determine these types. Figure 12.11 shows Individual Conditional Expectation (ICE) plots (Goldstein et al. 2015) for the production of SPMA, MA, UB, and HQ by workers as functions of AB concentration (as predicted by a *randomForest* analysis). Each of these ICE plots characterizes inter-individual heterogeneity

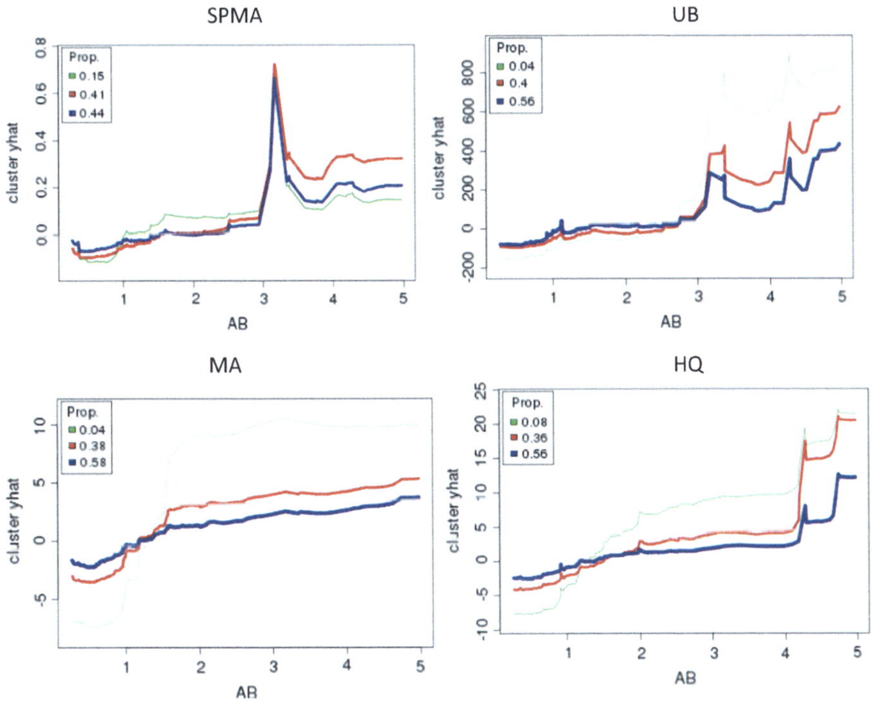

Fig. 12.11 Individual conditional expectation (ICE) plots for production of metabolites from low levels of air benzene exposure (AB). Fraction of worker population in each cluster is given by "Prop."(proportion) key in the upper left of each diagram. Vertical axes are scaled to show deviations from the overall population mean value. (Plots were generated by the CAT software using the ICEbox package in R (Goldstein et al. 2015). Plots control for individual gender, age, smoke, factory, weight, height, and AT by conditioning on their values in random forest analysis

by partitioning workers into 3 clusters based on their predicted AB-metabolite production functions. In contrast to PDPs showing how the *average* value of a dependent variable (e.g., quantity of a urinary metabolite) varies with an explanatory variable (e.g., AB in ppm), holding other variables at fixed values, an ICE plot disaggregates the average response by considering individual worker-level predictions, and then reaggregating them into distinct clusters for purposes of visualization (Goldstein et al. 2015). Figure 12.11 shows ICE plots for the four metabolites (SPMA, UB, MA, HQ) in Fig. 12.8, using 3 clusters in each plot. The proportion of workers in each cluster is indicated in the "Prop." box in the upper left of each plot. The clusters in Fig. 12.3 suggest that a relatively small proportion (4%) of workers produces more UB and MA on average (light green lines) than do other workers when exposed to higher AB concentrations (e.g., at AB = 4 ppm).

Figure 12.12 provides a more direct description of observed heterogeneity, without attempting to hold other factors fixed. In this plot, benzene exposure concentrations are rounded to the nearest whole number of ppm (0–5). For each of these 6 groups of AB concentrations, mean metabolite levels (dots) and standard deviations

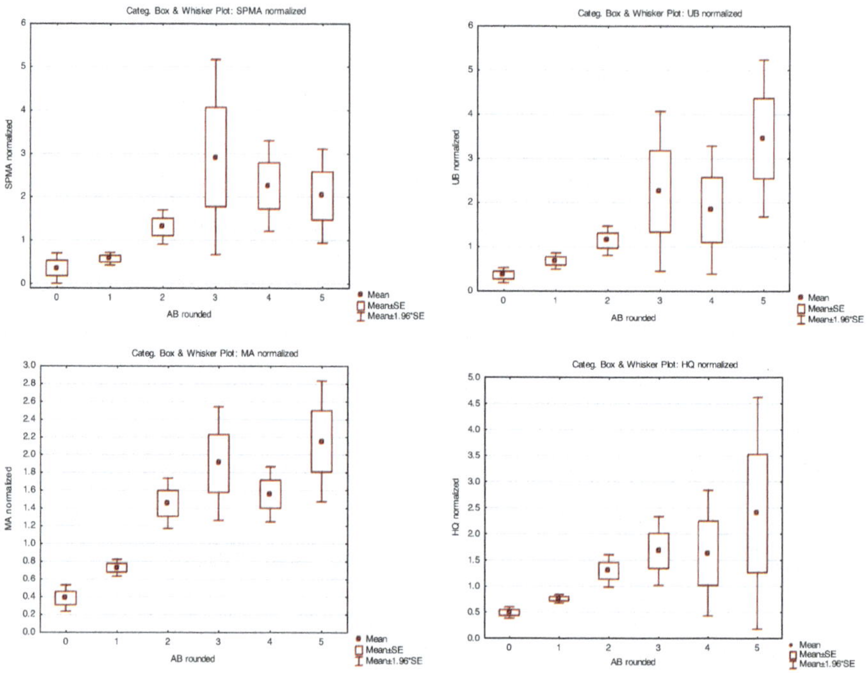

Fig. 12.12 Categorical box-and-whisker plots show substantial interindividual variability in production of metabolites (clockwise from upper left: normalized SPMA, UB, MA, HQ values, scaled so that population mean = 1 for each metabolite). Metabolite production is sub-linear at low doses (AB < 3 ppm) (i.e., successive increases in AB by 1 unit generate larger increases in metabolites)

(1 sd for the boxes, 2 sd for the whiskers) are displayed. Although this visualization does not capture the skewness of the underlying metabolite distributions (see histograms in Fig. 12.6), it does show that there is substantial inter-individual variability in metabolites produced by different individuals from approximately the same benzene exposure concentrations. The shapes of the plots at the lowest concentrations also show sub-linearity, i.e., the increases in metabolites produced are greater when AB increases from 1 to 2 than when it increases from 0 to 1.

Figure 12.13 depicts the joint values of multiple metabolites (UB and SPMA on the left, with a quadratic regression model (solid curve) and 95% normal confidence intervals shown as dashed curves); and UB, SPMA, and MA on the right), for workers exposed to 1 ppm of AM (rounded to the nearest ppm). These plots not only illustrate the wide inter-individual variability in metabolite levels for approximately the same AB exposure, but they also show that a few individuals have much higher levels of benzene metabolites than others. This interindividual variability is too great to be well modeled by a regression curve with a normally distributed, exposure-independent, additive error term. This highlights the value of using non-parametric methods, as in Figs. 12.8, 12.9, 12.10, and 12.11, to understand the conditional distributions of metabolites given AB levels.

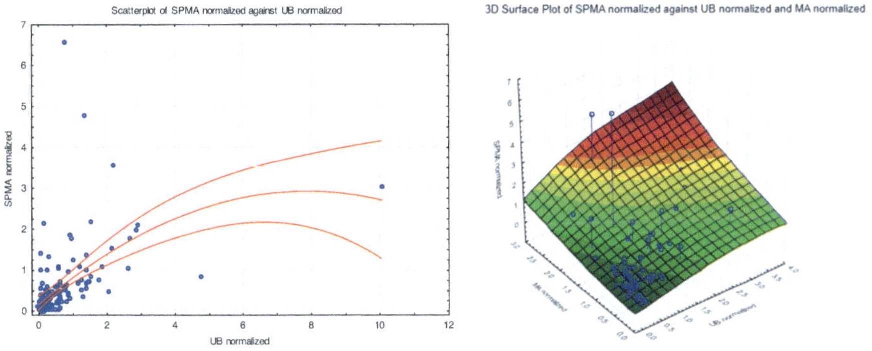

Fig. 12.13 Joint scatter plots for SPMA and UB (left) and MA, UB, and SPMA (right) for workers exposed to 1 ppm (rounded to the nearest ppm) of benzene in air (AB = 1)

Fig. 12.14 Women (upper curve) produce more SPMA than men at low concentrations (AB < 3 ppm) Vertical bars indicate approximate 95% confidence intervals

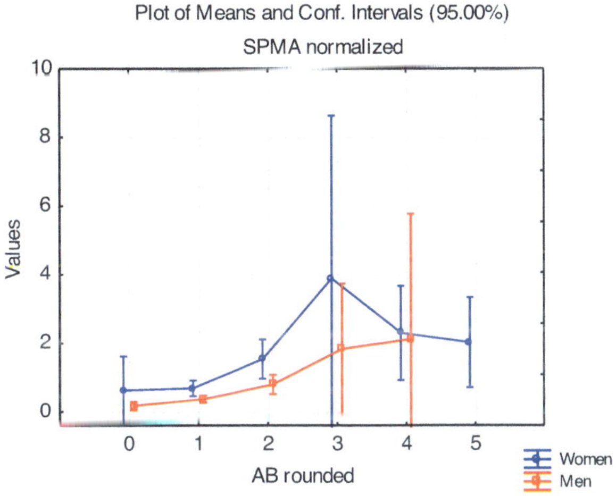

Consistent with earlier studies (Tranfo et al. 2018), Fig. 12.14 shows that women produce slightly more SPMA than men at low benzene concentrations; this explains the arrow from Gender to SPMA in the BN in Fig. 12.10. However, the effect of gender on metabolism is small compared to other inter-individual differences, as reflected in the wide vertical spread of metabolite levels for approximately the same AB concentrations (Fig. 12.12).

Intra-Individual Variability in Benzene Metabolism

The Tianjin factory worker contains repeated measurements of benzene exposure concentrations and urinary metabolite levels for some individual workers, with 263 data records for 177 individual workers. There were 4 measurement days (records in the data set) for 2 of the workers with low AB concentrations (<5 ppm); 3 records for each of another 11 of these workers; 2 records for 58 workers; and a single record for each of the remaining 106 workers. The repeated measurements provide an opportunity to study within-individual variability in exposure and metabolism.

Figure 12.15 plots the levels of SPMA, UB, and MA (top, middle, and bottom rows, respectively) at different AB levels, for the 2 individuals with 4 measurements each. (Note the very different vertical scales in these plots.) Worker 2 (right side) has close to twice the peak metabolite levels of Worker 1 (left side) for each

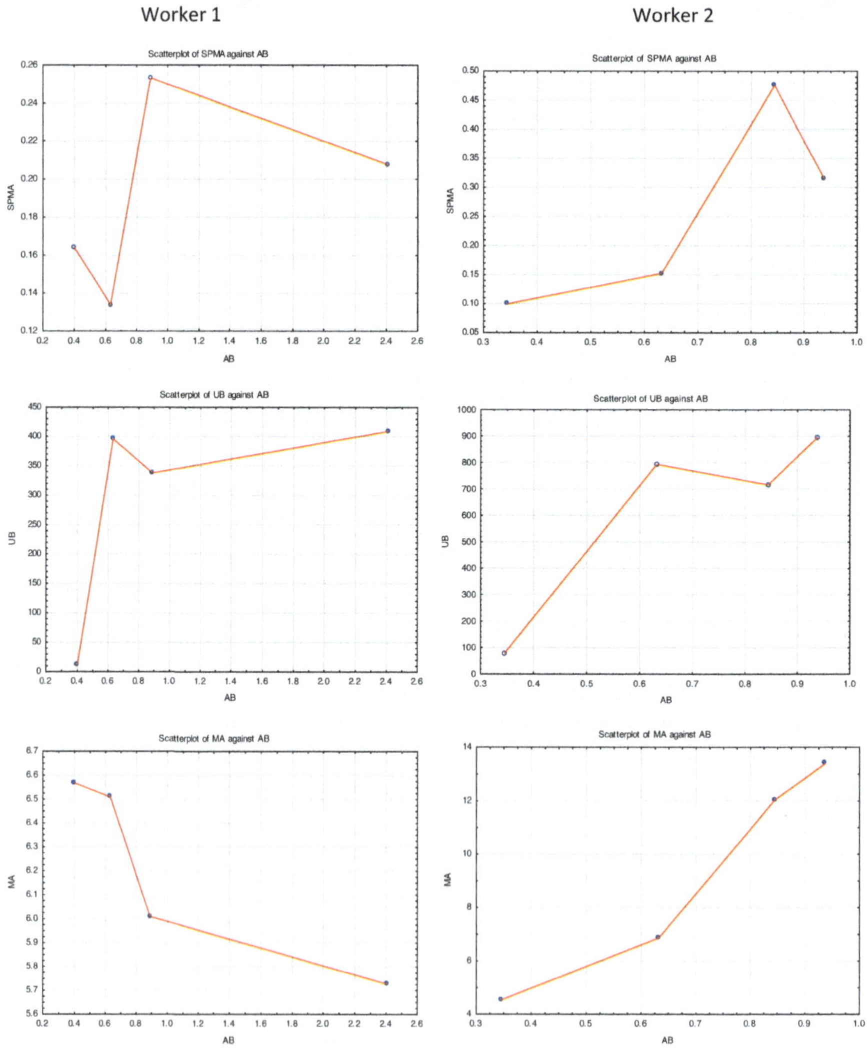

Fig. 12.15 Metabolites vs. AB (ppm) for 2 individual workers, each with 4 repeated measurements (Top row SPMA; middle row UB; bottom row MA)

metabolite, even though Worker 1 has a higher maximum benzene exposure concentration (2.4 ppm instead of less than 1 ppm). However, neither individual has uniformly higher levels of metabolites than the other across all four repeated measurements and AB levels (although Worker 2 does have higher levels of UB than Worker 1, even at lower AB exposures). While a study of only 2 individual workers with 4 repeated measurements each does not permit confident generalizations about intra-individual variability in benzene metabolism, the types of variability illustrated in Fig. 12.15 (including some curves in which metabolites decrease as AB increases, as in the slight decrease in MA shown in the lower left of Fig. 12.15) also occur for the 11 workers with 3 repeated measurements each. However, the effects of intra-individual variability in metabolism are difficult to discern, since the workers received different air benzene concentrations on different days. Despite these limitations, Fig. 12.14 suggests that, for the worker with relatively high metabolite formation (Worker 2), the levels of all three metabolites were much more than twice as great at AB concentrations between 0.8 ppm and 1 ppm (two right-most points in each plot) as at AB concentrations of about 0.35 ppm (leftmost point in each plot).

Levels of Benzene Metabolites and Markers in Relation to Increased Health Risks

Average levels of urinary benzene metabolites increase with increasing AB in the population of Chinese factory workers exposed to relatively low concentrations of benzene in air (AB < 5 ppm), even though there is considerable inter-individual variability in the quantity of metabolites formed for any given level of AB (Fig. 12.12). The question of what levels of metabolites (or of other markers with which they are correlated, such as 8-OHdG, various adducts, or micronuclei induction in peripheral blood lymphocytes) correspond to increased risk of MDS and AML has been much debated, with substantial differences in conclusions. For example:

- Vlaanderen et al. (2010) fit non-parametric (spline) smooth curves to heterogeneous data from 9 human studies of estimated cumulative benzene exposures and leukemia. They concluded that the best-fitting curves were linear or super-linear at low exposures, and that there was significantly increased risk of leukemia [relative risk (RR) = 1.14; 95% confidence interval (CI), 1.04–1.26] at estimated exposure levels as low as 10 ppm-years. Although these analyses did not adjust for errors in exposure estimates and did not address peak exposures, they are generally consistent both with earlier studies that found no threshold for estimated cumulative benzene exposures associated with increased risk of AML and other diseases (e.g., Glass et al. 2003) and also with subsequent reviews that accept the plausibility of leukemogenic effects of benzene even at low cumulative exposures, such as 0.5 ppm-years (Yoon et al. 2018).

- By contrast, Schnatter et al. (2012) combined data from 3 nested case-control studies of petroleum distribution workers from Australia, Canada, and the United Kingdom, and estimated that risk of MDS, but not of AML or other leukemias or myeloproliferative disorders, increased substantially (estimated odds ratio in a conditional logistic regression model was OR = 4.33) for workers with estimated cumulative benzene exposures > 2.9 ppm-years compared to workers with estimated cumulative benzene exposures < 0.35 ppm-years (highest vs. lowest tertiles, respectively). MDS risk was increased specifically in workers with estimated peak exposures >3 ppm. However, this analysis also used reconstructed exposures with unknown errors, and the error distributions were not modeled, potentially leading to misestimates of the true peak exposure concentrations that are associated with increased MDS risk. The authors suggest that MDS, rather than AML or other endpoints, may be the most relevant health risk at relatively low exposure concentrations.

The analyses in this chapter suggest a different perspective from either of these on the human health risks caused by low exposure concentrations of benzene, as follows.

- If risk of MDS and other diseases depend on chronic inflammation in the bone marrow (mediated by activation of the NLRP3 inflammasome) (Sallman and List 2019), then *cumulative exposure (ppm-years) is not an adequate dose metric* for predicting risk. Instead, risk is determined by the time until exposure causes chronic inflammation.
- PBPK models imply that, for a steady administered concentration of air benzene (AB), concentrations of metabolites in bone marrow increase until a dynamic equilibrium level is reached at which the rate of each metabolite's removal equals the rate of its addition to the marrow (via distribution from the liver and via local metabolism in the marrow). If exposure concentration of benzene in inhaled air is low enough, then the dynamic equilibrium concentration will be below the inflammasome activation threshold (or the smallest such threshold, if values are heterogeneous). Inflammation, and resulting inflammation-mediated increases in MDS and other disease risks, will not be triggered, and extending the duration of such a low-level exposure increases cumulative exposure (ppm-years) without increasing risk. It is not necessary to know which toxic metabolite(s) are involved in the pathogenesis of MDS and AML to reach this qualitative conclusion.
- Conversely, if the PBPK dynamic equilibrium for a given AB concentration is only slightly below the activation threshold, then twice this concentration administered for half as many years might quickly trigger inflammation, even though the original exposure (with the same ppm-years) did not. Cumulative exposure received after chronic inflammation starts may be nearly irrelevant if the chronic inflammation itself dominates subsequent risk.
- Even if inflammation is not the rate-limiting key event for increasing risk of MDS and AML, cumulative exposure is still a highly misleading metric for realistic past exposures involving AB concentrations in excess of about 20 ppm. Although we have focused so far on workers exposed to <5 ppm, Fig. 12.16

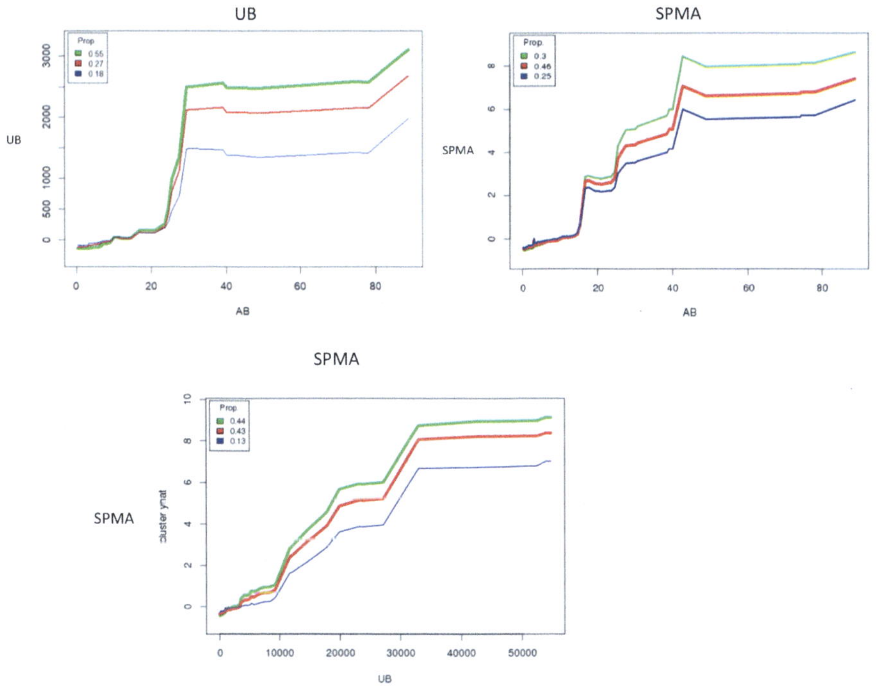

Fig. 12.16 (Top) Sub-linear production of benzene metabolites (UB on left; SPMA on right) by workers (partitioned into three clusters to indicate inter-individual heterogeneity), given AB. (Bottom) Prediction of SPMA from UB. (Plots control for individual gender, age, smoke, factory, weight, height, and AT by conditioning on their values in random forest analysis)

shows a wider perspective by plotting UB (in nM) and SPMA (μM) against AB for each of three clusters of workers in an individual conditional expectation (ICE) plot. The differences in metabolite production that we have been examining for AB < 5 ppm are almost negligible, on this expanded scale, compared to the main pattern of approximately linear metabolism at AB concentrations below about 15 ppm and a steep increase in production per unit of AB above that level, for all three clusters (light, medium, and heavy metabolizers).

- The bottom panel of Fig. 12.16 shows that SPMA can also be predicted from UB instead of AB. If UB is used instead of AB as a predictor of SPMA (along with the other variables in Fig. 12.16) in random forest analysis, the percentage of variance in SPMA that is explained increases from about 36% to about 41% (and to 47% if UB and AB are both used). A possible explanation is that urinary benzene (UB) reflects inter-individual variability in metabolism or in other exposure routes (e.g., dermal) that air benzene (AB) measurements do not address; differences in measurement error variances between UB and AB might also play a role.

- The steeply sub-linear (J-shaped) functions for metabolism in the top panels of Fig. 12.16 imply that a worker exposed to an average of 10 ppm-years consisting of 3 months a year of 40 ppm and 9 months per year of 0 ppm would have far higher average concentrations of metabolites (averaged over the year) than a similar worker exposed to 10 ppm throughout the year. Cumulative exposure and peak exposure both fail to capture the most relevant information for risk prediction, which is how long the worker spends exposed to AB concentrations sufficient to produce harmful (risk-increasing) concentrations of metabolites in the bone marrow.

From this perspective, mixed and contradictory results are to be expected from past studies and reviews that use cumulative exposure to benzene as a predictor of risk. The same cumulative exposure may be harmful or harmless, depending on how it is distributed over time.

The same reasoning implies that errors in exposure estimates can play a key role in determining the risk attributed to a given average exposure concentration. For example, the data and BN model predictions of Hack et al. (2010) indicate that AML risk increases for benzene concentrations above a level estimated as between 1 ppm and 10 ppm. For workers with occasional exposures to concentrations above 25 ppm, but estimated average exposures between 1 and 10 ppm, the risk may be dominated by metabolites formed during occasional exposures to higher concentrations (Fig. 12.16), although it is attributed to the lower average exposure concentrations. These considerations, if correct, imply that many previous analyses of benzene dose-response that use estimated cumulative exposure (ppm-years) to predict AML or MDS risk, without characterizing variability of exposure concentrations around their estimated values, lose essential information on which risk depends. Figure 12.16 suggests that a better understanding of benzene-associated risk can be achieved by considering the fraction of a worker's time spent exposed to relatively high concentrations (e.g., AB > 15 ppm), as this may dominate the contribution of exposure to formation of toxic metabolites.

Conclusions

Recent literature has raised the question of whether dose-response relationships for benzene are supra-linear, creating disproportionately high risks per ppm-year of exposure for exposure concentrations below 1 ppm (Lanphear 2017; Rappaport et al. 2010). We have examined this hypothesis using findings from recent literature on mode of action (MoA) and mechanistic considerations for benzene-induced myelodysplastic syndrome (MDS) and acute myeloid leukemia (AML) and by applying modern data science techniques to quantify and visualize relationships between air benzene and production of toxic metabolites in previously studied data for Tianjin factory workers. Our main conclusions from these analyses are as follows:

a. Physiologically-based pharmacokinetics (PBPK) modeling predicts that production of toxic metabolites is approximately linear (i.e., proportional to benzene concentration in inhaled air), or mildly sublinear, at low exposure concentrations (Fig. 12.3).
b. Bayesian network (BN) modeling suggests a sublinear (J-shaped) relationship between estimated air benzene (AB) concentrations and AML risk (Hack et al. 2010).
c. Data for Tianjin workers shows approximately linear metabolism at low concentrations (AB < 5 ppm), i.e., production of urinary metabolites is approximately proportional to concentration of benzene in inhaled air (Fig. 12.5). This linearity holds over nearly 2 orders of magnitude, from the lowest measured concentrations (about 0.2 ppm) to well over 10 ppm. At concentrations below 3 ppm, the production of metabolites is mildly sublinear (Fig. 12.12).
d. The production of metabolites at higher AB concentrations increases disproportionately before becoming saturated (Fig. 12.16). This implies that cumulative exposure (ppm-years) is not a sufficient summary of exposure from which to predict risk, which may help to explain conflicting conclusions in the literature.
e. The steeply sublinear shape ("elbow") at the left end of the metabolite-vs. AB curve (Fig. 12.16) also makes it important for risk assessments to consider the *variability* of exposure concentrations around estimated exposure concentrations. Analyses in which a mix of relatively high (e.g., AB > 15 ppm) and low (e.g., AB < 10 ppm) concentrations are summarized in a single cumulative exposure (ppm-year) estimate for each worker are likely to over-estimate low-concentration risk, by attributing to them adverse effects that are created primarily by the high concentrations.
f. If chronic inflammation via activation of the NLRP3 inflammasome is a critical event for induction of MDS by benzene (Sallman and List 2019), then PBPK considerations imply that sufficiently low AB concentrations will not create concentrations of toxic metabolites that exceed the NLRP3 inflammasome activation threshold, and hence they will not increase the risk of MDS, no matter how long the exposure duration is. This illustrates why ppm-years is not adequate as a predictor of risk: twice the ppm for half as many years might cause MDS even if lower concentrations for longer do not. Thus, the same ppm-years number can be harmless or harmful, depending on how AB concentrations are distributed over time.
g. If MDS is the most sensitive adverse health effect endpoint at relatively low exposure concentrations (Schnatter et al. 2012), then an exposure concentration threshold may exist for increases in risk due to benzene exposure. However, the use of cumulative exposures and reconstructed (uncertain) exposure estimates would obscure any such threshold. In addition, the substantial inter-individual variability in benzene metabolism (Figs. 12.11, 12.12, 12.13, and 12.14) implies that any exposure concentration threshold for AB has a distribution in a worker population.

Overall, the data and MoA literature we have examined do not support a hypothesis of supralinearity of risks at low benzene concentrations. However, they do suggest opportunities to improve benzene risk assessments by focusing on the proportion and timing of low and high exposure concentrations in each worker's exposure history, rather than combining these into a single cumulative exposure metric. For practical methods, pharmacokinetic (PBPK) modeling or PDP curves (such as Fig. 12.16) can be used to convert time series of approximate exposure concentrations to corresponding time series of corresponding benzene metabolites. Identifying the time until inflammation is triggered (as in Chap. 6), and using Bayesian networks developed from data (e.g., Fig. 12.10) to estimate increases in risk during periods of increased inflammatory markers, while controlling for other observed variables (e.g., age, sex, smoking, diet, polymorphisms, and other markers if available) appears to be a promising and practical approach for using current data science methods to advance benzene risk modeling.

References

Anderson LA, Pfeiffer RM, Landgren O, Gadalla S, Berndt SI, Engels EA. Risks of myeloid malignancies in patients with autoimmune conditions. Br J Cancer. 2009;100(5):822–8. https://doi.org/10.1038/sj.bjc.6604935.

Arnold SM, Angerer J, Boogaard PJ, Hughes MF, O'Lone RB, Robison SH, Schnatter AR. The use of biomonitoring data in exposure and human health risk assessment: benzene case study. Crit Rev Toxicol. 2013;43(2):119–53. https://doi.org/10.3109/10408444.2012.756455.

Bauer AK, Faiola B, Abernethy DJ, Marchan R, Pluta LJ, Wong VA, Roberts K, Jaiswal AK, Gonzalez FJ, Butterworth BE, Borghoff S, Parkinson H, Everitt J, Recio L. Genetic susceptibility to benzene-induced toxicity: role of NADPH: quinone oxidoreductase-1. Cancer Res. 2003;63(5):929–35.

Bechtold WE, Henderson RF. Biomarkers of human exposure to benzene. J Toxicol Environ Health. 1993;40(2-3):377–86.

Bogen KT. Inflammation as a cancer co-initiator: new mechanistic model predicts low/negligible risk at noninflammatory carcinogen doses. Dose-Response. 2019;17(2):1559325819847834. https://doi.org/10.1177/1559325819847834.

Campagna M, Satta G, Campo L, Flore V, Ibba A, Meloni M, Tocco MG, Avataneo G, Flore C, Fustinoni S, Cocco P. Biological monitoring of low-level exposure to benzene. Med Lav. 2012;103(5):338–46.

Carbonari D, Chiarella P, Mansi A, Pigini D, Iavicoli S, Tranfo G. Biomarkers of susceptibility following benzene exposure: influence of genetic polymorphisms on benzene metabolism and health effects. Biomark Med. 2016;10(2):145–63. https://doi.org/10.2217/bmm.15.106.

Carrieri M, Spatari G, Tranfo G, Sapienza D, Scapellato ML, Bartolucci GB, Manno M. Biological monitoring of low level exposure to benzene in an oil refinery: Effect of modulating factors. Toxicol Lett. 2018;2018:378. https://doi.org/10.1016/j.toxlet.2018.08.001.

Cox LA Jr. Modernizing the Bradford Hill criteria for assessing causal relationships in observational data. Crit Rev Toxicol. 2018;15:1–31. https://doi.org/10.1080/10408444.2018.1518404.

Cox LAT Jr. Risk analysis implications of dose-response thresholds for NLRP3 inflammasome-mediated diseases: respirable crystalline silica and lung cancer as an example. Dose-Response. 2019;17(2):1559325819836900. https://doi.org/10.1177/1559325819836900.

Cox LA, Schnatter AR, Boogaard PJ, Banton M, Ketelslegers HB. Non-parametric estimation of low-concentration benzene metabolism. Chem Biol Interact. 2017;278:242–55. https://doi.org/10.1016/j.cbi.2017.08.007.

Das M, Chaudhuri S, Law S. Benzene exposure--an experimental machinery for induction of myelodysplastic syndrome: stem cell and stem cell niche analysis in the bone marrow. J Stem Cells. 2012;7(1):43–59.

Du X, Jiang S, Zeng X, Zhang J, Pan K, Song L, Zhou J, Kan H, Sun Q, Zhao J, Xie Y. Fine particulate matter-induced cardiovascular injury is associated with NLRP3 inflammasome activation in Apo $E_{-/-}$ mice. Ecotoxicol Environ Saf. 2019;174:92–9. https://doi.org/10.1016/j.ecoenv.2019.02.064.

Fustinoni S, Campo L, Mercadante R, Consonni D, Mielzynska D, Bertazzi PA. A quantitative approach to evaluate urinary benzene and S-phenylmercapturic acid as biomarkers of low benzene exposure. Biomarkers. 2011;16(4):334–45. https://doi.org/10.3109/1354750X.2011.561499.

Galbraith D, Gross SA, Paustenbach D. Benzene and human health: a historical review and appraisal of associations with various diseases. Crit Rev Toxicol. 2010;40(Suppl 2):1–46. https://doi.org/10.3109/10408444.2010.508162.

Glass DC, Gray CN, Jolley DJ, Gibbons C, Sim MR, Fritschi L, Adams GG, Bisby JA, Manuell R. Leukemia risk associated with low-level benzene exposure. Epidemiology. 2003;14(5):569–77.

Goldstein A, Kapelner A, Bleich J, Pitkin E. Peeking inside the black box: visualizing statistical learning with plots of individual conditional expectation. J Comput Graph Stat. 2015;24(1):44–65. https://doi.org/10.1080/10618600.2014.907095.

Grigoryan H, Edmands WMB, Lan Q, Carlsson H, Vermeulen R, Zhang L, Yin SN, Li GL, Smith MT, Rothman N, Rappaport SM. Adductomic signatures of benzene exposure provide insights into cancer induction. Carcinogenesis. 2018;39(5):661–8. https://doi.org/10.1093/carcin/bgy042.

Gross SA, Paustenbach DJ. Shanghai health study (2001-2009): what was learned about benzene health effects? Crit Rev Toxicol. 2018;48(3):217–51. https://doi.org/10.1080/10408444.2017.1401581.

Guo X, Zhong W, Chen Y, Zhang W, Ren J, Gao A. Benzene metabolites trigger pyroptosis and contribute to haematotoxicity via TET2 directly regulating the Aim2/Casp1 pathway. EBioMedicine. 2019;47:578–89. https://doi.org/10.1016/j.ebiom.2019.08.056.

Hack CE, Haber LT, Maier A, Shulte P, Fowler B, Lotz WG, Savage RE Jr. A Bayesian network model for biomarker-based dose response. Risk Anal. 2010;30(7):1037–51. https://doi.org/10.1111/j.1539-6924.2010.01413.x.

Hamling JS, Coombs KJ, Lee PN. Misclassification of smoking habits: an updated review of the literature. World J Meta Anal. 2019;7(2):31–50. https://doi.org/10.13105/wjma.v7.i2.31.

Hirabayashi Y, Inoue T. Benzene-induced bone-marrow toxicity: a hematopoietic stem cell specific, aryl hydrocarbon receptor-mediated adverse effect. Chem Biol Interact. 2010;184(1 2):252–8. https://doi.org/10.1016/j.cbi.2009.12.022.

Hornung RW, Lanphear BP. The supralinear dose-response for environmental toxicants: a statistical artifact? Clin Toxicol. 2014;52(2):88–90. https://doi.org/10.3109/15563650.2013.87894.

Kerzic PJ, Irons RD. Distribution of chromosome breakpoints in benzene-exposed and unexposed AML patients. Environ Toxicol Pharmacol. 2017;55:212–6. https://doi.org/10.1016/j.etap.2017.08.033.

Knutsen JS, Kerger BD, Finley B, Paustenbach DJ. A calibrated human PBPK model for benzene inhalation with urinary bladder and bone marrow compartments. Risk Anal. 2013;33(7):1237–51. https://doi.org/10.1111/j.1539-6924.2012.01927.x.

Lanphear BP. Low-level toxicity of chemicals: no acceptable levels? PLoS Biol. 2017;15(12):e2003066. https://doi.org/10.1371/journal.pbio.2003066.

Lin LC, Chen WJ, Chiung YM, Shih TS, Liao PC. Association between GST genetic polymorphism and dose-related production of urinary benzene metabolite markers, trans, trans-muconic acid and S-phenylmercapturic acid. Cancer Epidemiol Biomark Prev. 2008;17(6):1460–9. https://doi.org/10.1158/1055-9965.EPI-08-0160.

McHale CM, Zhang L, Smith MT. Current understanding of the mechanism of benzene-induced leukemia in humans: implications for risk assessment. Carcinogenesis. 2012;33(2):240–52. https://doi.org/10.1093/carcin/bgr297.

McNally K, Sams C, Loizou GD, Jones K. Evidence for non-linear metabolism at low benzene exposures? A reanalysis of data. Chem Biol Interact. 2017;278:256–68. https://doi.org/10.1016/j.cbi.2017.09.002.

Meek ME, Klaunig JE. Proposed mode of action of benzene-induced leukemia: interpreting available data and identifying critical data gaps for risk assessment. Chem Biol Interact. 2010;184(1-2):279–85. https://doi.org/10.1016/j.cbi.2010.02.006.

Neumann HG. Risk assessment of chemical carcinogens and thresholds. Crit Rev Toxicol. 2009;39(6):449–61. https://doi.org/10.1080/10408440902810329.

Nourozi MA, Neghab M, Bazzaz JT, Nejat S, Mansoori Y, Shahtaheri SJ. Association between polymorphism of GSTP1, GSTT1, GSTM1 and CYP2E1 genes and susceptibility to benzene-induced hematotoxicity. Arch Toxicol. 2018;92(6):1983–90. https://doi.org/10.1007/s00204-017-2104-9.

Price PS, Rey TD, Fontaine DD, Arnold SM. A reanalysis of the evidence for increased efficiency in benzene metabolism at airborne exposure levels below 3 p.p.m. Carcinogenesis. 2012;33(11):2094–9. https://doi.org/10.1093/carcin/bgs257.

Qian S, Han Y, Shi Y, Xu W, Zhu Y, Jiang S, Chen Y, Yu Z, Zhang S, Yang Y, Yu K, Zhang S. Benzene induces haematotoxicity by promoting deacetylation and autophagy. J Cell Mol Med. 2019;23(2):1022–33. https://doi.org/10.1111/jcmm.14003.

Qu Q, Shore R, Li G, Jin X, Chen LC, Cohen B, Melikian AA, Eastmond D, Rappaport S, Li H, Rupa D, Waidyanatha S, Yin S, Yan H, Meng M, Winnik W, Kwok ES, Li Y, Mu R, Xu B, Zhang X, Li K. Validation and evaluation of biomarkers in workers exposed to benzene in China. Res Rep Health Eff Inst. 2003;115:1–72.

Rappaport SM, Kim S, Lan Q, Li G, Vermeulen R, Waidyanatha S, Zhang L, Yin S, Smith MT, Rothman N. Human benzene metabolism following occupational and environmental exposures. Chem Biol Interact. 2010;184(1-2):189–95. https://doi.org/10.1016/j.cbi.2009.12.017.

Rappaport SM, Kim S, Thomas R, Johnson BA, Bois FY, Kupper LL. Low-dose metabolism of benzene in humans: science and obfuscation. Carcinogenesis. 2013;34(1):2–9. https://doi.org/10.1093/carcin/bgs382.

Richardson DB. Multistage modeling of leukemia in benzene workers: a simple approach to fitting the 2-stage clonal expansion model. Am J Epidemiol. 2009;169(1):78–85. https://doi.org/10.1093/aje/kwn284.

Rushton L, Schnatter AR, Tang G, Glass DC. Acute myeloid and chronic lymphoid leukaemias and exposure to low-level benzene among petroleum workers. Br J Cancer. 2014;110(3):783–7. https://doi.org/10.1038/bjc.2013.780.

Sallman DA, List A. The central role of inflammatory signaling in the pathogenesis of myelodysplastic syndromes. Blood. 2019;133(10):1039–48. https://doi.org/10.1182/blood-2018-10-844654.

Sallman DA, Cluzeau T, Basiorka AA, List A. Unraveling the pathogenesis of MDS: the NLRP3 inflammasome and pyroptosis drive the MDS phenotype. Front Oncol. 2016;6:151. https://doi.org/10.3389/fonc.2016.00151.

Schnatter AR, Glass DC, Tang G, Irons RD, Rushton L. Myelodysplastic syndrome and benzene exposure among petroleum workers: an international pooled analysis. J Natl Cancer Inst. 2012;104(22):1724–37. https://doi.org/10.1093/jnci/djs411.

Slenter DN, Kutmon M, Hanspers K, Riutta A, Windsor J, Nunes N, Mélius J, Cirillo E, Coort SL, Digles D, Ehrhart F, Giesbertz P, Kalafati M, Martens M, Miller R, Nishida K, Rieswijk L, Waagmeester A, Eijssen LMT, Evelo CT, Pico AR, Willighagen EL. WikiPathways: a multifaceted pathway database bridging metabolomics to other omics research. Nucleic Acids Res. 2017;46(1):661–7.

Smith MT. Overview of benzene-induced aplastic anaemia. Eur J Haematol Suppl. 1996;60:107–10.

Snyder R. Leukemia and benzene. Int J Environ Res Public Health. 2012;9(8):2875–93. https://doi.org/10.3390/ijerph9082875.

Son MY, Deng CX, Hoeijmarkers JH, Rebel VI, Hasty P. A mechanism for 1,4-benzoquinone-induced genotoxicity. Oncotarget. 2016;7(29):46433–47. https://doi.org/10.18632/oncotarget.10184.

Teras LR, Diver WR, Deubler EL, Krewski D, Flowers CR, Switchenko JM, Gapstur SM. Residential ambient benzene exposure in the United States and subsequent risk of hematologic malignancies. Int J Cancer. 2019;145(10):2647–60. https://doi.org/10.1002/ijc.32202.

Titmarsh GJ, McMullin MF, McShane CM, Clarke M, Engels EA, Anderson LA. Community-acquired infections and their association with myeloid malignancies. Cancer Epidemiol. 2014;38(1):56–61. https://doi.org/10.1016/j.canep.2013.10.009.

Tranfo G, Pigini D, Paci E, Bauleo L, Forastiere F, Ancona C. Biomonitoring of urinary benzene metabolite SPMA in the general population in Central Italy. Toxics. 2018;6(3):E37. https://doi.org/10.3390/toxics6030037.

Vlaanderen J, Portengen L, Rothman N, Lan Q, Kromhout H, Vermeulen R. Flexible meta-regression to assess the shape of the benzene-leukemia exposure-response curve. Environ Health Perspect. 2010;118(4):526–32. https://doi.org/10.1289/ehp.0901127.

Waddell WJ. Critique of dose response in carcinogenesis. Hum Exp Toxicol. 2006;25(7):413–36.

Wang L, He X, Bi Y, Ma Q. Stem cell and benzene-induced malignancy and hematotoxicity. Chem Res Toxicol. 2012;25(7):1303–15. https://doi.org/10.1021/tx3001169.

Wei J, Wang H, Wang H, Wang B, Meng L, Xin Y, Jiang X. The role of NLRP3 inflammasome activation in radiation damage. Biomed Pharmacother. 2019;118:109217. https://doi.org/10.1016/j.biopha.2019.109217.

Weisel CP. Benzene exposure: an overview of monitoring methods and their findings. Chem Biol Interact. 2010;184(1-2):58–66. https://doi.org/10.1016/j.cbi.2009.12.030.

Yokley K, Tran HT, Pekari K, Rappaport S, Riihimaki V, Rothman N, Waidyanatha S, Schlosser PM. Physiologically-based pharmacokinetic modeling of benzene in humans: a Bayesian approach. Risk Anal. 2006;26(4):925–43.

Yoon JH, Kwak WS, Ahn YS. A brief review of relationship between occupational benzene exposure and hematopoietic cancer. Ann Occup Environ Med. 2018;30:33. https://doi.org/10.1186/s40557-018-0245-9.

Part III
Public Health Effects of Fine Particulate Matter Air Pollution

Chapter 13
Socioeconomic Correlates of Air Pollution and Heart Disease

Introduction

This chapter begins an investigation that occupies the rest of the book: examining how the causal analysis methods discussed in Part 2, along with the methodological points about interpretation of regression coefficients raised in earlier chapters, especially Chaps. 1, 2, 7, and 8, can be applied to the important public health topic of health risks caused by exposure to fine particulate matter (PM2.5) air pollution. The specific concept of causation of greatest interest is manipulative or interventional causation (Chap. 9), addressing how changing air pollution would change death and disease rates. This contrasts with other concepts of causation that seek to attribute some number of "premature" deaths or diseases per year to air pollution without addressing the extent, if any, to which these attributed deaths would be prevented by reducing pollution levels (see Chaps. 9 and 18).

This chapter begins the exploration of PM2.5-health risk associations and causality by examining how risk of heart disease depends on age, sex, smoking, income, education, marital status, and outdoor concentrations of fine particulate matter (PM2.5). To do so, it joins data available from the Centers for Disease Control and Prevention (CDC) behavioral risk factor surveillance (BRFSS) system for years 2008–2012 to US Environmental Protection Agency (EPA) data on county-specific concentrations of PM2.5 to quantify associations among these variables and to explore possible causal interpretations. Low income is identified as a direct cause of increased heart disease risk in this data set. The effect depends on age and sex: it is most pronounced for men under age 70 and for women under age 80. Income is significantly associated with all of the other variables examined and confounds the association between PM2.5 and heart disease risk. This association is significant in regression models that exclude income, but not in regression models that include it, both in the data set as a whole and in the subset of observations with PM2.5 < 15 μg/m^3. Applying causal directed acyclic graph (DAG) models and non-parametric

L. A. Cox Jr., *Quantitative Risk Analysis of Air Pollution Health Effects*, International Series in Operations Research & Management Science 299,
https://doi.org/10.1007/978-3-030-57358-4_13

model ensemble partial dependence plots from Chap. 9 confirms that higher incomes reduce heart disease risk, consistent with previous observations of socioeconomic gradients in health risks. They support interpretation of this as a robust causal relation apparent in non-parametric analyses, and hence independent of any specific parametric modeling assumptions.

Socio-economic variables are among the most important drivers of population health identified in many international studies. For example, the World Health Organization (WHO) lists social and economic environment as a leading determinant of health, followed by the physical environment and a person's individual characteristics and behaviors (WHO 2018). Chan et al. (2015) reported that, in Canada, low values of an index of socioeconomic status (SES) are associated with both fine particulate matter (PM2.5) exposure and low birth weights and related outcomes. In a Multi-Ethnic Study of Atherosclerosis and Air Pollution (MESA Air) in the United States, low neighborhood SES was associated with both higher PM2.5 exposures and increased risk of adverse cardiovascular health outcomes (Kaufman et al. 2016). A nine-country study in South America found that people with lower SES have increased risk of mortality from respiratory illnesses such as chronic obstructive pulmonary disease (COPD) (Romieu et al. 2012). A recent examination of significant statistical associations among socioeconomic variables, pollution levels, and self-reported histories of asthma, stroke, and heart attack in the United States found that higher income levels are associated with lower PM2.5 (fine particulate matter) exposure concentrations and with lower heart attack risks than are lower income levels; and that PM2.5 is significantly positively associated with heart attack risk when income is not included as a predictor, but not when it is included (Cox Jr. 2017). These patterns suggest that income may be an important confounder of PM2.5-heart attack associations. This chapter follows up on those observations by considering heart disease rather than heart attacks. It pays special attention to PM2.5-heart disease associations at exposure concentrations below 15 $\mu g/m^3$, the current secondary National Ambient Air Quality Standard (NAAQS). Its main goal is to quantify associations among socioeconomic variables (especially income), PM2.5 concentrations, and self-reported heart disease outcomes in a large population of adults 50 years or older for whom survey data are available from the Centers for Disease Control and Prevention (CDC) Behavioral Risk Factor Surveillance (BRFSS) System. A secondary goal is to examine possible causal interpretations of these associations using causal Bayesian networks and non-parametric machine-learning algorithms from Chap. 9.

Data Sources

The data set has been described previously (Cox Jr. 2017) and can be downloaded from http://cox-associates.com/CausalAnalytics/BRFSS.csv. Here, we briefly summarize its main features. Table 13.1 shows the layout of the data (the first 15 of 216,207 records) for individual respondents. The three categorical variables *Income*,

Table 13.1 Layout of the data, showing values of variables for the first 15 individual respondents

	1 State	2 County name	3 Year	4 Month	5 Age	6 Sex	7 Education	8 FluShot	9 HeartDisease Ever	10 Hispanic	11 Income	12 Marital status	13 Smoking	14 PM2.5
1	AZ	Apache	2010	1	95	0	4	0	0	0		3	1	1.8
2	AZ	Apache	2010	1	73	0	6	1	0	0	8	1	1	1.8
3	AZ	Apache	2010	1	66	0	5	1	0	0		1	0	1.8
4	AZ	Apache	2010	1	75	1	5	1	0	0	3	2	1	1.8
5	AZ	Apache	2010	1	70	0	4	1	0	0	4	3	1	1.8
6	AZ	Apache	2010	1	68	0	5	1	0	0	5	3	1	1.8
7	AZ	Apache	2010	1	72	0	6	1	0	0		1	0	1.8
8	AZ	Apache	2010	1	66	1	6	0	0	0	7	1	0	1.8
9	AZ	Apache	2010	1	52	0	5	1	0	0		3	1	1.8
10	AZ	Apache	2010	1	53	0	6	0	0	0	8	1	1	1.8
11	AZ	Apache	2010	1	75	0	5	1	0	0	3	3	0	1.8
12	AZ	Apache	2010	1	53	0	4	1	0	0	6	3	0	1.8
13	AZ	Apache	2010	1	63	0	5	1	0	0	7	1	0	1.8
14	AZ	Apache	2010	1	92	0	5	1	0	0		2	0	1.8
15	AZ	Apache	2010	1	69	0	4	0	0	0	3	2	0	1.8

Each column represents a variable, and each row contains the data for one respondent. The variables have various scale types. *Sex* (0 = female, 1 = male), *Smoking* (0 = no, 1 = yes), *Hispanic*, *Flu Shot*, and *HeartDiseaseEver* (0 = no, 1 = yes) are binary variables. *Income* and *Education* are ordered categorical variables, *Marital Status* is a nominal variable, and *PM2.5* is continuous

Education, and *Marital Status* have integer values for responses of 1–8, 1–6, and 1–6, respectively, with higher numerical values representing higher levels for *Income* and *Education*. *Smoking* is a binary variable that indicates whether a respondent reports having smoked at least 100 cigarettes (5 packs) during his or her life to date. The binary dependent variable *HeartDiseaseEver* indicates whether a doctor, nurse, or other health professional had ever told the respondent that s/he had heart disease, with answers coded as 1 for yes, 0 for no. The data consist of (a) Survey response data in the years 2008–2012 from over 228,000 individuals from the 15 most populous states, retrieved from the Centers for Disease Control and Prevention (CDC) Behavioral Risk Factor Surveillance (BRFSS) System (www.cdc.gov/brfss/questionnaires/state2013.htm); and (b) Environmental Protection Agency (EPA) data on PM2.5 concentrations for the counties in which these individuals lived at the time of the survey, retrieved from the US EPA website (www.epa.gov/airtrends/pm.html). Counties were used as the common key for merging annual average air pollution levels with individual response data. The BRFSS data consist primarily of either dichotomous (yes-no, which we coded as 1–0) variables for sex, smoking, Hispanic ethnicity, and self-reported history of ever having been diagnosed with heart disease; or categorical and ordered-categorical variables for age (50–99 years), income, education, and marital status. To these we joined a continuous variable, average PM2.5 concentrations in micrograms of fine particulate matter per cubic meter of air. We selected records (i.e., completed responses to the BRFSS survey) with complete data on *Age*, *Sex*, *HeartDiseaseEver*, *Smoking*, and county-specific PM2.5 for the year and month in which a respondent was surveyed. This reduced the number of records analyzed from 228,369 to 216,207. If values are not missing at random, the omission of records with incomplete data on one or more of these variables could result in a biased sub-sample of the full BRFSS data set. Hence, we focus on conditional relations among the variables in this sample (e.g., given values of the other variables, how likely is it that a respondent will report a history of heart disease?), and do not attempt to generalize beyond it.

Methods

We follow the methods and data-analytic strategy in Cox Jr. (2017), beginning with exploratory descriptive analyses that plot *HeartDiseaseEver* against explanatory variables such as age, income, sex, and average concentration of PM2.5 in the counties where respondents lived at the time of the survey. Approximate 95% confidence intervals (mean plus or minus 1.96 sample standard deviations) are indicated visually as vertical bars around the mean values shown in the interaction plots. Such exploratory data analyses can reveal nonlinear patterns of association among a few variables without requiring any parametric modeling assumptions. To quantify associations between multiple explanatory variables and health effects we used regression models. We supplemented these with non-parametric methods—a Bayesian network (BN) learning program (the *bnlearn* package in R) and causal partial dependence plots (PDPs) generated by the *dagitty* and *randomForest* package in R, as described in Chap. 9 and discussed further in connection with Figs. 13.9 and 13.10 to help make this chapter relatively self-contained—to discover, visualize, and quantify statistical dependence relations (represented by arrows between variables) and conditional independence relations (represented by a lack of arrows between variables) among all variables simultaneously.

As previously noted (Cox Jr. 2017), a challenge in this data set is that BRFSS respondents answered questions about whether they had *ever* been told that they had heart disease, but exposure concentrations are recorded only for the specific counties that they lived in at the time of the survey and only for the years 2008–2012. Responses from people who moved residences or who were diagnosed with heart disease when pollution conditions differed from those in the data set (2008–2012) might contribute irrelevant responses that dilute any observable relation between current pollutant concentrations and health effects. We addressed this challenge by checking that the relative ranking of counties by pollution levels is fairly stable over the 5 years of the study and is significantly associated with health effects; thus, although such dilution of associations probably occurs, there appears to be sufficient signal in this large data set to overcome the noise. Further details of these analytics methods and plan are given in Cox Jr. (2017).

Results

Dependence of Heart Disease on Age, Sex, and Smoking

Figure 13.1 shows an interaction plot describing how risk of ever having been diagnosed with heart disease varies with sex and age. The horizontal axis shows age categories; these represent ages rounded to the nearest decade (e.g., ages 65–75 are rounded to 70; ages 75–85 are rounded to 80, and so forth). The fractions of men (red curves with square data points) and women (blue curves with round data points) that report ever being told by a medical professional that they have heart disease are plotted on the vertical axes. Separate curves are shown for men and women. An approximate 95% confidence interval (vertical bar) is shown around each data point. As expected, risk increases with age, except among the relatively few respondents over age 95 (rounded to 100), who may represent an exceptionally disease-free subpopulation.

Risk increases with age more rapidly for men than for women. Figure 13.2 shows an analogous plot with effects of cigarette smoking, defined so that "No smoke' indicates fewer than 100 cigarettes (five packs) in a lifetime to date and "Smoked" indicates 100 cigarettes or more in a lifetime to date. For each age group 50–90 and for both sexes, smoking is associated with significantly increased risk of heart disease, as expected.

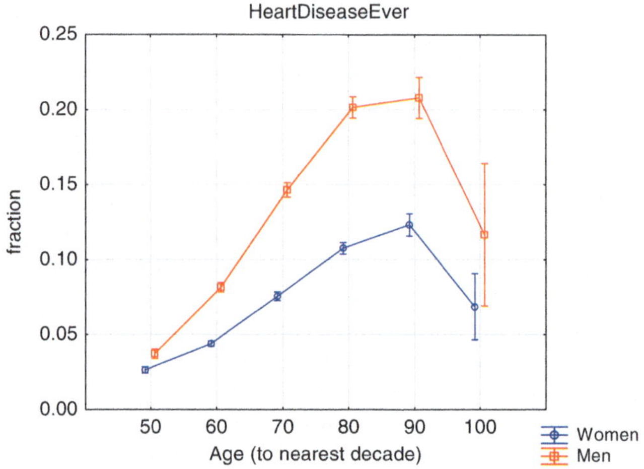

Fig. 13.1 Fraction of respondents reporting ever having had heart disease *vs.* age (horizontal axes, age in years rounded to the nearest decade). Vertical bars are 95% confidence intervals. Risk of heart disease increases with age for men and women aged 65–95. Men have higher risk of heart disease than women at every age

Statistical Dependence of Heart Disease on PM2.5 and Other Variables

Figure 13.3 shows how prevalence of self-reported heart disease increases with average PM2.5 concentration (rounded to the nearest integer) over the range from 5 to 15 μg/m^3, the range of greatest interest in this study. In a multiple linear regression model for the full data set, *Age*, *Sex*, *Smoking*, and *PM2.5* are all highly statistically significant predictors of heart disease risk (Table 13.2). In this main-effects only model, a 70-year old non-smoking man exposed to an average concentration of 15 μg/m^3 would have a cumulative probability of heart disease estimated as $-0.208 + 0.004*70 + 0.049 + 0.035*0 + 0.001*15 = 0.14$. If he were exposed to only 5 μg/m^3, he would have an estimated cumulative probability of heart disease of 0.13. Each μg/m^3 of exposure at these levels increases the conditional probability of lung disease by about 0.01.

However, *PM2.5* is no longer a significant predictor of heart disease risk if *Income* is included as a predictor (Table 13.3), consistent with previous findings for heart attack risk (Cox Jr. 2017). These results also hold for the subset of data with PM2.5 < 15 μg/m^3(202,846 out of 216,207 record) and if logistic regression is used instead of linear regression. (The odds ratios if *Income* is not included are 1.8 for *Sex*, 1.6 for *Smoking*, 1.05 for *Age*, and 1.01 for *PM2.5*, with all *P* values less than 0.01; with *Income* included, they are 2.0 for *Sex*, 1.6 for *Smoking*, 1.05 for *Age*, 0.88 for Income, and 1.00 for *PM2.5*, and all *P* values are less than 1E-16 except for PM2.5, which has a P value of 0.2.) Thus, income is a confounder of the association

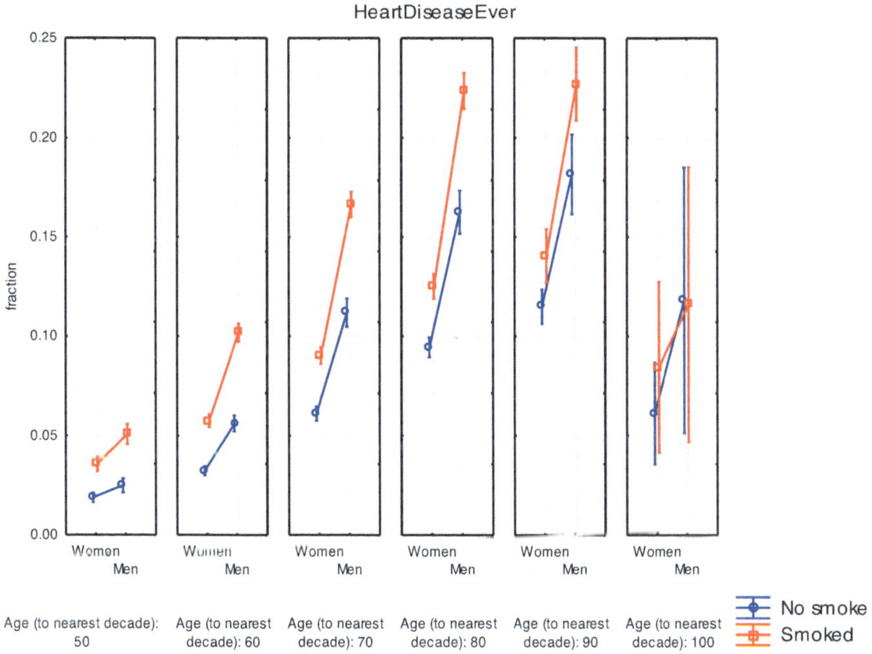

Fig. 13.2 Fraction of respondents reporting ever having had heart disease) *vs.* age (horizontal axes, age in years rounded to the nearest decade), sex (left side of each panel = women, right side = men) for non-smokers (blue circles) and smokers (red squares). Vertical bars are 95% confidence intervals. Heart disease risk is greater among smokers than among non-smokers for all age groups 50–90 and both sexes

between PM2.5 and heart disease risk, both overall and at relatively low (<15 μg/m^3) exposure concentrations.

Figures 13.4 and 13.5 examine this further by plotting heart disease risk for men and women against income category (Fig. 13.4) and plotting average PM2.5 concentration against income category for smokers and non-smokers (Fig. 13.5). (The BRFSS income codes are on a scale with code 2 for $10k–$15k annual income; code 4 for $20k–$25k annual income; and code 8 for greater than $75k annual income. Cases with missing data for income were skipped in making these plots, reducing the number of cases to 184,006.) Figure 13.4 shows that high income is strongly associated with reduced risk of heart disease, especially among women. Figure 13.5 shows that high income is also associated with lower average PM2.5 exposure concentrations, although the difference is proportionally much smaller. In addition, smokers tend to live in counties with slightly smaller average PM2.5 concentrations than non-smokers, although the difference is relatively small (Fig. 13.5).

Figure 13.6 puts these relations in perspective by showing that all of the binary variables, not just heart disease risk, vary significantly between the lowest and highest income categories. Understanding the interactions among these variables and age, education, and income clearly requires multivariate modeling.

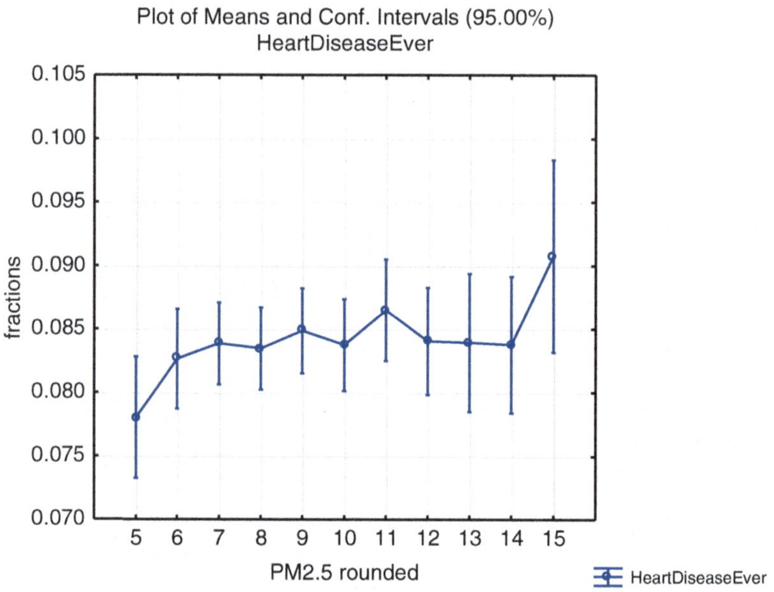

Fig. 13.3 Fraction of respondents reporting ever having had heart disease *vs.* average PM2.5 concentration for the county of respondent's residence at the time of the survey

Table 13.2 Regression Summary for dependent variable *HeartDiseaseEver*

N = 216,207	b*	Std. err. of b*	b	Std. err. of b	z	p-value
Intercept			−0.208	0.0041	−50.8	0.000000
Age	0.147	0.0021	0.004	0.0001	69.5	0.000000
Sex	0.086	0.0021	0.049	0.0012	40.3	0.000000
Smoking	0.064	0.0021	0.035	0.0012	30.0	0.000000
PM2.5	0.007	0.0021	0.001	0.0002	3.1	0.001666

PM2.5 is a highly statistically significant predictor of heart disease risk in this main-effects model

Table 13.3 Regression summary for dependent variable *HeartDiseaseEver* with income included

N = 184,006	b*	Std. err. of b*	b	Std. err. of b	z	p-value
Intercept			−0.1426	0.0053	−27.1	0.000000
Age	0.137	0.0024	0.0037	0.0001	58.4	0.000000
Sex	0.093	0.0023	0.0532	0.0013	40.2	0.000000
Smoking	0.059	0.0023	0.0330	0.0013	25.7	0.000000
PM2.5	0.003	0.0023	0.0003	0.0002	1.4	0.162194
Income	−0.069	0.0024	−0.0089	0.0003	−29.0	0.000000

PM2.5 is no longer asignificant predictor of heart disease risk in this model

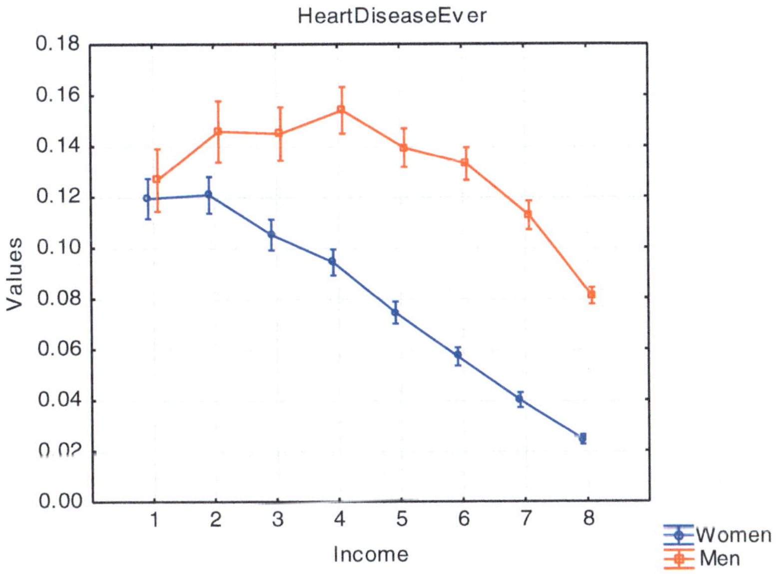

Fig. 13.4 Fraction of men and women reporting ever having had heart disease *vs.* respondent's income category. High income greatly reduces risk, especially in women

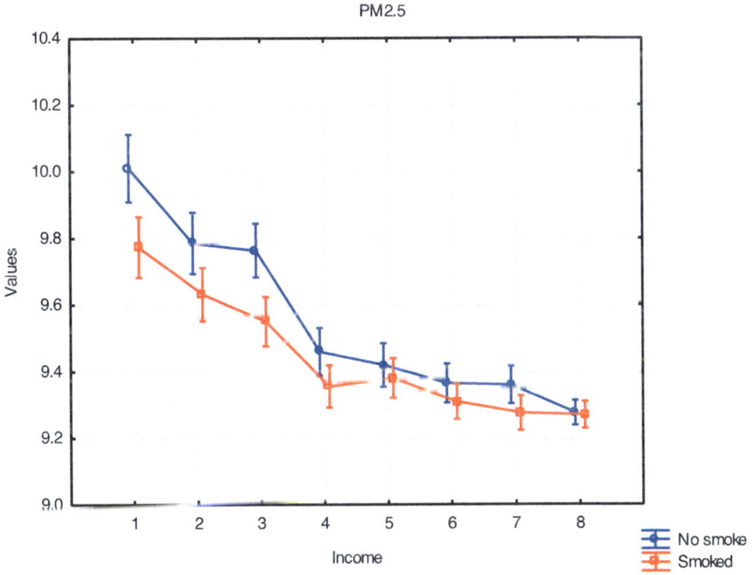

Fig. 13.5 Average PM2.5 concentration *vs.* respondent's income category and smoking status. For both smokers and non-smokers, PM2.5 exposure decreases with income

Age-Sex-Income Interactions in Predicting Risk of Heart Disease

The exploratory visualizations in Figs. 13.4 and 13.6 suggest the simple conclusion that risk of heart disease declines significantly with increasing income, as expected from the well-known socioeconomic gradient for health effects (e.g., Chetty et al. 2016). Multivariate analyses reveal that sex and age strongly modify this effect. Figure 13.7 fits a smooth response surface (using distance-weighted least squares) to age, income, and heart disease risk. The decreasing relation between income and risk is clear at young ages (e.g., under 70), but is reversed at older ages. Figure 13.8 provides a more detailed look at this interaction, disaggregating the respondents by age, sex, and income. It is clear that the beneficial effect of higher income on reducing heart disease risk (if these associations are causal) holds most clearly for women under age 80, corresponding to the two lower curves in the left panel (for ages less than 70 and between 70 and 80) and for men under 70, corresponding to the lowest curve in the right panel. This adds nuance to the often-noted "richer is healthier" finding for socioeconomic gradients. In this data set, it is clear that higher income accompanies lower risk of heart disease for most women and for men under 70, but not for older men.

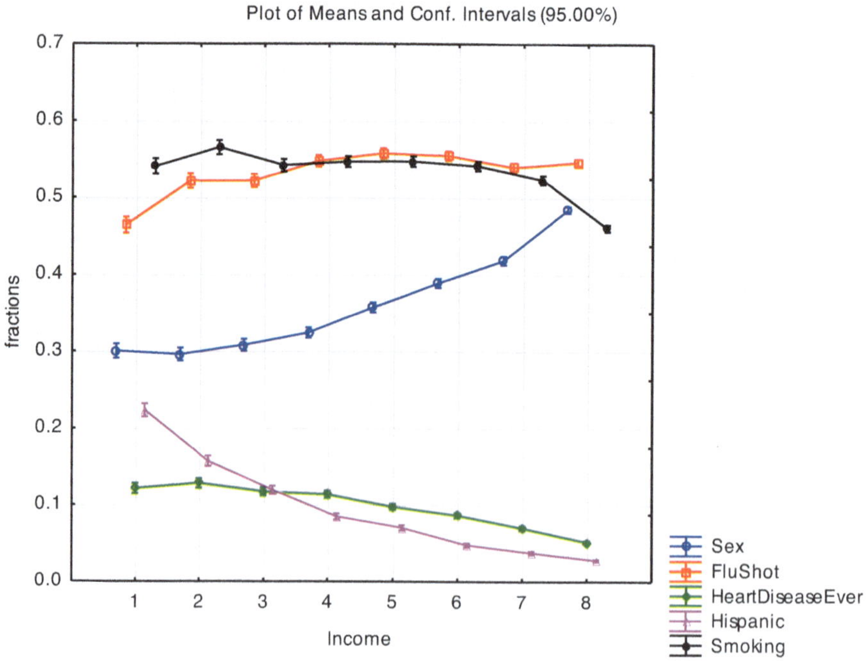

Fig. 13.6 Binary variables *vs.* respondent's income category. All five variables vary significantly with income (note the very short vertical 95% confidence interval bars)

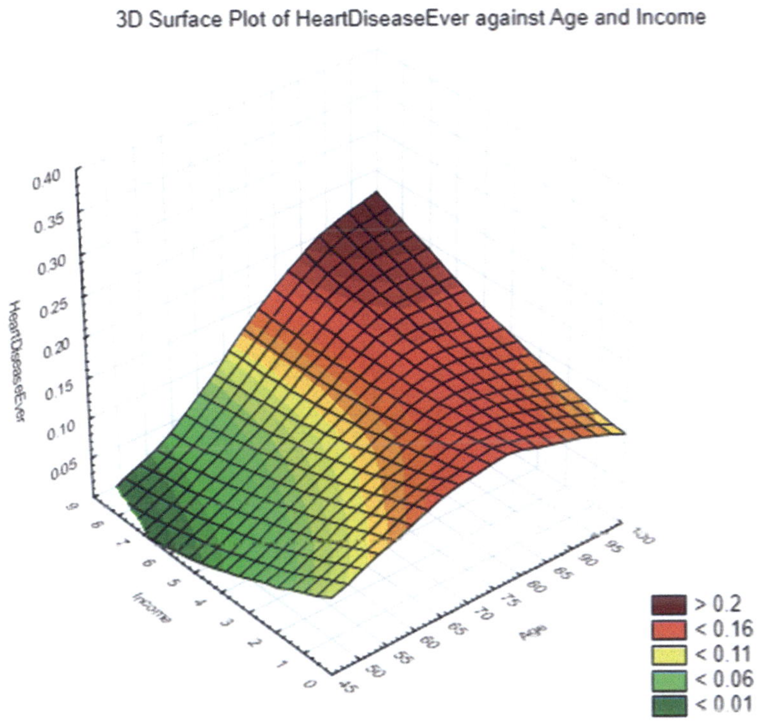

Fig. 13.7 Age and income interactions: high income reduces risk of heart disease by more at younger ages than at older ages

Predictive and Causal Analytics

To examine interactions among multiple predictors simultaneously, the analyses in this section apply some relatively computationally intensive machine learning algorithms to a random sample of size 5000 from the set of all 184,006 records with complete data for *Income* as well as for *Age*, *Sex*, *HeartDiseaseEver*, *Smoking*, and *PM2.5* for the year, month, and county of each respondent. For reproducibility, Appendix details the steps used in R to construct the sample and provides links to the data and to the Causal Analytics Toolkit (CAT) in-browser software used to perform these analyses. They can be reproduced and alternative analyses can be run using these links and either the same or different random samples of the data. The key conclusions are robust across multiple random samples, as can be verified by changing the random seed for the sampling step in Appendix.

Figure 13.9 shows a directed acyclic graph (DAG) model learned from the data via the *bnlearn* package in R, with all options set to their default values. As discussed further in Chaps. 9, 10, and 11, in this diagram, an arrow between two variables indicates that they are informative about each other, so that knowing the value of one helps to predict the value of the other. (By Bayes' Rule, or from information

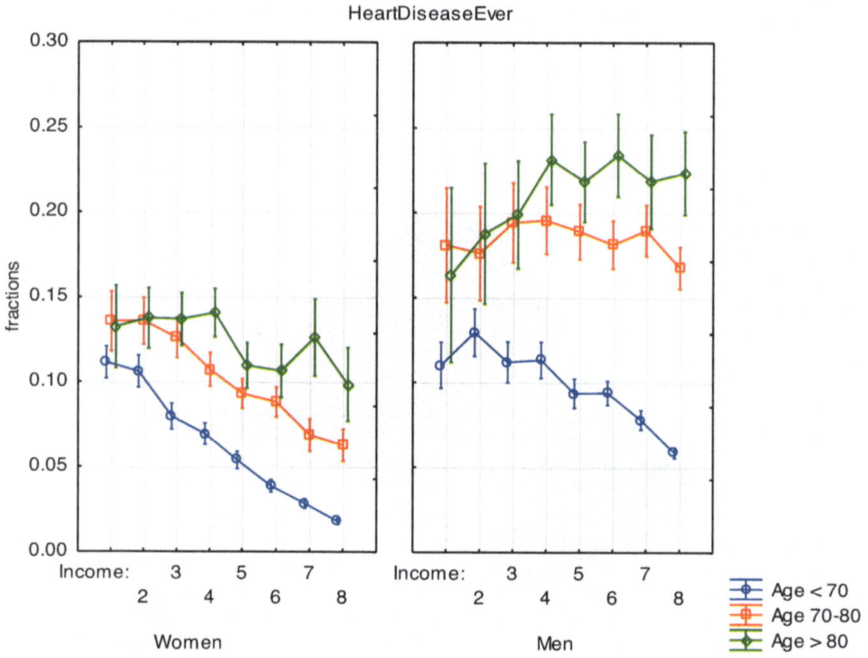

Fig. 13.8 Heart disease risk by age, sex, and income. Risk of heart disease decreases dramatically with income among women under 70, but not among men over 70

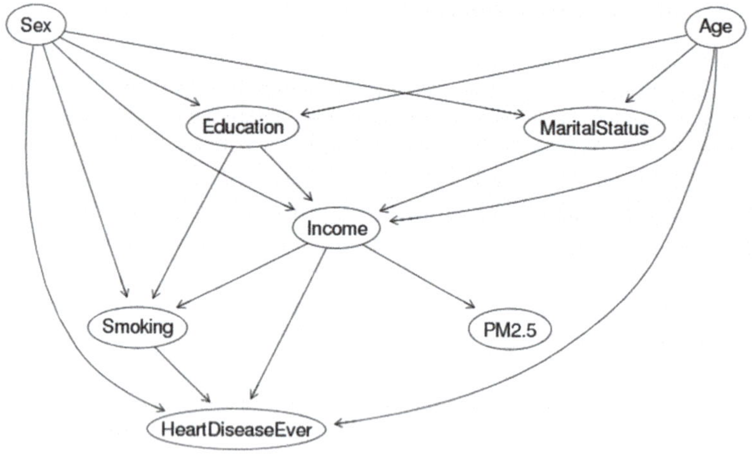

Fig. 13.9 A directed acyclic graph (DAG) model for information relations between variables. An arrow between variables indicate that they are informative about each other

theory identifying mutual information as a symmetric function of two random variables, such an arrow can run in either direction. One variable helps to predict another if conditioning on the former reduces the "uncertainty," measured as the expected conditional entropy, of the latter (Cover and Thomas 2006).) To orient the arrows to facilitate manipulative causal interpretation, we imposed the following knowledge-based constraints: *Sex* and *Age* were specified as sources with only outward-pointing arrows; and *HeartDiseaseEver* was specified as a sink with only inward-pointing arrows. Applying the CAT software with these constraints to the data then automatically generated the DAG model in Fig. 13.9.

Figure 13.9 shows that income is the least common ancestor of PM2.5 and heart disease. Sex, age, education, and marital status are also common ancestors, and hence they could appear as confounders in regression analyses, but Fig. 13.9 shows that their effects on PM2.5 are transmitted only through income. Thus, conditioning on income makes PM2.5 conditionally independent of these more remote ancestors. Interpretively, higher-income respondents tend to live in less polluted (lower PM2.5 concentration) counties and also, independently of PM2.5 concentration, higher-income respondents tend to have lower risks of heart disease for some settings of age, sex, and smoking. This pattern is consistent with the previous exploratory analyses using interaction plots.

Figure 13.10 shows a *causal partial dependence plot* for the predictive causal relation between income and heart disease risk. This combines the following set of ideas and algorithms (Chap. 9). Insofar as causes must be informative about their

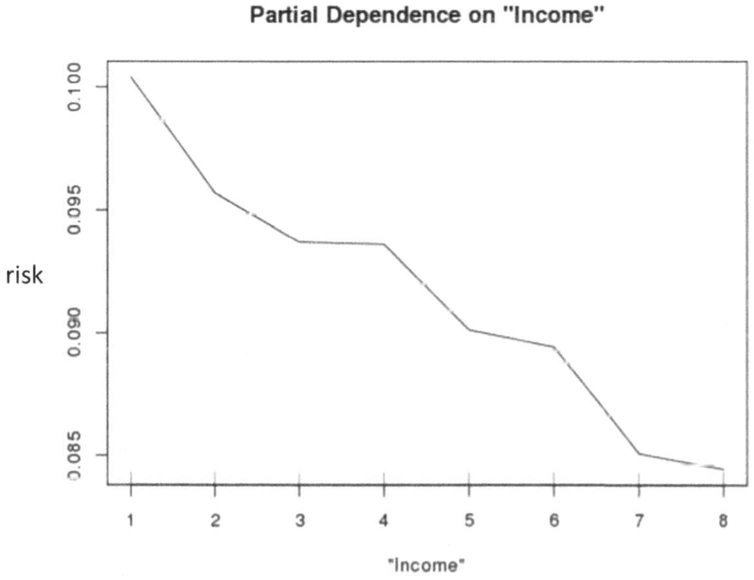

Fig. 13.10 A causal partial dependence plot (PDP) for the natural direct effect of income (horizontal axis) on absolute risk (i.e., probability) of self-reported heart disease (vertical axis), conditioning on variables in the adjustment set {*Age*, *Sex*, *Smoking*}

effects, one variable can be a direct cause of another in a DAG model such as Fig. 13.9 only if they are linked by an arrow. (Conversely, effects are not conditionally independent of their direct causes, even after conditioning on the values of other variables.) Estimates of the direct and total causal effects of one variable on another in a DAG model are obtained by conditioning on the variables in an *adjustment set*, which typically includes common ancestors of the cause and effect variables (e.g., confounders) but not common descendants (thus avoiding Berkson's selection bias).

For practical work, adjustment sets are computed by DAG-processing algorithms in packages such as *dagitty* (Textor et al. 2016). Given a cause such as income and an effect such as heart disease risk in a DAG model such as Fig. 13.9, adjustment sets can be automatically computed for estimating both the *natural direct effect* of the cause on the effect (i.e., how does the effect change as the cause changes, holding all other variables fixed at their current values?) and also the *total effect* (i.e., how does the effect change as the cause changes, allowing the effects on other variables to propagate via all directed paths from the cause to the effect?) For example, for the DAG model in Fig. 13.9 {*Age*, *Sex*, *Smoking*} is an adjustment set for estimating the natural direct effect of income on heart disease risk, and {*Age*, *Sex*, *Education*} is an adjustment set for estimating the total effect of income on heart disease risk, part of which is mediated by the effect of income on smoking.

Once an adjustment set has been identified (assuming that one exists for the causal effect of interest, which can then be identified from the data), the quantitative causal relationship between the cause and effect variables can be estimated by studying how the latter changes as the former is varied, conditioning on the values of the variables in the adjustment set. This is conveniently done using a partial dependence plot (PDP) generated by the *randomForest* package in R (see Chap. 9 for further discussion). This package automatically averages the predicted values for the effect variable over many (default value = 500) different classification and regression (CART) trees for each value of the cause variable, using each tree in the ensemble to condition on values of the variables in the adjustment set. A PDP for the dependence of an effect on a cause, conditioning on the variables in an adjustment set for this cause and effect, constitutes a *causal PDP*. The CAT software automates this sequence of analyses; it was used to generate the PDP in Fig. 13.10. Figure 13.10 can be interpreted as follows: Given the joint frequencies of sex, age, and smoking in this data set, the probability of ever having been diagnosed with heart attack for a randomly selected respondent declines from about 0.10 to about 0.085 as income code increases from 1 to 8. The total effect curve is very similar to the natural direct effect curve, declining from about 0.10 to about 0.083 as income code increases from 1 to 8. This slightly greater effect reflects a decline in smoking as income increases to accurately describe the lower prevalence of smoking among higher-income respondents. As expected from Fig. 13.9, there is no causal PDP for the direct or total effect of PM2.5 on heart disease risk, because PM2.5 is not a cause (parent or ancestor) of heart disease risk in this DAG model. Thus, selecting *PM2.5* and *HeartDiseaseEver* as the cause and effect variables and running CAT generates an error message that no adjustment set exists from which a causal impact can be estimated.

Discussion and Conclusions

Higher income is strongly and significantly associated with reduced prevalence of self-reported heart disease in the BRFSS data set, although both sex and age modify this association (Figs. 13.7 and 13.8). This finding is consistent with previous literature on socioeconomic health gradients (e.g., Chetty et al. 2016), although the interaction of income effects with sex and age adds additional detail. The observed pattern of multivariate associations is compatible with a causal impact of income on age-, sex- and smoking-specific heart disease risk in a directed acyclic graph (DAG) model (Figs. 13.9 and 13.10). Higher income is also statistically significantly associated with lower levels of PM2.5; thus, income is a potentially important confounder of PM2.5-heart disease associations. Whether PM2.5 is a significant predictor of heart disease is model-dependent. In linear and logistic regression models, PM2.5 is a significant predictor of heart disease if income is not included as a predictor, but not if it is (Tables 13.2 and 13.3). All socioeconomic and exposure variables examined are significantly associated with income (Figs. 13.4, 13.5, and 13.6), and hence they are all unconditionally associated with each other, but most of these associations have no causal interpretation, i.e., they vanish after conditioning on income (Fig. 13.9).

Overall, these results support low income as a probable cause of heart disease risk, especially for men under 70 years old and women under 80 years old. They highlight the importance of controlling for income in interpreting the associations of other variables, including PM2.5, with heart disease risk in cross-sectional data.

Appendix: Guide to Replication

For investigators who wish to replicate or extend the analyses presented in this chapter, the following instructions provide the required details.

1. The full BRFSS data set with PM2.5 data appended, for which Table 13.1 shows the first few records, can be downloaded from the following site: http://cox-associates.com/CausalAnalytics/BRFSS.csv
2. The data can be prepared for analysis via the following R script
   df<- read.csv("BRFSS.csv",stringsAsFactors=FALSE)
3. Linear and logistic regression models without income can be fit as follows:
   LinR <- lm(df$HeartDiseaseEver ~ df$Sex + df$Smoking + df$Age + df$PM2.5); summary(LinR)
   LogR <- glm(df$HeartDiseaseEver ~ df$Sex + df$Smoking + df$Age + df$PM2.5, family=binomial(link='logit')); exp(cbind(OR = coef(LR), confint(LR))); summary(LogR)
4. Linear and logistic regression models with income can be fit as follows:
   LinR <- lm(df$HeartDiseaseEver ~ df$Sex + df$Smoking + df$Age + df$PM2.5 + df$Income); summary(LinR)

LogR<- glm(df$HeartDiseaseEver ~ df$Sex + df$Smoking + df$Age + df$PM2.5 + df$Income, family=binomial(link='logit')); exp(cbind(OR = coef(LR), confint(LR))); summary(LogR)

5. For more computationally intensive analyses, a random sample from the full data set (small enough so that the *randomForest* algorithm can be used) can be created using the following commands:
   set.seed(1)
   brfss<- df[sample(1:nrow(df), 5000, replace = FALSE),]
   write.csv(brfss, "brfss2.csv")
6. Open the CAT software at http://cloudcat.cox-associates.com:8899/ (registering for free first if needed). Select the Data tab from the menu of commands, then browse to and upload the file "brfss2.csv". From the data tab, select the variables *HeartDiseaseEver, Income, Age, Sex, Smoking, PM2.5, Education, MaritalStatus*. To reproduce Fig. 13.9, select *Bayes* and specify *Age* and *Sex* and sources and *HeartDiseaseEver* as a sink. To reproduce Fig. 13.10, select *Income* as the cause and *HeartDiseaseEver* as the target effect in the interactive Bayesian network. Select direct or total to generate corresponding causal PDPs.

References

Chan E, Serrano J, Chen L, Stieb DM, Jerrett M, Osornio-Vargas A. Development of a Canadian socioeconomic status index for the study of health outcomes related to environmental pollution. BMC Public Health. 2015 Jul 28;15:714. https://doi.org/10.1186/s12889-015-1992-y. PMID: 26215141; PMCID: PMC4517649.

Chetty R, Stepner M, Abraham S, Lin S, Scuderi B, Turner N, Bergeron A, Cutler D. The association between income and life expectancy in the United States, 2001-2014. JAMA. 2016;315(16):1750–66. https://doi.org/10.1001/jama.2016.4226.

Cover TM, Thomas JA. Elements of information theory. 2nd ed. Hoboken: Wiley; 2006.

Cox LA Jr. Socioeconomic and air pollution correlates of adult asthma, heart attack, and stroke risks in the United States, 2010-2013. Environ Res. 2017;155:92–107. https://doi.org/10.1016/j.envres.2017.01.003.

Kaufman JD, Spalt EW, Curl CL, Hajat A, Jones MR, Kim SY, Vedal S, Szpiro AA, Gassett A, Sheppard L, Daviglus ML, Adar SD. Advances in understanding air pollution and CVD. Glob Heart. 2016;11(3):343–52. https://doi.org/10.1016/j.gheart.2016.07.004.

Romieu I, Gouveia N, Cifuentes LA, de Leon AP, Junger W, Vera J, Strappa V, Hurtado-Díaz M, Miranda-Soberanis V, Rojas-Bracho L, Carbajal-Arroyo L, Tzintzun-Cervantes G, HEI Health Review Committee. Multicity study of air pollution and mortality in Latin America (the ESCALA study). Res Rep Health Eff Inst. 2012;171:5–86.

Textor J, van der Zander B, Gilthorpe MS, Liskiewicz M, Ellison GT. Robust causal inference using directed acyclic graphs: the R package 'dagitty'. Int J Epidemiol. 2016;45(6):1887–94.

WHO (2018) Health impact assessment (HIA): the determinants of health. www.who.int/hia/evidence/doh/en/. Accessed 5 June 2018

Chapter 14
How Realistic Are Estimates of Health Benefits from Air Pollution Control?

Introduction: Do the Health Benefits from Cleaner Air Exceed Costs?

Chapter 13 established that income is an important confounder of some air pollution-associated health effects: low income increases health risks and is also associated with living in areas having higher air pollution levels. This raises the public health question: would reducing air pollution levels without addressing the other correlates of low income that might contribute to increased health risks, be effective in reducing health risks? Decades of regulatory science and political opinion have answered this question loudly and repeatedly in the affirmative. As discussed in subsequent chapters, practical experience has been much less encouraging, with substantial reductions in air pollution often making no clear contribution to causing improved public health. However, notwithstanding this experience, media, regulatory, and advocacy reports and recommendations continue to overwhelmingly assert that fine particulate matter (PM2.5, i.e., particulate matter smaller than 2.5 μm in diameter) in outdoor air kills people and causes serious health problems. Regulatory proposals to further decrease currently permitted levels of pollutants are supported by reference to large estimated or predicted health benefits from doing so. For example, in early 2011, the EPA released the results of its cost-benefit analysis of the 1990 Clean Air Act Amendments (CAAA). The assessment made two striking claims (EPA 2011a). (1) As of 2020, the CAAA would produce estimated health benefits valued at approximately two trillion (i.e., two thousand billion) dollars per year, compared to estimated compliance costs of only about $65 billion/year; and (2) The uncertainties in the cost-benefit analysis are small enough so that, "*The extent to which estimated benefits exceed estimated costs* and an in-depth analysis of uncertainties *indicate that it is extremely unlikely the costs of 1990 Clean Air Act Amendment programs would exceed their benefits under any*

L. A. Cox Jr., *Quantitative Risk Analysis of Air Pollution Health Effects*, International Series in Operations Research & Management Science 299,
https://doi.org/10.1007/978-3-030-57358-4_14

reasonable combination of alternative assumptions or methods identified during this study" (emphasis in original).

This chapter reexamines the assumptions, methods, and numbers used in reaching these encouraging conclusions, and seeks to improve upon them by more explicitly modeling key uncertainties. Although it is based on a paper published in 2012 (Cox Jr. 2012), little has changed in the basic approach commonly taken to regulatory benefits assessment or communication of estimated benefits to the public and to policy makers (see Chaps. 18 and 19). The result of this reexamination is a very different conclusion: that EPA's evaluation of health benefits is unrealistically high, by a factor that could well exceed 1000, and that it is therefore very likely that the costs of the 1990 CAAA exceed its benefits, plausibly by more than 50-fold. The reasoning involves re-examining specific uncertainties (including model uncertainty, toxicological uncertainty, confounder uncertainty, and uncertainty about what actually affects the timing of death in people) that were acknowledged qualitatively, but whose discrete contributions to uncertainty in health benefits were not quantified, in EPA's cost-benefit analysis.

Of greater methodological interest than the difference in conclusions is the difference in uncertainty analyses. If EPA's assessment of uncertainties were correct, then it should be impossible, or at least very improbable, for any plausible variations in assumptions to reverse the main conclusion that CAAA benefits greatly exceed its costs. Yet, this is precisely what the following sections show: plausible variations in assumptions easily reverse the sign of the estimated net benefits. The principal methodological challenge, therefore, is to diagnose the features of the uncertainty analysis that led to such a confident—but, we think, erroneous—conclusion, and to understand how to avoid such overconfidence in future risk-cost-benefit analyses.

The Main Logic of EPA's Benefit Assessment

Most of the benefit that EPA projects for 2020 is based on the estimated number and dollar value of mortalities that it expects will be prevented in that year by having cleaner air. The main calculation is simple:

$$\begin{aligned} \textit{Benefit} &= \textit{reduced number of deaths in}\, 2020 \\ &\quad \times \textit{value per statistical life saved} \\ &\approx (230{,}000\,\text{fewer deaths}) \\ &\quad \times (\$8{,}900{,}000\,\text{per life saved}) \\ &\approx \$2\,\text{trillion}. \end{aligned} \tag{14.1}$$

The rationales for these two input values are provided in detail in EPA's report. The 230,000 number for fewer deaths is given in Table 5-6, and is the estimated reduction in mortalities for all adults over 30 from reduced fine particulate matter (PM2.5) in air. The value per statistical life saved (VSL) in 2020, of $8,900,000, is given in EPA's Table 5-4.The actual calculation is more refined, as it includes discounting and additional sources of benefits (e.g., from reduced morbidities and improved visibility, as well as reduced mortalities from lower ozone levels), but these make little difference to the final answer, which is still close to two trillion dollars worth of benefits for 2020, and even more in later years. As EPA states, "These avoided deaths are valued at 1.8 trillion ($2006), with primary low and primary high bounds on this estimate of $170 billion to $5.5 trillion." (EPA 2011b). The reduction to $1.8 trillion reflects the passage of time before the estimated benefit is received. We will therefore focus on this simple calculation, and on the uncertainty analysis that led to these stated lower and upper bounds, as these calculations drive both the main quantitative results and the important qualitative conclusion that the value of health benefits almost certainly greatly exceeds the costs of compliance, which EPA estimates (Table 7-4) as $65 billion for 2020.

EPA's Input Estimates and Weibull Uncertainty Analysis

The upper and lower bounds for the two trillion dollar main estimate of benefits reflect uncertainties about the number of mortalities that will not occur in 2020 because of cleaner air, as well as uncertainty about the correct VSL value to use. The reduction in mortalities is estimated by multiplying estimated reductions in PM2.5 concentrations by an estimated concentration-response potency factor, called the C-R coefficient. As EPA (2011b) explains (p. 5–10), "We based the primary C-R coefficient estimate of the Second Prospective study on a Weibull distribution with a mean of 1.06% decrease in annual all-cause mortality per one μg/m^3. This mean is roughly equidistant between the results of the two most well-studied PM cohorts, the ACS cohort (0.58, as derived from Pope et al. 2002) and the Six Cities cohort (1.5, as derived from Laden et al. 2006), both of whose results have been robust to continued follow-up and extensive re-analysis." Likewise, for the VSL number, EPA again used a Weibull uncertainty distribution, noting (Table 5-4) that, "[The] Mean Value of Statistical Life (VSL) [is] based on 26 wage-risk and contingent valuation studies. A Weibull distribution, with a mean of $7 4 million (in 2006$), provided the best fit to the 26 estimates." The $7.4 million value is shown for 1990 income levels, and is updated to $8.9 million for 2020 income levels. These two Weibull distributions, for the C-R coefficient and VSL, together with estimated reductions in PM2.5 concentration levels due to CAAA, drive the 2 trillion dollar main benefit estimate and the upper and lower bounds around it.

Questioning the Logic: Does the Main Calculation Make Sense?

EPA's main analysis, as just summarized, has the commendable virtues of simple, transparent logic and well documented input values and reasoning. It invites open inspection by the scientific community, so that any important deficiencies can be noted and, if necessary, repaired. To such deficiencies, we now turn.

The logic of equation (14.1) requires estimating the reduction in mortalities in 2020 due to cleaner air (reduced PM2.5 concentrations) resulting from the CAAA. The total number of mortalities in 2020 *with* the CAAA is (or should be) calculated as follows: From the number of mortalities that would have occurred in 2020 without the CAAA (as estimated from population sizes and mortality rates), *subtract* the number of mortalities that are postponed until 2021 or later because of reduced mortality rates attributed to the CAAA; and *add* the number of mortalities that are postponed until 2020 from 2019 or earlier because of reduced mortality rates attributed to the CAAA. The difference between this number (i.e., the mortalities that occur in 2020 *with* the CAAA) and the original number (i.e., the mortalities that occur in 2020 *without* the CAAA) is the total change in mortalities in 2020 attributable to the CAAA. In other words,

$$\begin{aligned} &\textit{deaths in } 2020 \textit{ with CAAA} \\ &\quad = \textit{number that would have occurred without CAAA} \\ &\qquad - \textit{deaths postponed because of CAAA until after } 2020 \\ &\qquad + \textit{deaths postponed because of CAAA until } 2020 \textit{ from earlier} \end{aligned} \quad (14.2a)$$

Similarly,

$$\begin{aligned} &\textit{Change in deaths in } 2020 \textit{ attributable to CAAA} \\ &\quad = -\textit{deaths postponed because of CAAA until after } 2020 \\ &\qquad + \textit{deaths postponed because of CAAA until } 2020 \textit{ from before } 2020 \end{aligned} \quad (14.2b)$$

Unfortunately, EPA's calculation includes only the first term in the change (that is, the reduction term *-deaths postponed because of CAAA until after 2020*), but neglects the second term (that is, the increase term *deaths postponed because of CAAA until 2020 from before 2020*). EPA estimated that the average gain in life expectancy is less than a year (Table 5-8), so if mortality rates and population sizes changed relatively slowly between 2019 and 2021, the deaths postponed from 2020 to 2021 or later are approximately equal to the deaths postponed until 2020 from 2019 or earlier. Hence, *the net change in deaths in 2020 are approximately zero,* instead of the 230,000 reduction in mortality estimated by EPA.

More generally, as discussed further in Chaps. 18 and 19, calculating "deaths avoided" in any year is not the correct risk analysis expression to use to estimate the health benefits from a regulation or other intervention. A correct analysis (e.g., using age-specific hazard functions and life table analysis) shows that even a regulation that successfully decreases age-specific mortality rates, thereby increasing life expectancy, may not reduce the annual number of deaths in the population.

Example: Accounting for Deaths

The following simplified numerical example illustrates why a regulation that extends lives does not necessarily reduce deaths per year. For simplicity, suppose that, before the regulation, each individual lived to age 80, and then had a 50–50 chance of either dying at age 80, or of surviving for a further decade and dying at age 90. Suppose that the regulation prevents early deaths, so that all those who would have died at age 80 now die at age 90. (Those who would have died at age 90 still do so.) Now, how many people die each year before and after the regulation? (Assume a constant birth rate of N people per year, which is unaffected by the regulation.) Before the regulation, N people die each year (half of the cohort of N people born 80 years ago + half of those born 90 years ago, for a total of $N/2 + N/2 = N$.) After the regulation has fully taken effect, N people still die each year (all of the cohort born 90 years ago). Although half of the population lives longer, the number of deaths per year is unchanged.

Of course, even if the net deaths prevented in 2020 are zero (as will necessarily be true under steady-state conditions, and as will be approximately true if the population size is changing relatively slowly on a time scale of months to years), this would not imply that increasing life expectancy by several months has no value. But it does suggest that equation (14.1)'s emphasis on reduced mortalities in 2020 is the wrong way to capture this value. The problem is that the main projected health benefits are not (or should not be) deaths "prevented" or "avoided," or lives "saved," but rather life years gained (see Chaps. 18 and 19). EPA's own report[2] explains this well: "Avoided premature mortality is one of the more commonly cited results of benefits analyses for air pollution control. However, as noted in the valuation section of this chapter, *a more accurate description of the benefit of clean air is a reduction in the risk of mortality* for the exposed population over many years, which results in the extension of lives (sometimes referred to as 'lives saved'). *Other useful metrics of the benefit of cleaner air are the number of life years that are gained through the reduction of mortal risks*, and the number of years of life expectancy gained on average throughout the population."

Accordingly, we can re-do the calculation of benefits in equation (14.1), switching from deaths "prevented" or avoided (which, in reality, do not occur, since reducing PM2.5 does not confer immortality) to life-years gained. EPA (2011b, Table 5-8) estimates that, in 2020, almost two million (1,900,000) life-years will be gained because of cleaner air caused by CAAA. No value of a statistical life-year ($VSLY) gained is explicitly identified in EPA's report, and values in the literature vary widely, depending on the elicitation techniques used and the ages of the recipients. We tentatively use a $VSLY value of $150,000/life-year, which has previously been used for life-years gained by cancer patients through the year 2020 (Yabroff et al.

2008). (For comparison, if we assumed, simplistically, that the \$8,900,000 VSL number used in equation (14.1) consists of at least 50 equally-valued life-years, then the corresponding \$VSLY value would be at most \$8,900,000/50 = \$178,000. This ignores the important complexities of age-specific \$VSLY values and discounting, but suggests that the estimated \$VSLY value of \$150,000/year is not grossly inconsistent with EPA's \$VSL number.) Equation (14.1) can now be replaced with the following estimated value of health benefits based on estimated life-years gained:

$$\begin{aligned} Benefit &= \left(life - years\ gained\ in\, 2020\right) \\ &\quad \times \left(\$VSLY\ per\, life - year\ gained\right) \\ &= \left(1{,}900{,}000\,\text{life} - \text{years gained}\right) \\ &\quad \times \left(\$150{,}000\,/\,\text{life} - \text{year gained}\right) \\ &= 0.3\,\text{trillion}. \end{aligned} \tag{14.3}$$

Thus, simply replacing lives saved with life-years gained, and hence the \$VSL value in equation (14.1) with the \$VSLY value in equation (14.3), reduces the benefit estimate from about 2 trillion to about 0.3 trillion.

As it happens, a similar numerical conclusion results even if the logic of equation (14.1) is retained, but the ages at which projected benefits occur are taken into account. EPA estimates the median age of people whose lives are "saved" (i.e., who gain extra months of life) from cleaner air to be close to 80 years (EPA 2011b, Table 5-8). The \$VSL of \$8,900,000 is appropriate for a healthy young adult of age about 25 (Aldy and Viscusi 2007, Figure 1). The \$VSL for people aged close to 80 has been estimated to be approximately one sixth of the \$VSL for people aged about 25 (Murphy and Topel 2003, Figure 1). Thus, using \$VSL numbers that condition on the ages at which benefits are received would also reduce the estimated value of health benefits substantially, with a sixfold reduction reducing EPA's current estimate of \$1.8 trillion to a revised estimate of 0.3 trillion. Despite this numerical coincidence, however, equation (14.3) is logically (and perhaps ethically) preferable to equation (14.1), since, as EPA notes, the true benefit from reduced mortality rates is gain in expected life-years.

Two other ways of framing the simple analysis in this section may help to clarify the crucial role that conditioning on age should play in health effects benefits assessments. The first is that, in very round numbers, about two million life-years gained in 2020 can create about two trillion dollars worth of benefit only if each life-year gained is valued at about a million dollars (or, equivalently, at about \$80,000 per additional month of life). This is a much higher value than many octogenarians place on their own gains in life-years (Aldy and Viscusi 2007; Murphy and Topel 2003). Second, applying a \$VSL value appropriate for a 25-year old to people who are mostly over 75 (and a significant fraction of whom are over 100 years old, according to EPA's Table 5-8, since this group contributes as much to estimated life-years gained as the 40–49 year olds in 2020) suggests that cleaner air is being evaluated, perhaps inadvertently, as a sort of fountain of youth, restoring the full value of youth (e.g., with at least another 50 years of healthy life expectancy) to people who, in fact, have a median age close to 80 years (*ibid*). This appears to be overly optimistic.

Health economics studies of the costs and benefits of medical interventions routinely quantify health benefits in terms of the monetized value of quality adjusted life years (QALYs), per the recommendation of the U.S. Panel on Cost Effectiveness (Gold et al. 1996). QALY values of between $50k and $100k per year are common, and values up to $200k can be found, but these values are substantially less than the $1 million that EPA implicitly places on a life-year (without adjusting for quality). Thus, whether one uses monetized values of "lives saved," or life-years added, or quality-adjusted life-years added, EPA's valuation appears to be higher than others, for similarly aged (and perhaps similarly infirm) people, by approximately one order of magnitude.

Questioning the Major Premise: Will Reducing PM2.5 Concentrations Further Really Reduce Mortality Rates Further?

Even more important than uncertainty about the correct value of $VSL (or of $VSLY) is uncertainty about the correct value of the C-R coefficient that converts reductions in PM2.5 concentrations to corresponding projected reductions in mortality rates. If the C-R coefficient is zero, then the health benefit from further reducing PM2.5 concentrations is also zero. The assumption that the C-R coefficient is positive might be wrong. EPA's qualitative discussion of key uncertainties (Table 5-11 of EPA 2011b) explicitly identifies this crucial assumption, noting that the "Analysis assumes a causal relationship between PM exposure and premature mortality based on strong epidemiological evidence of a PM/mortality association. However, epidemiological evidence alone cannot establish this causal link." The possible impact of this assumption is identified as being a "potentially major" overestimate of benefits. However, EPA rates its confidence in its causal hypothesis as "High," arguing that, "The assumption of causality is suggested by the epidemiologic and toxicological evidence and is consistent with current practice in the development of a best estimate of air pollution-related health benefits. At this time, we can identify no basis to support a conclusion that such an assumption results in a known or suspected overestimation bias." EPA's quantitative uncertainty analysis goes further, implicitly expressing 100% confidence that this causal hypothesis is correct. The Weibull uncertainty distribution for the C-R coefficient implicitly assigns a probability of 100% to the C-R coefficient being positive, and hence a 0% probability to it being zero (or negative). Thus, the qualitative assessment of "high" confidence is implicitly mapped to a quantitative assessment of 100% certainty. Such complete certainty is unwarranted by available data and knowledge, as discussed next.

Is There Really a Significant Positive Association Between PM2.5 and Mortality Rates?

Being unable to identify a basis for a suspicion of an overestimation bias does not, of course, guarantee that there is no such bias. Nor, perhaps, should a (subjectively perceived) suggestion of the possibility of a causal relation in epidemiological and toxicological data, or consistency with current practices in developing best estimates, warrant a very confident conclusion—let alone 100% confidence—that a causal relation necessarily exists.

For toxicological data, both theory and empirical evidence indicate that the inflammation-mediated adverse health effects of fine particulates probably have thresholds, below which the C-R coefficient is zero (Stoeger et al. 2006; see also Chaps. 3, 4, 5, and 6). For example, low levels of exposure induce increased antioxidant (AOX) production in the lung, but higher levels induce levels of reactive oxygen species (ROS) that overwhelm the very limited capacity of the lung to increase AOX. Increased disease risks occur only when there is an imbalance that leads to a sustained high-ROS lung environment and other changes that disrupt homeostasis, which may not occur at or near currently permitted exposure levels.

For epidemiological data, the National Academy of Sciences and several expert statisticians have long questioned whether reported findings of significant positive C-R coefficients may be misinterpreting ambiguous data as showing evidence of a significant association where none actually exists (Clyde 2000; Koop and Tole 2004, 2006). This concern is based largely on awareness that multiple testing biases (which inflate the rate of false positives) can arise when pollution health effects investigators make numerous modeling choices (e.g., of which pollutant summaries, confounders, covariates, and interactions to include in the model; what logarithmic or other transformations to apply to them; lags in variables; model form specifications; treatment of exposure uncertainties, etc.) in arriving at the reported significant positive associations. Attempts to increase the objectivity of conclusions about PM-mortality rate associations, by considering all of the effect estimates from many (e.g., thousands of) computer-generated models that fit the available data approximately well, without applying human judgment to select or defend any particular model or conclusion, have found that the hypothesized associations may not exist. For example, some Bayesian model-averaging (BMA) analyses (averaging effects over multiple models, weighted by their likelihoods in light of the data) of time series studies have reported, contrary to findings based on subjective selection of modeling assumptions, that available time series data sets give no overall indication of a positive relation between PM2.5 and excess adverse health effects (Koop and Tole 2004, 2006). The probability that current or lagged PM2.5 values help to predict mortality rates (compared to predictions based solely on non-pollutant variables such as temperature and pressure) ranges from about 2% to 7% in some BMA studies (Koop and Tole 2004), and associations between pollutants (including particulate matter and ozone) and mortality rates have been reported as not significantly different from zero in some Canadian and U.S. investigations (Clyde 2000;

Koop and Tole 2004, 2006). However, the BMA analyses were conducted for time-series data, and thus are not directly comparable to the long-term cohort studies used by EPA. We do not know whether the conclusion that only a small fraction of models show any association of PM with mortality rates would hold in the two main studies (the Harvard Six Cities and American Cancer Society cohort study data) used in EPA's assessment (EPA 2011a).

Against such reports of apparently non-existent associations, some investigators whose methods and models have informed EPA's benefits assessment have argued that "Bayesian approaches in general and BMA in particular are not panaceas for model selection." (This is certainly true, as they are not model selection methods at all.) They further note that Bayesian approaches may be mistakenly applied, and therefore urge that "model selection should be informed by substantive knowledge about the environmental health processes influencing the outcome" (Thomas et al. 2007). This response appears to reject the BMA principle that model selection should not be used at all (since even the "best" single model is almost certain to be wrong); but to advocate instead the continued use of judgment, "informed by substantive knowledge," as a basis for selecting models. But this, in turn, risks filtering conclusions to conform with the modelers' *a priori* expectations about what should be found, thus reopening the possibility that confirmation bias (selecting modeling assumptions that are favored by prior expectations or expert opinions) and multiple selection bias might explain some reported positive associations. For example, imposing an *a priori* modeling assumption that mortality rates increase approximately in direct proportion to pollutant levels at low concentrations (e.g., in a Cox proportional hazards model), or that the C-R function is linear at low concentrations (and piecewise linear overall), makes it possible to conclude that the C-R curve "is linear, clearly continuing below the current U.S. standard of 15 μg/m^3" (Schwartz et al. 2008). In this way, substantive knowledge or *a priori* assumptions and model selection constraints can still be used to produce low-dose linear results and a positive C-R coefficient. EPA (2011b, Table 5-11, p. 5-42) identifies the "Assumption of a linear, no-threshold model for PM and ozone mortality" as a "Key uncertainty," but assesses as "High" its subjective confidence that this choice of model is correct, and as "Probably minor" the potential impact on overestimation of benefits of considering other model forms. This contrasts with time series studies that conclude that considering other models removes all significant positive associations between PM2.5 and mortality rates (Clyde 2000; Koop and Tole 2004, 2006). A possible explanation is that the one cohort study that EPA's table cites in support of these crucial conclusions (an EPA-(Health Effects Institute)-funded study of the ACS data (Krewski et al. 2009)) focused on models (random effects Cox proportional hazards and Poisson regression models) that imply low-dose linearity, rather than considering other (e.g., J-shaped or threshold) models that could have revealed absence of a positive C-R relationship at low exposure levels.

Even without further discussing, or seeking to resolve, how best to use computationally intensive statistical methods such as BMA to more fully characterize uncertainty about the C-R coefficient, it seems clear that there is significant uncertainty about whether a true association exists between ambient PM2.5 concentrations and mortality rates in some studies. EPA's qualitative caveat deserves emphasis, that the "Analysis assumes a causal relationship between PM exposure and premature mortality based on strong epidemiological evidence of a PM/mortality association, [but] epidemiological evidence alone cannot establish this causal link." Even this caveat should perhaps be tempered further, by acknowledging that calling the epidemiological evidence favoring a PM/mortality rate association "strong" may largely reflect unverified modeling assumptions (e.g., of a linear, no-threshold C-R relation at ambient levels). Frequent minority reports of statistically significant *negative* C-R coefficients, in models that allow this possibility, raise further questions about the coherence and reality of claimed statistical associations. For example, a recent review of associations between PM2.5 and all-cause and cause-specific mortality in 27 U.S. communities indicated *reduced* all-cause mortality rates at increased levels of PM2.5 for one third of the communities, including Birmingham, Dallas, Houston, Las Vegas, and Riverside (Franklin et al. 2007).

To understand the sensitivities of EPA's cost-benefit assessment to plausible alternative assumptions, it suffices to note that the question of whether a real (data-driven, rather than model selection-driven) association exists between PM2.5 concentrations and mortality rates remains open, apparently on fairly substantive grounds, with at least some BMA studies indicating rather strongly (e.g., with greater than 95% probability) that there is no such association (Koop and Tole 2004). Lack of consistency in effects across studies, and large, unexplained heterogeneity in apparent C-R functions (including some with negative slopes) across locations, add qualitative reasons to question whether a stable, coherent association exists. This uncertainty should be explicitly addressed and, if possible, quantified, as BMA analyses seek to do, rather than being tacitly ignored, e.g., by using only a single average value across all studies (Franklin et al. 2007) (perhaps with subjective confidence intervals) without explaining or resolving the observed negative relations.

Are Reported Statistical Associations Causal?

Even if a real (model-independent) statistical association exists, of course, it might not be causal. Many investigators have recognized that covariates and confounders (such as proportion of high-risk minorities in a city, or correlates of wealth or education or poverty, or residual confounding by weather variables, including high temperatures or barometric pressures, or changes and trends in these variables) might create significant statistical associations between pollution and mortality rates, even if the former does not necessarily cause the latter. Likewise, regressing some trend variables (e.g., declining cause-specific and overall mortality rates in the wake of innovations such as angioplasty, anti-smoking campaigns, and aspirin therapy) against others (such as declining ambient pollution levels in recent decades) can show strong, statistically significant associations between them, with reductions in exposure concentrations routinely being followed (as well as preceded) by reductions in mortality rates, even if there is no causal relation between them.

Such threats to valid causal interpretation of statistical associations are well recognized by many investigators in pollution health effects research, but approaches for trying to address them have varied widely. To meet such challenges and threats to valid causal inference more decisively, econometric and statistical tests for potential causation between variables have been extensively developed, using concepts such as conditional independence and Granger causality for multiple time series (Chap. 9). However, these methods have not yet been widely applied to PM2.5 and mortality data, with a few exceptions (e.g., Vitolo et al. 2018). Thus, the question of whether any statistical associations might be causal has not been settled using such relatively formal, objective tests, despite overwhelming opinion favoring this view by some vocal commentators, as discussed in Chaps. 1, 18, and 19. (As previously noted, inconsistencies across cities and studies, and large, unexplained heterogeneity in C-R coefficients estimated in different studies, including some significant negative ones, undermine the coherence of proposed causal interpretations.) Thus, the question of whether there is any positive *causal* relation between ambient levels of ozone and PM2.5 and mortality rates, is a second important, discrete, uncertainty that should be explicitly addressed in uncertainty analysis. It cannot be answered by arguing that there is a positive *statistical* association between them (at least under some model-selection assumptions), since statistical and causal relations have no necessary connection (e.g., omitted confounders can make statistical relations positive even if causal relations are zero or negative). Nor can it be settled by concluding, based on expertise, that the data are *consistent* with such a relationship (EPA 2011b), if the data are at least equally consistent with the opposite conclusion of no relationship (Clyde 2000; Koop and Tole 2004). We conclude that the question remains unsettled, and that it is appropriate for inclusion in uncertainty analysis, although it is often presented as a matter of settled science in media stories and by commentators relying on weight-of-evidence consensus judgments (Chaps. 1, 18, and 19).

EPA's benefit assessment EPA (2011b, p. 5–7, 5–10) recognized and addressed the possibility of confounding in its uncertainty analysis, as follows:

> Several other sources of uncertainty exist in the relationship between ambient pollution and the health outcomes, including model uncertainty, potential confounding by factors that are both correlated with the health outcome and each other, and potential misclassification of the study population exposures. …EPA recently conducted an expert elicitation (EE) study, which is the formal elicitation of subjective judgments, in order to more fully characterize the uncertainty surrounding the PM2.5/mortality C-R function. This study allowed experts to consider and integrate several sources of uncertainty in the form of a probability distribution for the C-R function…[W]e rely exclusively on longterm studies to quantify PM mortality effects. This is because cohort studies are able to discern changes in mortality rates due to long-term exposure to elevated air pollution concentrations. …Based on consultations with the Council's Health Effects Subcommittee (HES), the 812 Project Team developed a distribution of C-R function coefficients (i.e., the percent change in annual all-cause mortality per one μg/m^3 change in annual average PM2.5) for use in the PM-mortality C-R function for the Second Prospective study. This distribution is rooted in the epidemiological studies that most inform our understanding of the PM mortality C-R function, but reflects

> the broader findings of the EE study. We based the primary C-R coefficient estimate of the Second Prospective study on a Weibull distribution with a mean of 1.06% decrease in annual all-cause mortality per one μg/m³. This mean is roughly equidistant between the results of the two most well-studied PM cohorts, the ACS cohort (0.58, as derived from Pope et al. 2002) and the six cities cohort (1.5, as derived from Laden et al. 2006), both of whose results have been robust to continued follow-up and extensive re-analysis.

Thus, EPA explicitly recognized the possibility of confounding. They dealt with uncertainty about whether confounding, model uncertainty, or other sources of non-causal associations (such as exposure measurement error or misclassification) might have created an overall false-positive association between PM2.5 concentrations (C) and health responses (R) in past studies, by (a) Relying on two studies that reported significant positive associations; (b) Discarding, or leaving unexplained, studies that reported significant negative or non-significant C-R associations; and (c) Applying subjective expert judgments, encoded as a Weibull distribution (which cannot show zero or negative relationships), to conclude that a positive causal relation exists with 100% subjective confidence probability. (The Weibull distribution implicitly assigns a subjective probability of zero to the discrete possibility that fully controlling for all relevant confounders and trends would yield a negative or zero causal estimate of the percent decrease in annual all-cause mortality per μg/m³ of PM2.5.)

Whether or not their conclusion is correct, this *process*—considering only data and probability distributions that support or logically require, *a priori*, the conclusion of a positive relation between exposure and health effects—is surely not adequate for characterizing uncertainty about whether the underlying assumption that the true (not confounded) C-R relation is positive is correct. The results of expert elicitation, expressed in the form of a distribution for a (causally interpreted) C-R coefficient, are also of uncertain value as a guide to objective truth, insofar as (a) The elicited distribution does not match, and hence does not explain, the empirically observed distribution of C-R coefficients in meta-analyses of dozens of reputable studies, about a third of which are negative (Franklin et al. 2007); and (b) No one currently knows how or whether further reducing ambient concentrations of PM2.5 *per se*—as opposed to reducing other activities, conditions (e.g., temperature extremes), and pollutant constituents (e.g., various oxides and sulfates) that are strongly correlated with PM2.5—would cause any incremental reductions in mortality risks.

Experts, like other people, may have opinions on these matters. But one cannot elicit from experts knowledge that they do not have. In this case, correct "substantive knowledge" of causal relations showing how reducing current and future ambient PM2.5 concentrations would affect human health, or how and whether other confounders and their interactions affect the estimated C-R relations for PM2.5 is not yet available. We do not yet even know which components of PM2.5 (if any), possibly interacting with which other pollutants, cause adverse health effects under present ambient exposure conditions. This does not necessarily imply that no new action should be taken. But it does imply that, if PM2.5 *per se* is not involved in causing increased mortality rates, then further reductions in ambient PM2.5 levels may not produce any health benefits. This possibility should be acknowledged in uncertainty analysis of human health benefits by including a substantial discrete (non-zero) uncertainty about causation.

Do C-R Relations Have Thresholds or J-Shapes?

Even if historical data showed a clear, unambiguous, causal relation between concentration and response, with past declining levels of PM2.5 (or C) consistently causing proportionally declining levels of mortality rates (or R), it would be important to establish whether further reductions in C would produce further reductions in R. At some point, C might reach (or might already have decreased below) a threshold below which no further reductions in R are gained. Uncertainty about that threshold, and about whether current standards are already below it, should then be quantified. Or, conceivably, continuing reductions in C might eventually pass a point after which further reductions in C actually increase R, as suggested by a few of the U-shaped or J-shaped empirical C-R model curves fit to PM_{10} data for the twenty largest U.S. cities (Daniels et al. 2000) (This might also explain the previously noted disconcerting finding of negative C-R coefficients in approximately a third of cities examined (Franklin et al. 2007).) The probability of this contingency, and the conditional probability that we have already approached or passed the bottom of such a J shape if it exists (at least in some cities), should limit the certainty with which additional reductions in C can be projected to cause additional reductions in R.

Existing discussions in the air pollution health effects literature of the possibility of a threshold for the C-R function largely recapitulate the technical issues in discussions of whether a non-zero C-R coefficient exists at all. Those who favor a linear, no-threshold model argue that model selection criteria such as the Akaike Information Criterion (AIC) can be used to support it (Schwartz et al. 2008), insofar as other models tested typically do not give significantly better AIC scores. Those who are more agnostic about the linear no-threshold model point out that the AIC criterion is not appropriate for this use: in simulation experiments (for which the correct model is known) it only identified the correct model about half the time, and assuming a linear no-threshold model can greatly under- or over-estimate true mortality risks if non-linearities are present but are mistakenly assumed away based on AIC (Roberts and Martin 2006). Defenders of the linear no-threshold model further counter that population data exhibit no thresholds, but appear to be linear down to the lowest levels measured (Schwartz et al. 2008). Skeptics could respond that this pattern is precisely what one might expect if there really is a population threshold, but individual exposure estimates contain unmodeled errors, as in past studies; thus, the observed population C-R function is irrelevant to determining whether there is a true C-R threshold (Chap. 16). That errors in past exposure estimates cast doubt on inferences about health effects is starting to be more widely acknowledged (Sheppard et al. 2012), but no one has yet advanced a model that clearly fits most of the data, reconciles past inconsistencies, accounts for past measurement errors, and resolves the debate with an obviously correct C-R relation.

As in the case of uncertainty about whether a positive C-R coefficient exists, the role of uncertainty analysis is not to resolve whether threshold or J-shaped C-R relations exist (at at least some locations), but only to make sure that any significant uncertainties about these possibilities are identified and explicitly addressed in the presentation and evaluation of uncertain health benefits projected from reductions in exposure concentrations. However, EPA's quantitative uncertainty analysis implicitly assumes that there are *no* uncertainties about the correct form (linear, no-threshold with probability 100%) or sign (positive with 100% probability, zero or negative with 0% probability) of the C-R relation. Admitting that there are such uncertainties would substantially change both the uncertainty analysis and the evaluation of projected health benefits, as explained later.

How Much Smaller Will Future C-R Coefficients be than Past Ones?

Finally, suppose that a true, causal, linear C-R coefficient were to be confidently identified in historical data. Even this would not warrant a conclusion that *future* reductions in pollutant levels would achieve proportionate reductions in mortality rates and gains in life-years. Progress in medicine, including preventive medicine, is steadily shrinking the pool of highest-risk individuals whose lives might, under the above assumptions, be most prolonged by reducing pollutant levels. Long-term trends such as improved diet and exercise among adults, greater use of low-dose aspirin regimens and other medications to successfully prevent cardiovascular disease events, smoking reduction and cessation programs, and new treatment and therapeutic options (such as angioplasty) have led to a long-term trend of declining mortality risks from key diseases identified by EPA as being reduced by lower pollutant levels.

Some experts interpret epidemiological data as showing that "Short-term and long-term studies clearly indicate that relatively modest exposures to particulate matter in the ambient air are associated with increased morbidity and mortality due to coronary heart disease" (Simkhovich et al. 2009). Death from coronary heart disease (CHD), especially among the elderly and patients with pre-existing cardiovascular diseases, has been proposed as a specific mortality risk caused by high levels of particulate pollution (Simkhovich et al. 2008, 2009). The future relation between PM2.5 and mortality risks, therefore, may be affected by trends in CHD risk. These trends show very significant declines in CHD mortality risks over the past four decades, with further declines expected. For example, a recent review (Adabag et al. 2010) that quantified temporal trends in cardiovascular mortality risks in Minneapolis-St Paul, MN found that mortality rates for non-hospitalized men fell by about 2/3 between 1985 and 2008 (from about 150/100,000 in 1985 to about 50/100,000 in 2008). Similarly, for the United States as a whole, a recent World Health Organization (WHO) study (Capewell et al. 2010) found that "Since the late 1970s, age-adjusted CHD mortality rates have been halved in most industrialized countries, including the United States. …Approximately 44% of the substantial CHD mortality decline in the United States between 1980 and 2000 was attributable to changes in major risk factors [smoking, systolic blood pressure, total blood cholesterol, body mass index], and 47% to specific cardiological treatments. These findings resembled those from other industrialized countries. …In conclusion, implementing evidence-based policies to better control tobacco use and achieve a healthier diet across the population could potentially halve future CHD deaths in the United States." Although future reductions in risk will depend on future trends in diet, exercise, obesity, aging of the population, and medical interventions and treatments, there appear to be continued large opportunities for continued risk reductions in the United States (*ibid*). As such trends continue, the incremental benefits from additional pollution control might also be expected to shrink, as fewer people remain at high risk of mortality from pollution-associated diseases such as CHD. Thus, analysis of projected future benefits should include uncertainty about the size of future C-R coefficients, rather than assuming that C-R coefficients estimated from historical data will remain undiminished by current and future trends.

Discrete vs. Continuous Uncertainty Analysis

The major uncertainties about the health effects of reducing pollutant concentrations in air can be summarized as the following series of questions.

- *Is there a true (model-independent) positive C-R association?* What is the probability that there is a true (independent of model assumptions) statistical association between lower pollutant concentrations (C) and lower mortality rates (R)?
- *Is it causal?* Assuming that there is a true C-R association, is it causal? That is, will future reductions in PM2.5 and ozone cause corresponding reductions in future mortality rates? Or, to the contrary, is the association non-causal, e.g., reflecting past downward trends in both variables that do not represent a causal relation, and that do not support manipulating future mortality rates by changing future ambient pollutant concentrations? What is the probability that the C-R association (if there is one) is causal?
- *Is it linear non-threshold (or are present and future concentrations above any applicable threshold, or on the ascending part of any J-shaped C-R function)?* Assuming that a causal C-R relation exists in past data, for concentration levels prevalent then, how likely is it that a similar C-R coefficient will continue to hold in future, as ambient concentrations continue to fall? What is the probability that we are *not* yet at or below a nadir, threshold, or threshold-like nonlinearity, in the C-R function, which would cause future reductions in C to produce no further

reductions in R, and hence no incremental health benefits? Colloquially, we might wonder: How clean is clean enough to achieve the possible health benefits, and have we already reached a point past which no additional health benefits will accrue to additional reductions in concentrations? What is the probability that this is not the case?

- *Is it stable over time?* What is the probability that C-R coefficients will remain approximately the same, despite improvements in prevention and medical treatments for the diseases that are associated with pollutants? If the C-R coefficient does decline in future, as the at-risk population declines, then what is its new expected value?

A notable aspect of these uncertainties is that they are discrete: one can imagine answering yes or no to each one, if perfect information were available, and can therefore envision assigning a discrete probability (a number between 0 and 1) to each of these possibilities now, based on the imperfect information available now.

For purposes of plausible numerical illustration and sensitivity analysis, suppose that the quantitative probabilities for answers to these questions are as follows:

- *Probability that there is a true (model-independent) positive association* = 50%. As discussed previously, the data on associations are conflicting, with both positive and negative significant C-R associations being reported, and with results of individual studies appearing to be sensitive to choices of statistical methods and modeling assumptions. Bayesian Model Averaging (BMA) analyses that permit the possibility of no association between PM2.5 and mortality rates generally find little support for the hypothesis that such an association exists at ambient concentrations; the probability that it does exist was estimated as between about 2% and 7% in a recent BMA analysis (Koop and Tole 2004). Against this conclusion, many experts are convinced that the combined findings from multiple epidemiological studies "clearly indicate that relatively modest exposures to particulate matter in the ambient air are associated with increased morbidity and mortality due to coronary heart disease," (Simkhovich et al. 2009) and that the main remaining scientific challenge is to figure out why and how this occurs. It is not necessary to choose between these conclusions for purposes of illustrating how such uncertainties affect final benefits estimates. Instead, we will use a 50% probability of a true association as a point of departure for quantitative analysis, and then use sensitivity analysis to understand how changes in this probability affect conclusions.
- *Probability that a true association, if it exists, is causal* (and not explained away by trends, confounding, model specification errors, or the distinction between statistical and causal associations) $\leq$0.5. The inconsistencies and large unexplained differences in estimated C-R functions, signs, and magnitudes across studies, discussed in detail in the following chapters, might suggest a "more likely than not" conclusion that there is not a causal relation. A probability interval of $\leq$0.5 captures this uncertain probability. This is strengthened by some past reviews and conclusions, e.g. that, "Because the mortality risk estimates from important observational epidemiologic studies are extremely weak, derived from

studies unable to control for relevant confounding causes, and inconsistent by location, toxicologic and clinical information [it] is necessary to judge the likelihood and degree to which such findings are causal. Toxicologic data on typical forms of pollution-derived PM strongly suggest that current ambient concentrations in the U.S. are too small to cause significant disease or death. …The expectation that lives will be saved by reducing ambient PM2.5 in the U.S. is not supported by the weight of scientific evidence, although other bases for regulating PM may be justifiable" (Green and Armstrong 2003).

- *Probability that currently permitted concentrations are above any threshold* or nadir in the C-R function (if any; or probability that there is no such threshold or nadir) ≤0.5. Repeated finding of negative as well as positive C-R coefficients might suggest the possibility of a J-shaped function, and other reasons for expecting a threshold have previously been discussed (see also Chaps. 10 and 11). The inconsistencies and difficulties in identifying any clear positive C-R association (Green and Armstrong 2003) also suggest that we could now be below the clearly rising portion of a C-R function.
- *Expected reduction factor in C-R coefficient by 2020* ≤0.5, assuming that a linear no-threshold causal C-R coefficient exists. Improved medical diagnosis, intervention (e.g., with aspirin therapy, smoking cessation programs, or cholesterol-reducing measures, including changes in diet and exercise), and treatment, which are increasingly being used, can potentially reduce premature CVD mortalities by 80–90% or more, but an aging or more obese population over the next decade could modestly increase the C-R slope (Capewell et al. 2010). Based on these opposed trends, and in light of the past four decades of declining CVD risks, an overall reduction factor of 0.5 might not be implausible as a starting point for the reduction in any real C-R coefficient, although higher or lower values might be developed with additional modeling. We will use 0.5 as a point of departure for understanding the sensitivity of estimated benefits to this factor.

Each of these factors is conditioned on the outcomes of all of its predecessors being such that a positive linear no-threshold causal C-R coefficient exists.

With these rough estimates of factors (using the upper end of each uncertainty interval, e.g., 0.5 for the interval ≤ 0.5), the health benefits estimated by EPA would be adjusted as follows.

> (1.8 trillion initial estimate) × (1/6 reduction factor for $VSL if age or $VSLY is considered) × (0.5 probability that a true association exists) × (0.5 probability that a true association is causal, given that one exists) × (0.5 probability that ambient concentrations are above any thresholds or nadirs in the C-R function, given that a true causal C-R relation exists) × (0.5 expected reduction factor in C-R coefficient by 2020 due to improved medication and prevention of disease-related mortalities) = (1/6)*0.5*0.5*0.5*0.5 = $19 billion.

In round numbers, the combination of assumptions shown here reduces the estimated value of health benefits in 2020 100-fold, from two trillion dollars to $20 billion. Instead of being much greater than the estimated compliance costs of $65 billion, the estimated value of health benefits is substantially less. This conclusion is robust to changes in any of the above factors, in that replacing any of them with 1 (no reduction) would not change the conclusion that benefits are less than costs. Sensitivity analysis of this product is trivial: changing any factor by k-fold changes the entire product by k-fold, for any $k > 0$.

The point of departure values for this calculation were deliberately chosen realizing that they might be conservative (erring in the direction of over-stating C-R relations and health benefits), with all factors ≤ 0.5 being rounded up to 0.5 for purposes of this initial calculation. Two main changes that might increase realism would significantly reduce estimated benefits. One would be to accept at face value the BMA results indicating that a true positive association between present ambient levels of PM2.5 and mortality rates is unlikely (Koop and Tole 2004). This could reduce the point of departure for this probability by about a further order of magnitude, from 0.5 to about 0.05, and hence reduce the plausible high estimate for monetized value of annual health benefits from \$19 billion to \$1.9 billion. (A strong conviction that causality is very unlikely would have a similar effect.) The other main change would be to replace the 1/6 reduction factor for \$VSL with a reduction factor of 1/10 for \$VSLY (or value of a QALY), as discussed earlier. This would reduce annual benefits estimates by a further factor of 0.6 (e.g., from \$1.9B to approximately \$1B). Such numbers are well below the \$65B in annual compliance costs estimated by EPA. If these two alternative values were used as points of departure, then the conclusion that expected benefits are less than expected costs would be robust to changes that increased any or all of the remaining factors to 100%.

These calculations are all based on expected values (albeit for upper bounds, if the points of departure are indeed conservative). For a risk-averse individual or society, however, the economic value of an uncertain benefit is *less* than its expected value, by an amount that depends on the risk premium for uncertainty. A further reduction in the evaluation of the uncertain benefits (by perhaps a further factor of 2, depending on the risk premium) would be needed to take into account risk aversion for uncertain gains.

Such further adjustments and refinements, while possibly producing a more realistic answer, might also invite inessential controversy. Our main conclusion, that the 1990 CAAA benefits plausibly amount to only a small fraction of their direct compliance costs (the only costs considered in the \$65 billion estimate), does not depend on choosing very small probabilities as points of departure, nor on greatly revising EPA's value judgments about \$VSL or \$VSLY, nor on incorporating risk aversion for uncertain benefits. Rather, it is a robust result of the combination, in series, of multiple substantial uncertainties. The result is robust in the sense that no particular one of these uncertainties (and no small subset, if the suggested smaller points of departure are adopted) is essential for reaching this conclusion.

However, the conclusion that the 1990 CAAA human health benefits are very probably substantially smaller than compliance costs offers a sharp contrast to EPA's more reassuring conclusion that "*The extent to which estimated benefits exceed estimated costs and an in-depth analysis of uncertainties indicate that it is extremely unlikely the costs of 1990 Clean Air Act Amendment programs would exceed their benefits under any reasonable combination of alternative assumptions or methods identified during this study.*" What explains the difference? The key is that *EPA's uncertainty analysis ignores discrete uncertainties*, such as those we have enumerated, in favor of a continuous, non-negative probability distribution—the Weibull distribution—that puts zero probability density on zero or negative numbers as possible values. Applied to the crucial C-R coefficient, the Weibull uncertainty distribution implies 100% confidence that the coefficient is positive, notwithstanding the numerous studies that have reported zero or negative coefficients. This framing of the uncertainty analysis only allows one to ask (and experts to answer) *how large* are the assumed benefits, but not to question (or provide opinions on) whether they exist. Thus, the major discrete uncertainty about whether positive benefits exist at all—the main focus of our analysis—is simply assumed away in EPA's Weibull uncertainty distribution.

In this context, EPA's conclusion that "it is extremely unlikely the costs of 1990 Clean Air Act Amendment programs would exceed their benefits under any reasonable combination of alternative assumptions or methods identified" is no surprise. It is little more than an assertion of a tautology, similar to, "Assuming that I am right, it is extremely unlikely that any reasonable combination of alternative assumptions would show that I am wrong." But this style of uncertainty analysis fails to cast any light on uncertainty about the major premise. The not-implausible numbers offered above suggest that that probability of zero human health benefits might well exceed $1 - (0.5*0.5*0.5) = 87.5\%$ based solely on uncertainties about whether there is a positive, causal (not due to confounding or coincidence of trends), non-threshold C-R relation for current and future ambient pollutant concentrations (assuming, conservatively, that there is a 50% probability for each of these three uncertainties—if smaller points of departure are used, then the probability of zero human health benefits exceeds 99%). Thus, assuming that the probability of zero (or negative) human health benefits is zero substantially understates the true uncertainty, and hence substantially overstates the uncertain benefits of the CAAA. As noted by other commentators in a different context (chemical risk assessments), "Fundamentally, the EPA should replace risk values that are built on science-policy assumptions with risk estimates that acknowledge underlying uncertainties" (Gray and Cohen 2012).

The use of a Weibull uncertainty distribution also conceals other discrete uncertainties, arising from blinkered analysis and failure of imagination. For example, suppose that the CAAA creates more clear days and bright skies. How, if at all, would this affect peak temperatures during the summer, and mortality rates among people without air conditioners? How, if at all, would it affect the future burden of skin cancer? The cost-benefit analysis does not say, and the Weibull uncertainty analysis implicitly disregards all such uncertainties due to drivers of possible costs and benefits not already identified in the analysis. Or suppose that, contrary to our expectations, the CAAA does prolong the last year of life for a substantial fraction of elderly patients suffering from heart, lung, and circulatory diseases. How would this affect health care costs and living costs, and the benefits of longer life, as evaluated by patients and their families? These costs and uncertainties are likewise implicitly assumed away in EPA's analysis, but considering them might further increase the probability that the net benefits of the CAAA are negative, and further increase the estimated ratio of costs to benefits, beyond the roughly 65-to-1 value (or more) suggested by some of our calculations.

Conclusions

EPA's cost-benefit assessment indicates that the CAAA offers the United States an admirable return on investment: an expenditure of only $65 billion per year secures health benefits valued at nearly two trillion dollars per year. If this is credible, any policy maker would be foolish to refuse it. EPA's uncertainty analysis suggests that the large excess of estimated costs over estimated benefits is not only credible, but virtually certain: it is deemed to be "extremely unlikely" that any reasonable changes in assumptions would reverse the sign of net benefits.

However, these promising conclusions depend essentially on the use of a Weibull distribution in the uncertainty analysis, which tacitly assumes away any possibility of serious, discrete errors or uncertainties in the key assumptions—especially about whether future reductions in PM2.5 and ozone will necessarily cause future increases in life expectancy. Yet, this is exactly the uncertainty that a large, data-driven literature questions; this literature is explored in the following chapters. Numerous reported findings of unexplained negative and highly heterogeneous positive C-R coefficients call into question the assumption that any such single, positive, causal coefficient exists.

Taking into account such discrete uncertainties suggests that the probability of achieving positive benefits could well be small (e.g., less than 12.5%, or less than 1%, if some BMA results (Koop and Tole 2004) are used) so that EPA's uncertainty analysis assuming that this probability is 100% may be more misleading than informative. A policy maker to whom the CAAA costs and benefits are presented as: "Pay $65 billion for a chance of probably much less than 20% to achieve benefits of about $300 billion," as our numbers suggest, might well decide that the cost is too great for the uncertain benefits. Even if all of our suggested specific numbers are rejected and replaced by more carefully developed ones, it seems that considering discrete uncertainties has important effects on the analysis, and that such uncertainties should be presented to policy makers if they are to make well-informed decisions to serve the public interest.

The calculations presented here do not necessarily prove beyond doubt that the CAAA costs far more than it produces in benefits. But they make this possibility very likely, and hence well worth including, instead of assuming away, in any uncertainty analysis of CAAA cost-benefit comparisons. Being scrupulously clear about what we do and do not know is a prerequisite for better-informed risk management decision-making and more useful risk-cost-benefit analysis of regulations.

Scientific questions about what health effects might be caused at different exposure levels of fine particulate air pollution, and whether the concentration-response relation has thresholds or other nonlinear characteristics at low concentrations, remain fascinating and important topics for research. They invite application of the best available methods of causal analysis and modeling. The remainder of this book makes a start at responding to this scientific challenge.

References

Adabag AS, Luepker RV, Roger VL, Gersh BJ. Sudden cardiac death: epidemiology and risk factors. Nat Rev Cardiol. 2010;7:216–25.

Aldy JE, Viscusi WK. Age differences in the value of statistical life: revealed preference evidence. Rev Environ Econ Policy. 2007;1(2):241–60.

Capewell S, Ford ES, Croft JB, Critchley JA, Greenlund KJ, Labarthe DR. Cardiodiovascular risk factor trends and potential for reducing coronary heart disease mortality in the United States of America. Bull World Health Organ. 2010;88(2):81–160.

Clyde M. Model uncertainty and health effect studies for particulate matter. Environmetrics. 2000;11(6):745–63.

Cox LA Jr. Reassessing the human health benefits from cleaner air. Risk Anal. 2012;32(5):816–29. https://doi.org/10.1111/j.1539-6924.2011.01698.x.

Daniels MJ, Dominici F, Samet JM, Zeger SL. Estimating particulate matter-mortality dose-response curves and threshold levels: an analysis of daily time-series for the 20 largest US cities. Am J Epidemiol. 2000;152(5):397–406.

EPA. The benefits and costs of the clean air act from 1990 to 2020: summary report. Washington, D.C.: U.S. EPA, Office of Air and Radiation; 2011a.

EPA. The benefits and costs of the clean air act from 1990 to 2020. Full report. Washington, D.C.: U.S. EPA, Office of Air and Radiation; 2011b.

Franklin M, Zeka A, Schwartz J. Association between PM2.5 and all-cause and specific-cause mortality in 27 US communities. J Expo Sci Environ Epidemiol. 2007;17(3):279–87.

Gold M, Siegel JE, Russell LB, Weinstein MC. Cost-effectiveness in health and medicine. New York: Oxford University Press; 1996.

Gray GM, Cohen JT. Rethink chemical risk assessments. Nature. 2012;389:27–8.

Green LC, Armstrong SR. Particulate matter in ambient air and mortality: toxicologic perspectives. Regul Toxicol Pharmacol. 2003;38(3):32.

Koop G, Tole L. Measuring the health effects of air pollution: to what extent can we really say that people are dying from bad air? J Environ Econ Manag. 2004;47:30–54.

Koop G, Tole L. An investigation of thresholds in air pollution–mortality effects. Environ Model Softw. 2006;21(12):1662–73.

Krewski D, Jerrett M, Burnett RT, et al. Extended follow-up and spatial analysis of the American Cancer Society study linking particulate air pollution and mortality. Res Rep Health Eff Inst. 2009;140:5–114.

Laden F, Schwartz J, Speizer FE, Dockery DW. Reduction in fine particulate air pollution and mortality: Extended follow-up of the Harvard Six Cities study. Am J Respir Crit Care Med. 2006 Mar 15;173(6):667–72. https://doi.org/10.1164/rccm.200503-443OC. Epub 2006 Jan 19. PMID: 16424447; PMCID: PMC2662950.

Murphy K, Topel R. The economic value of medical research. In: Murphy K, Topel R, editors. Measuring the gains from economic research. an economic approach. Chicago: University of Chicago Press; 2003. p. 9–40.

Pope CA 3rd, Burnett RT, Thun MJ, Calle EE, Krewski D, Ito K, Thurston GD. Lung cancer, cardiopulmonary mortality, and long-term exposure to fine particulate air pollution. JAMA. 2002 Mar 6;287(9):1132–41. https://doi.org/10.1001/jama.287.9.1132. PMID: 11879110; PMCID: PMC4037163.

Roberts S, Martin MA. The question of nonlinearity in the dose-response relation between particulate matter air pollution and mortality: can Akaike's Information Criterion be trusted to take the right turn? Am J Epidemiol. 2006;164(12):1242–50.

Schwartz J, Coull B, Laden F, Ryan L. The effect of dose and timing of dose on the association between airborne particles and survival. Environ Health Perspect. 2008;116(1):64–9.

Sheppard L, Burnett RT, Szpiro AA, Kim S-Y, Jerrett M, Pope CA, Brunekreef B. Confounding and exposure measurement error in air pollution. Air Qual Atmos Health. 2012;5(2):203–16.

Simkhovich BZ, Kleinman MT, Kloner RA. Air pollution and cardiovascular injury epidemiology, toxicology, and mechanisms. J Am Coll Cardiol. 2008;52(9):719–26.

Simkhovich BZ, Kleinman MT, Kloner RA. Particulate air pollution and coronary heart disease. Curr Opin Cardiol. 2009;24(6):604–9.

Stoeger T, Reinhard C, Takenaka S, Schroeppel A, Karg E, Ritter B, Heyder J, Schulz H. Instillation of six different ultrafine carbon particles indicates a surface area threshold dose for acute lung inflammation in mice. Environ Health Perspect. 2006;114(3):328–33.

Thomas DC, Jerrett M, Kuenzli N, Louis TA, Dominici F, Zeger S, Schwartz J, Burnett RT, Krewski D, Bates D. Bayesian model averaging in time-series studies of air pollution and mortality. J Toxicol Environ Health A. 2007;70(3-4):311–5.

Vitolo C, Scutari M, Ghalaieny M, Tucker A, Russell A. Modeling air pollution, climate, and health data using Bayesian Networks: a case study of the English regions. Earth Space Sci. 2018;5:76–88. https://doi.org/10.1002/2017EA000326.

Yabroff KR, Bradley CJ, Mariotto AB, Brown ML. Estimates and projections of value of life lost from cancer deaths in the United States. J Natl Cancer Inst. 2008;100(24):1755–62.

Chapter 15
Do Causal Concentration-Response Functions Exist?

Introduction

Concentration-response (C-R) functions relating concentrations of pollutants in ambient air to mortality risks or other adverse health effects provide the basis for many public health risk assessments, benefits estimates for clean air regulations, and recommendations for revisions to existing air quality standards (Schwartz et al. 2002). The assumption that C-R functions relating levels of exposure and levels of response estimated from historical data usefully predict how future changes in concentrations would change risks has seldom been carefully tested. This chapter critically reviews literature on C-R functions for fine particulate matter (PM2.5) and mortality risks, building on the ideas and methods from Chaps. 2, 7 and 8 (especially that regression coefficients do not necessarily have valid causal interpretations) and Chaps. 9, 10, 11, 12, and 13 (especially that Bayesian networks and partial dependence plots can provide useful additional information about dependence relationships that help to identify, constrain, and quantify potential direct and total causal relationships, modelled via invariant causal conditional probability tables). We find that most of the historically influential papers in the literature on PM2.5 and mortality risks only describe historical statistical associations rather than valid causal models for predicting effects of interventions that change concentrations. The few papers that explicitly attempt to model causality rely on unverified statistical modeling assumptions, casting doubt on their predictions about effects of interventions. Modern causal inference algorithms for observational data (Chap. 9) have been little used in C-R modeling to date. Applying these methods to publicly available data from Boston and the South Coast Air Quality Management District around Los Angeles shows that C-R functions estimated for one do not hold for the other. Changes in month-specific PM2.5 concentrations from one year to the next do not help to predict corresponding changes in average elderly mortality rates in either location. Thus, the assumption that estimated statistical C-R relations predict effects

L. A. Cox Jr., *Quantitative Risk Analysis of Air Pollution Health Effects*,
International Series in Operations Research & Management Science 299,
https://doi.org/10.1007/978-3-030-57358-4_15

of pollution-reducing interventions may not be true. Better causal modeling methods are needed to better predict how reducing air pollution would affect public health.

Causal vs. Associational Concentration-Response Relations

A concentration-response (C-R) curve shows levels of adverse health responses in exposed populations on its vertical axis and levels of ambient concentrations of a pollutant on its horizontal axis. Such curves, which are usually upward-sloping, have been widely used to predict the public health impacts of proposed reductions in air pollutants (Pope et al. 2015; Schwartz et al. 2002). These predictions are made by assuming that reducing air pollution will reduce adverse health effects, moving both exposure and response variables leftward and downward along the C-R curve. Thus, a C-R curve is commonly given both of the following two interpretations:

1. *Associational interpretation.* A C-R curve shows the estimated response (*R*) for each estimated level of exposure concentration (*C*), based on historical data. The association between C and R is typically described by a regression model. Such a model may also include trends, season, weather, and other variables, including co-pollutants (Schwartz et al. 2002). We call this the *associational* interpretation of a C-R curve. It describes C-R associations between levels of *C* and levels of *R* in historical data. Common measures of C-R associations include relative risk (*RR*), the regression coefficient for *C* as a predictor of *R*, and quantities derived from them such as the odds ratio (*OR*), population attributable risk, etiologic fraction, or global burden of disease estimates.
2. *Causal interpretation.* The C-R curve shows what the estimated response *would become* if the exposure concentration were to be fixed at different levels. Thus, it also shows how the response would change if concentration were changed. In the terminology of Chap. 9, this is a manipulative causal interpretation of the curve.

As illustrated by the examples in Table 15.1, these two interpretations are usually conflated, leading to causal interpretations of associations. For example, causal claims such as that "The magnitude of the *association* suggests that controlling fine particle pollution *would result* in thousands of fewer early deaths per year" (Schwartz et al. 2002) have remained very common in the epidemiological literature on air pollution health effects, and in regulatory risk assessments based on this literature, even though, as discussed next, there is no necessary relation between the magnitude or direction of a historical C-R association and the effects on *R* of reducing *C*.

In principle and in practice, there may be no single curve satisfying both interpretations. To again use the example of aspirin: finding a positive association between daily consumption of aspirin and heart attack risk in an elderly population, perhaps because elderly people at higher risk of heart attacks are more likely to be prescribed an aspirin regimen to reduce that risk, would imply nothing about how

Table 15.1 Examples of association and causation being conflated in the literature on PM2.5 health effects (all emphases added)

Interpretation in literature	Comments
"We observed statistically significant and robust *associations* between air pollution and mortality… these results suggest that fine-particulate air pollution, or a more complex pollution mixture associated with fine particulate matter, *contributes to* excess mortality in certain U.S. cities" (Dockery et al. 1993)	Associations do not suggest a contribution to excess mortality unless they are causal
"The magnitude of the *association* suggests that controlling fine particle pollution *would result* in thousands of fewer early deaths per year" (Schwartz et al. 2002)	Associations do not allow prediction of results from changes in exposure concentrations unless the associations represent manipulative causal relations
"We examined the *association* between PM(2.5) and both all-cause and specific-cause mortality… Our findings describe the magnitude of the *effect* on all-cause and specific-cause mortality, the modifiers of this association, and suggest that PM(2.5) may pose a *public health risk* even at or below current ambient levels" (Franklin et al. 2006)	An association with mortality is not an effect on mortality. A C-R association does not suggest that exposure poses a public health risk, unless the association is causal
"Residential ambient air pollution exposures were *associated* with mortality… our study is the first to assess the *effects* of multiple air pollutants on mortality with fine control for occupation within workers from a single industry" (Hart et al. 2011)	Associations with mortality are not effects on mortality
"Each increase in PM2.5 (10 μg/m³) was *associated* with an adjusted increased risk of all-cause mortality (PM2.5 average on previous year) of 14%... These results suggest that further public policy efforts that reduce fine particulate matter air pollution are likely to have continuing *public health benefits*" (Lepeule et al. 2012)	Associations do not suggest that public policy efforts that reduce exposure are likely to create public health benefits unless the associations reflect manipulative causation
"Ground-level ozone (O3) and fine particulate matter (PM2.5) are *associated* with increased risk of mortality. We quantify the *burden* of modeled 2005 concentrations of O3 and PM2.5 on health in the United States. … Among populations aged 65–99, we estimate nearly 1.1 million *life years lost from PM2.5 exposure*… Among the 10 most populous counties, the percentage of deaths *attributable to PM2.5* and ozone ranges from 3.5% in San Jose to10% in Los Angeles. These results show that despite significant improvements in air quality in recent decades, recent levels of PM2.5 and ozone still pose a *nontrivial risk* to public health" (Fann et al. 2012)	In the absence of manipulative causation, statistical associations between pollutant levels and mortality risks do not quantify effects caused by exposure on burden of disease or on life-years lost or on deaths, nor do they indicate a risk to public health

(continued)

Table 15.1 (continued)

Interpretation in literature	Comments
"Ambient fine particulate matter (PM2.5) has a large and well-documented global *burden of disease*. Our analysis uses high-resolution (10 km, global-coverage) concentration data and cause-specific *integrated exposure-response (IER) functions developed for the Global Burden of Disease 2010* to assess how regional and global improvements in ambient air quality *could reduce attributable mortality from* PM2.5. Overall, an aggressive global program of PM2.5 mitigation in line with WHO interim guidelines *could avoid* 750,000 (23%) of the 3.2 million deaths per year currently (ca. 2010) attributable to ambient PM2.5" (Apte et al. 2015)	The Global Burden of Disease IER functions are based on relative risk measures of association. They do not allow prediction or assessment of "how… improvements on ambient air quality could reduce attributable mortality" or avoid deaths unless the underlying relative risks represent (manipulative) causal relations
"We use a high-resolution global atmospheric chemistry model combined with *epidemiological* concentration response *functions* to investigate *premature mortality attributable to* PM2.5 in adults ≥30 years and children < 5 years. …[A]pplying worldwide the EU annual mean standard of 25 μg/m(3) for PM2.5 could *reduce global premature mortality* due to PM2.5 exposure by 17%… Our results reflect the need to adopt stricter limits for annual mean PM2.5 levels globally… *to substantially reduce premature mortality* in most of the world" (Giannadaki et al. 2016)	Epidemiological exposure concentration-response associations and estimates of PM2.5-attributable mortalities based on them do not imply that reducing PM2.5 would reduce mortality, or allow such reductions to be predicted, unless the associations represent manipulative causal relations
"*Relative risks* were derived from a previously developed exposure-response model. …Nationally, the *population attributable mortality fraction* of PM2.5 for the four disease causes was 18.6% (95% CI, 16.9-20.3%). … Aggressive and multisectorial intervention strategies are urgently needed to *bring down the impact o*f air pollution on environment and health" (Lo et al. 2016)	Relative risks and population attributable mortality fractions are measures of exposure-response associations. Such associations do not imply that interventions to reduce exposures would reduce risks of adverse responses unless there is a manipulative causal relation between them
"In the US Medicare population from 2000 to 2012, short-term exposures to PM2.5 and warm-season ozone were *significantly associated with* increased risk of mortality. This risk occurred at levels below current national air quality standards, suggesting that these *standards may need to be reevaluated*" (Di et al. 2017)	Significant exposure-mortality associations do not suggest that standards may need to be reevaluated unless the associations reflect manipulative causality rather than confounding, biases, etc.

changing daily consumption of aspirin would *change* future heart attack risks. The data only address associations between historical levels, which do not reveal the impacts of future changes. The historical association between their levels may be positive and the future relationship between changes in their levels negative, so that reducing aspirin consumption would increase risk, even though historical levels of aspirin consumption and risk are positively correlated. More generally, finding a positive C-R association in historical data does not necessarily imply anything about how changing C would change R. It is *not* generally true that if a C-R model describes past data values for C and R, then it also predicts how changing exposure concentration from a current level to a new level will change the average response.

Example: C-R Associations do not Necessarily Provide Valid Causal Predictions

To illustrate, consider a simplified setting in which daily mortality rate, R and average daily exposure concentration, C have been measured for several years and the following associational C-R model perfectly fits the data:

$$R = \mathrm{C} + 50\left(\text{Model 1: Associational C} - \text{R relation}\right)$$

That is, each additional unit of exposure concentration is associated with an additional unit of daily mortality, which on these scales corresponds to a 2% increase above the zero-exposure baseline mortality rate of 50. Do these data imply or suggest that reducing C would reduce R? Not necessarily, because an associational model does not necessarily represent causal mechanisms (McClellan 2016; Moolgavkar 2016). Other variables may affect how or whether R changes when C changes. For example, suppose that an unmeasured third variable, T, such as minimum daily temperature, affects C and R as described by the following structural equations:

$$C = 50 - 0.5T, \text{for } 0 \le T \le 100$$

$$R = 150 - C - T, \text{for } 0 \le C + T \le 150\left(\text{Model 2 : Causal C} - \text{R relation}\right)$$

These equations have the following explicit causal interpretation: if the value of a variable on the right side of an equation is changed, then the value of the dependent variable on the left side will change to restore equality. Thus, changes propagate from right to left through these equations. The causal C-R relation is then very simple: decreasing C by one unit via an exogenous intervention would *increase* R by one unit (since the two are *negatively* related, with a C-R coefficient of -1 for the causal impact of changes in C on changes in R). The associational C-R relation is simply Model 1. (Solving the first equation for T (yielding $T = 100 - 2C$) and substituting into the second equation to eliminate T yields the reduced-form C-R equation: $R = C + 50$.) Model 1 correctly reveals that days with one unit less of C have historically been associated with one unit *less* of R (i.e., the two are *positively* associated), even though Model 2 correctly reveals that reducing C by one unit would *increase* R by one unit. Thus, the causal and associational C-R relations are quite different. The associational model would be entirely valid for predicting how many deaths would occur on days with different exposure concentrations in the absence of interventions, but only the causal model can be used to correctly predict how changing *C* would change *R*. In reality, of course, many C-R models include temperature, but other unmeasured, unmodeled, or mis-modeled variables can drive wedges between associational and causal C-R relations, so care should be taken to distinguish between these potentially quite different interpretations of C-R relations. Such care has not been widely exercised in the literature reviewed next.

The importance of distinguishing between association and causation in interpreting C-R functions has recently been emphasized in comments on estimated human health benefits from reducing ambient concentrations of fine particulate matter (PM2.5) air pollution (North 2016; Cox Jr. 2016; Frey 2016; McClellan 2016; Moolgavkar 2016; Smith 2016). As detailed in Chap. 14, in studies that have been influential for estimating quantitative public health benefits attributed to PM2.5 reductions, the EPA has assumed a causal relation between PM2.5 reductions and reductions in both acute and chronic mortality for purposes of estimating health and economic benefits, while acknowledging in technical appendices that these causal assumptions are uncertain and are not clearly established by epidemiological data. This chapter seeks to contribute to a sounder and less assumption-dependent scientific basis for understanding the probable human health consequences of changing pollution levels, by challenging the common practice of treating association as if it were causation for purposes of air pollution C-R modeling and by drawing attention to different concepts of causation and to less assumption-based methods for assessing causal impacts,.

One specific purpose of this chapter is to critically review how C-R functions have been developed and used in public health risk assessments and risk management policy recommendations, focusing on fine particulate matter (PM2.5) and mortality risks as a prominent example. Because interpretation of C-R functions for purposes of risk management decision-making is typically based on consideration of the wider context provided by multiple studies, rather than relying on analyses of any single data set, we focus on patterns of data analysis and causal interpretation that are prevalent across multiple studies. A second purpose of this chapter is to examine the extent to which associational C-R curves estimated from publicly available data permit accurate prediction of changes in health effects, *R*, based on changes in ambient concentrations of air pollution, *C*. To this end, we examine available data for the Boston and Los Angeles areas, recently identified in C-R studies as locations where further reducing PM2.5 concentrations is predicted to create substantial health benefits (Cromar et al. 2016; Schwartz et al. 2016). A third purpose of this chapter is to examine how well C-R relations estimated for elderly people in Boston apply across the continent in Los Angeles. Finding the same C-R curve in such different geographic areas might suggest that it represents a predictively useful causal relation (consistent with the invariant causal prediction (ICP) property from Chap. 9). Conversely, finding that different C-R curves hold in different locations or at different times in the same location would be evidence to the contrary.

The remainder of this chapter is organized around these three purposes. The following section first examines how C-R functions have been developed and used in the epidemiological literature attributing adverse health effects to fine particulate matter. It critically discusses methods used in this literature to draw causal inferences and compares them to other methods for causal inference—what we call *information-based methods*—developed in the past century in genetics, engineering, economics and econometrics, statistics, systems biology, physics, computer science and artificial intelligence, machine learning, and other fields (Pearl 2009a). The next section applies these information-based causal inference methods to two publicly available data sets for PM2.5 and elderly mortality in the Northeast (Boston) and Southwest (Los Angeles air basin) and compares the results to each other and to results from associational (regression) modeling. Finally, we discuss limitations of the illustrative results presented and the information-based methods reviewed and acknowledge their rich and deep grounding and intellectual history. We conclude that current methods of information-based causal analysis used in other fields can provide a valuable complement to other techniques used in air pollution epidemiology. Applying them suggests that *causal C-R functions, as they are usually described, understood, and applied, may not exist.* Rather, published C-R functions quantify statistical associations that do not necessarily predict correctly how changing exposure concentrations would affect risk of adverse health responses. To better achieve this predictive goal, information-based methods can be used to address aspects of causality that are not well addressed by associational and potential outcomes methods.

Critical Review and Synthesis of Literature on C-R Relationships for PM2.5

Many papers have applied estimated C-R regression relationships to quantify human health risks associated with fine particulate matter (PM2.5) and to project human health benefits from further reducing ambient PM2.5 levels. Various criticisms of statistical uncertainties, limitations, and biases in C-R regression estimates have been offered (e.g., Krstić et al. 2016; Young and Xia 2013). The following paragraphs refer to selected papers, including recent ones and ones frequently cited in regulatory risk assessments, to illustrate the following key themes:

1. Associational and causal C-R functions have commonly been conflated in the literature, usually without explicit discussion or careful consideration of what causation means (Maldonado 2013).
2. There are many possible non-causal explanations for reported positive C-R associations (see Chap. 2). These are usually not systematically addressed in the air pollution health effects literature using appropriate methods such as multiple-bias analysis (Greenland 2005).
3. Recent attempts to address causality for PM2.5 and mortality have relied on modeling assumptions of unknown or doubtful validity, such as that there are no omitted confounders, that interactions (e.g., between temperature and PM2.5) are negligible, or that simplistic parametric statistical models correctly describe

the data-generating process. They leave unaddressed other crucial assumptions, such as that individuals with different exposures are otherwise exchangeable, or that individual-level and population-level causation are consistent (Maldonado 2013).
4. Modern information-based algorithms for causal inference can be applied to PM2.5 health effects data, although few studies have yet done so.

Many Important Past Papers Equate Associational and Causal C-R Relations

Current risk assessments, benefits assessments, and recommendations for revising standards for criteria pollutants build on a decades-old scientific literature that routinely makes the following three implicit assumptions:

a. Associational and causal C-R functions can be modeled by the same curve;
b. This C-R curve can be estimated quantitatively from relevant data via regression models or other associational methods such as odds ratios, relative risks, attributable risks, and burden of disease estimates, perhaps augmented with human judgments based on the Sir Austin Bradford Hill considerations (Hill 1965) or other weight-of-evidence (WoE) considerations such as the strength, consistency, temporality, and biological plausibility of associations (Fedak et al. 2015).
c. C-R regression coefficients or other associational measures estimated from one set of locations and times can be applied to exposure concentrations for other locations and times to estimate excess mortalities caused by air pollution and the potential human health benefits that would be caused by reducing it (e.g., Chen et al. 2013). In effect, this simply assumes the Invariant Causal Prediction (ICP) property from Chap. 9 without testing and verifying it.

Table 15.1 gives examples of statements from articles that make one or more of these assumptions. Examples of associational and causal language are italicized (all emphases added). Brief comments in the right column note where stated causal interpretations do not follow from the associations presented. These articles include many that have been cited by EPA and others in deliberations on PM2.5 risks and regulations in recent decades, continuing up to the present.

Such examples can be multiplied many-fold. Published papers on health effects attributed to pollutants typically move freely between associational and causal interpretations of C-R associations. They routinely present positive C-R associations as implying or suggesting that reducing air pollution would improve public health. But this is true only if the C-R associations are in fact causal—more specifically, only if they describe manipulative causality (Woodward 2013), as opposed to associational or counterfactual or predictive causation. To make this chapter relatively self-contained, these concepts are reviewed later; they are also discussed in more detail in Chap. 9 and its references. None of these papers establishes that the presented C-R associations do in fact describe (manipulative) causation, so none is suitable for predicting the effects on *R* of changing *C* by reducing air pollution. Of course, even if many published causal conclusions do not follow from their stated associational premises, this does not necessarily imply that either the premises or the conclusions are mistaken. It only implies that the methods used to draw causal conclusions are not reliable: they do not generate conclusions that logically follow from the premises. They cannot be depended on to yield correct conclusions and may even yield contradictory results in the hands of different investigators (Dominici et al. 2014; Glaeser 2006), although they might also yield correct conclusions on some important occasions, as in the example of the Bradford-Hill considerations applied to cigarette-smoking and lung cancer.

There are Many Potential Non-causal Explanations for Positive C-R Associations

If most positive C-R associations—or at least those satisfying commonly discussed weight-of-evidence considerations such as strength, consistency, specificity, temporality, and biological plausibility (Hill 1965; Höfler 2005; Fedak et al. 2015)—were in fact usually causal, then the distinction between associational and causal C-R relations might be of little more than academic interest. But there are many possible non-causal explanations for such C-R associations. As discussed in Chap. 2, depending on the details of the study design used, non-causal C-R associations can arise for any of the following reasons.

- *Coincident historical trends.* Both *C* and *R* might be falling over time, but not because either causes the other. An example discussed later is a study of mortality risks before and after coal burning bans (Clancy et al. 2002). Non-stationary time series (e.g., statistically independent random walks) often exhibit significant correlations in the absence of causation because they each independently tend to have trends over any interval of observation. To the extent that they are all correlated with time, they tend to be correlated with each other—the problem known as "spurious regression" (e.g., Yule and Udny 1926). "History," meaning trends or events that cause effects to change following interventions, but not because of the interventions, has been recognized and controlled for as one of the standard "threats to internal validity" of causal inferences in social statistics and quasi-experimental studies since the 1960s (Shadish et al. 2002). These ideas have had limited, but useful, impact on improving causal inference in health risk research (Slack and Draugalis 2001).
- *Omitted confounders.* Many data sets that identify positive C-R associations do not include potential socioeconomic confounders such as income, education, and occupation of those exposed, or societal factors such as stress and behavioral and societal factors (Valberg 2003), or time-varying confounders such as wind speed (Urban and Kyselý 2014; Morabito et al. 2014) or winter colds and flus. Such

unmeasured variables might exert far larger effects on mortality than environmental exposures and differ systematically between more- and less-exposed areas.

- *Residual confounding*. This arises when a confounding variable is incompletely controlled for, e.g., by using discrete levels of a continuous confounding variable such as age or temperature, which allows for exposure and response variables to remain confounded *within* the discrete levels; or by using relatively stiff splines or inflexible parametric statistical models to partially control for continuous variables such as recent daily temperatures. For example, Wang et al. (2016) note that "Many studies have reported the associations between long-term exposure to PM2.5 and increased risk of death. However, to our knowledge, none has used a causal modeling approach or controlled for long-term temperature exposure, and few have used a general population sample." To address these gaps, they explain that "We *estimated the causal effects* of long-term PM2.5 exposure on mortality and tested the effect modifications by seasonal temperatures, census tract–level socioeconomic variables, and county-level health conditions… Specifically, we *estimated the association* between long-term exposure to PM2.5 and mortality while *controlling for geographical differences using dummy variables* for each census tract in New Jersey, a state-wide time trend *using dummy variables for each year* from 2004 to 2009, and *mean summer and winter temperatures for each tract in each year*". In addition to the usual problem of conflating "we estimated the causal effects" with "we estimated the association," using a single dummy variable for each year and mean summer and winter temperatures leaves room for substantial residual confounding of exposure-response relations by time and by temperature variations within the summer and winter seasons.
- *Modeling biases* provide other well-known non-causal sources of association. As discussed in Chap. 2, these include biases from omitted variables, omitted error terms or classification error probabilities for predictors, biases in data selection or coding, model form selection biases, and model functional form specification errors. They are sometimes addressed via "multiple-bias modeling" (Greenland 2005; Höfler et al. 2007). They can greatly affect conclusions. As illustrated in our example with Models 1 and 2, different choices about which predictors to include on the right side of a regression model can change the sign, as well as the magnitude, of the C-R regression coefficient. Such findings have led some prominent researchers to conclude that "There is a growing consensus in economics, political science, statistics, and other fields that the associational or regression approach to inferring causal relations—on the basis of adjustment with observable confounders—is unreliable in many settings" (Dominici et al. 2014). Sources of modeling bias can also interact in subtler ways. For example, suppose that mortality rate and exposure concentration depend only on temperature, but not on each other, via the following structural equations:

$$mortality\ rate = 100 - (0.1 * temperature)^2$$

$$\textit{concentration} = 100 - \textit{temperature}, \text{for } 0 < \textit{temperature} < 100.$$

If a large data set is created by sampling temperature values from the interval from 0 to 100°, with each observed value of temperature and concentration differing from its true value by a small random error independently and identically uniformly distributed between −2 and 2, then fitting a linear regression model for mortality rate as a function of observed concentration and temperature will show that mortality rate is significantly *associated* only with concentration, and not with temperature, even though by construction mortality rate *depends* only on temperature, and not on concentration. Fitting a mis-specified (linear) regression model yields a statistically significant C-R regression coefficient in the absence of a causal relation. Of course, in this simple example, regression diagnostics (e.g., plotting the data and residuals) would reveal the need for a nonlinear (quadratic) term rather than a linear term for temperature, allowing the incorrect conclusions to be avoided. But for the large and complex epidemiological models commonly used in practice, regression diagnostics and multiple-bias analyses that would allow effects of model specification errors and omitted errors-in-variables to be corrected are often not presented, potentially allowing modeling biases to affect results in unquantified ways (Greenland 2005).

Some of these possible non-causal explanations are sometimes mentioned in papers reporting positive C-R associations. However, they are seldom systematically listed and refuted by data, leaving readers uncertain about whether reported associations reflect reliable truths about the world or are only artifacts of modeling choices (Greenland 2005; Dominici et al. 2014; Glaeser 2006).

That potential non-causal explanations for C-R associations occur commonly in practice makes it important to report model diagnostics and to address such rival explanations before interpreting reported C-R associations as having causal significance or policy relevance (Greenland 2005). Yet, influential papers such as those in Table 15.1 often do little more than argue that smoking or selection biases are unlikely to explain the full C-R association before presenting a causal interpretation and urging policy-relevant recommendations based on it.

Recent Papers Draw Causal Conclusions from Observational Data by Making Unverified Assumptions

Encouragingly, the more recent literature has started to address the question of causation more explicitly. This is most often done by introducing unverified modeling assumptions to justify interpreting regression coefficients, associations, and differences of means or proportions as if they were measures of causal impacts. The following types of studies and modeling assumptions have been used to support important causal claims about C-R associations for PM2.5 and mortality.

- *Intervention studies*, such as a widely cited study of effects on mortality risks of coal burning bans in Dublin County, Ireland (Clancy et al. 2002), assume that if health risks change *following* an intervention, then the change is *caused by* the intervention. This assumption was tested a decade after the original Dublin study in an updated study that compared mortality rates in areas affected and not affected by the bans (Dockery et al. 2013). In contrast to the original study, which had been used to justify policy decisions to extend coal-burning bans in Ireland based on a much-publicized belief that cleaner air had been found to cause reduced mortality, the updated study using control groups concluded that the bans had produced no detectable reductions in total mortality rates (Dockery et al. 2013). As explained by Zigler and Dominici et al. (2014), "However, even when studying an abrupt action, threats to causal validity can arise, as illustrated in extended analyses of the Dublin coal ban that revealed that long-term trends in cardiovascular health spanning implementation of the ban—not the coal ban itself—contributed to apparent effects on cardiovascular mortality." Since the 1960s, the "one-group pretest-posttest design" used in the original study has been identified by social statisticians as inappropriate for causal inferences, since it leaves uncontrolled the threat of coincident historical change, as well as other threats to valid causal inference (Campbell and Stanley 1963, p. 7). By contrast, a pretest-posttest control group design is appropriate for causal inference (*ibid*). It can show that a large reduction in particulate pollution had no detectable effect on total mortality, as in Dublin, if that is the case; or it can provide strong evidence that high pollution levels cause excess mortalities if mortality rate spikes where and when air pollution spikes—such as in Donora in 1948, or London in 1952—but not otherwise, e.g., in the same cities and months a year earlier or later than the high pollution episode, or in other cities with similar temperatures, humidity, influenza rates, etc. but without the high air pollution.
- *Instrumental variable* (IV) studies assume that a variable (called an "instrument") directly affects exposure but not response and is itself unaffected by unmeasured confounders (Baiocchi et al. 2014). How response varies as the instrument varies can then be used to estimate the effect on responses of the changes in exposure associated with variations in the instrument, without any effects from unmeasured confounders. For example, Schwartz et al. (2016) noted that, while many time series studies have established associations of daily pollution variations with daily deaths, most of these associations have been found at relatively high exposure concentrations, and "causal modeling approaches are also lacking." To address this lack, they developed an IV approach that "combined height of the planetary boundary layer and wind speed, which impact concentrations of local emissions, to develop the instrument for PM2.5, BC, or NO2 variations that were independent of year, month, and temperature." They "conclude that there is a causal association of local air pollution with daily deaths at concentrations below EPA standards. The estimated attributable risk in Boston exceeded 1800 deaths during the study period, indicating important public health benefits can follow from further control efforts." However, the conclusions of such IV analyses depend on the validity of the assumptions that the instrument

affects exposure but not response, and that it is unaffected by omitted confounders. Whether these assumptions are valid is usually unknown. As noted by O'Malley (2012), "IV analyses make strong assumptions that cannot be conclusively tested by the data" and different modeling choices, e.g., about how to treat lagged values, can yield very different results. Schwartz et al. (2016) assume that their instrument "is unlikely to be correlated with other causes of death," but the validity of this assumption is unknown, especially since wind speed, a component of the instrument, has been found to be correlated with cardiovascular mortality (Urban and Kyselý 2014), elderly mortality (Morabito et al. 2014), and stroke symptom onset (Kim et al. 2016).

- *Regression discontinuity designs* (RDDs) assume that if exposures and effects are significantly different for people with values of some characteristic (such as age) above *vs.* below an arbitrary threshold (such as an age threshold for legal drinking), then the difference in effects is caused by the difference in exposures, rather than by other differences. This assumption will be mistaken if the difference in effects is instead caused by other differences between people above and below the threshold (as might happen if the legal drinking age is also the legal age threshold for other activities such as gambling, smoking, driving without supervision, employment in certain occupations, and so forth). In air pollution health effects research, RDDs have been used to attribute differences in life expectancies between people living north and south of a river to differences in air pollution rather than to other regional differences in exercise or other variables, but this attribution remains an untested assumption (Chen et al. 2013). RDDs typically yield biased effects estimates unless the functional form of the C-R relationship is correctly specified.
- *Counterfactual* and *potential outcome* models, including *difference-in-differences* models, assume that differences between observed responses to observed exposure concentrations and unobserved model-predicted responses to "counterfactual" exposure concentrations are caused by the differences between the observed and counterfactual exposures, rather than by errors in the model or by systematic differences in other factors such as distributions of income, location, and age between the more- and less-exposed individuals. For example, Wang et al. (2016) assume that death count (the potential outcome) depends on predictors via the parametric model

$$\ln\left(E\left(Y^{a}{}_{c,t}\right)\right) = \beta_0 + \beta_1 a + \beta_2 Z_c + \beta_3 U_t + \beta_4 W_{c,t} + \ln\left(P_c\right),$$

- where
- $Y^{a}{}_{c,t}$ = number of deaths that would occur in the population of census tract c were itexposed to a in year t;
- Z_c represents spatial confounders that vary among census tracts but not over the time period of the study;
- U_t represents confounders that vary over time but not among census tracts;
- $W_{c,t}$ represents confounders that vary over time and among census tracts, and

- $\ln(P_c)$ is the natural log of the population in census tract *c*.

If the assumed additive form of the right side is correct, then the change in deaths from one year to the next within a census tract should depend only on time-varying terms, and the difference in these changes (i.e., the "difference in differences") in deaths across census tracts should depend only on factors such as exposure that differ over time and between census tracts. Assuming that these differences-in-differences are caused only by differences in exposures, the authors conclude that "Under the assumption of the difference-in-differences approach, we identified a causal effect of long-term PM2.5 exposure on mortality that was modified by seasonal temperatures and ecological socioeconomic status." But the validity of the model assumptions is unknown. Counterfactual causality is based on estimating what *would have* happened had exposure conditions been different. Since what would have happened is not observed, models are used to guess at what would have happened; if these guesses are wrong, then their causal conclusions may be wrong. In competitive evaluations (Hill 2016), potential outcomes methods exhibit 20-fold greater errors and biases than other methods of causal analysis, as discussed below, reflecting the fact that they are sensitive to models of unknown validity. Commendably, such limitations have been meticulously noted in some recent papers advocating and applying potential outcome methods. For example, Zigler et al. (2012) state that "Our analysis of the CAAA estimated that 1991 non-attainment designations for PM10 did causally reduce Medicare mortality in 2001, and that there are important causal pathways through which this effect occurred without affecting average ambient concentrations of PM10 or O3 during 1999–2001. …Our results are predicated on the belief that after adjusting for demographic characteristics in 2000–2001… and preregulation pollution levels, *there are no unmeasured factors relevant to air quality and mortality that differ systematically between attainment and nonattainment areas*. …Furthermore, *we used a relatively restrictive (exponential) spatial decay function* that was indexed by a single parameter, but more flexible (e.g. anisotropic) spatial covariance functions could provide better fit to pollution data and should be explored." More recently, Zigler et al. (2016) concluded on the basis of potential outcomes modeling that "all-cause Medicare mortality and respiratory-related hospitalization rates were causally reduced in areas designated as nonattainment for PM10 during 1990–1995 compared with the rates that would have occurred without the designation," but then responsibly noted that "A key limitation of our analysis is the fact that we estimated the effect of the nonattainment designation by regarding all monitoring locations in a nonattainment area as 'treated' However, nonattainment designations… sometimes resulted in no action at all. … [In addition, the] prospect of unmeasured confounding remained a threat to the validity of our results." Such expositions deserve credit for highlighting the fact that the conclusions reached are not necessarily valid, insofar as they depend on assumptions that are not necessarily correct. Crucial assumptions, such as that unmeasured confounding does not invalidate the conclusions, are usually left untested in the counterfactual/ potential outcomes approach, and many of its ablest practitioners suggest that they are inherently untestable. Thus, for example, Petersen and van der Laan (2014) write that "In sum, the flexibility of a structural causal model allows us to avoid many (although not all) unsubstantiated assumptions and thus facilitates specification of a causal model that describes the true data-generating process. Alternative causal models differ in their assumptions about the nature of causality and make fewer untestable assumptions. …[A] formal causal framework can provide a tool for defining a statistical estimation problem that comes as close as possible to addressing the motivating scientific question, given the data and knowledge currently available, while remaining transparent regarding the additional assumptions required to endow the resulting estimate with a causal interpretation." Clearly stating the untested assumptions (called "convenience-based assumptions" by Petersen and van der Laan, and distinguished by them from "real knowledge") that have been used to draw causal conclusions from observational data is certainly an admirable part of the program of counterfactual and potential outcomes causal research. However, the net result is that causal inferences and effects estimates in this framework are no more certain than the untested assumptions that support them. In practice, the approach often ends up interpreting associations predicted from regression models as if they were causal, while making (and explicitly stating) the convenience-based assumptions needed to warrant such an interpretation. This leaves open the practical question of whether the assumptions and causal conclusions are valid.

- *Predictive causality methods* assume that if exposure helps to predict a response, then exposure might be a cause of response. For example, *Granger causality* between an exposure and a response time series (Kleinberg and Hripcsak 2011) relies on the idea that causes help to predict their effects. Technically, variable X is a Granger-cause of variable Y if the future of Y is not conditionally independent of the history of X, given the history of Y (see Chap. 9). Thus, nicotine-stained fingers can be a Granger cause of lung cancer, helping to predict it, even if cleaning one's fingers would have no effect on future lung cancer risk: a predictive cause need not be a manipulative cause. Granger causality does not protect against causal associations from omitted confounders, nor does predictive causality have any necessary implications for manipulative causality. These points are often ignored or misunderstood in recent epidemiology papers on air pollution health effects. For example, Schwartz et al. (2016) state that that "We also used Granger causality to assess whether omitted variable confounding existed… Granger causality… argues that omitted covariates that are correlated with time varying exposure and outcome are as likely to be correlated with tomorrow's exposure as yesterday's exposure." This is neither a correct description of Granger causality, nor a valid test for omitted confounders. For example, consider a time series in which cold snaps occur only rarely, and for a single day at a time; people then turn on their heaters and PM2.5 concentrations spike on the day of the cold snap and the following day; flu-like symptoms occur one day later (i.e., 2 days after the cold snap day); and finally both temperature and PM2.5 then return to their usual distribution of values. If PM2.5 and flu-like symptoms are measured and cold snaps are an omitted covariate, there would be no justification for assuming that cold snaps "are as likely to be correlated with tomorrow's exposure [i.e., PM2.5 on the day after flu-like symptoms are observed] as yesterday's exposure [i.e., PM2.5 on the day before flu-like symptoms are observed]." To the contrary, occurrence of cold snaps could be perfectly correlated with yesterday's spikes in PM2.5 and yet have no correlation with tomorrow's random values of PM2.5. Even if Granger causality had been correctly established, the Schwartz et al. (2016) conclusion "that there is a causal association of local air pollution with daily deaths at concentrations below EPA standards… indicating important public health benefits can follow from further control efforts" would not follow: finding a causal association based on Granger causality or IV has no necessary implications for effects of control efforts, because such causal associations do not necessarily reflect manipulative causality (Woodward 2013). Just as showing that nicotine-stained fingers help to predict lung cancer would not imply that cleaning fingers would reduce risk of lung cancer, so showing that polluted air is a Granger-cause of mortality would not imply that cleaning air would reduce risk of mortality.

- *Attributable risk* and *burden-of-disease* studies, as previously discussed, assume that if responses are greater among people with higher exposures, then this difference is caused by the difference in exposures, and could be reduced by reducing it. Typically, this assumption is made without any careful justification: it simply confuses association with causation. Examples are widespread, e.g., Fann et al. (2012), Lepeule et al. (2012), Schwartz et al. (2016), and Lo et al. (2016).
- *Computer-based simulation modeling studies* assume that simulated impacts in a computer model predict real-world responses. Many recent studies simulate health impacts of changes in exposures by applying C-R impact factors or functions to simulated changes in exposure distributions. Validation of such models for specific applications is crucial for determining their usefulness and trustworthiness, but in practice they are usually used without such validation. As one prominent example, the United States Environmental Protection Agency (EPA) provides a publicly available computer program, BenMAP, that quantifies the subjective judgments of selected experts about C-R relations. The technical documentation for BenMAP (United States Environmental Protection Agency 2015, p. 60–61) explicitly and repeatedly states "no causality included" in summarizing health impact associations based on expert judgments. The text further explains that "Experts A, C, and J indicated that they included the likelihood of causality in their subjective distributions. However, the continuous parametric distributions specified were inconsistent with the causality likelihoods provided by these experts. Because there was no way to reconcile this, we chose to interpret the distributions of these experts as unconditional and ignore the additional information on the likelihood of causality." BenMAP's health functions for long-term impacts should not be interpreted causally without carefully addressing these caveats. Nonetheless, many investigators present the results of BenMAP calculations as if they yielded "Estimated Excess Morbidity and Mortality Caused by Air Pollution" (Cromar et al. 2016) despite the BenMAP documentation and even though BenMAP has not been validated as a causal model. Similar comments apply to the increasingly widespread use of statistical and computational

models to infer changes in mortality from changes in exposure concentration based on previously estimated C-R slope factors, without any attempt to validate manipulative causation (e.g., Lin et al. 2016).

Although most current literature on formal causal modeling of PM2.5-associated health effects depends on unverified assumptions, several recent papers strongly emphasize their authors' subjective confidence and conviction that associations can and should be interpreted causally. For example, one recent paper asserts that, in Boston, "the association between PM2.5 and deaths is almost certainly causal," even while noting that the supporting analysis "relies on the untestable assumption of no unmeasured confounding" or related untestable assumptions (Schwartz et al. 2015). This leaves open the question of whether the untested assumptions are correct. It also ignores a substantial statistical literature on methods for testing the assumption of no unmeasured confounding (Marra et al. 2014).

Many papers that express confidence in causal interpretations of epidemiological associations cite animal and *in vitro* studies as providing additional evidence that supports the hypothesis of causality (e.g., Schwartz et al. 2016). Adducing such data, or more general discussions and conjectures about plausible biological mechanisms by which exposures might cause the effects attributed to them, as evidence that epidemiological associations should be interpreted causally is problematic for several reasons. As noted by McClellan (2016) in discussing one such paper, "much of this discussion is quite simplistic and, indeed, naïve with regard to the actual complexity of disease processes." For example, premises such as that exposure leads to increased levels of reactive oxygen species (ROS) in the lung and that elevated levels of ROS in the lung are found in certain lung diseases do not necessarily constitute valid evidence that exposure increases risk of lung diseases, as elevated ROS levels also occur in response to exposure in healthy people and animals and do not necessarily or usually indicate any pathological mechanism at work (Cox Jr. 2011). More fundamentally, disease processes take place *within individuals*, while epidemiological C-R functions, relative risk ratios, and regression coefficients describe associations at the level of *populations* of individuals. Discussion of causal mechanisms, factor interactions, and confounding at the individual level may have no clear implications for analogous phenomena at the population level unless the frequency distribution of heterogeneous individual types within the populations is well understood (Maldonado 2013). Yet, papers that adduce mechanistic evidence in the context of epidemiological data seldom address this crucial methodological point, implicitly assuming that human populations can be treated as if they were composed of homogeneous individuals, with mechanistic discussions at the individual level being simply scaled up to apply to populations.

A more sober assessment is that *all* the papers we have reviewed rely on implicit or explicit untested modeling assumptions—typically, one or more of those discussed above, such as that changes following an intervention were caused by it, that statistical or computer simulation models used to predict counterfactual outcomes do so correctly, that instrumental variables are valid, and that unmeasured confounders do not exist—to justify their claims about causation. All address statistical counterfactual or predictive causation rather than manipulative causation, although manipulative causation is what risk managers and policy makers need to be informed about to predict effects of proposed interventions such as regulations or coal burning bans. None of these papers demonstrates a manipulative causal C-R relationship between changes in exposure concentrations and changes in total mortality risks. Some papers refute previous statistical causal claims (e.g., Dockery et al. 2013). Thus, it appears that the literature on adverse public health effects that are solidly proved to be caused by PM2.5 exposure and to be preventable by reducing PM2.5 exposure is still in its infancy, despite decades of association-based and assumption-based studies and repeated (but not logically valid) causal claims based on such assumption. Sound studies that address manipulative causality are still very much needed.

Opportunities Remain to Apply Modern Causal Inference Algorithms

The literature reviewed on PM2.5 health effects can be summarized as follows.

1. Past key papers do not distinguish clearly between association and causation, or directly address manipulative causation. As pointed out by Wang et al. (2016): "Many studies have reported the associations between long-term exposure to PM2.5 and increased risk of death. However, to our knowledge, none has used a causal modeling approach." Similarly, Schwartz et al. (2016) noted that "While many time series studies have established associations of daily pollution variations with daily deaths… causal modeling approaches are also lacking."
2. The previous section identified a few studies that seek to apply assumption-based causal modeling approaches such as intervention studies, instrumental variable models, counterfactual or potential outcomes models, predictive causation models, burden of disease models, and computer simulation models. However, such studies are rare compared to the large number that deal only with associations. All of them rely on untested modeling assumptions, of unknown validity, to justify interpreting associations as causation.
3. More importantly, none of the papers reviewed clearly distinguishes between statistical counterfactual or predictive causation and manipulative causation. None demonstrates that manipulating exposure concentrations would change future mortality risks, or that it has done so in the past (once statistical errors in causal interpretations are corrected, as in the Dublin coal burning ban studies discussed by Clancy et al. (2002) and Dockery et al. (2013)). Frequent claims that reducing exposure concentrations would reduce mortality appear to be based on a widespread misconception that one type of causation implies others, despite counterexamples such as nicotine-stained fingers being a Granger cause but not a manipulative cause of lung cancer.

In short, the preceding review finds that the very large literature making policy-relevant causal claims and predictions about the human health effects of reducing air pollution is not well supported by analyses appropriate for manipulative causation. Causal claims about the predicted health effects of reducing air pollution are typically derived by conflating causation with association and conflating manipulative causation with other types of causation.

Fortunately, as discussed in Chap. 9 and illustrated in Chaps. 10, 11, 12, and 13, an extensive technical literature provides algorithms for computer-aided causal discovery, modeling and analytics that offer constructive means to address the preceding limitations of existing C-R functions. These algorithms have emerged from a confluence of research in economics and econometrics, neuroscience, bioinformatics and systems biology, physics, engineering, statistics and applied probability, philosophy and formal logic, computer science, machine learning, and artificial intelligence (Pearl 2009b). They clearly distinguish among manipulative, counterfactual, predictive, and other (e.g., exogeneity-based or dynamic transition-based) forms of causation and enable causal inferences that do not require untestable assumptions or subjective weight-of-evidence judgments (Pearl 2010).

This technical literature on algorithms and methods for causal inference from observational data is already large and is growing rapidly though articles published in sources such as *The Journal of Causal Inference*, *Artificial Intelligence*, *Neural Information Processing Workshops on Causality*, *Uncertainty in Artificial Intelligence* (UAI) conference proceedings, and documentation of algorithms implemented in R and Python. From comparisons of the competitive performance of dozens of causal inference algorithms on benchmark problems over the past decade, it is possible to distill a short guide to the principles and algorithms that generally work best in practice. Many top-performing methods in current causal analytics competitions for observational data use some or all the following principles (Hill 2016; Bontempi and Flauder 2015), which are discussed in more detail in Chap. 9.

- *Information principle*: Causes provide information that helps to predict their effects and that cannot be obtained from other variables. This principle creates a bridge between well-developed computational statistical and machine learning methods for identifying informative variables that improve prediction of dependent variables such as health effects, and the needs of causal inference. It allows techniques of predictive analytics to be applied to screen variables for potential causation. In practice, the information principle is usually implemented via algorithms that identify *conditional independence* among variables (Pearl 2009a). By this criterion if effect Y is conditionally independent of exposure variable X, given the values of other variables, then X is not eligible to be a cause of Y. Granger-causality is the special case of this principle in which the future of Y must not be conditionally independent of the history of X, given the history of Y. Although conditional independence tests do not establish manipulative causation, they provide a useful screen for potential manipulative causes, insofar as they provide a condition that is usually *necessary* (but not *sufficient*) for manipulative causation.

For example, discovering that nicotine-stained fingers are a Granger cause of lung cancer would allow nicotine-staining to be identified as a *potential* manipulative cause of lung cancer, but we know that smoking is the actual manipulative cause.

In practice, statistical dependencies among variables can be discovered and displayed by algorithms that generate Bayesian networks or classification and regression trees (CART models) (Young and Xia 2013) from data, such as the *bnlearn* or *party* packages package in R, respectively. A Bayesian network (BN) shows arrows between pairs of variables that are statistically dependent even after conditioning on all other variables. Arrows between variables are absent if they are conditionally independent of each other after conditioning on other variables (e.g., Pearl 2009b; Rottman and Hastie 2014). The structure of such networks can often be learned directly from large data sets, rather than having to be specified as a hypothesis or based on expert knowledge, by using algorithms that test for conditional independence and quantify conditional probability dependencies (Frey et al. 2003; Aliferis et al. 2010; *bnlearn* package documentation by M. Scutari, www.bnlearn.com/). A CART tree (or, more generally, a "recursive partitioning" tree) shows combinations of values of predictors (or of ranges of their values) that yield significantly different conditional distributions for the dependent variable, and in this sense provide useful information for predicting it. Such trees can be used to help learn Bayesian network structures and to quantify their conditional probability tables from data (Frey et al. 2003; Aliferis et al. 2010).

- *Propagation of changes principle:* Changes in causes help to explain and predict changes in their effects (Friston et al. 2013; Wu et al. 2011). This applies the information principle to changes in variables over time. It can often be visualized in terms of changes propagating along links (representing statistical dependencies) in a Bayesian network or other network model (e.g., Friston et al. 2013). The goal of causal analysis in C-R modeling is to predict how changing exposure would change health effects, so studying how changes propagate among variables over time is of great interest. In practice, examining propagation of changes from exposure to response variables requires longitudinal data and analysis of information flows from lagged to present values of variables.
- *Nonparametric analyses*. Multivariate non-parametric methods, most commonly, classification and regression trees (CART) algorithms, are used to identify information dependencies among variables (e.g., Halliday et al. 2016). If no significant change occurs in the conditional empirical cumulative distribution function of a dependent variable as the value of an explanatory variable varies, for any combination of values of the remaining variables, then this lack of dependence does not support a conclusion that the explanatory variable is a cause of the dependent variable. The dependent variable is then *conditionally independent* of the explanatory variable, given the values of other variables. Effects are not conditionally independent of their direct causes. The useful fact that CART trees can also be used to test for conditional independence, with the dependent variable being conditionally independent of variables not in the tree, given the variables that are in it, at least as far as the tree-growing algorithm can discover,

only became widely appreciated and applied in causal analysis and machine-learning after 2000 (e.g., Frey et al. 2003; Aliferis et al. 2010). CART trees can be automatically generated using freely available tree-growing recursive partitioning algorithms in R packages such as *party* or *rpart* or Python *scikit-learn*.

- *Model ensembles*. Rather than relying on any single statistical model, the top-performing causal analytics algorithms typically fit hundreds of nonparametric models to subsets of the data(e.g., CART trees grown on random subsets of predictors to help de-correlate their predictions) (Hernandez et al. 2015; Furqan and Siyal 2016). Averaging predictions of how the dependent variable depends on other variables over an ensemble of models usually yields estimates with lower bias and error variance than any single predictive model. Computational statistics packages such as the *randomForest* package in R automate construction, validation, and predictive analytics for such model ensembles and present results in simple graphical forms. For example, partial dependence plots show how predicted values of a dependent variable change as a single predictor is systematically varied leaving all other variables with their empirical distributions (http://scikit-learn.org/stable/auto_examples/ensemble/plot_partial_dependence.html). Applied to C-R functions, such partial dependence plots show how *R* varies with *C* when accounting for the effects on *R* of all other variables via an ensemble of CART trees. *If* this dependency represents manipulative causality, then the partial dependency plot indicates how the conditional expected value of *R* should be expected to change when *C* is manipulated, given the empirical joint distribution of other measured predictors on which R also depends. This is an important *if*. Even state-of-the-art causal inference algorithms typically do *not* allow manipulative causality to be inferred confidently in the absence of interventions. Thus, humility about what can be accomplished using only observational data is prudent. But Bayesian Networks, Random Forest ensembles, and partial dependence plots do allow *possible* causal dependencies—meaning dependencies that satisfy the information principle—to be identified and quantified from data. They also have several important practical advantages, including
    - *Automating variable-selection and coding*, thus reducing opportunities for p-hacking and confirmation bias to affect the analysis and conclusions;
    - *Detecting and modeling high-order interactions* among variables. This can be done using trees representing conjunctions of predictor ranges that lead to significantly different conditional distributions of the dependent variable, with the depth of the tree corresponding to the order of interactions considered;
    - *Coping with model uncertainties* by using ensembles of models; and
    - *Avoiding model selection and misspecification biases* by using non-parametric methods to avoid having to make parametric regression modeling assumptions.

As noted by Hernandez et al. (2015), "Random forests (RF)… is a popular method for dealing with high-dimensional data, mainly because of its computational speed and high accuracy. It is a non-parametric method and so does not make any major distributional assumptions about the data. RF automatically allows for non-linear interaction effects, a desirable property in many high-dimensional datasets.... However, as RF is a machine learning algorithm and does not use a statistical model it cannot provide probability-based uncertainty intervals as in a Bayesian setting." Hybrid algorithms such as Bayesian Additive Regression Trees using Bayesian Model Averaging (BART-BMA) have very recently been proposed as a promising approach to combine these strengths of RFs with the ability of Bayesian methods to generate posterior uncertainty intervals (*ibid*; Chipman and McCulloch 2016). Such hybrid methods improve on traditional Bayesian Model Averaging (BMA) for parametric regression models by using nonparametric trees and allowing the forms of dependencies, rather than just the predictors selected, to be varied and averaged over. However, although they may well turn out to represent a lasting and valuable advance, these comparatively recent Bayesian tree ensemble approaches are not yet as mature and well-tested as pure RF ensemble methods, which will therefore be used for the computations in this chapter.

The design principles for causal inference algorithms just described—the information principle, propagation of changes, use of nonparametric estimates, and averaging over ensembles of predictions—characterize *information-based approaches* to causal inference. They emphasize discovery of which variables in a data set are informative about, and hence help to predict, which others, without regard for statistical associations and without depending on any specific parametric modeling assumptions. Information-based approaches are very different from associational or assumption-based ones. For example, two variables can be highly informative about each other even if the statistical correlation between them is zero, as in the case of $Y = X^2$ where X is uniformly distributed between -1 and 1. Conversely, two variables Y and Z can be strongly correlated even if they are conditionally independent of each other, so that neither provides information that helps to predict the other, given the values of other variables. An example is the system $Y = X^2$ and $Z = X^2$, with the value of X determining the values of both Y and Z. In this case, Y and Z will be perfectly correlated with each other (but not with X), although neither causes the other and X causes both. Information-based methods complement the assumption in potential outcomes or counterfactual models, that differences in causes make effects *differ* from what they otherwise would have been, with the idea that differences in causes help to *predict* differences in their effects. This should be true even when using nonparametric and model ensemble methods to make the predictions, including testing for conditional independence and quantifying probabilistic dependencies. Whether one variable helps to predict another can be tested using observational data without making hypothetical modeling assumptions about what would have happened had exposures or other conditions been different from those observed, and without relying on specific parametric modeling assumptions of unknown validity, by using nonparametric methods such as Bayesian Network learning, tree-growing algorithms, and ensemble methods.

Empirically, causal inference algorithm performance and validation results from a causal inference competition (Hill 2016) yielded the following quantitative results for comparison of a tree-based algorithm to a counterfactual/potential outcomes algorithm (Inverse Probability of Treatment Weighting (IPTW)), averaged over 20 challenge data sets for which the correct data-generating processes were known to the competition organizers, but not to the causal inference algorithm designer competitors who submitted algorithms to estimate these known causal relationships.

- *Bias:* Nonparametric regression tree methods had a bias of -0.007 in estimating the causal effects from data, compared to a −0.15 bias, about 20-fold greater, for the IPTW counterfactual causality algorithm.
- *Error:* The nonparametric regression tree algorithm yielded a root mean-squared prediction error of 0.02 compared to 0.41 for the IPTW algorithm, i.e., again about a 20-fold difference.
- *Coverage probabilities and uncertainty interval lengths:* The nonparametric regression tree algorithm had smaller uncertainty intervals and larger coverage probabilities than the IPTW algorithm.

It thus appears that the nonparametric tree-based approach can complement counterfactual approaches in at least some cases, not only requiring fewer modeling assumptions, but also providing better performance.

An Example: C-R Modeling of Annual Changes in PM2.5 and Elderly Mortality Rates in Boston and Los Angeles

This section moves beyond a critical review of the literature on C-R functions and causal inference algorithms by applying information-based causal inference principles and algorithms to publicly available data for Boston (Suffolk County, MA) and Los Angeles (South Coastal Air Quality Management District (SCAQMD), CA) area data. The goal is to explore what can be learned about associational and predictive causal C-R functions by these methods, recognizing that, despite their extensive development in the journals and literatures previously cited, information-based causal analysis methods have not yet been widely used or well vetted in air pollution health effects research and epidemiology. Accordingly, our exploration is cautious, seeking only to illustrate the insights produced by applying these algorithms to two example data sets, rather than to reach definitive general conclusions. However, the state-of-the-art in computational statistical software makes these methods easy to apply, and it is more instructive to do so than only to discuss the literature about them.

Data

We use two data sets to illustrate information-based causal analytics methods, one from the Boston area (Suffolk County, MA) for 2000–2013 and the other from California's SCAQMD, which contains Los Angeles, for 2007–2010. The data fields for both areas are similar. Table 15.2 shows the layout of the data for the SCAQMD, from Lopiano et al. (2015). (The full SCAQMD data is available as the "LA" data set bundled with the free in-browser CAT software at http://cloudcat.cox-associates.com:8899/.)

The full data set has 1461 rows of data, one for each day from January 1 of 2007 through December 31, 2010. The variables (columns) in Table 15.2, and their data sources, are as follows:

- Calendar variables *year*, *month*, and *day* identify when the data were collected. Each row of data represents one day of observations.
- *mortality75* is a count variable giving the number of deaths among people aged at least 75 dying on each day, as recorded by the California Department of Health. (This variable was originally named *AllCause75*, but we renamed it as *mortality75*.)
- *PM2.5* is the daily average ambient concentration of fine particulate matter (PM2.5) in micrograms per cubic meter of air, as recorded by the California Air Resources Board (CARB) at www.arb.ca.gov/aqmis2/aqdselect.php.
- The three meteorological variables *tmin* = minimum daily temperature, *tmax* = maximum daily temperature, and *MAXRH* = maximum relative humidity, are from ORNL (http://cdiac.ornl.gov/ftp/ushcn_daily/) and the US Environmental Protection Agency (EPA) www3.epa.gov/ttn/airs/airsaqs/detaildata/download-aqsdata.htm.

Lopiano et al. (2015) and the above sources provide further details on these variables. For example, for the *Mortality75* variable, Lopiano et al. explain that elderly mortality counts consist of "The total number of deaths of individuals… 75+ years of age with group cause of death categorized as AllCauses… Note accidental deaths

Table 15.2 Layout of data for PM2.5 concentration, weather, and elderly mortality ("*mortality75*") variables in the LA-SCAQMD

Year	Month	Day	Mortality75	PM2.5	tmin	tmax	MAXRH
2007	1	1	151	38.4	36	72	68.8
2007	1	2	158	17.4	36	75	48.9
2007	1	3	139	19.9	44	75	61.3
2007	1	4	164	64.6	37	68	87.9
2007	1	5	136	6.1	40	61	47.5
2007	1	6	152	18.8	39	69	39
2007	1	7	160	19.1	41	76	40.9
2007	1	8	148	13.8	41	83	33.7
2007	1	9	188	14.6	41	84	37.5
2007	1	10	169	39.6	41	78	63.2
2007	1	11	160	19.2	37	66	85.9
2007	1	12	160	22.3	31	56	67.2
2007	1	13	166	11.7	27	55	40.4
2007	1	14	157	20.8	24	56	34

were excluded from our analyses." The definitions of the populations covered and the death categories used are taken from the cited sources, but it is clear that average PM2.5 concentrations at monitor sites do not apply in detail to each individual, any more than the weather conditions describe each individual's exposure to temperature and humidity. Rather, these aggregate variables should be interpreted only as providing data from which we can study whether days with lower recorded PM2.5 levels, or lower recorded minimum temperatures, relative humidity, and so forth, also have lower mortality, and, if so, whether predictive causality or other causal relationships hold between them. In addition to these variables, we also include the derived variable *time*, defined as the total number of months since the start of the data set. Month and year are each treated as categorical variables, but time is a continuous variable that allows longer-term trends to be modeled.

The Boston data are very similar, but with *Dewpoint* replacing *MAXRH* as a measure of humidity, and years ranging from 2000–2013. PM2.5 data were obtained from the US EPA Air Quality System (AQS) website (www.epa.gov/ttn/airs/airsaqs/) for central-site monitoring locations in the Greater Boston Area; daily quality controlled local climatological data (QCLCD) were downloaded from NOAA; and individual-level mortality records were obtained from the Massachusetts Department of Public Health.

We focus on elderly people (≥75 years old) because past literature has pointed to the elderly as being especially susceptible to health harm from PM2.5 (Goto et al. 2016), with considerable conjecture about causal mechanisms such as PM2.5 increasing oxidative stress and cardiovascular disease risks (Wang et al. 2016). Past literature has also noted that "in a time-series study in Boston [moving] the time scale from days to months (i.e., 60 days) increased the estimated PM effect" (Laden et al. 2006, citing earlier work by Schwartz et al.). We therefore aggregate the PM2.5 and elderly mortality data to the monthly level, so that our C-R functions will describe relations between monthly averages of daily PM2.5 concentrations and monthly averages of daily mortality counts for people 75 years old or older. The key question for C-R modeling with these choices of endpoints and time scales is whether months with higher average PM2.5 have correspondingly higher elderly mortality rates after controlling for other factors. For studying how well changes in PM2.5 levels help to predict changes in elderly mortality rates, we follow Wang et al. (2016) in using a changes over a one-year time interval. The key research question is how well differences in PM2.5 from one month to the same month a year later in a given location (Boston or the SCAQMD) predict corresponding differences in elderly mortality counts.

To give a visual impression of the data, Fig. 15.1 plots several of the variables for the 48 months from January, 2007 (denoted as 200701) through December, 2010. Monthly averages of daily minimum and maximum temperatures, PM2.5 concentrations, and mortality counts are shown. Since the SCAQMD has a population close to 20 times larger than Boston's, the Boston mortality counts are multiplied by 20 in Fig. 15.1 to put them on the same scale. It is clear from visual inspection of Fig. 15.1 that elderly mortality decreases as temperature increases, but it is much less clear whether PM2.5 variations help to predict elderly mortality variations. Answering that question is a challenge for C-R analysis.

Methods and Analytic Plan

Our analytic plan is as follows.

1. *Identify conditional independence and dependencies* among variables in Table 15.2 using nonparametric Bayesian network learning algorithms and regression trees. This step screens for possible causal relations using the previously discussed information principle that variable *X* is a potential cause of variable *Y* only if *X* provides information that helps to predict *Y* and that cannot be obtained from other sources.
2. *Repeat step 1 for changes in variables over a time interval of one year.* This implements the propagation-of-changes principle that changes in causes should help to predict changes in their effects.
3. *Quantify the C-R dependence (if any) between PM2.5 concentration and elderly mortality*, without further assessing whether it is causal, using a partial dependence plot generated by the *randomForest* package in R. As discussed in the previous section, this algorithm averages the results of hundreds of nonparamet-

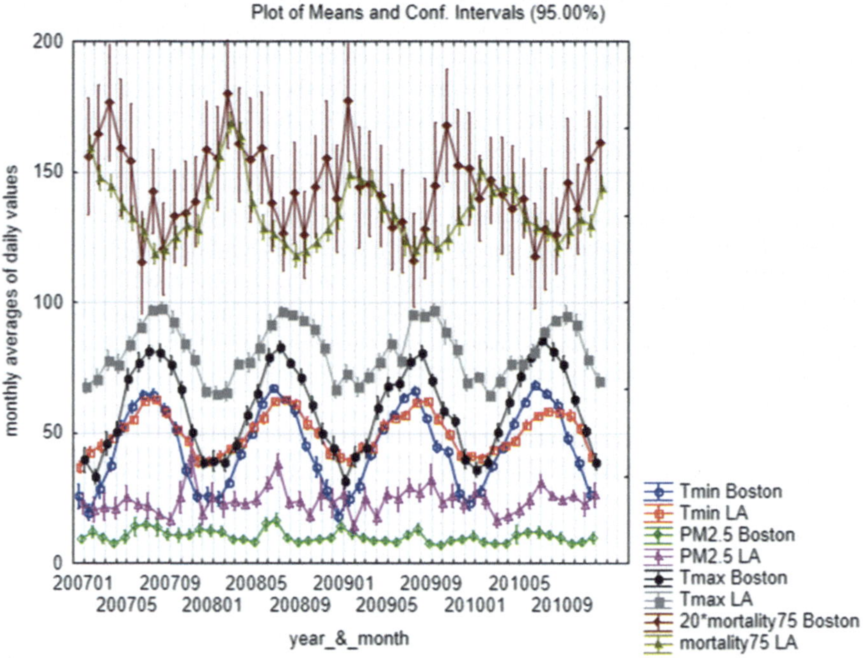

Fig. 15.1 Monthly time series of selected variables, 2007–2010

ric regression trees to estimate how daily mortality count varies as PM2.5 is swept over its full range of values. The *randomForest* package documentation contains details of this process (https://cran.r-project.org/web/packages/randomForest/randomForest.pdf).
4. Compare the results from steps 1–3 to those from multiple linear regression analysis.

All analyses were performed using free computational statistics packages from the CRAN repository for the R project, https://cran.r-project.org/, as follows:

- Bayesian Network learning algorithms were run using the R package *bnlearn*, (www.bnlearn.com/) with all settings at their default values. This implements the information principle by using nonparametric machine learning algorithms to discover conditional independence and dependencies among variables.
- Regression trees were grown using the R package *party* (https://cran.r-project.org/web/packages/party/party.pdf) to further explore and visualize multivariate dependency and conditional independence relationships.
- Random Forest model ensembles were generated by the R package *randomForest*, https://cran.r-project.org/web/packages/randomForest/randomForest.pdf) to quantify associations between two variables, controlling for the levels of other variables using multiple nonparametric models.
- Additional R packages (*car*, *MASS*, *leaps*, *MSBVAR*) were used for parametric regression analyses and Granger causality testing.

To facilitate easy replication and interpretation by investigators not familiar with these packages, we accessed all R packages and displayed results using the Causal Analysis Toolkit (CAT), available at http://cloudcat.cox-associates.com:8899/. The analyses can be replicated by clicking on the "Analyze" button on CAT. Details on CAT are provided in an appendix to Cox Jr. (2016) and in a User's Guide at www.cox-associates.com/downloads/. Details of all algorithms and supporting statistical theory are given in the online documentation for the corresponding R packages.

Results and Discussion

This section first discusses results for Boston and then compares them to results for the LA (SCAQMD) air basin. As a point of departure, we note that if the goal were simply to find regression models in which PM2.5 is significantly positively associated with elderly mortality after controlling for selected other variables via regression, it could easily be accomplished. Table 15.3 shows an example of such a regression model for the Boston data, with a highly significant regression coefficient ($p = 1.7\text{e-}9$) of 0.17 for PM2.5, implying that each 10 μg/m^3 increase in PM2.5 is associated with an average of 1.7 extra deaths per day, about a 22% increase, since the mean death count is 7.675 deaths per day. Of course, as discussed earlier, such associations do not necessarily reveal anything about causation, since the association could be explained by non-causal factors such as omission of time as a potential confounder, in this example. Once time is included, so that the downward secular trends in PM2.5 and mortality as functions of time (elapsed months) can be modeled, PM2.5 no longer appears as a statistically significant predictor of elderly mortality in multiple linear regression ($p = 0.25$ instead of 1.7E-9), indicating that it appears as a highly statistically significant predictor in Table 15.3 only because it is acting there as a surrogate for the omitted variable *time*. Applying the Granger causality test using the *granger.test* function in the R package *MSBVAR* shows that PM2.5 and elderly mortality are Granger-causes of each other, meaning that past and current values of each are significant predictors of the future values of the other, even after conditioning on its own past values and current value. Again, this indicates likely confounding by one or more other variables, such as time and/or temperature, that are correlated with both PM2.5 and elderly mortality, thereby making each informative about the other in bivariate analyses when the confounders are

Table 15.3 Example of a multiple linear regression model for Boston with a significant positive C-R coefficient for the association between PM2.5 and elderly mortality

CAT_linear (mortality75,PM2.5,Tmax,Tmin,Dewpoint,month)				
Dependent, variable: moratlity75				
Residuals				
Min	1Q	Median	3Q	Max
–1.4800	–0.4655	–0.0493	0.4449	1.9475
Coefficients				
	Estimate	Std. error	t value	Pr (>|t|)
(Intercept)	8.37718	0.77798	10.77	<2e-16***
PM2.5	0.17307	0.02696	6.42	1.7e-03***
Tmax	–0.01625	0.03747	–0.43	0.665
Tmin	–0.05547	0.06282	–0.88	0.379
Dewpaint	0.00464	0.03384	0.14	0.891
month10	1.13162	0.59401	1.91	0.059
month11	0.47998	0.42894	1.12	0.265
month2	0.01960	0.29656	0.07	0.947
month2	–0.07521	0.27538	–0.27	0.785
month3	0.33780	0.33338	1.01	0.313
month4	0.91830	0.47917	1.92	0.057
month5	0.88112	0.64737	1.36	0.176
month6	0.56257	0.83408	0.67	0.501
month7	0.63506	0.96497	0.71	0.479
month8	0.98317	0.95095	1.03	0.303
month9	1.11129	0.80365	1.38	0.169

Signif. codes: 0 ‘***’ 0.001 ‘**’ 0.01 ‘*’ 0.05 ‘.’ 0.1 ‘ ’ 1
Residual standard error: 0.712 on 152 degrees of freedom
Multiple R-squared: 0.596, adjusted R-squared: 0.556
F-statistic: 14.9 on 15 and 152 DF, p-value: <23–16

omitted. In both cases, statistical significance of an association does not imply substantive or causal significance.

Figure 15.2 shows the structure of a Bayesian network (BN) learned from the same Boston data by the R package *bnlearn*. In this network, an arrow between two variables shows that they are not conditionally independent of each other. Absence of arrows between two variables shows that they are conditionally independent of each other given the values of other variables. The directions of the arrows do *not* necessarily reflect causality. They simply indicate one way to decompose the joint probability distribution of all variables so that it can be computed from the marginal distributions of the input variables, those with only outward-pointing arrows, and conditional probabilities for the values of each other variable, given the values of the variables that point into it (Pearl 2009b). Nonetheless, the Bayesian network is a useful guide to possible predictive causality, insofar as the information principle implies that one variable can be a cause of another only if they are adjacent (linked by an arrow) in the Bayesian network. Figure 15.2 indicates that the passage of time is informative about both elderly mortality and PM2.5 (both decline with time), providing a reason that they are Granger-causes of each other.

If the *bnlearn* algorithms for discovering conditional independence and dependencies were completely accurate and trustworthy oracles, we would be done:

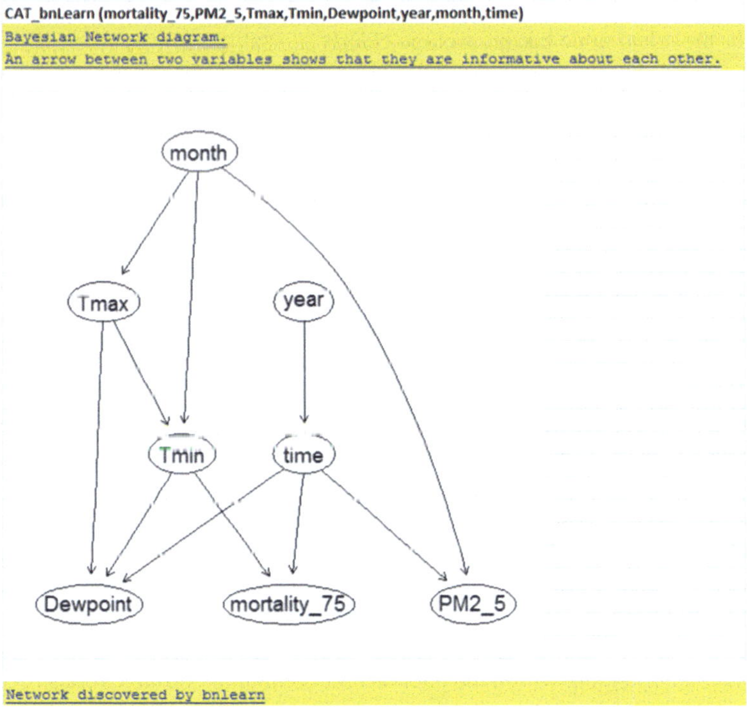

Fig. 15.2 Structure of Bayesian Network for Boston data discovered by R package *bnlearn*

Fig. 15.2 would imply that PM2.5 concentration is not a predictive cause of elderly mortality risk on the chosen time scale of months, and thus is also unlikely to be a manipulative cause. The highly statistically significant ($p < 1.7E-9$) regression coefficient for PM2.5 as a predictor of elderly mortality in Table 15.3 would then be an example of a statistical effect but not a causal one. (In this case, fitting a linear model for elderly mortality counts introduces model specification errors that can be reduced by including PM2.5 or other month-dependent variables, thus making PM2.5 a useful statistical predictor even if it has no causal impact on mortality.) However, no algorithms for learning Bayesian network structures are perfectly accurate, and it is therefore worth continuing the analysis assuming that a C-R relationship might exist and simply be too weak to have been detected by the *bnlearn* algorithms.

Figure 15.3 shows a regression tree grown on the same Boston data as in Fig. 15.2 and Table 15.3 using the *party* package in R. This tree identifies combinations of ranges of values for multiple variables that lead to significantly different conditional distributions for the elderly mortality (*mortality75*) dependent variable. The variables selected by the tree-growing algorithm as informative for predicting elderly mortality include *time* and *Tmin*, in agreement with the Bayesian network in Fig. 15.2, but also include PM2.5, unlike the Bayesian network. To read the regression tree in Fig. 15.3, note that each shaded "leaf" node at the bottom of the tree shows the conditional mean value of the dependent variable, elderly mortality, given the ranges of values for the variables in the path leading to that leaf. For example, the right-most leaf node has an average elderly mortality count of 6.646 deaths per day for $n = 53$ months satisfying with $Tmin > 48.677°$ and $time > 45$ months. The intermediate nodes show the p values from F tests for rejecting the null hypothesis that the conditional distributions of elderly mortality are not different from each other on the left and right of each split. Figure 15.3 shows that months with warmer temperatures have lower average daily elderly mortality counts and that mortality rates have decreased over time. Relatively cold months with an average daily minimum temperature below 48.677° and average PM2.5 concentrations above 13.271 $\mu g/m^3$ have elevated elderly mortality.

Although individual trees are often not robust to perturbations in the data (e.g., growing trees on multiple random subsets of the data may produce many different trees), averaging results over ensembles of hundreds of trees for random subsets of the data using the *randomForest* R package yields the relatively robust C-R partial dependence plot in Fig. 15.4. The same plot is shown on two different vertical scales, the left one emphasizing the variations that occur over their narrow range, and the right showing that the plot is almost flat when considered on a vertical scale that includes the origin. As average daily PM2.5 concentrations for specific months range over twofold, from below 10 $\mu g/m^3$ to above 20 $\mu g/m^3$, corresponding average daily elderly mortality counts range from about 7.64 to about 7.75 deaths per day, or roughly 1.01-fold. Although this variation could be due to omitted confounders or residual confounding within the intervals for continuous variables constructed by the tree-growing algorithm, no test performed so far rules out the possibility of a genuine predictive causal effect.

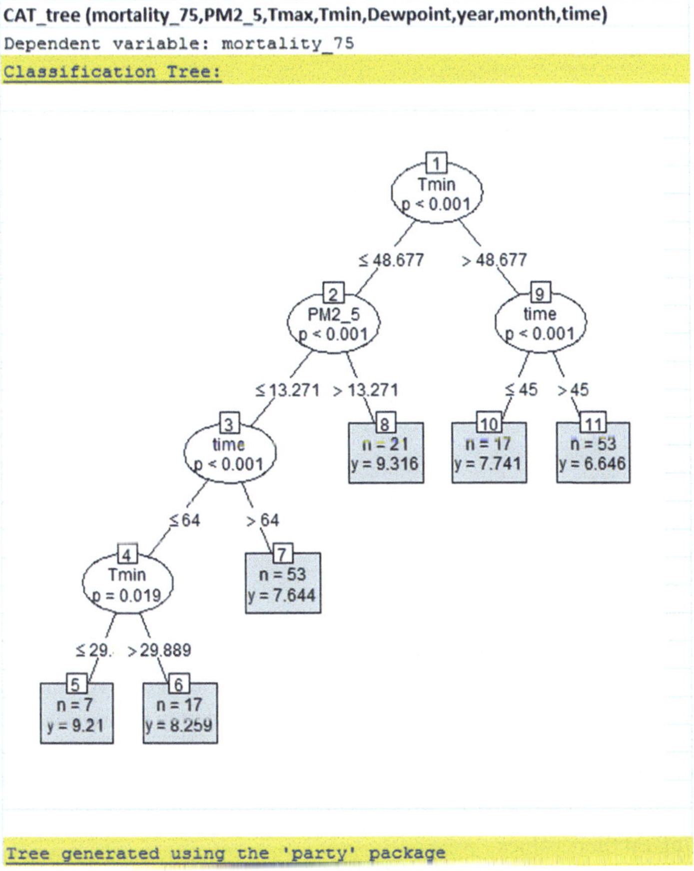

Fig. 15.3 A regression tree for elderly mortality in the Boston data

However, when the same analyses are repeated using *changes* in variables over a 1-year period, the results are quite different. Only change in *Tmin* is a significant predictor of change in elderly mortality, being adjacent to it in the Bayesian network for changes. Change in PM2.5 is also adjacent to change in *Tmin*, so change in temperature could confound any association between change in PM2.5 and change in mortality. Regression trees do not identify any significant predictors of change in elderly mortality, but plotting the empirical cumulative distribution function (ECDF) of change in mortality (between a month and that same month a year later) conditioned on the upper and lower quartiles of values for change in *Tmin* shows that increases of more than 2.45° in *Tmin* are associated with reductions in elderly mortality risks (a leftward-shifted ECDF curve), while decreases in *Tmin* by 2.19° or more are associated with increases in elderly mortality risks (rightward-shifted ECDF curves). By contrast, for PM2.5 the ECDFs for the top and bottom PM2.5 quartiles are not shifted in either direction but cross and re-cross each other. Thus,

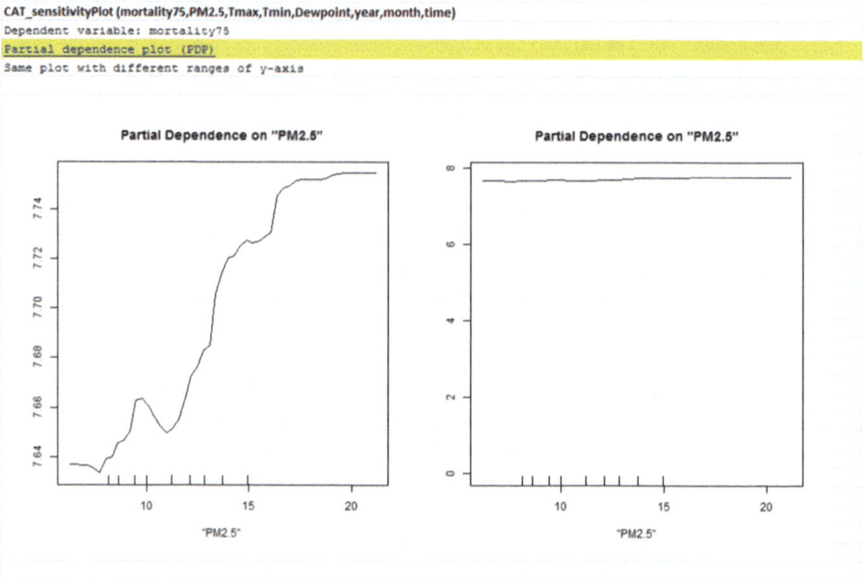

Fig. 15.4 A C-R partial dependence plot for elderly mortality vs. PM2.5 in Boston

in this case, the Bayesian network algorithm proves more sensitive than regression trees in detecting what appears to be a genuine dependency relation between temperature and elderly mortality, and also perhaps more accurate than classification trees in identifying no significant dependency between PM2.5 and elderly mortality rate once other variables such as temperature and time have been accounted for by conditioning on their values.

Finally, we examine how well the C-R curve estimated in Fig. 15.4 for Boston applies to the SCAQMD air basin. Figure 15.5 shows the main result. As average daily PM2.5 concentration in different months ranges from under 20 μg/m^3 to over 40 μg/m^3, average daily mortality counts stay essentially flat at 134.5 $\pm$ 0.5 deaths per day, with the slight residual C-R association between them being negative. PM2.5 and elderly mortality are adjacent in the Bayesian network for the SCAQMD, so the possibility of a slightly negative causal relation is not ruled out by these data. Ideally, confidence bands or uncertainty intervals would be provided with such plots, but, as noted by Hernandez et al. (2015), "However, as RF [Random Forests] is a machine learning algorithm and does not use a statistical model it cannot provide probability-based uncertainty intervals as in a Bayesian setting." Instead, we simply note that the PM2.5-elderly mortality association is positive in Fig. 15.4 and negative in Fig. 15.5, so that both cannot be correct descriptions of one and the same causal relationship: the invariant causal prediction (ICP property from Chap. 9 does not hold.

Comparing Figs. 15.4 and 15.5, it is clear that estimated C-R functions cannot simply be transferred from one city or region to another. Table 15.4 confirms this for

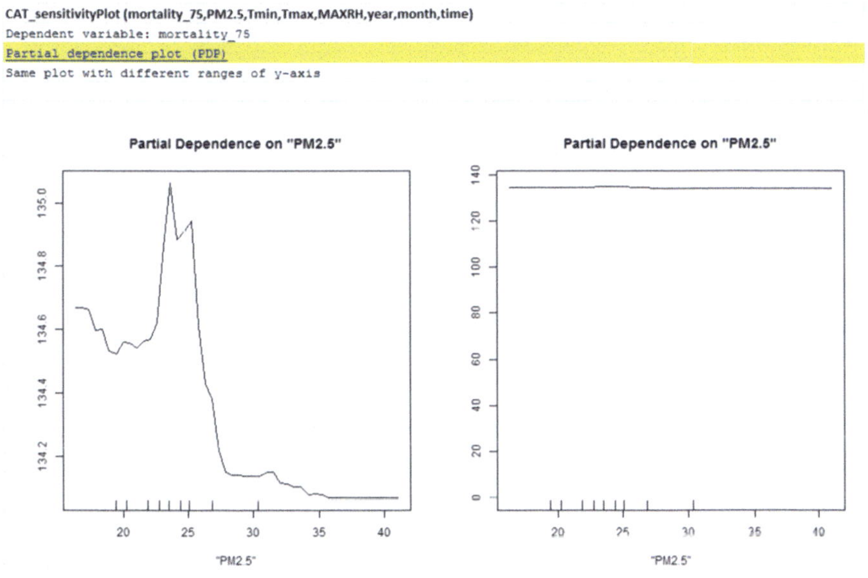

Fig. 15.5 A C-R partial dependence plot for elderly mortality vs. PM2.5 in the SCAQMD air basin in southern California

regression. In contrast to Table 15.3 for Boston, in a multiple linear regression model for the SCAQMD air basin, *Tmin* has a significant negative association with elderly mortality for the LA region, but PM2.5 has no significant association with mortality. In a different regression model with only *PM2.5*, *MAXRH*, and *year* included as predictors, the C-R association between PM2.5 and elderly mortality is actually negative, −0.633, with $p < 0.10$, suggesting that confounding by *Tmin* and *month* can affect the regression coefficient for PM2.5 if they are omitted from the model. Regression tree analysis identifies only *Tmin* as a predictor of elderly mortality, with average daily mortality counts of 147.5 deaths per day when *Tmin* is below 45.4°, compared to 120.6 deaths per day when *Tmin* exceeds 58.1°. In analysis of changes over a one year period, however, neither regression tree analysis nor Bayesian network learning identified any dependencies among changes that helped to predict changes in elderly mortality, including changes in *Tmin*, in contrast to Boston.

Taken as a whole, these results suggest that the common practice of transferring C-R regression coefficients or associations estimated from data for one region to exposure changes estimated for a different area, as in the papers of Apte et al. (2015), Cromar et al. (2016), Lo et al. (2016), and many others, is not well justified. Whether because of differences in PM2.5 composition or differences in covariate distributions between populations or because associations in one area simply do not reflect stable causal laws useful for predicting impacts in another, it appears that there is no single C-R function that correctly describes historical PM2.5-elderly

Table 15.4 Example of a multiple linear regression model for the SCAQMD air basin with a significant negative coefficient for the association between *Tmin* and elderly mortality

CAT_linear (mortality_75,PM2.5,Tmin,MAXRH,month)				
Dependent variable: mortality_75				
Residuals				
Min	1Q	Median	3Q	Max
−8.044	−2.956	−0.296	1.778	16.436
Coefficients				
	Estimated	Std. error	t value	Pr (>|t|)
(Intercept)	197.2808	20.6510	9.55	5e-11***
FM2.5	0.1868	0.1956	0.96	0.3464
Tmin	−1.1148	0.4988	−2.25	0.0323*
MAXRH	−0.0634	0.1623	−0.39	0.6985
month10	−9.1743	7.3792	−1.24	0.2225
month11	−14.2395	4.6642	−3.05	0.0045**
month12	−9.7250	4.0597	−2.40	0.0224*
month2	4.8563	4.3763	1.11	0.2752
month3	0.7252	4.4807	0.16	0.8724
month4	−7.3510	5.3307	−1.38	0.1772
month5	−9.0287	7.0615	−1.28	0.2100
month6	−6.3543	8.9425	−0.71	0.4823
month7	−8.0882	10.9937	−0.74	0.4671
month8	−7.0913	11.1635	−0.64	0.5297
month9	−6.8509	10.2753	−0.67	0.5096

Signif. codes: 0 '***' 0.001 '**' 0.01 '*' 0.05 '.' 0.1 ' ' 1
Residual standard error: 5.21 on 33 degrees of freedom
Multiple R-squared: 0.874, adjusted R-squared: 0.82
F-statistic: 16.3 on 14 and 33 DF, p-value: 6.15e-11

mortality associations for both the Boston and the Los Angeles regions, controlling for the effects of other variables as in Figs. 15.4 and 15.5. Nor does there appear to be a C-R function that allows changes in month-specific mortality risks to be predicted based on changes in month-specific PM2.5 concentrations, as no dependencies were found between these changes. These analyses do not support the usual assumption that a single stable C-R function exists that can be used for these purposes.

Study Uncertainties, Limitations and Extensions

No Proof of Manipulative Causality from Observational Data

The methods and conclusions illustrated in Figs. 15.2, 15.3, 15.4, and 15.5 and our review of frameworks for causal inference and modeling of C-R functions have several limitations. First, the information-based approaches illustrated in Figs. 15.2,

15.3, 15.4, and 15.5 only address whether conditions that are typically *necessary* for manipulative causation hold, such as whether exposures help to predict mortality rates after conditioning on observed confounders. These conditions do not provide *sufficient* conditions for inferring or quantifying manipulative causation based on observational data. They cannot, and are not intended to, "prove" manipulative causation based on observational data. Rather, they can be viewed as screening tests that establish whether, in the data analyzed, exposures are found to significantly predict health effects (and changes in exposure are found to significantly predict changes in responses), even after conditioning on other variables. If so, then it might be prudent to tentatively assume, for purposes of risk management, that C-R relations that have passed these screening tests are causal, while acknowledging that the tests stop short of definitively proving that exogenously changing (i.e., manipulating) future exposures would change future responses.

In the presence of unmeasured confounders, causality is difficult to establish for a relatively weak health risk factor such as ambient air pollution. However, consistency of the hypothesis of causation with available data can be established or refuted more easily (Chap. 9). Even when confounders are not measured, their effects can often be detected as unexplained but significant associations between observed responses in disjoint subpopulations, such as between mortality rates among men over 75 and women under 75. Conditions for estimating effects of unmeasured ("latent") variables and for detecting causal paths despite unmeasured confounders have been developed (e.g., Spirtes et al. 1995). Focusing on necessary conditions that can be tested using available data helps to avoid the admitted philosophical and statistical difficulties of defining necessary and sufficient conditions for establishing manipulative causation, and also the need to make strong, untested assumptions to justify stronger causal conclusions. The price of this simplicity and testability is that the conclusions reached are only about predictive causation and not necessarily about manipulative causation.

No Elucidation of Causal Pathways or Explanatory Causal Mechanisms Mediating C-R Relationships

A second important limitation of the methods we have reviewed and illustrated is that they focus on whether there is evidence that a predictive causal C-R relation is present, and on quantifying it if so, but not on explaining how it works. This non-explanatory approach is sometimes referred to as "black box" causal analysis (Imai et al. 2011). Its goals are far more modest than those of automated causal discovery algorithms in general (Cooper et al. 2015; Spirtes and Zhang 2016). Current research in causal discovery seeks to provide algorithms and principles to support the common scientific ambition "to understand the mechanisms by which variables came to take on the values they have (that is, to find a generative model), and to predict what the values of those variables would be if the naturally occurring mechanisms were subject to outside manipulations" (Spirtes 2010)—in other words, to address causal explanation, to reveal how causes produce their effects, and to make accurate predictions based on understanding of manipulative causation. This is part of the philosophical agenda *of causal realism*, succinctly described as follows (Lewis-Beck et al. 2003):

There are two broad types of theories of causation: the Humean theory ("causation as regularities") and the causal realist theory ("causation as causal mechanism"). …Consider these various assertions about the statement, "X caused Y":

- *X* is a necessary and/or sufficient condition of Y.
- If *X* had not occurred, Y would not have occurred.
- The conditional probability of Y given X is different from the absolute probability of *Y* ($P(Y|X) \neq P(Y)$).
- *X* appears with a non-zero coefficient in a regression equation predicting the value of *Y*.

  There is a causal mechanism leading from the occurrence of *X* to the occurrence of *Y*.
- The central insight of causal realism is that the final criterion is in fact the most fundamental. According to causal realism, the fact of the existence of underlying causal mechanisms linking *X* to *Y* accounts for each of the other criteria; the other criteria are symptoms of the fact that there is a causal pathway linking *X* to *Y*. … Causal realism insists, finally, that empirical evidence must be advanced to assess the credibility of the causal mechanism that is postulated between cause and effect. …A causal mechanism is a sequence of events or conditions, governed by lawlike regularities, leading from the *explanans* to the *explanandum*.

The challenge of elucidating causal pathways and processes describing *how* changes in exposures propagate through sequences of law-like biological, chemical, or other mechanisms to affect health, thus yielding explanations of C-R causal relations that can be used to predict accurately the health effects in individuals or populations of manipulating exposures, has been addressed via extensions of the principles and algorithmic approaches discussed in previous sections (Pearl 2014; Chap. 9). Although a full discussion of these extensions is beyond the intended scope of this chapter, it is worth recognizing that (a) Detecting and quantifying causal C-R relations in populations stops well short of the mechanistic explanations addressed by advanced causal discovery algorithms; but (b) The framework of information-based causal inference, including causal graphs (e.g., causal Bayesian Networks), conditional independence tests, and conditional dependence relationships quantified by partial dependence plots, can be extended to address explanation and description of causal pathways and mechanisms.

A traditional approach to elucidating causal pathways in epidemiology within a potential outcomes framework is causal *mediation analysis* (Chap. 9) For example, Imai et al. (2011) present algorithms for estimating the "indirect effect" of an exposure or treatment variable on an outcome or response variable that is transmitted via an observed mediator, as well as for estimating the "direct effect" transmitted via other pathways not involving that mediator. They note that, as with other potential outcomes analyses, their estimation procedures require strong, unverifiable assumptions. As explained by Keele et al. (2015), "This gives causal mediation analysis the character of observational studies, where confounding between [the mediator and potential outcomes] must be ruled out 'on faith' to some extent." However, they make a constructive advance by presenting nonparametric estimation and sensitivity analysis algorithms, implemented in an R package, that quantify how large the errors introduced by unobserved pre-treatment confounders might be. The problem of unobserved post-treatment confounders remains open, however.

Despite considerable effort and ingenuity invested in mediation analysis, many past applications are now widely understood to be incorrect or misleading due to failure to distinguish clearly between *conditioning* on values of variables and *manipulating* the values of those variables (Pearl 2014; Spirtes and Zhang 2016). For example, Pearl (2014) comments that much mediation analysis, including "principal stratification's mishandling of mediation" and other technical methods that he characterizes as products of "a century of blunders and confusions," stem from misguided efforts to use *conditioning* on a particular value of a mediator as a proxy for *holding* the value of a mediator fixed at that value. This confuses "seeing" with "doing," in Pearl's evocative terminology.

Even technically sophisticated algorithms developed and used for mediation analysis and estimation of causal impacts in recent decades, such as principal stratification analysis (Frangakis and Rubin 2002; Vanderweele 2011), and G-estimation of causal effects in the presence of time-varying confounding (Robins et al. 1992), do not overcome the fundamental problem that conditioning on observed values is quite distinct from setting those values (Pearl 2014). It is now widely appreciated that effects estimated from mean differences in responses conditioned on different values of an exposure need not coincide with the effects that would be caused by changing exposure from one value to another. To the contrary, conditioning or stratifying on some variables can create spurious C-R associations that have no causal interpretation. Algorithms for determining which subsets of variables to condition on to obtain unbiased estimates of direct causal effects of exposure concentration on response when causal DAGs are known have recently become available in special-purpose causal analytics software (Textor et al. 2011; Textor 2015). Conditioning or stratifying on variables without knowledge of the correct DAG model can be misleading for predicting how changing exposures would change outcomes; as previously noted, observing that heart attack risks are higher among elderly people with higher levels of aspirin consumption would not warrant any causal conclusions about how changing aspirin consumption would change heart attack risks.

Attempts to stratify individuals to more clearly reveal the effects that would be caused by changing exposures via pathways involving specified mediating variables do not solve the problem that conditioning is not manipulation. Like other data analysis strategies (such as matching, propensity scores, and instrumental variables) that try to approximate randomized assignments by conditioning on observed data, stratification methods do not address the key point that statistical relationships among observed levels of variables do not necessarily reveal how changing one would change others. Some experts have concluded that principal stratification "is of some use in assessing 'direct effects' [but] it is not the appropriate tool for assessing 'mediation.' There is nothing within the principal stratification framework that corresponds to a measure of an 'indirect' or 'mediated' effect" (Vanderweele 2011). Such assessments suggest that it is worthwhile to consider other approaches to elucidating causal pathways and mechanisms.

A complementary approach to causal explanation and prediction of effects of future interventions emphasizes the operation of sequences of causal mechanisms *within individuals* (Dash et al. 2013; Maldonado 2013). Causal discovery algorithms can be applied to data from cells, signaling pathways, biochemical and physiological processes, etc. within individuals (Lagani et al. 2016; Faes et al. 2015; Schiatti et al. 2015). Paying careful attention to the temporal sequences in which different observed variables respond to an exogenous change in the value of just one or a few of them can help to reveal the structure of causal networks and estimate the magnitudes and time delays in transmission of causal impacts among time series variables (Mohammad and Nishida 2011; Schiatti et al. 2015). Even without such exogenous shocks, quantifying predictive relations among variables using information-theoretic algorithms allows causal network structure and estimates of effect sizes and delays to be derived from time series data on multiple causally related variables under fairly general conditions (Sun et al. 2015). These approaches for elucidating causal mechanisms and pathways *within individuals*, developed largely in artificial intelligence (Dash et al. 2013) and systems biology (Lagani et al. 2016), complement the statistical strategies discussed and illustrated in previous sections for estimating differences in average responses across *subpopulations* of individuals with different exposures. Information-based algorithms can help identify which variables might cause which others, and show the shape and magnitude of C-R relations in partial dependence plots, but additional algorithms can also help to quantify the timing of transmission of causal information among variables, adding further resolution to the description of causal processes (*ibid*).

Methods of causal inference that focus on explanatory mechanisms can also strongly complement black-box statistical causal analysis by allowing valid causal conclusions to be drawn from studies of relatively few individuals if the studies suffice to elucidate the mechanisms of pathogenesis. Once it is understood how an exposure causes a response, this understanding can be used to make predictions and to generalize beyond the specific individuals and experimental or clinical conditions used to gain the initial understanding. For example, the discovery that smoking induces chronic unresolved inflammation in the lungs of susceptible smokers, followed by increased cell death and regenerative hyperplasia of the alveolar epithelium, provides a possible mechanistic basis for understanding (or, if it were not already known, predicting) that smoking will also increase risk of chronic obstructive lung disease (COPD) and lung cancer in these smokers (Cox Jr. 2011). This causal prediction could be arrived at without the need to study changes in COPD and lung cancer in large populations over time as smoking habits change. Similarly, suppose that it is known from toxicological and clinical data that inhalation of long, thin amphibole asbestos fibers increases repetitive injury of mesothelial tissue, resulting in (a) localized inflammation and release of TNF-α from alveolar macrophages attracted to the site of tissue injury and inflammation; and (b) upregulation of the expression of TNF-α receptors on mesothelial cells at the site, thus inducing mesothelial cells with DNA damage to survive and proliferate (via an NF-κB signaling pathway) instead of undergoing apoptosis and being safely removed (Yang et al. 2008). Such mechanistic knowledge, when available and correct, can be used to explain and predict health risks, such as the relatively high relative potency of amphibole fibers in causing malignant mesothelioma compared to the lower potency of cleavage fragments and mineral particles that do not have these biological activities (Bareinboim and Pearl 2013). Conversely, mechanistic understanding can also help to identify when epidemiological associations are unlikely to be causal. For example, for PM2.5, Green et al. (2002) suggested that health effects *attributed* to PM2.5 exposure were unlikely on toxicological grounds to be *caused* by PM2.5 exposure, because most PM2.5 lacks the potency needed to induce such effects at realistic exposure concentration levels. Thus, mechanistic understanding can complement epidemiological data in showing how and why some exposures cause adverse effects while others do not.

Unresolved Ambiguity and Geographic Heterogeneity of PM2.5 Exposure Metrics and Unresolved Negative Studies

A familiar but important limitation of efforts to quantify C-R functions for PM2.5 is that "PM2.5" does not denote a well-defined, unique substance. Instead it refers to a heterogeneous collection of fine particulate matters. This raises the possibility that PM2.5 in one location might exert toxic effects while PM2.5 in a different location might not, simply because of differences in the composition of PM2.5 between the two locations. In this case, the concept of a causal C-R function that can be estimated from one or more data sets and then applied to estimate health benefits from reducing PM2.5 at other locations fails not because of such subtleties as that historical C-R associations are not valid predictive models, but simply because the same value of *C* is likely to have different meanings in different locations, measuring different things, and hence having different effects on *R*. Under such conditions, the existence of a single predictively useful C-R function for PM2.5 should not be expected.

There is empirical support for this concern, as well as a sub-literature that speculates about which specific components of PM2.5 might exert adverse health effects. For occupational exposures to a specific form of particulate matter, carbon black, Dell et al. (2006) applied standardized mortality ratio (SMRs) and Cox Proportional hazards regression modeling to quantify C-R association between lung cancer and respiratory disease mortality and cumulative inhalable carbon black exposure among over 6000 US carbon black workers. They found that no consistent C-R association between cumulative exposure to inhalable carbon black and respiratory disease mortality. On the public health side, Chay et al. (2003) "found that [1970

Clean Air Act] regulatory status is associated with large reductions in [total suspended particulates] pollution but has little association with reductions in either adult or elderly mortality" in the United States; the authors interpreted these negative findings cautiously, noting limitations in their study. Building on this work, Obenchain and Young (2016) concluded that reducing air pollution did not reduce deaths. Young and Xia (2013) found that there is geographic heterogeneity in air quality-mortality associations across the United States, with no effect of PM2.5 on life expectancy in the western United States. Young and Xia (2013) and Krstić (2013) both critically re-analyze previous reports (Pope et al. 2002) associating PM2.5 with reduced life expectancy in the United States and highlight the importance of adequate control for potentially significant confounding factors and the need to consider influential outliers, specific variable-attributable effects, and geographical heterogeneity. Cox Jr and Popken (2015) examined longitudinal data on PM2.5 and mortality rates in 100 U.S. cities and concluded that there were many cities with statistically significant PM2.5-mortality associations, mostly (but not entirely) positive, but no clear evidence that PM2.5 is a Granger (predictive) cause of mortality (see Chap. 17). These studies looked for effects of PM2.5 reductions on mortality rates on time scales of years to decades. On a time scale of days to weeks, Zu et al. analyzed data from a natural experiment in which forest fires sent daily average PM2.5 concentrations above 60 μm/m^3 in Boston and above 80 μm/m^3 in New York for 3 days in 2002. They concluded that these spikes in PM2.5 did not produce "any discernible increase in daily mortality subsequent to the dramatic increase in ambient PM2.5 levels." On a wider spatial scale, an innovative study by Greven et al. (2011) used estimated "local coefficients" to try to reduce effects of unmeasured confounders in a very large spatio-temporal dataset (the Medicare Cohort Air Pollution Study), which included individual-level information on time of death and age on a population for over 18 million people between 2002 and 2006. They concluded that "Based on the local coefficient alone, we are not able to demonstrate any change in life expectancy for a reduction in $PM_{2.5}$." Outside the U.S., on a time scale of years to decades, as previously discussed, Dockery et al. (2013) reported that substantial (roughly 45% to 70%) reductions in black smoke in Ireland were accompanied by no detectable decreases in all-cause or cardiovascular mortality rates, correcting earlier reports to the contrary. Such negative findings, discussed further in Chaps. 16 and 17, clearly challenge the many assumption-based calculations, such as those in Table 15.1, that confidently predict that further reducing PM2.5 will substantially reduce all-cause mortality rates.

More generally, the fact that PM2.5 has fallen dramatically in many locations without detectably affecting mortality rates (Chap. 17), contrary to projections from global burden of disease (GBD) and other models, invites explanation. One possible explanation is that C-R functions estimated by regressing past levels of mortality against past levels of PM2.5 exposure concentrations are not valid predictive causal models. Another possible explanation is that the composition of the PM2.5 exposures (or of the exposed populations) in locations such as Boston, New York, California and the western United States, Ireland, and other sites of negative studies may differ from the composition of PM2.5 implicitly assumed in models that project substantial mortality reduction benefits from further reducing PM2.5. Since the composition of PM2.5 to which these models apply is unspecified, their relevance to PM2.5 found in other specific locations is unknown. Without attempting to further resolve issues of geographic and compositional heterogeneity, and consequent limitations on the applicability of mortality reduction benefits calculations that assume that a single C-R function applies everywhere, it seems clear that such a universally applicable function might not exist simply because C lacks a unique, consistent definition in terms of causally relevant agents.

Generalizability of Causal C-R Functions Across Studies and Applications

Apart from the challenges that arise from aggregating disparate substances under the heading "PM2.5,"a more general issue for C-R functions is *generalizability* across studies—that is, the extent to which a C-R function estimated under one set of circumstances can be applied to others to correctly predict the impacts on responses of changing exposure concentrations. This problem has been studied since the 1960s in the quasi-experimental design literature on how to establish the *external validity* of causal inferences (Shadish et al. 2002). It has been addressed more recently within the information-based causal analytics framework of directed acyclic graphs (DAGs), conditional independence relationships, and conditional probability tables (e.g., Lee and Honavar 2013; Bareinboim and Pearl 2013). As a simple example, consider a causal DAG model $X \rightarrow Y \leftarrow Z$ in which X is a manipulable decision variable such as exposure concentration, Y is a health variable of interest such as mortality rate, and Z is a vector of covariates such as age and sex. Suppose that a unique causal model specifying the probability distribution for Y values given X and Z values can be identified and estimated from some mix of observational and experimental or clinical data, yielding a conditional probability table (CPT) specifying the conditional probability of each possible value of Y for each set of input values for X and Z. This CPT can be denoted in symbols by $P(Y = y \mid X$ is set to value x, Z has value $z)$ (or, more briefly, by $P(y \mid do(x), z)$), standing for the conditional probability that Y has value y given that X is set to value x and that Z has value z. (The use of "*do*" to distinguish variables whose values are set by a decision-maker from variables whose values are only observed is due to Pearl.) If this causal model is correct, and Y indeed depends only on X and Z, then the CPT can be transported to new settings in which the population distribution of covariate values is quite different from their distribution in the study or studies used to estimate the CPT. The CPT then represents a (probabilistic) causal law that should be invariant across settings, even though the joint frequency distribution of covariates, and hence the local C-R function for the average value of Y as a function of x, may differ in different populations.

In general, the C-R function $E(Y \mid do(x))$ will be different for different distributions of Z, implying that no single C-R function holds in different settings, simply because the C-R function estimated for one population reflects the distribution of covariates in that population as well as the causal law that determines risk from x for any given set of covariate values. Thus, contrary to widespread current practice, it would be misguided to believe that C-R effects estimated from one study can or should be applied unchanged elsewhere, or that effects of pollutants on human health estimated for a specific population and set of conditions in one country, such as China, "can be applied to other countries, time periods, and settings" (Chen et al. 2013). Conditioning appropriately on *Z* provides so-called *transport formulas* for deriving the correct C-R function for a new population from the previously estimated CPT and the joint distribution of *Z* in the new population, if both are known. (Specifically, for our simple example, the probability that the mortality rate for a randomly selected individual will become *y* if the pollutant concentration is set to *x* is given by the probability identity: $P(y \mid do(x)) = \Sigma_z P(y \mid do(x), z)P(z)$, where $P(z)$ is the probability or relative frequency of covariates z in the target population for which causal impacts are to be estimated and $P(y \mid do(x), z)$is the causal CPT, which may have been developed from previous studies elsewhere. The population C-R function for this target population is then given by the identity $E(Y \mid do(x)) = \Sigma_y P(y \mid do(x))*y$.) Various important generalizations of this idea have been worked out in the recent epidemiological (Hernan and Vanderweele 2011; Schwartz et al. 2011) and computer science literatures, including methods for using causal relationships estimated from experiments and observations in several different source environments to estimate causal impacts of interventions from observational data for a target environment, when this is possible, and otherwise determining that it is not possible (Hernan and Vanderweele 2011; Lee and Honavar 2013; Bareinboim and Pearl 2013). However, such transport formulas have not been developed and applied for causal C-R functions in the PM2.5 health effects literature to date, making it inappropriate to apply (i.e., transport) C-R functions estimated from studied populations to different populations, as encouraged by programs such as BenMAP (United States Environmental Protection Agency 2015). This appears to be an area where simply being aware of, and applying, modern causal analytics methods such as transportability conditions and formulas might be able to improve epidemiological and public health risk assessments practice relatively quickly by enabling practitioners to adjust estimated causal functions to apply to different locations. On the other hand, to properly adjust for differences in distributions of response-related covariates across populations, considerable relevant biological knowledge may be necessary to identify the covariates that should enter the transport formulas, and enough data on the joint distributions of those covariates in the source and target populations must be available for the transport formulas to be applied.

Consideration of Effects on Different Time Scales

Our illustrative analyses used months as the time scale, essentially asking whether knowing average daily PM2.5 within a month helps to predict average daily mortality that month after conditioning on other variables such as temperature. Since daily values of the variable are available, the presence of effects on shorter time scales than months can also be examined. Similar to the results for monthly data in the previous section, Bayesian Network, classification tree, and partial dependence plot analyses of daily data again show that daily mortality rates depend on temperature but not on PM2.5. Including values of variables lagged by 1–7 days as predictors shows that today's elderly mortality depend not only on recent elderly mortality rates in the preceding 3 days (autocorrelation), but also on recent daily minimum temperatures (over at least the most recent 4 days). Elderly mortality does not depend directly on current or lagged PM2.5 values, but PM2.5 values are autocorrelated and lagged values of PM2.5, maximum relative humidity, and daily minimum temperature depend on each other (over a window of at least 4 days). These findings with daily resolution of the time scale confirm that elderly mortality is found to depend directly on daily minimum temperature but not directly on PM2.5 (same-day or lagged). The additional resolution from using daily data reveals transient autocorrelations and cross-correlations among these variables, and also relative humidity, over a period of a few days that are not apparent at the monthly level of aggregation, suggesting the importance of controlling for several days of temperature and humidity as confounders of the PM2.5-elderly mortality association. With time steps of months, these more detailed transients can be ignored. However, the existence of complex transients and interactions among variables on a time scale of days to weeks highlights the importance of carefully choosing time windows for case-crossover designs or dynamic regression models, since simply comparing exposure levels on nearby days with and without a death (e.g., Schwartz 2004) may misattribute to differences in daily exposure concentrations effects that are actually caused by lagged values of other variables that are correlated with such concentrations. For the Boston data, there are many days with missing data, especially in earlier years, so no analogous analyses are possible. Aggregating data to the monthly level resolves this problem, as the vast majority of days in each month have data from which monthly average values of all variables can be calculated. In addition, as previously noted, time steps of months may bring out more clearly than daily data the lasting effects of changes in pollution levels, temperature, or other variables on average daily mortality rates (Laden et al. 2006).

No Discussion of the Foundations and Deep Grounding of Methods

Our review and application of information-based causal inference algorithms has deliberately emphasized principles and algorithms that have led to high performance in competitive benchmarking tests, while skipping over centuries of previous work. As noted by Pearl (2014), "Traditional statisticians fear that, without extensive reading of Aristotle, Kant and Hume, they are not well equipped to tackle the subject of causation, especially when it involves claims based on untested assumptions." Even the relatively short history of computational approaches to causal analysis of data, which is only about a century old, can be intimidating. Some of its key milestones are as follows:

- 1920s: *Path analysis* was introduced and developed by geneticist Sewell Wright (1921). This was the first approach to use directed acyclic graph (DAG) models in conjunction with quantitative analysis of statistical dependencies and independencies to clarify the distinction between correlation and causality. They have been so used ever since. Although Wright's path analysis was restricted to linear models, it can be seen as a forerunner of the Bayesian networks introduced some 70 years later, which generalize path coefficients to *conditional probability tables* (CPTs). These allow for non-parametric estimation of arbitrary (possibly non-linear) probabilistic dependencies among variables by specifying the conditional probabilities for the possible values of a variable, given each combination of values for the variables that point into it in a DAG model. In practice, this conditional probability distribution or table at a node of a DAG model can be represented relatively efficiently as a classification tree for the node's value, given the values of its parents (inputs) in the DAG, rather than by explicitly listing all possible combinations of input values (Frey et al. 2003). Path analysis and closely related linear structural equations models (SEMs) were extensively developed by social scientists and statisticians in the 1960s and 1970s and became primary tools of causal analysis in the social sciences in those decades (Blalock 1967; Kenny 1979).
- 1950s: *Structural equations models (SEMs)* were developed as tools for causal analysis. For example, polymath and Nobel Laureate Herbert Simon defined causal ordering of variables in systems of structural equations (Simon 1953) and applied conditional independence and exogeneity criteria for distinguishing between direct and indirect effects and between causal and spurious correlations in econometrics and other fields (Simon 1954).
- 1960s: *Quasi-experiments* were introduced, standard threats to valid causal inference in observational studies were identified and listed, and statistical designs and tests for overcoming them in observational studies were devised, most notably by social statisticians Campbell and Stanley (1963). These methods were extended and applied to evaluation of the success or failure of many social and educational interventions in the 1960s and 1970s, leading to a large body of techniques for program evaluation. The methods of data analysis and causal analysis developed for quasi-experiments, which consist largely of enumerating and refuting potential non-causal explanations for observed associations, have subsequently been extensively applied to "natural experiments" in which changes affect a subset of a population, allowing a quasi-experimental comparison of changes in responses in the affected subpopulation to contemporaneous changes in responses in the unaffected (control) subpopulation. These methods make it unnecessary to depend on already collected data on historical C and R levels, and instead allow potentially valid causal inferences about how changes in C change R.

- 1965: *Hill considerations for causality introduced.* In 1965, Sir Austin Bradford Hill, expressing doubt that any valid algorithmic approach for causal discovery could exist, and without reference to the computational causal inference methods introduced since the 1920s in genetics, social statistics, econometrics, and engineering (Wiener 1956), introduced a set of intuitive "considerations" to help humans make judgments about causality based on associations (Hill 1965). These considerations, discussed in Chap. 9, stand apart from the rest of the history of causal analysis methods, being neither greatly influenced by nor greatly influencing the technical developments that have led to successful current algorithms for causal discovery and inference. They have been enormously influential in epidemiology and regulatory risk assessment, however, where they have encouraged efforts to use judgment and weight-of-evidence deliberations to interpret associations causally. Some attempts have been made to link Hill's considerations to counterfactual causality (Höfler 2005), but they play no role in current causal analysis algorithms, and the rates of false positives and false negative causal conclusions reached with their help have not been quantified. In Hill's judgment, "What they can do, with greater or less strength, is to help us to make up our minds on the fundamental question—is there any other way of explaining the set of facts before us, is there any other answer equally, or more, likely than cause and effect?" As a psychological aid to help epidemiologists, risk assessors and regulators to make up their minds, Hill's considerations have proved effective, but their performance as a guide for drawing factually correct conclusions about causality—especially manipulative causality—from observational data is less clear. Chapter 9 provides further discussion and synthesis of the Hill considerations with the information-based approaches to causality discussed earlier.
- 1970s: *Conditional independence tests and predictive causality tests for time series* were developed to identify predictive causal relationships between time series, most notably by Nobel Laureate econometrician Clive Granger and colleague Christopher Sims, building on earlier ideas by mathematician and electrical engineer Norbert Wiener (1956). Granger (or Granger-Sims) tests for predictive causality have been extended to multiple time series and applied and generalized by neuroscientists analyzing observations of neural firing patterns in the brain (Friston et al. 2013; Furqan and Siyal 2016; Wibral et al. 2013).
- 1980s: *Counterfactual and potential outcomes techniques* were proposed for estimating average causal effects of treatments in populations, largely by statistician Donald B. Rubin and colleagues, building on work by statistician Jerzey Neyman in 1923. Over the course of four decades, specific computational methods put forward in this framework to quantify average causal effects in populations, usually by trying to use observations and assumptions to estimate what would have happened if treatments or exposures had been randomly assigned, have included matching on observed covariates (Rubin 1974), Bayesian inference (Rubin 1978), matching with propensity scores (Rosenbaum and Rubin 1983), potential outcomes models with instrumental variables (Angrist et al. 1996), principal stratification (Zhang and Rubin 2003), and mediation analysis (Rubin 2004). These methods have been influential in epidemiology, where they

have been presented as suitable for estimating average effects caused by treatments or interventions. But they have also been criticized within the causal analysis community as being needlessly obscure, reliant on untestable assumptions, and prone to give biased, misleading, and paradoxical results in practice, in part because they do not necessarily estimate genuine (manipulative) causal effect (e.g., Pearl 2009b). From this perspective, the useful contributions of the potential outcomes framework can be subsumed into and clarified by methods of structural equations modeling (*ibid*).
The 1980s also saw the introduction of classification and regression trees (CART) methods (Breiman et al. 1984). These would eventually provide nonparametric tests for conditional independence, useful for learning Bayesian network structures from data (Frey et al. 2003). They also provided the base nonparametric models for Random Forest ensembles and related non-parametric ensemble algorithms now widely used in machine learning (Furqan and Siyal 2016).
- 1990s: *Probabilistic graphical models* were developed in great detail and given clear mathematical and conceptual foundations (Pearl 1993). These included Bayesian networks and causal graph models, together with inference algorithms for learning them from data and for using them to draw causal inferences and to estimate the sizes of effects caused by interventions. These methods are most prominently associated with the Turing Award-winning work of computer scientist Judea Pearl and his coauthors. They grew out of the intersection of artificial intelligence and statistics. They provide a synthesis and generalization of many earlier methods, including structural equations modeling (both linear and nonlinear), probabilistic causation, manipulative causation, predictive (e.g., Granger) causation, counterfactual and potential outcomes models, and directed acyclic graph (DAG) models, including path analysis. Conditional independence tests and quantification of conditional probabilistic dependencies play key roles in this synthesis, as set forth in landmark books by Pearl (2000) and Koller and Friedman (2009). The full, careful development of probabilistic graphical models and algorithms created what appears to be a lasting revolution in representing, understanding, and reasoning about causality in a realistically uncertain world (North 2019).
- 2000–Present: *Causal discovery and inference algorithms* for learning causal DAG models from data and for using them to draw causal inferences and to quantify or place bounds on the sizes of impacts caused by different interventions have been extensively developed, refined, tested, and compared over the past two decades. Important advances included clarifying which variables in a DAG model must and must not be conditioned on to obtain unbiased estimates of causal impacts in known DAG models (Textor 2015; Shpitser and Pearl 2008), as well as transportability formulas for applying causal relationships discovered and quantified in one or more learning settings to a different target setting(Hernan and Vanderweele 2011; Lee and Honavar 2013; Bareinboim and Pearl 2013).

Recent years have also seen substantial generalizations of earlier methods. For example, transfer entropy, a nonparametric generalization of Granger causality, quantifies the rates of directed information flows among time series variables. Introduced by physicist Thomas Schreiber (2000) and subsequently refined and extended by workers in computational finance and neuroscience (Wibral et al. 2013), transfer entropy and closely related methods appear to be promising for creating algorithms to discover causal DAG structures and quantitative dependency relationships and time lag characteristics from observations of multiple time series.

Even such an abridged list of milestones makes clear that causal analytics is now a large and deep field with a host of interrelated technical concepts and algorithms supported by a confluence of insights and methods from statistics, social statistics and program evaluation, electrical engineering, economics and econometrics, physics, computer science, computational finance, neuroscience, and other fields. Any brief survey must therefore be relatively superficial; full treatments run into thousands of pages (e.g., Koller and Friedman 2009), and even documentation for R packages implementing the key ideas can be hundreds of pages. This deep grounding of current information-based causal analytics methods and algorithms in nearly a century of computational methods backed by centuries of philosophizing about causality might well inspire a prudent humility (Pearl 2014). Yet, for the practitioner with limited time and a need to draw sound causal inferences from data, two relatively recent developments make even superficial understanding of key ideas and software packages highly useful. The first is that many formerly distinct causal analysis methods have now been synthesized and unified within the framework of information-theoretic methods and directed acyclic graphs. This framework brings together ideas from potential outcomes and counterfactual causation, predictive causality, DAG modeling, and manipulative causality (Pearl 2000, 2010). The second is the success of the object-oriented software paradigm in platforms such as R and Python. Modern software enables and encourages encapsulation of technical implementation details so that only key ideas and behaviors of software objects need be understood to use them correctly. This allows users with only a superficial understanding of exactly what a software package does and how it works to use it appropriately to do valuable tasks. For example, a user who understands only that causes must be informative about their effects, and that this can be indicated graphically by arrows between variables showing which ones are identified as being informative about each other and which are conditionally independent of each other, can use this limited understanding to interpret correctly the results of sophisticated algorithms. As a practical matter, making tools such as Bayesian network learning algorithms, classification trees, and partial dependency plots widely available and easy to apply can complement insights from regression-based and other associational and counterfactual methods to reveal and quantify potential causal relationships in observational data.

Summary and Conclusions

The critical review in this chapter of the literature on concentration-response (C-R) relations for fine particulate matter and mortality has identified the following challenges:

- C-R functions that describe historical associations do not necessarily predict how changing C would change R. This is partly because associations may not represent manipulative causal relations, as when positive associations between baby aspirin consumption and heart attack risk, or between nicotine-stained fingers and subsequent risk of lung cancer, do not allow a valid prediction that reducing one would reduce the other.
- Almost all of the existing literature on PM2.5-mortality C-R functions deals with associations and not with causality (Wang et al. 2016).
- The few papers that do attempt to model causality in C-R relations for PM2.5 exposure and mortality fail to distinguish among counterfactual, predictive, and manipulative causality. Most of these papers follow a counterfactual approach that relies heavily on unverified modeling assumptions about unobserved potential outcomes, as discussed further in Chap. 17. This amounts to assuming, rather than showing, that associations in regression models can be interpreted causally. Arguably, manipulative causation should be the gold standard for discussions of causality and the public health effects that would be caused by changing PM2.5 exposures. Although even state-of-the-art causal inference algorithms cannot definitively establish manipulative causality from observational data, they can identify several measures of predictive causality (Granger causality, information relations in DAG models, and so forth) that provide a valuable screen for evidence of potential manipulative causation.
- A large literature on modern causal inference algorithms for observational data (Chap. 9) has yet to be applied to C-R modeling. The most successful methods in this literature, as assessed in competitive challenges, emphasize the information criterion (a necessary but not sufficient condition) that causes should be informative about their effects. Changes in causes should also be informative about changes in their effects. These principles can be applied to observed data using freely available R packages for predictive analytics. Ensembles of non-parametric CART trees used to detect and quantify information about responses provided by exposure concentrations are among the most successful and most mature current algorithmic approaches used in computational causal inference.
- Applying these methods to publicly available data from the Northeast (Boston) and Southern California(Los Angeles) shows that C-R associations found in Boston do not hold in the SCAQMD air basin. Quite significant (e.g., twofold) changes in average PM2.5 concentrations do not help to predict changes in average elderly mortality rates in either location, at least on the time scales of monthly averages separated by a one-year lag used in our example analyses.

We conclude that well-defined causal C-R functions do not necessarily exist for PM2.5 and elderly mortality, at least on the time scales considered. Different statistically significant associations hold in different areas, but they do not necessarily correspond to the theoretical construct of a causal C-R function that can be estimated at one location and applied to another to approximate how changes in concentrations would affect changes in public health. It seems highly desirable for future work to distinguish more clearly among statistical association, predictive causal, and manipulative causal C-R functions than past research has done. The fundamental premise that C-R functions exist that can predict the public health effects caused by reductions in pollutant concentrations needs to be carefully reexamined and tested, as it does not appear to hold in general.

References

Aliferis CE, Statnikov A, Tsamardinos I, Mani S, Koutsoukos XS. Local causal and markov blanket induction for causal discovery and feature selection for classification part I: algorithms and empirical evaluation. J Mach Learn Res. 2010;11:171–234.

Angrist JD, Imbens GW, Rubin DB. Identification of causal effects using instrumental variables. J Am Stat Assoc. 1996;91(434):444–55.

Apte JS, Marshall JD, Cohen AJ, Brauer M. Addressing global mortality from ambient PM2.5. Environ Sci Technol. 2015;49(13):8057–66.

Baiocchi M, Cheng J, Small DS. Instrumental variable methods for causal inference. Stat Med. 2014;33(13):2297–340.

Bareinboim E, Pearl J (2013) Meta-transportability of causal effects: a formal approach. In: Proceedings of the Sixteenth International Conference on Artificial Intelligence and Statistics

Blalock HM. Causal inferences in nonexperimental research. Chapel Hill: UNC Press; 1967.

Bontempi G, Flauder M. From dependency to causality: a machine learning approach. J Mach Learn Res. 2015;16:2437–57.

Breiman L, Friedman J, Olshen R, Stone C. Classification and regression trees. Belmont: Wadsworth; 1984.

Campbell DT, Stanley JC. Experimental and quasi-experimental designs for research. Boston: Houghton Mifflin Company, 1963.

Chay K, Dobkin C, Greenstone M. The clean air act of 1970 and adult mortality. J Risk Uncertain. 2003;27:279–300.

Chen Y, Ebenstein A, Greenstone M, Li H. Evidence on the impact of sustained exposure to air pollution on life expectancy from China's Huai River policy. Proc Natl Acad Sci U S A. 2013;110(32):12936–41. https://doi.org/10.1073/pnas.1300018110.

Chipman H, McCulloch R (2016) Package 'BayesTree'. https://cran.r-project.org/web/packages/BayesTree/BayesTree.pdf

Clancy L, Goodman P, Sinclair H, Dockery DW. Effect of air-pollution control on death rates in Dublin, Ireland: an intervention study. Lancet. 2002;360(9341):1210–4.

Cooper GF, Bahar I, Becich MJ, Benos PV, Berg J, Espino JU, Glymour C, Jacobson RC, Kienholz M, Lee AV, Lu X, Scheines R, Center for Causal Discovery Team. The center for causal discovery of biomedical knowledge from big data. J Am Med Inform Assoc. 2015;22(6):1132–6.

Cox LA Jr, Popken DA. Has reducing PM2.5 and ozone caused reduced mortality rates in the United States? Ann Epidemiol. 2015;25(3):162–73.

Cox LA Jr. A causal model of chronic obstructive pulmonary disease (COPD) risk. Risk Anal. 2011;31(1):38–62.

Cox LA Jr. Rethinking the meaning of concentration–response functions and the estimated burden of adverse health effects attributed to exposure concentrations. Risk Anal. 2016;36(9):1770–9.

Cromar KR, Gladson LA, Perlmutt LD, Ghazipura M, Ewart GW. American Thoracic Society and Marron Institute Report. Estimated excess morbidity and mortality caused by air pollution above American Thoracic Society-Recommended Standards, 2011-2013. Ann Am Thorac Soc. 2016;13(8):1195–201.

Dash D, Voortman M, de Jongh M. Sequences of mechanisms for causal reasoning in artificial intelligence. In: Proceeding IJCAI '13 Proceedings of the twenty-third international joint conference on artificial intelligence. New York: AAAI Press; 2013. p. 839–45.

Dell LD, Mundt KA, Luippold RS, Nunes AP, Cohen L, Burch MT, Heidenreich MJ, Bachand AM, International Carbon Black Association. A cohort mortality study of employees in the U.S. carbon black industry. J Occup Environ Med. 2006;48(12):1219–29.

Di Q, Wang Y, Zanobetti A, Koutrakis P, Dominici F, Schwartz JD. Association of short-term exposure to air pollution with mortality in older adults. J Am Med Assoc. 2017;318(24):2446–56.

Dockery D, Pope C, Xu X, et al. An association between air pollution and mortality in six US cities. N Engl J Med. 1993;329:1753–9.

Dockery DW, Rich DQ, Goodman PG, Clancy L, Ohman-Strickland P, George P, Kotlov T, HEI Health Review Committee. Effect of air pollution control on mortality and hospital admissions in Ireland. Res Rep Health Eff Inst. 2013;176:3–109.

Dominici F, Greenstone M, Sunstein CR. Science and regulation. Particulate matter matters. Science. 2014;344(6181):257–9.

Faes L, Porta A, Nollo G. Algorithms for the inference of causality in dynamic processes: application to cardiovascular and cerebrovascular variability. Conf Proc IEEE Eng Med Biol Soc. 2015;2015:1789–92.

Fann N, Lamson AD, Anenberg SC, Wesson K, Risley D, Hubbell BJ. Estimating the national public health burden associated with exposure to ambient PM2.5 and ozone. Risk Anal. 2012;32(1):81–95.

Fedak KM, Bernal A, Capshaw ZA, Gross S. Applying the Bradford Hill criteria in the 21st century: how data integration has changed causal inference in molecular epidemiology. Emerg Themes Epidemiol. 2015;12:14.

Frangakis CE, Rubin DB. Principal stratification in causal inference. Biometrics. 2002;58:21–9.

Franklin M, Zeka A, Schwartz J. Association between PM2. 5 and all-cause and specific-cause mortality in 27 US communities. J Expo Sci Environ Epidemiol. 2006;17:279–87.

Frey HC. Dose-response models are conditional on determination of causality. Risk Anal. 2016;36(9):1751–4.

Frey L, Fisher D, Tsamardinos I, Aliferis CF, Statnikov A, (2003). Identifying Markov blankets with decision tree induction. In: Proceedings of the third IEEE international conference on data mining, Melbourne, FL Nov 19–22 2003. pp 59–66

Friston K, Moran R, Seth AK. Analysing connectivity with Granger causality and dynamic causal modelling. Curr Opin Neurobiol. 2013;23(2):172–8.

Furqan MS, Siyal MY. Random forest Granger causality for detection of effective brain connectivity using high-dimensional data. J Integr Neurosci. 2016;15(1):55–66.

Giannadaki D, Lelieveld J, Pozzer A. Implementing the US air quality standard for PM2.5 worldwide can prevent millions of premature deaths per year. Environ Health. 2016;15(1):88.

Glaeser EL(2006) Researcher incentives and empirical methods. NBER Technical Working Paper No. 329 October

Goto D, Ueda K, Ng CFS, Takami A, Ariga T, Matsuhashi K, Nakajima T. Estimation of excess mortality due to long-term exposure to PM2.5 in Japan using a high-resolution model for present and future scenarios. Atmos Environ. 2016;140:320–32.

Green LC, Crouch EA, Ames MR, Lash TL. What's wrong with the National Ambient Air Quality Standard (NAAQS) for fine particulate matter (PM(2.5))? Regul Toxicol Pharmacol. 2002;35(3):327–37.

Greenland S. Multiple-bias modelling for analysis of observational data. J R Stat Soc A. 2005; 168(2):267–306.

Greven S, Dominici F, Zeger S. An approach to the estimation of chronic air pollution effects using spatio-temporal information. J Am Stat Assoc. 2011;106:396–406.

Halliday DM, Senik MH, Stevenson CW, Mason R. Non-parametric directionality analysis - extension for removal of a single common predictor and application to time series. J Neurosci Methods. 2016;268:87–97.

Hart J, Garshick E, Dockery D, Smith T, Ryan L, Laden F. Long-term ambient multi-pollutant exposures and mortality. Am J Respir Crit Care Med. 2011;183:73–8.

Hernan M, Vanderweele T. On compound treatments and transportability of causal inference. Epidemiology. 2011;22:368.

Hernandez B, Raftery AE, Pennington SR, Parnell AC (2015) Bayesian additive regression trees using Bayesian model averaging. https://arxiv.org/pdf/1507.00181.pdf

Hill AB. The environment and disease: association or causation? Proc R Soc Med. 1965;58:295–300.

Hill J (2016) Atlantic causal inference conference competition: is your SATT where it's at? http://jenniferhill7.wixsite.com/acic-2016/competition

Höfler M. The Bradford Hill considerations on causality: a counterfactual perspective. Emerg Themes Epidemiol. 2005;2:11.

Höfler M, Lieb R, Wittchen HU. Estimating causal effects from observational data with a model for multiple bias. Int J Methods Psychiatr Res. 2007;16(2):77–87.

Imai K, Keele L, Tingley D, Yamamoto T. Unpacking the black box of causality: learning about causal mechanisms from experimental and observational studies. Am Polit Sci Rev. 2011;105:4.

Keele L, Tingley D, Yamamoto T. Identifying mechanisms behind policy interventions via causal mediation analysis. J Policy Anal Manage. 2015;34(4):937–63.

Kenny DA. Correlation and causality. New York: Wiley; 1979.

Kim J, Yoon K, Choi JC, Kim H, Song JK. The association between wind-related variables and stroke symptom onset: a case-crossover study on Jeju Island. Environ Res. 2016;150:97–105.

Kleinberg S, Hripcsak G. A review of causal inference for biomedical informatics. J Biomed Inform. 2011;44(6):1102–12.

Koller D, Friedman N. (2009). Probabilistic Graphical Models: Principles and Techniques. The MIT Press. Cambridge, MA USA.

Krstić G. A reanalysis of fine particulate matter air pollution versus life expectancy in the United States. J Air Waste Manag Assoc. 2013;63(2):133–5.

Krstić G, Krstić NS, Mauricio Z-B. The br2—weighting Method for Estimating the Effects of Air Pollution on Population Health. J Mod Appl Stat Methods. 2016;15(2):42. https://doi.org/10.22237/jmasm/1478004000.

Laden F, Schwartz J, Speizer FE, Dockery DW. Reduction in fine particulate air pollution and mortality: extended follow-up of the Harvard Six Cities study. Am J Respir Crit Care Med. 2006;173:667–72.

Lagani V, Triantafillou S, Ball G, Tegnér J, Tsamardinos I. Probabilistic computational causal discovery for systems biology. In: Geris L, Gomez-Cabrero D, editors. Uncertainty in biology: a computational modeling approach, vol. 17. Cham: Springer; 2016.

Lee S, Honavar V (2013) m-Transportability: transportability of a causal effect from multiple environments. In: Proceedings of the twenty-seventh AAAI conference on artificial intelligence. www.aaai.org/ocs/index.php/AAAI/AAAI13/paper/viewFile/6303/7210

Lepeule J, Laden F, Dockery D, Schwartz J. Chronic exposure to fine particles and mortality: an extended follow-up of the Harvard Six Cities study from 1974 to 2009. Environ Health Perspect. 2012;120:965–70.

Lewis-Beck MS, Bryman A, Liao TF. Encyclopedia of social science research methods. Thousand Oaks: Sage Publications; 2003.

Lin H, Liu T, Fang F, Xiao J, Zeng W, Li X, Guo L, Tian L, Schootman M, Stamatakis KA, Qian Z, Ma W. Mortality benefits of vigorous air quality improvement interventions during the periods of APEC blue and parade blue in Beijing. Environ Pollut. 2016;220:222–7.

Lo WC, Shie RH, Chan CC, Lin HH. Burden of disease attributable to ambient fine particulate matter exposure in Taiwan. J Formos Med Assoc. 2016;116(1):32–40.

Lopiano KK, Smith RL, Young SS (2015) Air quality and acute deaths in California, 2000–2012. https://arxiv.org/abs/1502.03062

Maldonado G. Toward a clearer understanding of causal concepts in epidemiology. Ann Epidemiol. 2013;23(12):743–9.

Marra G, Radice R, Missiroli S. Testing the hypothesis of absence of unobserved confounding in semiparametric bivariate probit models. Comput Stat. 2014;29:715.

McClellan RO. Providing context for ambient particulate matter and estimates of attributable mortality. Risk Anal. 2016;36(9):1755–65.

Mohammad Y, Nishida T. Discovering causal change relationships between processes in complex system. Los Alamitos: Institute of Electrical & Electronics Engineers (IEEE); 2011.

Moolgavkar SH. Fine particulate matter pollution and mortality. Risk Anal. 2016;36(9):1766–9.

Morabito M, Crisci A, Messeri A, Capecchi V, Modesti PA, Gensini GF, Orlandini S. Environmental temperature and thermal indices: what is the most effective predictor of heat-related mortality in different geographical contexts? Sci World J. 2014;2014:961750.

North DW. Introduction to special issue on air pollution health risks. Risk Anal. 2016;36(9):1688–92. https://doi.org/10.1111/risa.12707.

North DW. Book reviews mega-review: causality books. Risk Anal. 2019;39(7):1647–54.

Obenchain RL, Young SS. Local control strategy: simple analyses of air pollution data can reveal heterogeneity in longevity outcomes. Risk Anal. 2016;37(9):1742–53.

O'Malley AJ. Instrumental variable specifications and assumptions for longitudinal analysis of mental health cost offsets. Health Serv Outcome Res Methodol. 2012;12(4):254–72.

Pearl J. Comment: graphical models, causality, and intervention. Stat Sci. 1993;8:266–9.

Pearl J. Causality: models, reasoning, and inference. 1st ed. New York: Cambridge University Press; 2000.

Pearl J. Causal inference in statistics: an overview. Stat Surveys. 2009a;3:96–146. https://doi.org/10.1214/09-SS057.

Pearl J. Causality: models, reasoning and inference. 2nd ed. New York: Cambridge University Press; 2009b.

Pearl J. An introduction to causal inference. Int J Biostat. 2010;6(2):7.

Pearl J. Reply to commentary by Imai, Keele, Tingley, and Yamamoto concerning causal mediation analysis. Psychol Methods. 2014;19(4):488–92.

Petersen ML, van der Laan MJ. Causal models and learning from data: integrating causal modeling and statistical estimation. Epidemiology. 2014;25(3):418–26.

Pope CA, Burnett RT, Thun MJ, et al. Lung cancer, cardiopulmonary mortality, and long-term exposure to fine particulate air pollution. JAMA. 2002;287:1132–41.

Pope CA, Cropper M, Coggins J, Cohen A. Health benefits of air pollution abatement policy: role of the shape of the concentration-response function. J Air Waste Manag Assoc. 2015;65(5):516–22.

Robins JM, Blevins D, Ritter G, Wulfsohn M. G-estimation of the effect of prophylaxis therapy for Pneumocystis carinii pneumonia on the survival of AIDS patients. Epidemiology. 1992;3:319–36.

Rosenbaum PR, Rubin DB. The central role of the propensity score in observational studies for causal effects. Biometrika. 1983;70(1):41–55. https://doi.org/10.1093/biomet/70.1.41.

Rottman BM, Hastie R. Reasoning about causal relationships: inferences on causal networks. Psychol Bull. 2014;140(1):109–39. https://doi.org/10.1037/a0031903.

Rubin DB. Estimating causal effects of treatments in randomized and nonrandomized studies. J Educ Psychol. 1974;66(5):688–701. https://doi.org/10.1037/h0037350.

Rubin DB. Bayesian inference for causal effects: the role of randomization. Ann Stat. 1978;6:34–58.

Rubin DB. Direct and indirect causal effects via potential outcomes (with discussion). Scand J Stat. 2004;31:161–70.

Schiatti L, Nollo G, Rossato G, Faes L. Extended Granger causality: a new tool to identify the structure of physiological networks. Physiol Meas. 2015;36(4):827–43.
Schreiber T. Measuring information transfer. Phys Rev Lett. 2000;85(2):461–4. https://doi.org/10.1103/PhysRevLett.85.461.
Schwartz J. The effects of particulate air pollution on daily deaths: a multicity case crossover analysis. Occup Environ Med. 2004;61:956–61.
Schwartz J, Laden F, Zanobetti A. The concentration-response relation between PM(2.5) and daily deaths. Environ Health Perspect. 2002;110(10):1025–9.
Schwartz S, Gatto NM, Campbell UB. Transportabilty and causal generalization. Epidemiology. 2011;22(5):745–6.
Schwartz J, Austin E, Bind MA, Zanobetti A, Koutrakis P. Estimating causal associations of fine particles with daily deaths in Boston. Am J Epidemiol. 2015;182(7):644–50.
Schwartz J, Bind MA, Koutrakis P. Estimating causal effects of local air pollution on daily deaths: effect of low levels. Environ Health Perspect. 2016;125(1):23–9.
Shadish W, Cook T, Campbell D. Experimental and quasi-experimental designs for generalized causal inference. Boston: Houghton Mifflin; 2002.
Shpitser I, Pearl J. Complete identification methods for the causal hierarchy. J Mach Learn Res. 2008;9:1941–79.
Simon HA. Causal ordering and identifiability. In: Hood WC, Koopmans TC, editors. Studies in econometric method. Cowles commission forresearch in economics monograph No. 14. New York: Wiley; 1953. p. 49–74.
Simon HA. Spurious correlation: a causal interpretation. J Am Stat Assoc. 1954;49(267):467–79.
Slack MK, Draugalis JR. Establishing the internal and external validity of experimental studies. Am J Health Syst Pharm. 2001;58(22):2173–81.
Smith AE. Inconsistencies in risk analyses for ambient air pollutant regulations. Risk Anal. 2016;36(9):1737–44.
Spirtes P. Introduction to causal inference. J Mach Learn Res. 2010;11:1643–62.
Spirtes P, Zhang K. Causal discovery and inference: concepts and recent methodological advances. Appl Inform. 2016;3:3.
Spirtes P, Meek C, Richardson T. Causal inference in the presence of latent variables and selection bias. In: UAI'95 proceedings of the eleventh conference on uncertainty in artificial intelligence. San Francisco: Morgan Kaufmann Publishers Inc.; 1995. p. 499–506.
Sun J, Taylor D, Boll EM. Causal network inference by optimal causation entropy. SIAM J Appl Dyn Syst. 2015;14(1):73–106.
Textor J (2015) Drawing and analyzing causal DAGS with DAGitty. http://www.dagitty.net/manual-2.x.pdf
Textor J, Hardt J, Knüppel S. DA Gitty: a graphical tool for analyzing causal diagrams. Epidemiology. 2011;22(5):745. https://doi.org/10.1097/EDE.0b013e318225c2be.
United States Environmental Protection Agency (2015) BenMAP environmental benefits mapping and analysis program – community edition. User's manual appendices. www.epa.gov/sites/production/files/2015-04/documents/benmap-ce_user_manual_appendices_march_2015.pdf
Urban A, Kyselý J. Comparison of UTCI with other thermal indices in the assessment of heat and cold effects on cardiovascular mortality in the Czech Republic. Int J Environ Res Public Health. 2014;11(1):952–67.
Valberg PA. Possible noncausal bases for correlations between low concentrations of ambient particulate matter and daily mortality. Nonlinearity Biol Toxicol Med. 2003;1(4):521–30. https://doi.org/10.1080/15401420390271137.
Vanderweele TJ. Principal stratification--uses and limitations. Int J Biostat. 2011;7(1):28.
Wang Y, Kloog I, Coull BA, Kosheleva A, Zanobetti A, Schwartz JD. Estimating causal effects of long-term PM2.5 exposure on mortality in New Jersey. Environ Health Perspect. 2016;124(8):1182–8.

Wibral M, Pampu N, Priesemann V, Siebenhühner F, Seiwert H, Lindner M, Lizier JT, Vicente R. Measuring information-transfer delays. PLoS One. 2013;8(2):e55809. https://doi.org/10.1371/journal.pone.0055809.

Wiener N. The theory of prediction. In: Beckenbach EF, editor. Modern mathematics for engineers. New York: McGraw-Hill; 1956.

Woodward J (2013) Causation and manipulability. In: Edward N. Zalta (ed) The Stanford encyclopedia of philosophy. http://plato.stanford.edu/archives/win2013/entries/causation-mani/

Wright S. Correlation and causation. J Agric Res. 1921;20:557–85.

Wu MH, Frye RE, Zouridakis G. A comparison of multivariate causality based measures of effective connectivity. Comput Biol Med. 2011;41(12):1132–41.

Yang H, Testa JR, Carbone M. Mesothelioma epidemiology, carcinogenesis and pathogenesis. Curr Treat Options in Oncol. 2008;9(2-3):147–57.

Young SS, Xia JQ. Assessing geographic heterogeneity and variable importance in an air pollution data set. Sta Anal Data Mining. 2013;6(4):375–86.

Yule A, Udny G. Why do we sometimes get nonsense-correlations between time-series? -- A study in sampling and the nature of time-series. J R Stat Soc. 1926;89(1):1–63.

Zhang JL, Rubin DB. Estimation of causal effects via principal stratification when some outcomes are truncated by "death". J Educ Behav Stat. 2003;28:353–68. https://doi.org/10.3102/10769986028004353.

Zigler CM, Dominici F, Wang Y. Estimating causal effects of air quality regulations using principal stratification for spatially correlated multivariate intermediate outcomes. Biostatistics. 2012;13(2):289–302.

Zigler CM, Kim C, Choirat C, Hansen JB, Wang Y, Hund L, Samet J, King G, Dominici F, HEI Health Review Committee. Causal inference methods for estimating long-term health effects of air quality regulations. Res Rep Health Eff Inst. 2016;187:5–49.

Chapter 16
How Do Exposure Estimation Errors Affect Estimated Exposure-Response Relations?

Introduction

Associations between fine particulate matter (PM2.5) exposure concentrations and a wide variety of undesirable outcomes are routinely reported. Adverse outcomes associated with PM2.5 range from autism, auto theft, and COVID-19 mortality to elderly mortality, suicide, and violent crime. Many influential articles argue that reducing National Ambient Air Quality Standards for PM2.5 is desirable to reduce these outcomes; of late, it has become fashionable to explicitly interpret estimated statistical associations causally, implying that reducing PM2.5 would reduce associated adverse outcomes, typically by making untested modeling assumptions to justify interpreting association as causation. Yet, other studies have found that reducing particulate pollution dramatically, by as much as 70% and dozens of micrograms per cubic meter, has not detectably affected all-cause mortality rates even after decades, despite strong, statistically significant positive exposure concentration-response (C-R) associations between them (Zigler and Dominici 2014; see also Chap. 17).

This chapter examines whether this disconnect between association and causation might be explained in part by ignored estimation errors in estimated exposure concentrations. We use EPA air quality monitor data from the Los Angeles area of California to examine the shapes of estimated C-R functions for PM2.5 when the true C-R functions are assumed to be step functions with well-defined response thresholds. The estimated C-R functions mistakenly show risk as smoothly increasing with concentrations even well below the response thresholds, thus incorrectly predicting substantial risk reductions from reductions in concentrations that do not affect health risks. We conclude that ignored estimation errors obscure the shapes of true C-R functions, including possible thresholds, possibly leading to unrealistic predictions of the changes in risk caused by changing exposures. Instead of estimating improvements in public health per unit reduction (e.g., per 10 μg/m^3 decrease)

L. A. Cox Jr., *Quantitative Risk Analysis of Air Pollution Health Effects*, International Series in Operations Research & Management Science 299, https://doi.org/10.1007/978-3-030-57358-4_16

in average PM2.5 concentrations, it may be essential to consider how interventions change the distributions of exposure concentrations.

Interpreting PM2.5-Health Effect Associations

Numerous influential studies have reported positive statistical associations between concentrations of fine particulate matter (PM2.5) in air and risk of adverse health effects, especially mortality. An important and common policy-relevant conclusion is that such associations suggest that substantial public health benefits might be achieved by further reducing National Ambient Air Quality Standards (NAAQS) for PM2.5. A recent high-profile paper in this tradition (Di et al. 2017) summarized the PM2.5-morality association in a large (National Medicare) data set and its significance for national environmental policy as follows:

> The US Environmental Protection Agency is required to reexamine its National Ambient Air Quality Standards (NAAQS) every 5 years, but evidence of mortality risk is lacking at air pollution levels below the current daily NAAQS in unmonitored areas and for sensitive subgroups. ...Daily PM2.5 and ozone levels in a 1-km × 1-km grid were estimated... [and] daily exposures were calculated for every zip code in the United States. Each short-term increase of 10 μg/m3 in PM2.5 (adjusted by ozone) ... [was] statistically significantly associated with a relative increase of 1.05% (95% CI, 0.9–1.15%) ... in daily mortality rate... Absolute risk differences in daily mortality rate were 1.42 (95% CI, 1.29–1.56)... per 1 million persons at risk per day. There was no evidence of a threshold in the exposure-response relationship. In the US Medicare population from 2000 to 2012, short-term exposures to PM2.5 and warm-season ozone were significantly associated with increased risk of mortality. This risk occurred at levels below current national air quality standards, suggesting that these standards may need to be reevaluated. (Di et al. 2017)

Mortality is not the only adverse health effect that has been associated with particulate matter. As the US EPA prepares to review the PM2.5 NAAQS in 2018, there has been an outpouring of recent studies associating PM2.5 with adverse outcomes, including(but not limited to) Alzheimer's (Calderón-Garcidueñas et al. 2016), anemia (Honda et al. 2017), asthma (Young et al. 2014), autism (Morales-Suárez-Varela et al. 2017), anxiety (Pun et al. 2017), baldness (Rajput 2015), birth defects and cleft palates (Zhou et al. 2017), bladder cancer (Yeh et al. 2017), burglaries (Mapou et al. 2017), cardiac arrest (Dennekamp et al. 2010), cardiovascular mortality and morbidity (Kim et al. 2017; Thurston et al. 2016), carpal tunnel syndrome in dialysis patients (Weng et al. 2017), cognitive impairment (Tallon et al. 2017a), coronary artery disease (McGuinn et al. 2017), COVID-19 mortality risks (Wu et al. 2020), criminal assaults (Mapou et al. 2017), dandruff (Rajput 2015), diabetes (Bai et al. 2018; Renzi et al. 2017), dementia (Chen et al. 2017), depression (Sram et al. 2017), dermatitis (Oh et al. 2018), genotoxicity in cheek cells from artisanal cashew nut roasting workers (de Oliveira Galvão et al. 2017), hair loss (Rajput 2015), heart attacks (Liu et al. 2018), congenital heart defects (Zhang et al. 2016),impotence (Tallon et al. 2017b), infections (Yadav et al. 2003), infertility (Xue and Zhang 2018), influenza (Ma et al. 2017), IQ deficits in youth (Wang et al. 2017), insomnia or low sleep efficiency (ATS 2017), juvenile delinquency and aggression (Younan et al. 2017), kidney disease (Bowe et al. 2018), kidney cancer (Turner et al. 2017), liver cancer (Pedersen et al. 2017), liver cancer patient mortality rates (Deng et al. 2017), lung cancer (Huang et al. 2017), decreased lung function among people who feel that their neighborhoods are poorly maintained (Malecki et al. 2018), memory loss in adults (Sram et al. 2017),metabolic syndrome (Shamy et al. 2017), obesity (Mazidi and Speakman 2017), pneumonia (Yadav et al. 2003), psychiatric stress in veterans (Mehta et al. 2015), sleep-disordered breathing (Shen et al. 2018), substance abuse (Younan et al. 2017), suicide (Ng et al. 2016), theft of motor vehicles (Mapou et al. 2017), and violent crime (Herrnstadt and Muehlegger 2015). Particularly troubling is that associations of PM2.5 with mortality and morbidities are reported as having no safe levels or thresholds in exposure-response relations, even at concentrations at or below current standards, as noted by Di et al. (2017) for mortalities.

Associations between particulate matter (including black smoke) and mortality risk have been recognized for decades (e.g., Clancy et al. 2002). Frustratingly, some previous interventions (e.g., coal-burning bans) that attempted to reduce mortality and morbidities by reducing exposure concentrations of particulates have shown that even reducing mean air concentrations of particulate matter by up to 70% (35 μg/m^3) caused no detectable change in total mortality rates after more than two decades (Dockery et al. 2013). This suggests that the relation between statistical exposure-response associations and manipulative causation that would allow responses to be reduced by reducing exposures is more subtle than was initially hoped and expected (Zigler and Dominici 2014; Harvard School of Public Health 2002). (That the bans saved lives is still mistakenly claimed in Ireland to support nation-wide extensions of regulations; see Chap. 17.) In the United States, clear, significant statistical associations between PM2.5 concentrations and mortality rates have not implied that even substantial historical reductions in PM2.5 lead to detectable reductions in mortality rates (Cox Jr and Popken 2015). Association is not causation, inviting the question: Why not? How can there be a strong, consistent, robust statistical PM2.5-mortality exposure-response relation without a causal link whereby reducing PM2.5 reduces mortality rates? The PM2.5-mortality association seems to satisfy most of the traditional Bradford Hill considerations for associative causation and is routinely presented in the abundant literature on air pollution healthy effects as if this also implied manipulative causation (Cox Jr. 2017). Why doesn't it?

The remainder of this chapter examines in detail one possible explanation for the discrepancy between (a) statistical (association-based) exposure-response curves for mortalities or morbidities estimated from epidemiological data; and (b) causally effective exposure-response curves, for which reducing exposure reduces response as described by the curve. That explanation is *unmodeled measurement or estimation error* in estimates of exposure. As explained by Rhomberg et al. (2011):

> It is well known among statisticians that random errors in the values of independent variables (such as exposure in exposure-response curves) may tend to bias regression results. For increasing curves, this effect tends to flatten and apparently linearize what is in truth a

> steeper and perhaps more curvilinear or even threshold-bearing relationship. The degree of bias is tied to the magnitude of the measurement error in the independent variables. It has been shown that the degree of bias known to apply to actual studies is sufficient to produce a false linear result, and that although nonparametric smoothing and other error-mitigating techniques may assist in identifying a threshold, they do not guarantee detection of a threshold. The consequences of this could be great, as it could lead to a misallocation of resources towards regulations that do not offer any benefit to public health. (Rhomberg et al. 2011)

It is easy to see how a threshold-like exposure concentration-response (C-R) relation might decouple statistical associations from manipulative causation. As a trivial hypothetical example, suppose that in a certain location, coal-fired power plants are used only on days when it is unusually hot or cold and air conditioning or electric heat are being widely used. Average pollution concentrations are 24 μg/m^3 on days when the power plants are used and 8 μg/m^3 on other days. If there is an exposure concentration threshold of 10 μg/m^3 for adverse health responses to pollution, then cutting in half all daily exposure concentrations would not affect health responses. Although the bimodal distribution of exposure concentrations would shift leftward, neither part would cross the threshold. But if exposure estimation errors "flatten and apparently linearize" the observed (i.e., estimated) exposure-response (C-R) curve, as Rhomberg et al. describe, then the failure of the 50% reduction in mean exposure concentration to change health outcomes might appear puzzling. In reality, of course, neither a bimodal distribution of exposure concentrations nor a sharp threshold in the true (causal) population exposure concentration-response relation may be present, and yet exposure estimation error might still distort the shape of the estimated C-R curve, leading to model-predicted reductions in risk from lowering exposures quite different from what actually occurs.

The following sections quantify the real-world distributions of errors in exposure estimates for PM2.5 concentrations using publicly available data for 10 air quality monitoring stations in Southern California. They examine the extent to which real exposure estimation errors for PM2.5 concentrations would mask or distort a threshold in the causal C-R curve if one existed in the relevant range of concentrations, as envisioned by Rhomberg et al. (2011). A goal is to understand the general relation between exposure estimation errors and estimated C-R curves when the true C-R curve has a threshold. This understanding should be equally applicable to different mortality and morbidity endpoints, from autism to violent crime.

Data

The US EPA provides data on daily average PM2.5 concentrations at many monitor locations in the US, along with their locations (latitudes and longitudes). These data are available atwww.epa.gov/outdoor-air-quality-data/download-daily-data. Table 16.1 shows the first few records (rows) of data retrieved from this site for the year 2016 for Los Angeles County. Explanations of variables are provided at www.epa.gov/outdoor-air-quality-data/about-air-data-reports. The records shown in

Table 16.1 Daily PM2.5 concentrations at an air quality monitoring site (AQS) in Los Angeles county

Date	AQS_ site_ID	POC	Daily mean PM2.5 concentration	Units	Daily_ AQI_ value	Daily_ OBS_ count	Percent_ complete	AQS_ Parameter_ code	AQS_ parameter_ desc	CBSA_ code	CBSA_name	State_ code	State	County_ code	County	Site_ latitude	Site_ longitude
1/1/2016	60370002	1	6	ug/m^3 LC	25	1	100	88101	-Local Con	31080	Long Beach-	6	California	37	Los Angeles	34.1365	−117.92391
1/4/2016	60370002	1	8.7	ug/m^3 LC	36	1	100	88101	-Local Con	31080	Long Beach-	6	California	37	Los Angeles	34.1365	−117.92391
1/7/2016	60370002	1	0.7	ug/m^3 LC	3	1	100	88101	-Local Con	31080	Long Beach-	6	California	37	Los Angeles	34.1365	−117.92391
1/10/2016	60370002	1	12.2	ug/m^3 LC	51	1	100	88101	-Local Con	31080	Long Beach-	6	California	37	Los Angeles	34.1365	−117.92391
1/13/2016	60370002	1	11	ug/m^3 LC	46	1	100	88101	-Local Con	31080	Long Beach-	6	California	37	Los Angeles	34.1365	−117.92391
1/16/2016	60370002	1	16	ug/m^3 LC	59	1	100	88101	-Local Con	31080	Long Beach-	6	California	37	Los Angeles	34.1365	−117.92391
1/19/2016	60370002	1	14.4	ug/m^3 LC	56	1	100	88101	-Local Con	31080	Long Beach-	6	California	37	Los Angeles	34.1365	−117.92391

Source: www.epa.gov/outdoor-air-quality-data/download-daily-data

Lower quartiles range from 6 μg/m^3 for site 1–9.15 for site 3. Upper quartiles range from 11 μg/m^3 for site 6–16 μg/m^3 for site 3

Table 16.1 are for a specific air quality monitoring site (AQS) (ID 60370002) at the location shown in the latitude and longitude columns at the right of the table.

We select the LA area because publicly available data show both (a) A clear, positive, statistically significant exposure concentration-response (C-R) association between PM2.5 and daily elderly mortality rates for the corresponding air basin (e.g., a significant positive regression coefficient in a Poisson regression model for daily mortality count as a function of daily high and low temperatures, humidity, and PM2.5 concentrations); and yet (b) No detectable predictive causal relation between changes in PM2.5 and changes in mortality (Cox Jr. 2017). We use data from the LA area to examine what role exposure estimation error might play in creating a strong statistical C-R association in the absence of a strong causal impact of changes in exposure on changes in response.

Figure 16.1 shows the locations (latitudes and longitudes) of AQS monitoring sites in the "Los Angeles-Long Beach-Anaheim, CA" core-based statistical area (CBSA). Ten AQS sites, indicated by a black rectangle enclosing them in Fig. 16.1, are clustered relatively closely together. Figure 16.2 shows a closer view of these ten locations and assigns them numbers 1–10 for future reference. Data from these sites will be used next to quantify errors from imputing to each location the PM25 concentrations measured at nearby monitors.

Table 16.2 shows the layout of the data. For each site (column), and for each day (row) in 2016 on which data were collected for that site, the daily average PM2.5 concentration recorded for that site is shown in the table. Not all sites record PM2.5 concentrations every day, as indicated by the missing values in the table.

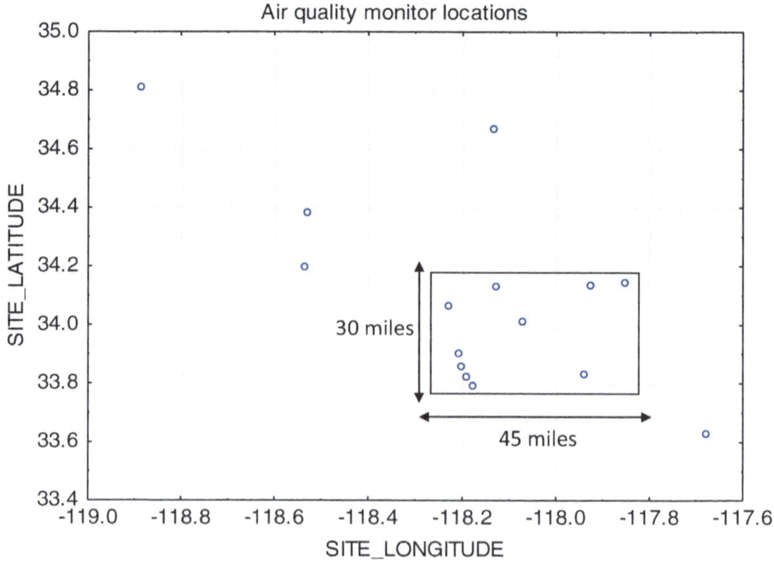

Fig. 16.1 AQS station locations and 10 stations selected for analysis

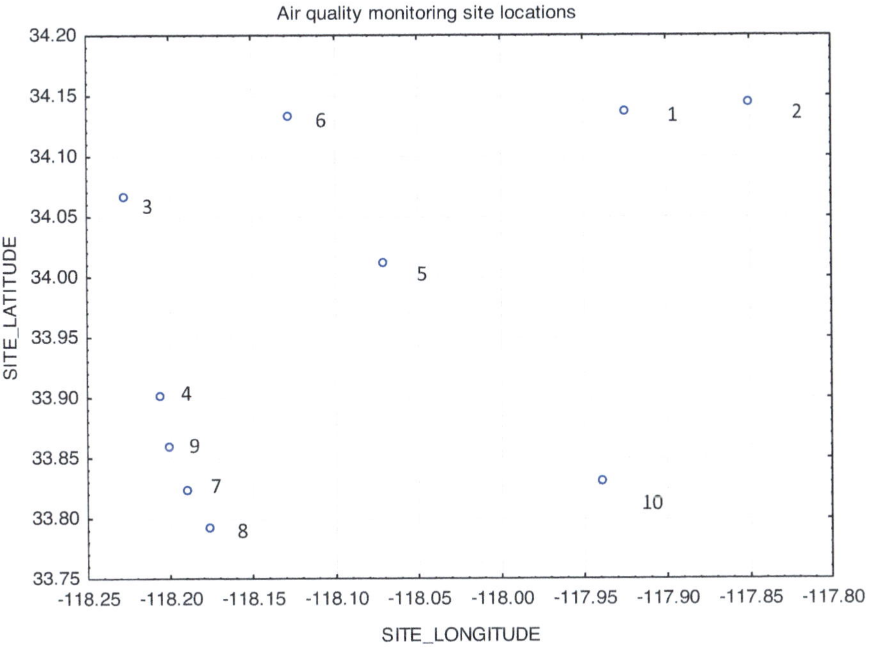

Fig. 16.2 Locations of ten AQS monitoring sites in Los Angeles County used in subsequent data analyses. This is an expansion of the 45 mile × 30 mile rectangle in Fig. 16.1

Methods

Two ways to estimate individual exposures to air pollutants are to assign to each individual on each day either the average daily concentration measured at the nearest monitoring site to the individual's residence, or the average daily concentration for a surrounding geographic area (e.g., a county, zip code, SMSA, etc.) estimated by averaging measurements from several monitors included within its boundaries. To understand the quantitative distribution of errors in the imputed exposure values produced by these methods, we use them to estimate PM2.5 concentrations at different AQS monitoring sites and then compare the estimated values to the true values recorded at those sites. Basic non parametric descriptive analytics methods (scatter plots, interaction plots, and non-parametric (smoothing) regression) and correlations characterize how informative exposure concentration measurements at one location are about exposure concentrations at other locations.

To see how exposure estimation errors distort estimated exposure-response curves, we first consider their effects on estimation of a sharp exposure-response threshold function—that is, a step function that has probability 0 of response for daily exposure concentrations below a threshold, such as 10 μg/m^3, and probability 1 of response for daily exposures concentrations at or above that threshold value. (The response in this case would have to be a morbidity, rather than mortality, to

Table 16.2 Layout of the daily PM2.6 concentrations data for the 10 AQS sites in Los Angeles County in Fig. 16.2

1 Month	2 Day	3 PM2.5 site 1	4 PM2.5 site 2	5 PM2.5 site 3	6 PM2.5 site 4	7 PM2.5 site 5	8 PM2.5 site 6	9 PM2.5 site 7	10 PM2.5 site 8	11 PM2.5 site 9	12 PM2.5 site 10
1	1	6.0	8.9	11.9	28.0	17.6	17.7	22.0	30.8	26.5	7.1
1	2		10.5	12.6				23.5	22.4	23.8	17.1
1	3		10.5	19.0				25.6	25.1	30.2	17.1
1	4	8.7	13.4	10.9	13.2	14.3	10.0	10.5	11.7	12.8	9.4
1	5		11.2	7.8				5.1	6.6	7.2	5.8
1	6		7.5	5.4				6.1	5.4	8.3	8.3
1	7	0.7	4.5	4.4		5.2	2.7	4.5	3.9	5.4	5.6
1	8		9.7	7.0				11.2	9.7	13.6	14.2
1	9		12.7	10.8				12.5	10.2	15.1	11.8
1	10	12.2	14.1	15.6		16.9	17.5	7.2	7.2	9.2	11.8

Lower quartiles range from 6 $\mu g/m^3$ for site 1–9.15 for site 3. Upper quartiles range from 11 $\mu g/m^3$ for site 6–16 $\mu g/m^3$ for site 3 avoid killing everyone when a response occurs.) Next, less extreme cases are considered for which exposure concentrations in excess of a threshold increases risk above its non-zero background level. In reality, of course, different individuals might have different thresholds (or, more generally, different sigmoid individual exposure-response curves), but understanding how exposure estimation error affects estimates of these known step-function exposure-response curves illuminates more general cases in which mixtures of individual step-function responses are affected.

The statistical analyses and visualizations reported here were performed using the Statistica 12 commercial statistics package, R, and the free R-based Causal Analytics Toolkit (CAT) software (https://cox-associates.com/).

Results

Variability Around Average Exposure Concentration Values

Based on mutual proximity, the monitoring sites in Fig. 16.2 can be partitioned into the following clusters: {1, 2}, {3, 5, 6}, {4, 7, 8, 9}, and {10}. Figure 16.3 shows how average PM2.5 concentrations vary throughout the year for each of these sites. The three panels in Fig. 16.3 correspond to the first three spatial clusters. Site 10 is added to the cluster {4, 7, 8, 9} because its monthly PM2.5 values are similar to theirs.

The vertical axes in these interaction plots represent the monthly averages of the daily concentrations of PM2.5 recorded for each site. Time is aggregated by months, so that each data point (1/month) represents the average of 1 month of daily values, except for the scatterplot in the lower right, where each dot represents a value for 1 day and site. (Several dots may be superimposed.) The vertical bars around the monthly averages in the other three plots represent approximate 95% confidence intervals (monthly mean concentration plus or minus 1.96 sample standard deviations).In the lower right panel, a non-parametric (loess smooth) curve is fit through the scatterplot, and the vertical scatter around it indicates the spread of values among days and sites around the monthly average value.

These plots reveal systematic relations between values measured at different locations. In the upper left panel, average PM2.5 concentrations are less for site 1 than for site 2 in January-May, and are greater for site 1 than for site 2 during the rest of the year. These differences are statistically significant for some individual months (April, October), as indicated by non-overlapping confidence intervals. In the upper right panel, site 3 has higher average PM2.5 values than site 6 throughout the year. The lower left panel shows many significant differences between monthly concentrations at different sites (e.g., between sites 4 and 9). The lower right panel indicates that there are wide scatters of PM2.5 values at different sites and on different days around the monthly averages. Figure 16.4 makes the same point using days instead of months as the time steps. Each dot represents a PM2.5 value for an

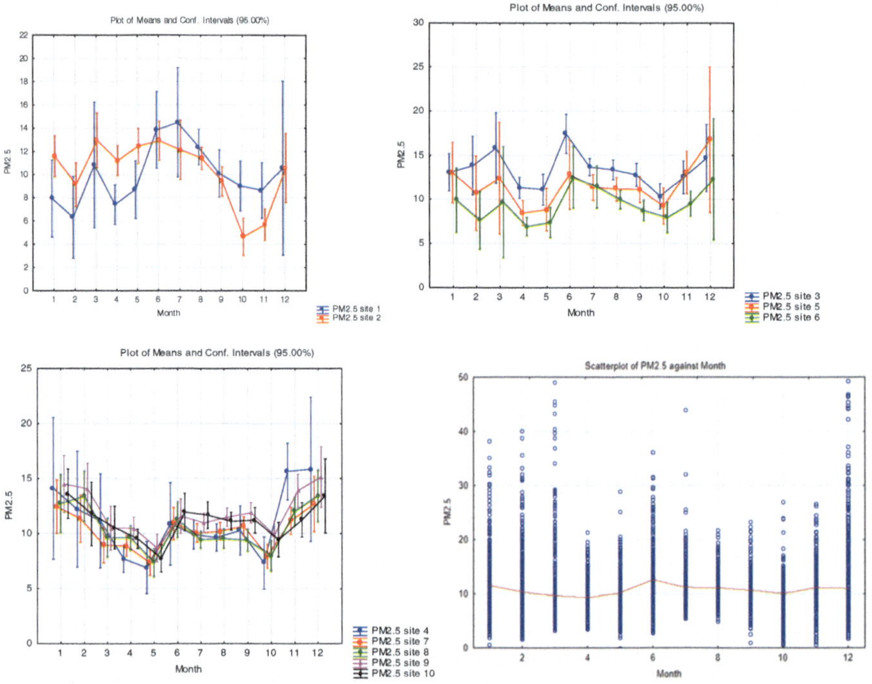

Fig. 16.3 Monthly average PM2.5 concentrations (μg/m³), by AQS site and month. Lower quartiles range from 6 μg/m³ for site 1–9.15 for site 3. Upper quartiles range from 11 μg/m³ for site 6 to 16 μg/m³ for site 3

individual AQS monitoring site on a specific day, and the numbers on the horizontal axis indicate month and day using the code 100*Month + day, so that 412, for example, would be the 12th day of the 4th month of the year, i.e., April 12. The wide scatter of recorded PM2.5 values around the daily average (indicated approximately by the smooth curve) is clear, as is the fact that it varies throughout the year: PM2.5 values in excess of 30 μg/m³ occur frequently in December-March, for example (when elderly mortality rates are also highest in this area (Cox Jr. 2017)), but not at all in August-November (when elderly mortality rates are also lower). Such distributional details are lost when only average concentrations are considered.

That some sites have systematically higher PM2.5 values than others suggests that discussions of how risk varies with *average* PM2.5 values may have limited relevance to understanding health risks. For example Di et al. (2017) emphasize that "short-term exposures to PM2.5 and warm-season ozone were significantly associated with increased risk of mortality. This risk occurred at levels below current national air quality standards, suggesting that these standards may need to be reevaluated." However, the claim that "This risk occurred at levels below current national air quality standards" refers only to *estimated average* levels of exposure concentrations (averaged at the level of zip codes), not to the variable actual levels of exposure concentrations that individuals received. In light of the variability illustrated in

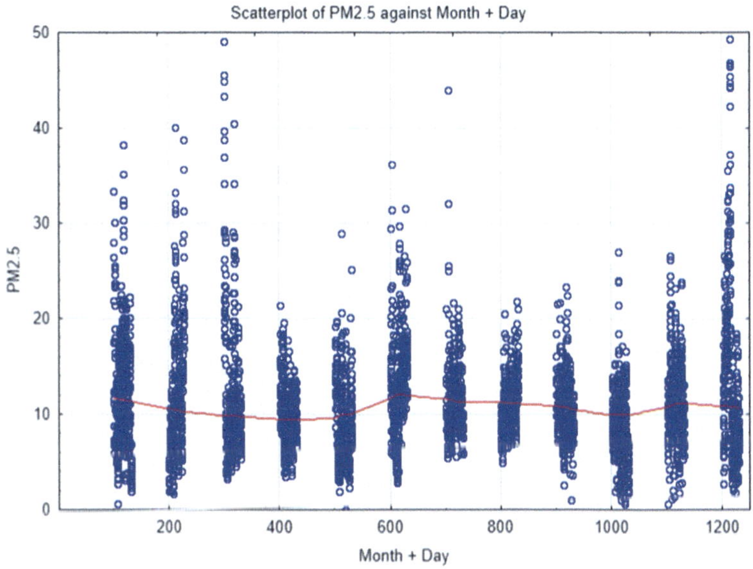

Fig. 16.4 Daily PM2.5 measurements at different AQS sites in Los Angeles County. The horizontal axis indicates month and day using 201 for February 1, 1011 for October 11, etc. Lower quartiles range from 6 μg/m³ for site 1–9.15 for site 3. Upper quartiles range from 11 μg/m³ for Site 6–16 μg/m³ for site 3

Fig. 16.4, it seems plausible that on a day when a zip code has an estimated average PM2.5 concentration "below current national air quality standards," there might well be many sub-areas with concentrations 3 or 4 times greater than the average value, well above the current standards. The lower right panel of Fig. 16.3 shows a similar finding on a monthly time scale. How deaths would change if no areas had concentrations above the current NAAQS on any given day or month is not clear from such variable data. In short, substantial variability among measured concentrations at different sites on the same day implies that estimated average daily concentrations for an entire area do not reveal either the relatively high exposure concentrations that occur within the area or the full distribution of concentrations, which may be needed to understand causal exposure-response functions.

Effects of Using Average Concentrations or Concentrations at One Location as Surrogates for Concentrations at Nearby Locations

PM2.5 concentrations on the same day are generally strongly and positively correlated with each other across AQS sites, although the sizes of correlations vary considerably for different pairs of sites. Figure 16.5 shows the numerical correlations

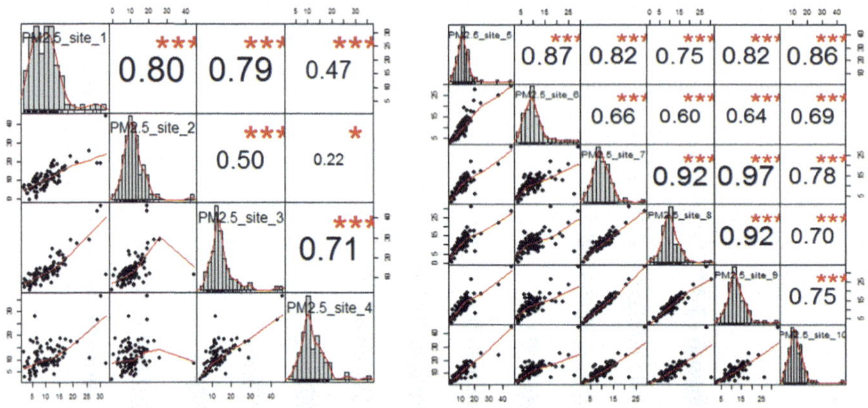

Fig. 16.5 Scatterplot matrices for sites 1–4 (left) and 5–10 (right) with histograms (main diagonal), nonparametric smoothing (kernel density) regression overlays, absolute correlations (numbers) and approximate significance levels (asterisks: * = 0.05, ** = 0.01, *** = 0.001). (The reported significance levels do not adjust for non-normal data)

(numbers above the main diagonals) and corresponding scatter plots and superimposed smooth regression curves (plots below the main diagonals) for sites 1–4 (left panel) and 5–10 (right panel). The frequency histograms of concentrations on the main diagonals show long right tails. For completeness, Table 16.3 shows all pairwise Pearson's correlations between daily PM2.5 concentrations at different sites. All correlations are for days with PM2.5 measurements at all AQS sites.

Spatial proximity is a useful but imperfect predictor of strength of correlation. For example, concentrations at site 1 are correlated more strongly with concentrations at sites 6 and 3, some distance away, than with concentrations at the nearby site 2, which is relatively poorly correlated with other sites. Correlation is only part of the story, however: shifting a monitor's concentrations up or down by any fixed amount does not change the correlation numbers, but may greatly affect how well its concentrations approximate concentrations at other locations.

To understand the effects on estimated exposure concentration-response (C-R) functions of using PM2.5 concentrations from nearby monitors, or the average value of PM2.5 over all monitors in the areas, as proxies for the concentrations at a given location, we will first consider the extreme case of a step-function C-R function, the same for all individuals, with a response probability of 0 for days with average PM2.5 concentrations no greater than 10 micrograms per cubic meter and a response probability of 1 for days with average PM2.5 concentrations above 10. This is not intended to be realistic—e.g., it leaves out delays between exposure and response and gradual increases of response probability with increasing exposure concentrations—but is meant only to clarify the effect of exposure estimation error on reconstruction of a C-R function having a simple known form. Figure 16.6 shows the C-R functions estimated for each of the 10 AQS monitoring locations when average daily PM2.5 exposure concentration, averaged over all ten locations, is used as the exposure variable on the x axis and the daily response probability on the y axis

Table 16.3 Correlations among PM2.5 daily concentrations at different sites

Variable	Means	Std. dev.	PM2.5 site 1	PM2.5 site 2	PM2.5 site 3	PM2.5 site 4	PM2.5 site 5	PM2.5 site 6	PM2.5 site 7	PM2.5 site 8	PM2.5 site 9	PM2.5 site 10	PM2.5 average
PM2.5 site 1	10.1	5.5	1.00	0.80	0.80	0.44	0.78	0.87	0.48	0.39	0.42	0.67	0.82
PM2.5 site 2	10.5	5.5	0.80	1.00	0.67	0.23	0.50	0.56	0.39	0.35	0.31	0.54	0.59
PM2.5 site 3	13.4	6.5	0.80	0.67	1.00	0.69	0.90	0.39	0.72	0.65	0.66	0.78	0.91
PM2.5 site 4	11.0	5.6	0.44	0.23	0.69	1.00	0.84	0.56	0.93	0.90	0.95	0.73	0.86
PM2.5 site 5	11.7	5.7	0.78	0.50	0.90	0.84	1.00	0.37	0.81	0.74	0.80	0.85	0.96
PM2.5 site 6	9.5	4.8	0.87	0.66	0.89	0.66	0.87	1.00	0.65	0.58	0.61	0.69	0.90
PM2.5 site 7	10.3	4.4	0.48	0.39	0.72	0.93	0.81	0.65	1.00	0.93	0.97	0.82	0.88
PM2.5 site 8	10.5	4.7	0.39	0.35	0.65	0.90	0.74	0.58	0.93	1.00	0.93	0.79	0.82
PM2.5 site 9	12.0	4.8	0.42	0.31	0.66	0.95	0.80	0.61	0.97	0.93	1.00	0.77	0.86
PM2.5 site 10	11.1	5.2	0.67	0.54	0.78	0.73	0.85	0.69	0.32	0.79	0.77	1.00	0.87
PM2.5 average	11.2	4.6	0.82	0.59	0.91	0.86	0.96	0.90	0.38	0.82	0.86	0.87	1.00

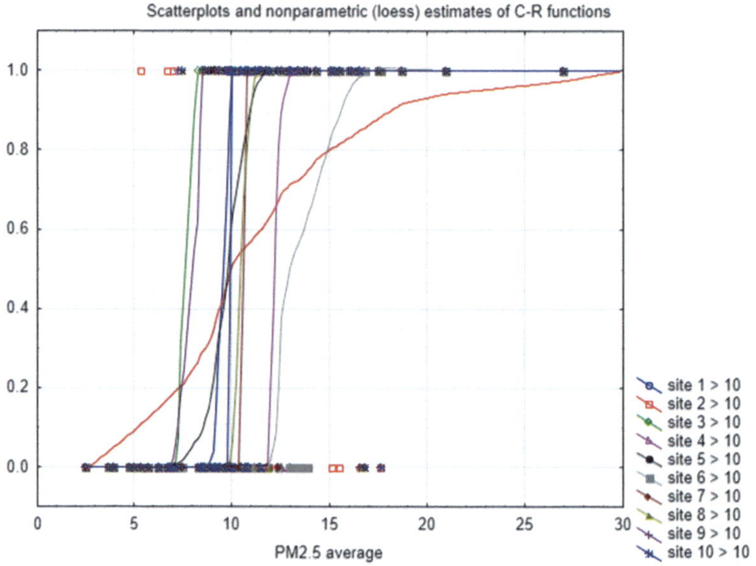

Fig. 16.6 Estimated C-R functions at the 10 AQS sites in Los Angeles County if the true C-R function is a step function from 0 to 1 at PM2.5 = 10 μg/m³. Lower quartiles of concentrations range from 6 μg/m³ for site 1–9.15 for site 3. Upper quartiles range from 11 μg/m³ for site 6–16 μg/m³ for site 3

is either 1 or 0 at each location, depending on whether the true (site-specific and day-specific) concentration there is or is not greater than 10. The estimated C-R functions in Fig. 16.6 are obtained by fitting nonparametric smooth (loess) regression curves to the average PM2.5 concentration from all ten sites on each day and the corresponding site-specific and day-specific response probabilities (0 or 1) for each site. Non-parametric smoothing is used to avoid potential parametric model specification errors or model selection biases. There are between 115 days of data (for Site 4) and 366 days of data (for sites 3 and 8) for each site, depending on monitoring frequency.

In the absence of exposure estimation errors, all ten curves in Fig. 16.6 would be the same: they would all jump from 0 for PM2.5 less than 10 to 1 for PM2.5 greater than 10. But because of the differences between the true concentrations and the imputed (average) concentrations, the estimated C-R functions at a large fraction of the sites fall far from this ideal. The estimated C-R curve for site 2 provides an especially poor approximation to the correct C-R curve, increasing gradually over the entire range from about 2–30 μg/m³, consistent with the warnings of Rhomberg et al. (2011). In addition, many of the estimated C-R functions, while providing better approximations to step functions, are shifted rightward or leftward with respect to the true test C-R function, i.e., they increase from 0 to 1 at concentrations ranging roughly from 8 to 18 μg/m³ instead of jumping at the correct value of 10 μg/m³.

Figure 16.7 shows the composite estimated C-R function obtained from the site-specific estimated C-R functions in Fig. 16.6 by plotting the fraction of the population responding on a given day as a function of the average PM2.5 concentration for that day (averaged over all 10 sites), assuming that 10% of the total population is located at each AQS site. As suggested by Rhomberg et al. (2011), this estimated C-R function is a smoothly increasing sigmoid curve, even though the underlying true C-R function that it is estimating is a sharp threshold (step) function. The estimated C-R function starts increasing at about 5 μg/m^3 even though the true C-R function does not increase until 10 μg/m^3, and the estimated C-R function is still well below 1 at 15 μg/m^3, even though the true C-R function jumps to 1 at 10 μg/m^3.

Figures 16.6 and 16.7 show the distorted C-R function created by exposure estimation errors when the exposure estimated for each location is the average PM2.5 concentration for the whole area, i.e., the average concentration over all 10 AQS sites.

Similar distortions occur when the PM2.5 concentrations measured at one monitoring site are used as surrogates for the concentrations at other locations. Figure 16.8 shows an example very similar to Fig. 16.7 except that the estimated concentration is now the concentration at site 5 (the most centrally located site of the 10 sites in Fig. 16.2) instead of the average concentration over all sites. The resulting sigmoid estimated C-R function is similar to that in Fig. 16.7, but somewhat wider; again, it is not a very close approximation to the underlying true C-R function, which steps from 0 to 1 at 10 μg/m^3.

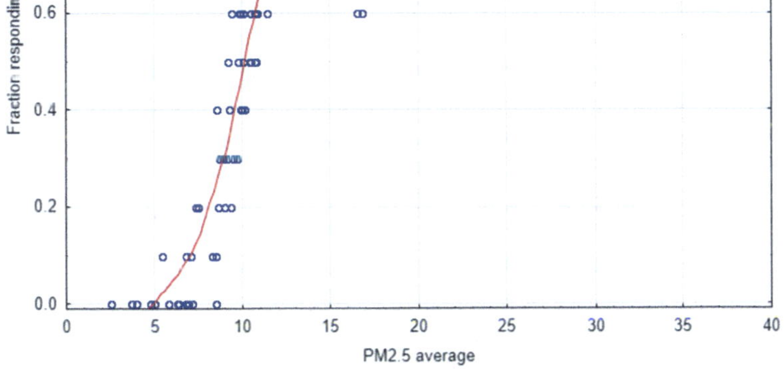

Fig. 16.7 Estimated C-R function if all ten AQS sites are weighted equally and average PM2.5 concentration for all AQS sites is used as an exposure surrogate. Lower quartiles range from 6 μg/m^3 for site 1–9.15 for site 3. Upper quartiles range from 11 μg/m^3 for site 6–16 μg/m^3 for site 3

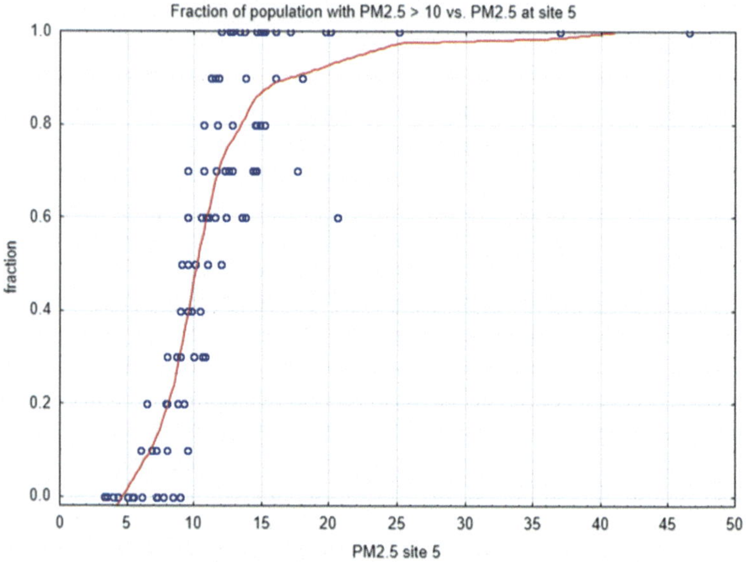

Fig. 16.8 Estimated C-R function if all ten AQS sites are weighted equally and PM2.5 concentration at site 5 is used as an exposure surrogate

The step function used so far is extreme, representing an all-or-nothing response threshold. A more realistic test function for elderly mortality would have a non-zero background mortality rate on the order of 0.001 expected deaths per day for a person with a remaining life expectancy of a few years. Suppose that exposure to 10 μg/m^3 or more were to increase this rate by 2%, to 0.00102/person/year. Figure 16.9 shows an estimated C-R function fit to simulated data from a population of ten million people, one million at each of the 10 AQS sites, where the number of deaths at each site on any given day has a binomial distribution with parameters $n = 1{,}000{,}000$ and $p = 0.001$ if the daily PM2.5 concentration recorded for that site does not exceed 10μg/m^3, and has a binomial distribution with parameters $n = 1{,}000{,}000$ and $p = 0.00102$ otherwise. A nonparametric (loess) regression curve is fit to the simulated data (using the scatter.smooth function in R; for replication, the random number generator for the simulation was initialized using set.seed(1234)). The effect of exposure estimation errors in this case is to obscure the threshold at 10 μg/m^3. Instead, the estimated C-R function appears to increase smoothly and approximately linearly with concentration, starting at the lowest exposure levels, with no evidence of a threshold.

This pattern is consistent with the warning by Rhomberg et al. (2011) that "It has been shown that the degree of bias known to apply to actual studies is sufficient to produce a false linear result, and that although nonparametric smoothing and other error-mitigating techniques may assist in identifying a threshold, they do not guarantee detection of a threshold."It is also consistent with the report by Di et al. (2017) that "[R]esponses were almost linear, with no indication of a mortality risk threshold

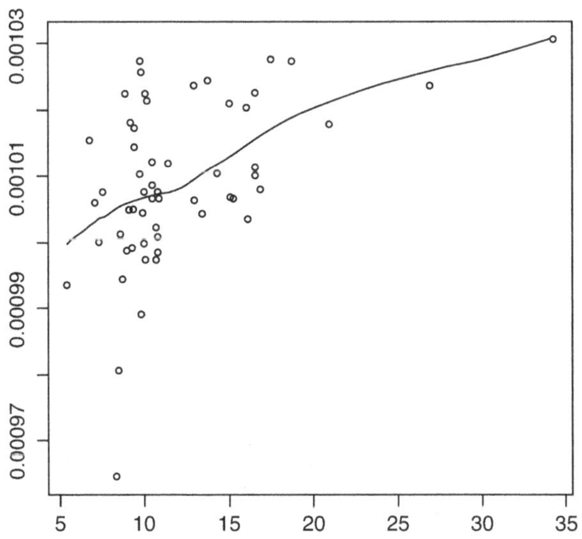

Fig. 16.9 Estimated C-R function if PM2.5 concentrations above 10 increase daily mortality from 1 death per 1000 person-days to 1.02 deaths per 1000 person-days in a population of 10 million people uniformly distributed among the ten AQS sites, using average PM2.5 concentration for all AQS sites as an exposure surrogate. Lower quartiles of site-specific PM2.5 concentration distributions range from 6 μg/m³ for site 1–9.15 for site 3. Upper quartiles range from 11 μg/m³ for site 6–16 μg/m³ for site 3

at very low concentrations," bearing in mind that the "very low concentrations" referred to are only estimates based on zip code-level average values, and that the distributions of true concentrations—some of which were probably much higher than the averages, as illustrated in Fig. 16.4—were not considered.

Possible Implications for Differences between Associational and Causal C-R Functions

A C-R function based on average values, such as each of the curves in Figs. 16.7, 16.8, and 16.9, might seem to suggest that changing the average PM2.5 concentration would cause the response rate to change as described by the curve. This causal interpretation is often made, but it is unjustified in general: drawing a regression curve through a scatterplot has no necessary implications for how changing one variable would change the other (Cox Jr. 2017). The effects of measurement errors illustrated in Figs. 16.6, 16.7, 16.8, and 16.9 raise a distinct but related methodological point: without specifying how a pollution control measure affects the *distribution* of exposure concentrations, rather than just the *average* exposure concentration, it may be impossible, even in principle, to determine what change in responses (if any) the intervention will cause. This is because the same change in average concentration may correspond to very different changes in distributions of concentrations that have very different implications for changes in responses.

As a simple example, consider again the case of a bimodal distribution with half of days exposing people to 24 μg/m^3 and the other half exposing them to 8 μg/m^3, for an average of 16 μg/m^3. Now suppose that average exposure is reduced by 5 μg/m^3, from 16 μg/m^3 to 11 μg/m^3, and that the true C-R function is a step function at 10 μg/m^3. What will be the effect of this reduction on health risks? The correct answer depends on how the distribution of concentrations has changed. It cannot be uniquely determined from the information given. If the reduction is accomplished by reducing each day's concentration by 5 μg/m^3, then there is no impact on health risks, as previously noted: a reduction from 24 to 19 leaves exposures above 10, and a reduction from 8 to 3 leaves exposure below 10, so neither reduction affects health risks described by a C-R step function at 10 μg/m^3 (or anywhere else strictly between 8 and 19 μg/m^3 or above 24 μg/m^3 or below 3 μg/m^3). On the other hand, if the same reduction in average concentration is accomplished by reducing the exposure for half of the high-exposure days from 24 μg/m^3 to 4 μg/m^3 (giving an average of $0.25*24 + 0.25*4 + 0.5*8 = 11$ μg/m^3), then risk on these days will be reduced from the high (above-threshold) to the low (below-threshold) level. If the reduction is accomplished by increasing the concentrations on low-exposure days from 8 μg/m^3 to 11 μg/m^3 but reducing the concentrations on high-exposure days from 24 μg/m^3 to 11 μg/m^3, then the reduction in average exposure concentration from 16 μg/m^3 to 11 μg/m^3 will be accompanied by an increase in risk, with twice as many people experiencing the higher risk level after the reduction. Thus, the same 5 μg/m^3 reduction in *average* exposure concentration can create an increase, decrease, or no change in health risk, depending on how the underlying distribution of concentrations changes. Such examples illustrate that using only average PM2.5 concentrations can lose essential information, and that the overall shape of the frequency distribution of PM2.5 exposure concentrations must be considered (e.g., using a few quantiles, such as the deciles) to accurately understand and interpret population concentration-response data.

Many papers express their main findings in the form of predictions that each reduction of exposure concentration by a certain amount (customarily 10 μg/m^3 for PM2.5) would produce a certain change in health risks (e.g., Schwartz et al. 2008; Shi et al. 2016; Wang et al. 2016; Di et al. 2017). Such papers typically do not specify how the underlying distribution of concentrations is assumed to change when the average exposure is decreased. Hence they fail to engage with the fact that this change in distribution drives change in risk, and that the same average reduction in exposure concentration can have a wide range of implications for changes in risk, perhaps including increases, decreases, and no change. The usual justification is that estimated C-R functions are approximately linear and non-threshold; if the true C-R function is also linear and non-threshold, then average change in response is simply proportional to average change in exposure concentration, and hence there is no need to be explicit about how the underlying distribution of concentrations changes. The significance of the findings in Figs. 16.6, 16.7, 16.8, and 16.9 is that true C-R functions need not resemble estimated C-R functions that ignore exposure estimation errors. A linear no-threshold model for an estimated C-R function can correspond to a highly nonlinear (step function) true C-R function. The literature that conflates true and estimated C-R functions leaves unaddressed the true shape of the C-R function and the possibly crucial effects of proposed interventions on the shapes of concentration distributions.

Discussion and Conclusions

It is common practice in air pollution health effects research, as in many other areas of applied epidemiology (e.g., occupational health studies using job exposure matrices), to use surrogate measures or estimates for unknown individual-level exposure concentrations and to analyze and refer to the surrogate measures or estimates as if they were accurate measures of true concentrations. Common surrogate measures include concentrations at nearby monitors and averaged concentrations from multiple monitors in an area. Although we have focused on the recent paper of Di et al. (2017) as a motivating example, many other influential papers present estimated exposures as if they were true exposures. For example,

- Shi et al. (2016) state that "PM2.5 was associated with increased mortality. … [We] estimated significant acute and chronic effects of PM2.5 exposures below current EPA standards. These findings suggest that improving air quality below current standards may benefit public health." Shi et al. refer to *estimates* of *average* PM2.5 concentrations (in this case, predicted via a model from satellite aerosol optical depth estimates on a 1 km × 1 km grid) as simply "PM2.5 exposures" without differentiating between estimated and true PM2.5 exposures and without modeling the errors in the estimates or the uncertainties and variability in the true values. Other research has estimated a root mean squared prediction error of about 5.5 μg/m^3 for such estimates in Mexico City (Just et al. 2015), but aerosol optical depth appears to contribute little predictive power to accurate estimation of PM2.5 on urban or sub-regional scales in the eastern United States (Paciorek et al. 2012). Shi et al. do not quantify what fraction of people experienced exposure concentrations above current standards on days when the estimated average was below current standards, but the long tails of concentration distributions in our Fig. 16.4 suggest the importance of such distributional considerations.
- Wang et al. (2016) state that "We estimated the causal effects of long-term PM2.5 exposure on mortality." Again, the "PM2.5 exposure" referred to is not actual exposure, but the output of "A 3-stage statistical modeling approach for predicting daily PM2.5" containing unknown errors. (The authors address model quality using correlation coefficients, but correlation coefficients are insensitive to even very large biases.)
- A decade earlier, Schwartz et al. (2008) wrote that "Understanding the shape of the concentration–response curve for particles is important for public health, and

lack of such understanding was recently cited by U.S. Environmental Protection Agency (EPA) as a reason for not tightening the standards. Similarly, the delay between changes in exposure and changes in health is also important in public health decision making. We addressed these issues using an extended follow-up of the Harvard Six Cities Study. … We found that the concentration–response curve is linear, clearly continuing below the current U.S. standard of 15 μg/m^3, and that the effects of changes in exposure on mortality are seen within 2 years." However, the PM2.5 concentrations referred to are actually only very rough estimates of average concentrations: "Each participant's exposure to air pollution each year was defined by that year's concentrations of PM2.5 in that participant's city. Concentrations of PM2.5 were measured at a centrally located air-monitoring station in each community starting in 1979 and ending in 1986–1988 depending on the city. …We calculated city-specific annual mean PM2.5 concentrations as the average of four quarterly means of daily data for each available year. For years before sampling, PM2.5 values were assumed equal to the earliest sampling year." In light of our results, such imputation of PM2.5 concentrations based on average values can lead to non-threshold, linear-looking estimated C-R curves even if the true curve has a sharp threshold; thus, such analyses should not be interpreted as revealing the shape of the true C-R function. Statements such as "the concentration–response curve is linear, clearly continuing below the current U.S. standard" should be clearly understood to refer only to the *estimated* C-R function, which maybe very different from the true C-R function.

These examples can be multiplied many times over, as the literature that treats estimated exposure concentrations as if they were error-free measurements is voluminous (Cox Jr. 2017). The key methodological points are the same for each: estimated C-R functions may be increasing at relatively low doses and lack thresholds even if the true C-R function has neither of these properties.

Our examples and illustrations have used simple step-functions as counter-examples, since these are the most extreme cases for testing whether estimation errors can obscure the shape of a true C-R relation, making even a step function look smooth, non-threshold, and approximately linear over the relevant range of concentrations, as cautioned by Rhomberg et al. (2011). Figures 16.6, 16.7, 16.8, and 16.9 support the conclusion that realistic exposure estimation errors do indeed create such distortions. Whether strongly nonlinear, threshold-like C-R functions are biologically plausible is a separate question from the methodological one of whether ignoring exposure estimation errors can conceal even such strong nonlinearities. However, the following considerations suggest that strong nonlinearities or thresholds in PM2.5 dose-response functions at or below current ambient levels are a realistic possibility that deserves to be considered carefully.

- Epidemiological data on the true shape of the low-dose concentration-response relation for PM2.5 are ambiguous, in part because true exposures are uncertain. Thus, while some previous expert commentators on the possible existence of a threshold concluded that "There also appears to be a monotonic (e.g., linear or log-linear) concentration-response relationship between PM2.5 and mortality risk observed in cohort studies that extends below present-day regulations of 15 μg/m^3 for mean annual levels, without a discernable 'safe' threshold," others warned that "At the human population level, however, various sources of variability and uncertainty tend to smooth and 'linearize' the concentration-response function (such as the low data density in the lower concentration range, possible influence of measurement error, and individual differences in susceptibility to air pollution health effects)… even though likely mechanisms include nonlinear processes for some key events. …The results from [selected studies] support no-threshold log-linear models, but issues such as the possible influence of exposure error and heterogeneity of shapes across cities remain to be resolved." Other experts concluded that "The available epidemiological database on daily mortality and morbidity does not establish either the presence or absence of threshold concentrations for adverse health effects. … For the purpose of estimating public health impacts, the Panel favored the primary use of an assumed threshold of 10 μg/m^3." (United States Environmental Protection Agency (US EPA) 2010)

- The fact that estimated C-R functions are expected not to have thresholds, whether or not true (i.e., without estimation error) C-R functions do, implies that not finding a threshold in estimated C-R functions provides no useful evidence about whether the underlying true C-R functions have thresholds.
- Recent findings that pathological responses to PM2.5 exposures are mediated by chronic inflammation involving the NLRP3 inflammasome, which requires sufficiently large and sustained exposure concentrations and durations to trigger assembly and activation, suggest that careful attention to biological thresholds for PM2.5 health effects is warranted (Wang et al. 2017; Xu et al. 2018).

Our finding that estimated and true C-R relations may be very different when realistic exposure error estimates are considered implies that frequent previous epidemiological findings of estimated C-R functions that appear to be approximately low-dose linear and no-threshold (e.g., Schwartz et al. 2008; Shi et al. 2016; Di et al. 2017) do not address or resolve the question of whether the true C-R function describing response probabilities caused by different exposure concentrations has a threshold, or perhaps a population distribution of thresholds. The question appears to be well worth pursuing further. In addition, although this chapter has focused exclusively on effects of exposure estimation errors in obscuring the true shape of the low-dose concentration-response function for PM2.5, this does not imply that resulting uncertainty about effect thresholds is the only reason to question whether ambient levels of PM2.5 cause the adverse effects often attributed to them. Additional uncertainties about causality arise from epidemiological studies that rigorously distinguish between association and causation (Chap. 17), as well as from animal data and biological mechanistic data on thresholds for inflammation-mediated diseases (Chaps. 3, 4, 5, and 6). Finally, even if outdoor ambient PM2.5 concentrations were to be quantified with little error, personal exposures to PM2.5 still depend largely—often predominantly—on indoor concentrations of PM2.5 and on occupational exposures, which can differ widely even for neighboring homes and among members of the same household.

The methodological points here are not intended to be new. Statisticians have frequently warned against treating averages as if they were actual known values, and this warning has been widely popularized, e.g., in work on the "flaw of averages" (Savage 2009). As previously acknowledged, Rhomberg et al. (2011) warned more specifically that ignoring exposure estimation uncertainty in quantitative estimation and regression modeling of C-R functions can yield distorted estimates that appear to be smooth and non-threshold, and possibly linear, even if the true function being estimated has clear thresholds. Our findings confirm these warnings for PM2.5 concentration estimates and estimation of C-R functions. We found that the differences between the true daily average PM2.5 values recorded for 10 AQS monitoring sites and the estimated values imputed from measurements at nearby sites (either the average over all 10 sites, as in Figs. 16.7 and 16.9; or the value at individual sites, such as site 5 in Fig. 16.8) were large enough to smooth out and to make sigmoid-appearing (Figs. 16.7 and 16.8) or approximately linear-looking (Fig. 16.9) estimates even of step-function C-R functions with well-defined concentration thresholds. The practical implication of these findings for PM2.5 in the LA area examined is that finding an estimated C-R function that appears to be linear no-threshold down to the lowest concentrations considered is not inconsistent with the possibility that the true-but-unknown C-R function has a response threshold well above these low levels. The key question of what the true C-R function looks like remains to be addressed, but addressing it usefully will probably require distinguishing between true and estimated exposure concentrations.

References

ATS (2017) American Thoracic Society. Air pollution may disrupt sleep. ScienceDaily. www.sciencedaily.com/releases/2017/05/170522080830.htm

Bai L, Chen H, Hatzopoulou M, Jerrett M, Kwong JC, Burnett RT, van Donkelaar A, Copes R, Martin RV, van Ryswyk K, Lu H, Kopp A, Weichenthal S. Exposure to ambient ultrafine particles and nitrogen dioxide and incident hypertension and diabetes. Epidemiology. 2018;29(3):323–32. https://doi.org/10.1097/EDE.0000000000000798.

Bowe B, Xie Y, Li T, Yan Y, Xian H, Al-Aly Z. Particulate matter air pollution and the risk of incident CKD and progression to ESRD. J Am Soc Nephrol. 2018;29(1):218–30. https://doi.org/10.1681/ASN.2017030253.

Calderón-Garcidueñas L, Avila-Ramírez J, Calderón-Garcidueñas A, González-Heredia T, Acuña-Ayala H, Chao CK, Thompson C, Ruiz-Ramos R, Cortés-González V, Martínez-Martínez L, García-Pérez MA, Reis J, Mukherjee PS, Torres-Jardón R, Lachmann I. Cerebrospinal fluid biomarkers in highly exposed PM2.5 urbanites: the risk of Alzheimer's and Parkinson's diseases in young Mexico City residents. J Alzheimers Dis. 2016;54(2):597–613. https://doi.org/10.3233/JAD-160472.

Chen H, Kwong JC, Copes R, Hystad P, van Donkelaar A, Tu K, Brook JR, Goldberg MS, Martin RV, Murray BJ, Wilton AS, Kopp A, Burnett RT. Exposure to ambient air pollution and the incidence of dementia: a population-based cohort study. Environ Int. 2017;108:271–7. https://doi.org/10.1016/j.envint.2017.08.020.

Clancy L, Goodman P, Sinclair H, Dockery DW. Effect of air-pollution control on death rates in Dublin, Ireland: an intervention study. Lancet. 2002;360(9341):1210–4.

Cox LA Jr, Popken DA. Has reducing fine particulate matter and ozone caused reduced mortality rates in the United States? Ann Epidemiol. 2015;25(3):162–73. https://doi.org/10.1016/j.annepidem.2014.11.006.

Cox LAT Jr. Do causal concentration-response functions exist? A critical review of associational and causal relations between fine particulate matter and mortality. Crit Rev Toxicol. 2017;47(7):603–31. https://doi.org/10.1080/10408444.2017.1311838.

de Oliveira Galvão MF, de Queiroz JD, Duarte ESF, Hoelzemann JJ, de André PA, Saldiva PH, Menezes Filho JA, Batistuzzo de Medeiros SR. Characterization of the particulate matter and relationship between buccal micronucleus and urinary 1-hydroxypyrene levels among cashew nut roasting workers. Environ Pollut. 2017;220(Pt A):659–71. https://doi.org/10.1016/j.envpol.2016.10.024.

Deng H, Eckel SP, Liu L, Lurmann FW, Cockburn MG, Gilliland FD. Particulate matter air pollution and liver cancer survival. Int J Cancer. 2017;141(4):744–9. https://doi.org/10.1002/ijc.30779.

Dennekamp M, Akram M, Abramson MJ, Tonkin A, Sim MR, Fridman M, Erbas B. Outdoor air pollution as a trigger for out-of-hospital cardiac arrests. Epidemiology. 2010;21(4):494–500. https://doi.org/10.1097/EDE.0b013e3181e093db.

Di Q, Dai L, Wang Y, Zanobetti A, Choirat C, Schwartz JD, Dominici F. Association of short-term exposure to air pollution with mortality in older adults. JAMA. 2017;318(24):2446–56. https://doi.org/10.1001/jama.2017.17923.

Dockery DW, Rich DQ, Goodman PG, Clancy L, Ohman-Strickland P, George P, Kotlov T, HEI Health Review Committee. Effect of air pollution control on mortality and hospital admissions in Ireland. Res Rep Health Eff Inst. 2013;176:3–109.

Harvard School of Public Health (2002) Press release: ban on coal burning in Dublin cleans the air and reduces death rates. www.hsph.harvard.edu/news/press-releases/archives/2002-releases/press10172002.html

Herrnstadt E, Muehlegger E (2015) Air pollution and criminal activity: evidence from Chicago microdata. NBER Working Paper No. w21787. Available at SSRN: https://ssrn.com/abstract=2703192

Honda T, Pun VC, Manjourides J, Suh H. Anemia prevalence and hemoglobin levels are associated with long-term exposure to air pollution in an older population. Environ Int. 2017;101:125–32. https://doi.org/10.1016/j.envint.2017.01.017.

Huang F, Pan B, Wu J, Chen E, Chen L. Relationship between exposure to PM2.5 and lung cancer incidence and mortality: a meta-analysis. Oncotarget. 2017;8(26):43322–31.

Just AC, Wright RO, Schwartz J, Coull BA, Baccarelli AA, Tellez-Rojo MM, Moody E, Wang Y, Lyapustin A, Kloog I. Using high-resolution satellite aerosol optical depth to estimate daily PM2.5 geographical distribution in Mexico City. Environ Sci Technol. 2015;49(14):8576–84.

Kim H, Kim J, Kim S, Kang SH, Kim HJ, Kim H, Heo J, Yi SM, Kim K, Youn TJ, Chae IH. Cardiovascular effects of long-term exposure to air pollution: a population-based study with 900 845 person-years of follow-up. J Am Heart Assoc. 2017;6(11):e007170. https://doi.org/10.1161/JAHA.117.007170.

Liu H, Tian Y, Cao Y, Song J, Huang C, Xiang X, Li M, Hu Y. Fine particulate air pollution and hospital admissions and readmissions for acute myocardial infarction in 26 Chinese cities. Chemosphere. 2018;192:282–8. https://doi.org/10.1016/j.chemosphere.2017.10.123.

Ma JH, Song SH, Guo M, Zhou J, Liu F, Peng L, Fu ZR. Long-term exposure to PM2.5 lowers influenza virus resistance via down-regulating pulmonary macrophage Kdm6a and mediates histones modification in IL-6 and IFN-β promoter regions. Biochem Biophys Res Commun. 2017;493(2):1122–8. https://doi.org/10.1016/j.bbrc.2017.09.013.

Malecki KM, Schultz AA, Bergmans RS. Neighborhood perceptions and cumulative impacts of low level chronic exposure to fine particular matter ($PM_{2.5}$) on cardiopulmonary health. Int J Environ Res Public Health. 2018;15(1):E84. https://doi.org/10.3390/ijerph15010084.

Mapou AEM, Shendell D, Ohman-Strickland P, et al. Environmental factors and fluctuations in daily crime rates. J Environ Health. 2017;80(5):8–23.

Mazidi M, Speakman JR. Ambient particulate air pollution (PM2.5) is associated with the ratio of type 2 diabetes to obesity. Sci Rep. 2017;7(1):9144. https://doi.org/10.1038/s41598-017-08287-1.

McGuinn LA, Ward-Caviness C, Neas LM, Schneider A, Di Q, Chudnovsky A, Schwartz J, Koutrakis P, Russell AG, Garcia V, Kraus WE, Hauser ER, Cascio W, Diaz-Sanchez D, Devlin RB. Fine particulate matter and cardiovascular disease: comparison of assessment methods for long-term exposure. Environ Res. 2017;159:16–23. https://doi.org/10.1016/j.envres.2017.07.041.

Mehta AJ, Kubzansky LD, Coull BA, Kloog I, Koutrakis P, Sparrow D, Spiro A, Vokonas P, Schwartz J. Associations between air pollution and perceived stress: the Veterans Administration Normative Aging Study. Environ Health. 2015;14:10. https://doi.org/10.1186/1476-069X-14-10.

Morales-Suárez-Varela M, Peraita-Costa I, Llopis-González A. Systematic review of the association between particulate matter exposure and autism spectrum disorders. Environ Res. 2017;153:150–60. https://doi.org/10.1016/j.envres.2016.11.022.

Ng CF, Stickley A, Konishi S, Watanabe C. Ambient air pollution and suicide in Tokyo, 2001-2011. J Affect Disord. 2016;201:194–202. https://doi.org/10.1016/j.jad.2016.05.006.

Oh I, Lee J, Ahn K, Kim J, Kim YM, Sun Sim C, Kim Y. Association between particulate matter concentration and symptoms of atopic dermatitis in children living in an industrial urban area of South Korea. Environ Res. 2018;160:462–8. https://doi.org/10.1016/j.envres.2017.10.030.

Paciorek CJ, Liu Y, HEI Health Review Committee. Assessment and statistical modeling of the relationship between remotely sensed aerosol optical depth and PM2.5 in the eastern United States. Res Rep Health Eff Inst. 2012;167:5–83.

Pedersen M, Andersen ZJ, Stafoggia M, Weinmayr G, Galassi C, Sørensen M, Eriksen KT, Tjønneland A, Loft S, Jaensch A, Nagel G, Concin H, Tsai MY, Grioni S, Marcon A, Krogh V, Ricceri F, Sacerdote C, Ranzi A, Sokhi R, Vermeulen R, Hoogh K, Wang M, Beelen R, Vineis P, Brunekreef B, Hoek G, Raaschou-Nielsen O. Environ Ambient air pollution and primary liver cancer incidence in four European cohorts within the ESCAPE project. Respiration. 2017;154:226–33. https://doi.org/10.1016/j.envres.2017.01.006.

Pun VC, Manjourides J, Suh H. Association of ambient air pollution with depressive and anxiety symptoms in older adults: results from the NSHAP study. Environ Health Perspect. 2017;125(3):342–8. https://doi.org/10.1289/EHP494.

Rajput R. Understanding hair loss due to air pollution and the approach to management. Hair Ther Transplant. 2015;5:133. https://doi.org/10.4172/21670951.1000133.

Renzi M, Cerza F, Gariazzo C, Agabiti N, Cascini S, Di Domenicantonio R, Davoli M, Forastiere F, Cesaroni G. Air pollution and occurrence of type 2 diabetes in a large cohort study. Environ Int. 2017;112:68–76. https://doi.org/10.1016/j.envint.2017.12.007.

Rhomberg LR, Chandalia JK, Long CM, Goodman JE. Measurement error in environmental epidemiology and the shape of exposure-response curves. Crit Rev Toxicol. 2011;41(8):651–71. https://doi.org/10.3109/10408444.2011.563420.R.

Savage SL. The flaw of averages: why we underestimate risk in the face of uncertainty. Hoboken: Wiley; 2009.

Schwartz J, Coull B, Laden F, Ryan L. The effect of dose and timing of dose on the association between airborne particles and survival. Environ Health Perspect. 2008;116(1):64–9. https://doi.org/10.1289/ehp.9955.

Shamy M, Alghamdi M, Khoder MI, Mohorjy AM, Alkhatim AA, Alkhalaf AK, Brocato J, Chen LC, Thurston GD, Lim CC, Costa M. Association between exposure to ambient air particulates and metabolic syndrome components in a Saudi Arabian Population. Int J Environ Res Public Health. 2017;15(1):E27. https://doi.org/10.3390/ijerph15010027.

Shen YL, Liu WT, Lee KY, Chuang HC, Chen HW, Chuang KJ. Association of $PM_{2.5}$ with sleep-disordered breathing from a population-based study in Northern Taiwan urban areas. Environ Pollut. 2018;233:109–13. https://doi.org/10.1016/j.envpol.2017.10.052.

Shi L, Zanobetti A, Kloog I, Coull BA, Koutrakis P, Melly SJ, Schwartz JD. Low-concentration PM2.5 and mortality: estimating acute and chronic effects in a population-based study. Environ Health Perspect. 2016;124(1):46–52.

Sram RJ, Veleminsky M Jr, Veleminsky M Sr, Stejskalová J. The impact of air pollution to central nervous system in children and adults. Neuro Endocrinol Lett. 2017;38(6):389–96.

Tallon LA, Manjourides J, Pun VC, Mittleman MA, Kioumourtzoglou MA, Coull B, Suh H. Erectile dysfunction and exposure to ambient air pollution in a nationally representative cohort of older Men. Environ Health. 2017a;16(1):12. https://doi.org/10.1186/s12940-017-0216-6.

Tallon LA, Manjourides J, Pun VC, Salhi C, Suh H. Cognitive impacts of ambient air pollution in the National Social Health and Aging Project (NSHAP) cohort. Environ Int. 2017b;104:102–9. https://doi.org/10.1016/j.envint.2017.03.019.

Thurston GD, Ahn J, Cromar KR, Shao Y, Reynolds HR, Jerrett M, Lim CC, Shanley R, Park Y, Hayes RB. Ambient particulate matter air pollution exposure and mortality in the NIH-AARP diet and health cohort. Environ Health Perspect. 2016;124(4):484–90. https://doi.org/10.1289/ehp.1509676.

Turner MC, Krewski D, Diver WR, Pope CA, Burnett RT, Jerrett M, Marshall JD, Gapstur SM. Ambient air pollution and cancer mortality in the cancer prevention study II. Environ Health Perspect. 2017;125(8):087013. https://doi.org/10.1289/EHP1249.

United States Environmental Protection Agency (US EPA) (2010) Summary of expert opinions on the existence of a threshold in the concentration-response function for pm2.5-related mortality technical support document (TSD) June 2010 Compiled by: U.S. Environmental Protection Agency Office of Air Quality Planning and Standards Health and Environmental Impact Division Air Benefit-Cost Group Research Triangle Park, North Carolina. https://www3.epa.gov/ttnecas1/regdata/Benefits/thresholdstsd.pdf

Wang Y, Kloog I, Coull BA, Kosheleva A, Zanobetti A, Schwartz JD. Estimating causal effects of long-term PM2.5 exposure on mortality in New Jersey. Environ Health Perspect. 2016;124:1182–8. https://doi.org/10.1289/ehp.1409671.

Wang H, Song L, Ju W, Wang X, Dong L, Zhang Y, Ya P, Yang C, Li F. The acute airway inflammation induced by $PM_{2.5}$ exposure and the treatment of essential oils in Balb/c mice. Sci Rep. 2017;7:44256. https://doi.org/10.1038/srep44256.

Weng CH, Hu CC, Yen TH, Huang WH. Association between environmental particulate matter and carpal tunnel syndrome in patients undergoing hemodialysis. Kidney Blood Press Res. 2017;42(5):827–36. https://doi.org/10.1159/000484422.

Wu X, Nethery RC, Sabath BM, Braun D, Dominic F. Exposure to air pollution and COVID-19 mortality in the United States: a nationwide cross-sectional study. BMJ. 2020;200:54502. https://doi.org/10.1101/2020.04.05.20054502.

Xu F, Qiu X, Hu X, Shang Y, Pardo M, Fang Y, Wang J, Rudich Y, Zhu T. Effects on IL-1β signaling activation induced by water and organic extracts of fine particulate matter ($PM_{2.5}$) in vitro. Environ Pollut. 2018;237:592–600. https://doi.org/10.1016/j.envpol.2018.02.086.

Xue T, Zhang Q. Associating ambient exposure to fine particles and human fertility rates in China. Environ Pollut. 2018;235:497–504. https://doi.org/10.1016/j.envpol.2018.01.009.

Yadav AK, Kumar K, Kasim A, et al. Visibility and incidence of respiratory diseases during the 1998 haze episode in Brunei Darussalam. Pure Appl Geophys. 2003;160:265–77.

Yeh HL, Hsu SW, Chang YC, Chan TC, Tsou HC, Chang YC, Chiang PH. Spatial analysis of ambient $PM_{2.5}$ exposure and bladder cancer mortality in Taiwan. Int J Environ Res Public Health. 2017;14(5):E508. https://doi.org/10.3390/ijerph14050508.

Younan D, Tuvblad C, Franklin M, Lurmann F, Li L, Wu J, Berhane K, Baker LA, Chen JC (2017) Longitudinal analysis of particulate air pollutants and adolescent delinquent behavior in southern california. J Abnorm Child Psychol. https://doi.org/10.1007/s10802-017-0367-5

Young MT, Sandler DP, DeRoo LA, Vedal S, Kaufman JD, London SJ. Ambient air pollution exposure and incident adult asthma in a nationwide cohort of U.S. women. Am J Respir Crit Care Med. 2014;190(8):914–21. https://doi.org/10.1164/rccm.201403-0525OC.

Zhang B, Liang S, Zhao J, Qian Z, Bassig BA, Yang R, Zhang Y, Hu K, Xu S, Zheng T, Yang S. Maternal exposure to air pollutant PM2.5 and PM10 during pregnancy and risk of congenital heart defects. J Expo Sci Environ Epidemiol. 2016;26(4):422–7. https://doi.org/10.1038/jes.2016.1.

Zhou Y, Gilboa SM, Herdt ML, Lupo PJ, Flanders WD, Liu Y, Shin M, Canfield MA, Kirby RS. Maternal exposure to ozone and $PM_{2.5}$ and the prevalence of orofacial clefts in four U.S. states. Environ Res. 2017;153:35–40. https://doi.org/10.1016/j.envres.2016.11.007.

Zigler CM, Dominici F. Point: clarifying policy evidence with potential-outcomes thinking--beyond exposure-response estimation in air pollution epidemiology. Am J Epidemiol. 2014; 180(12):1133–40.

Chapter 17
Have Decreases in Air Pollution Reduced Mortality Risks in the United States?

Introduction: Using Data from Natural Experiments to Understand Causality

An aim of applied science in general and of epidemiology in particular is to draw sound causal inferences from observations. For public health policy analysts and epidemiologists, this includes drawing inferences about whether historical changes in exposures have actually caused the consequences predicted for, or attributed to, them. The example of the Dublin coal-burning ban mentioned in Chaps. 15 and 16 suggests that accurate evaluation of the effect of interventions is not always easy, even when data are plentiful. Students are taught to develop hypotheses about causal relations, devise testable implications of these causal hypotheses, carry out the tests, and objectively report and learn from the results to refute or refine the initial hypotheses. For at least the past two decades, however, epidemiologists and commentators on scientific methods and results have raised concerns that current practices too often lead to false-positive findings and to mistaken attributions of causality to mere statistical associations (Lehrer 2012; Sarewitz 2012; Ottenbacher 1998; Imberger et al. 2011). Formal training in epidemiology may be a mixed blessing in addressing these concerns. As discussed in Appendix C of Chap. 9, concepts such as "attributable risk," "population attributable fraction," "burden of disease," "etiologic fraction," and even "probability of causation" are solidly based on relative risks and related measures of statistical association; they do not necessarily reveal anything about predictive, manipulative, structural, or explanatory (mechanistic) causation (Greenland and Brumback 2002). Limitations of human judgment and inference, such as confirmation bias (finding what we expect to find), motivated reasoning (concluding what it pays us to conclude), and overconfidence (mistakenly believing that our own beliefs are more accurate than they really are), do not spare health effects investigators. Experts in the health effects of particular compounds are not always also experts in causal analysis, and published causal conclusions are often unwarranted, with a pronounced bias toward finding "significant" effects where

L. A. Cox Jr., *Quantitative Risk Analysis of Air Pollution Health Effects*, International Series in Operations Research & Management Science 299,
https://doi.org/10.1007/978-3-030-57358-4_17

none actually exists (false positives) (Lehrer 2012; Sarewitz 2012; Ioannidis 2005; The Economist 2013).

This chapter applies methods of causal hypothesis-testing (Granger causality tests and conditional independence tests) to assessment of the effects caused by past changes, for the important practical problem of assessing improvements in public health risks caused by past reductions in air pollution concentrations. To do so, we take advantage of the fact that between years 2000 and 2010, air pollutant levels in counties throughout the United States changed significantly, with fine particulate matter (PM2.5) declining over 30% in some counties, and ozone (O3) exhibiting large variations from year to year. This history provides an opportunity to compare county-level changes in average annual ambient pollutant levels to corresponding changes in all-cause and cardiovascular disease (CVD) mortality rates over the course of a decade. This chapter examines data from these "natural experiments" of changing pollutant levels for 483 counties in the 15 most populated U.S. states using quantitative methods for causal hypothesis testing, such as conditional independence and Granger causality tests. We shall see that the hypothesis of a significant statistical association between air pollution and mortality rates is well supported, but that the hypothesis of a predictive causal relation between them is not. For example, no significant positive associations are found between *changes* in PM2.5 or O3 levels and corresponding changes in disease mortality rates between 2000 and 2010, nor for shorter time intervals of 1–3 years.

Dominici et al. (2014) recently noted that "[A]nalyses of observational data have had a large impact on air-quality regulations and on the supporting analyses of their accompanying benefits, [but] associational approaches to inferring causal relations can be highly sensitive to the choice of the statistical model and set of available covariates that are used to adjust for confounding. … There is a growing consensus… that the associational or regression approach to inferring causal relations—on the basis of adjustment with observable confounders—is unreliable in many settings." They demonstrate that the choice of regression model can result in either statistically significant positive or statistically significant negative associations between air pollutant levels and mortality rates. This implies that implicit modeling choices can greatly affect – or even determine—the results presented to decision-makers and the public. Table 17.1 provides some examples of important policy-relevant conclusions and doubts about their validity from the recent air pollution health effects literature.

To overcome this difficulty, Dominici et al. (2014) proposed the use of *quasi-experiments* (QEs), or natural experiments, in which outcomes are compared between a treatment and control group are compared, but without random assignment or other determination of the treatment status by the researcher. As an example, they cite the Dublin coal-burning ban study discussed in Chap. 1, reporting significantly lower mortality rates in the 6 years following a ban on coal-burning in Dublin County, Ireland compared to the six years prior to the ban (Clancy et al. 2002). This proposal to use QEs to better assess causal relations between pollution levels and health effects has been hailed as "a paradigm-shifting solution" (Harvard Law Today 2014). Yet, since QEs were first introduced in social statistics in the 1960s, expert practitioners

Table 17.1 Some conflicting claims about health effects known to be caused by air pollution

Pro (causal interpretation or claim)	Con (counter-interpretation or claim)
"Epidemiological evidence is used to quantitatively relate $PM_{2.5}$ exposure to risk of early death. We find that UK combustion emissions cause 13,000 premature deaths in the UK per year, while an additional 6000 deaths in the UK are caused by non-UK European Union (EU) combustion emissions" (Yim and Barrett 2012)	"[A]lthough this sort of study can provide useful projections, its results are only estimates. In particular, although particulate matter has been associated with premature mortality in other studies, a definitive cause-and-effect link has not yet been demonstrated" (NHS 2012)
"[A]bout 80,000 premature mortalities [per year] would be avoided by lowering PM2.5 levels to 5 μg/ m^3 nationwide" in the U.S. 2005 levels of PM2.5 caused about 130,000 premature mortalities per year among people over age 29, with a simulation-based 95% confidence interval of 51,000–200,000 (Fann et al. 2012)	"Analysis assumes a causal relationship between PM exposure and premature mortality based on strong epidemiological evidence… However, epidemiological evidence alone cannot establish this causal link" (EPA (U.S. Environmental Protection Agency) 2011, Tables 6–11). Significant negative associations have also been reported between PM2.5 (Krstić 2010) and short-term mortality and morbidity rates, as well as between levels of some other pollutants (e.g., NO2 (Kelly et al. 2011) and ozone (Powell et al. 2012)) and short-term mortality and morbidity rates
"Some of the data on the impact of improved air quality on children's health are provided, including… the reduction in the rates of childhood asthma events during the 1996 Summer Olympics in Atlanta, Georgia, due to a reduction in local motor vehicle traffic" (Buka et al. 2006). "During the Olympic Games, the number of asthma acute care events decreased 41.6% (4.23 vs. 2.47 daily events) in the Georgia Medicaid claims file," coincident with significant reductions in ozone and other pollutants (Friedman et al. 2001)	"In their primary analyses, which were adjusted for seasonal trends in air pollutant concentrations and health outcomes during the years before and after the Olympic Games, the investigators did not find significant reductions in the number of emergency department visits for respiratory or cardiovascular health outcomes in adults or children." In fact, "relative risk estimates for the longer time series were actually suggestive of increased ED [emergency department] visits during the Olympic Games" (Health Effects Institute (HEI) 2010)
"An association between elevated PM10 levels and hospital admissions for pneumonia, pleurisy, bronchitis, and asthma was observed. During months when 24-h PM10 levels exceeded 150 μ/m^3, average admissions for children nearly tripled; in adults, the increase in admissions was 44%" (Pope 1989)	"Respiratory syncytial virus (RSV) activity was the single explanatory factor that consistently accounted for a statistically significant portion of the observed variations of pediatric respiratory hospitalizations. No coherent evidence of residual statistical associations between PM_{10} levels and hospitalizations was found for any age group or respiratory illness" (Lamm et al. 1994)

(continued)

Table 17.1 (continued)

Pro (causal interpretation or claim)	Con (counter-interpretation or claim)
"Reductions in respiratory and cardiovascular death rates in Dublin suggest that control of particulate air pollution could substantially diminish daily death....Our findings suggest that control of particulate air pollution in Dublin led to an immediate reduction in cardiovascular and respiratory deaths" (Clancy et al. 2002) "The results could not be more clear, reducing particulate air pollution reduces the number of respiratory and cardiovascular related deaths immediately" (Harvard School of Public Health 2002)	Mortality rates were already declining long before the ban, and occurred in areas not affected by it. "Serious epidemics and pronounced trends feign excess mortality previously attributed to heavy black-smoke exposure" (Wittmaack 2007). "Thus, a causal link between the decline in mortality and the ban of coal sales cannot be established" (Pelucchi et al. 2009). "In contrast to the earlier study, there appeared to be no reductions in total mortality or in mortality from other causes, including cardiovascular disease, that could be attributed to any of the bans. That is, after correcting for background trends, similar reductions were seen in ban and non-ban areas" (Health Effects Institute (HEI) 2013)

Source: Adapted from Cox Jr. et al. (2013)

have recognized that "in many quasi-experiments, one is most often left with the question: 'Are there alternative explanations for the apparent causal association?'" (Harris et al. 2006). Such alternative explanations, or threats to the internal validity of causal inferences for the studied populations, are discussed in Chap. 1. They must be refuted before valid causal inferences can be drawn from QEs (Campbell and Stanley 1966; Maclure 1991; Rothman and Greenland 2005). For example, to be valid, the conclusion that a ban on coal-burning *caused* an immediate reduction in all-cause and cardiovascular mortality (Harvard School of Public Health 2002) would have had to refute plausible alternative explanations. Including a relevant historical or contemporaneous control group (using a pretest-posttest design or a nonequivalent control group design, respectively, in QE terminology) would have allowed the elimination of non-causal explanations, such as that (a) mortality rates were already declining before the ban, and continued to do so without significant change during and afterward for reasons unrelated to the ban (the "History" threat to internal validity, in QE terminology); or (b) mortality rates declined at the same rate in areas not affected by the ban as in areas affected by it. For the Dublin study, both possibilities (a) and (b) proved to be true, so that no valid conclusions about the impact of the ban on all-cause or cardiovascular mortality rates can be drawn (Wittmaack 2007; Pelucchi et al. 2009). Indeed, upon reanalysis using relevant control groups, no effect of the ban on these outcomes could be detected (Health Effects Institute (HEI) 2013). Yet, as Dominici et al, rightly note, natural experiments occur frequently and if properly analyzed, can provide crucial policy-relevant insights into causality (or lack thereof) in observed exposure-response relations. In the U.S. for example, geographic heterogeneity in the rates at which pollutant levels have declined in different regions has created many natural experiments for assessing the effects of these changes on public health over time.

To take advantage of these natural experiments, the following sections compare changes in PM2.5 and O3 levels from 2000 to 2010 to corresponding changes in all-cause and CVD age-specific mortality rates over the same interval, for hundreds of counties in the 15 largest states in the U.S. Treating county as the unit of observation, as in the Dublin study and many others where individual-level exposure data are not available, invites application of longitudinal designs and methods in which each county's history of pollution levels and mortality rates serves as its own control group for purposes of determining how subsequent changes in pollution are associated with subsequent changes in mortality rates (Campbell and Stanley 1966). Using repeated observations on the same counties over time also allows the effects of unmeasured (and possibly unknown) confounders to be largely controlled for as changes in pollutant levels and mortality rates are calculated—the basic strategy of panel data analysis (Angrist and Pischke 2009). The goal of our analysis is to understand the extent to which historical associations between pollutant levels and mortality rates reflect a clear causal relation, rather than merely coincident trends, the effect of confounders, or modeling choices.

Table 17.2, which repeats information from Chap. 8 for convenience, lists several quantitative methods for causal hypothesis testing, modeling, and analysis that have been extensively developed and applied over the past six decades (Cox Jr. 2013). Chapter 2 provides much more detail, and various advantages of these techniques, as compared to qualitative causal criteria (Rothman and Greenland 2005) such as the traditional Hill considerations and other weight-of-evidence and associational methods, are well explained and illustrated in the references for Table 17.2 (e.g., Greenland and Brumback 2002), along with their limitations (e.g., Freedman 2004). Prominent among these advantages is the development of empirically testable implications of causal hypotheses, such as conditional independence implications, timing implications, information-theoretic implications, and exogeneity implications, with conditional probability distributions of some variables being determined by the values of others.

These testable implications capture the inherent asymmetry inherent in the notion of causation, unlike correlations or other symmetric measures of association. They can be tested statistically using publically available standard computer codes, such as those in R and Python/NumPy. This enables different investigators, perhaps with very different prior beliefs, to reach the same conclusions from the same data. This points the way toward greater objectivity and definitiveness in determining via such tests the extent to which data do or do not support causal hypotheses, based on their testable implications.

Other reasons why modern methods of quantitative causal analysis should be (and increasingly are) included among current approaches in the epidemiologist's tool kit are discussed in modern epidemiology textbooks and monographs (e.g., Hernan and Robins 2011) and in the references to Table 17.2. The purpose of this chapter is not to further review these methods, but to apply those that are most useful to the air pollution and mortality rate records in the United States.

Table 17.2 Some formal methods for modeling and testing causal hypotheses

Method and references	Basic idea	Appropriate study design
Quasi-experimental design and analysis (Campbell and Stanley 1966)	Can control group comparisons refute alternative (non-causal) explanations for observed associations between hypothesized causes and effects, e.g., coincident trends and regression to the mean? If so, this strengthens causal interpretation	Observational data on subjects exposed and not exposed to interventions that change the hypothesized cause(s) of effects
Conditional independence tests (Freedman 2004, Friedman and Goldszmidt 1998; Chap. 9)	Is hypothesized effect (e.g., cardiovascular disease (CVD) mortality rate) statistically independent of hypothesized cause (e.g., PM2.5 concentration), given (i.e., conditioned on) the values of other variables, such as education and income? If so, this undermines causal interpretation	Cross-sectional data; Can also be applied to multi-period data (e.g., in dynamic Bayesian networks)
Panel data analysis (Angrist and Pischke 2009; Stebbings Jr. 1978)	Are changes in exposures followed by changes in the effects that they are hypothesized to help cause? If not, this undermines causal interpretation; if so, this strengthens causal interpretation Example: Are reductions in PM2.5 levels followed (but not preceded) by corresponding changes in CVD mortality rates?	Panel data study: Collect a sequence of observations on same subjects or units of observation (e.g., counties) over time
Granger causality test (Eichler and Didelez 2010)	Does the history of the hypothesized cause improve ability to predict the future of the hypothesized effect? If so, this strengthens causal interpretation; otherwise, it undermines causal interpretation Example: Can CVD mortality rates be predicted better from time series histories of PM2.5 levels and mortality rates than from the time series history of mortality rates alone?	Time series data on hypothesized causes and effects
Intervention analysis and change point analysis (Helfenstein 1991; Gilmour et al. 2006)	Does the best-fitting model of the observed data change significantly at or following the time of an intervention? If so, this strengthens causal interpretation Do the quantitative changes in hypothesized causes predict and explain the subsequently observed quantitative changes in hypothesized effects? If so, this strengthens causal interpretation Example: Do mortality rates fall faster in counties where pollutant levels fall faster than in other counties?	Time series observations on hypothesized effects, and knowledge of timing of intervention(s) Quantitative time series data for hypothesized causes and effects

(continued)

Table 17.2 (continued)

Method and references	Basic idea	Appropriate study design
Counterfactual and potential outcome models (Moore et al. 2012)	Do exposed individuals have significantly different response probabilities than they would have had if they had not been exposed? Example: Do people have lower mortality risk after historical exposure reductions than they would have had otherwise?	Cross-sectional and/or longitudinal data, with selection biases and feedback among variables allowed
Causal network, path analysis, and structural equations models of change propagation (Hack et al. 2010; Chaps. 9 and 15)	Do changes in exposures (or other causes) create a cascade of changes through a network of causal mechanisms (represented by equations), resulting in changes in the effect variables? Example: Do relatively large variations in daily levels of fine particulate matter (PM2.5) air pollution create corresponding variations in markers of oxidative stress in the lungs?	Observations of variables in a dynamic system out of equilibrium
Negative controls (for exposures or for effects) (Lipsitch et al. 2010)	Do exposures predict health effects better than they predict effects that cannot be caused by exposures? Example: Do pollutant levels predict cardiovascular mortality rates better than they explain car accident mortality rates? If not, this weakens causal interpretation of the CVD associations	Observational studies

Data and Methods

Cause specific mortality rates, by county and age group, were downloaded from the Centers for Disease Control and Prevention (CDC) Wonder "Compressed Mortality, 1999–2010" database (Centers for Disease Control and Prevention (CDC) 2014). To create a geographically diverse sample, mortality rates were extracted at the county level for the 15 largest states in the U.S. (California, Texas, New York, Florida, Illinois, Pennsylvania, Ohio, Georgia, Michigan, North Carolina, New Jersey, Virginia, Washington, Massachusetts, Arizona) representing approximately 65% of the total U.S. population. We extracted mortality rates (per hundred thousand person-years) for all causes of death, and then created three disease subcategories:(1) diseases of the circulatory system (International Classification of Diseases, 10th revision codes [ICD-10]I00-I99), (2) all external causes of death(ICD-10 codes V01-Y89) and (3) total disease-related mortalities (all causes of death excluding external causes). The dependent variables shown in subsequent tables thus included the following:

- *CVRatePer100K*—Mortality rate (per 100,000 people per year) due to all heart/circulatory diseases

- *ExtRatePer100K*—Mortality rate due to external causes (used as a negative control). (To investigate whether the methods used can detect causal known relationships, we also used a positive control in which a known causal effect was simulated, as discussed later for Table 17.7.)
- *ACRatePer100K*—Mortality rate due to all disease-related(non-external) causes

Most of our analyses were restricted to ages 65+ years, as they have the highest CVD mortality rates. Age was categorized as 65–74 years, 75–84 years, and 85+ years.

County-level air quality data for PM2.5 (daily 24-h mean) and O3 (daily maximum 8-h moving average) were downloaded from the U.S. Environmental Protection Agency Air Quality System (AQS) for all monitors located in each county (n = 483) of the fifteen states listed above (EPA (U.S. Environmental Protection Agency) 2011). Data were obtained for the years 2000–2010. The two pollutant measures were summarized as county-level annual averages in our analyses.

The mortality and air quality data were merged by state/county and year. The resulting merged data file contained data for 483 distinct counties from 2000–2010, although not all counties collected both ozone and PM2.5 data for all years. These merged data files are freely available from the authors upon request.

Statistical Analysis Methods

The methods in Table 17.2 that are most useful for the air pollution and mortality rate data sets just described include conditional independence tests, longitudinal comparisons of changes in death rates and changes in pollution levels, Granger causality tests, and negative controls comparing presumably non-causal associations between longitudinal changes in accident and other "external" (non-disease) death rates and changes in pollutant levels to associations between changes in disease mortality rates and changes in pollutant levels. These are described in the following paragraphs. All statistical computations were carried out using the *Statistica 12.5* statistical computing environment, with the exception of the Granger causality tests, described below. Other methods in Table 17.2, such as change-point analysis and intervention analysis for an intervention that occurs at a single point in time (e.g., closing a steel mill or banning coal-burning in Dublin) are less relevant for these data, since both changes in PM2.5 and changes in mortality rates occurred gradually over a decade, rather than abruptly from before to after some intervention.

Association-Based Methods: Correlation and Regression

Although not methods of causal analysis, association-based methods such as correlation and regression analysis are widely used in air pollution health effects research (Dominici et al. 2014). We used these methods also to test whether applying them in this data set produced similar results to past studies. Intuitively, the absence of any association might be interpreted to suggest that causation is unlikely (Hill 1965; Ward 2009). We used Pearson product-moment linear correlation coefficients and linear regression coefficients as measures of linear association, since past research suggests an approximately linear association of PM2.5 and O3 with mortality (e.g., Lepeule et al. 2012).

Conditional Independence Tests

If a statistically significant association between exposure and response variables is found, e.g., based on linear correlation and regression tests, then an important screening test for potential causation is the *conditional independence* test (Chap. 9): does a significant association remain even after conditioning on potential confounders, such as age or year? For example, if a significant association between PM2.5 and CVD mortality were hypothesized to be due to confounding by year (because both PM2.5 and CVD mortality rates both declined with time, even if one did not cause the other), then one could condition on year (i.e., holding it fixed at a given value, such as 2010), and test whether the conditional association vanishes within the subset of records with that value (e.g., with Year = 2010).

To avoid biasing results by manual selection of variables to condition on, we relied on automated backward stepwise variable selection in our multiple regression models. This is a standard—but deservedly controversial—technique. We do not advocate it for general use, as it over-fits models to data, producing excess false positives in simple settings. We therefore have used it only as a readily available automated approach that may be more familiar and easily available than alternatives such as Bayesian Model Averaging, but we have also verified the main conclusions using multiple disjoint random samples of the data (20% cross-validation), to guard against the defects of backward stepwise selection. The backward stepwise selection procedure uses successive F tests to determine whether dropping individual variables (e.g., O3 concentration) from the set of potential explanatory variables significantly decreases the ability of the model to predict values of the dependent variable (e.g., CVD mortality risk). If not, i.e., if the F test indicates that the dependent variable is conditionally independent of a potential explanatory variable (such as O3), given the values of other variables in the model, then that variable is automatically dropped from the final set of explanatory variables. Despite its flaws, use of this technique reduces subjectivity in choosing explanatory variables. We used the default settings in *Statistica* (e.g., *p* values of 0.05 to define significant associations).

Correlations Among Changes Over Time

Perhaps the most important screening test we use for potential causality is examining whether changes in an exposure help predict and explain changes in a response. A frequent confusion in epidemiology is to interpret the *slope* of a concentration-response relation as indicating the future *change* in response (e.g., mortality rates) that would be caused by a unit change in future exposure concentration. This is incorrect, since many concentration-response associations are not entirely causal (e.g., due to confounders or modeling biases). Rather than using slopes of cross-sectional regression lines as proxies for causal impacts, we directly tested whether there were significant positive correlations and regression coefficients between longitudinal changes in county-specific PM2.5 and O3 levels from 2000 to 2010, and corresponding longitudinal changes in county-specific and age-specific mortality rates; and whether counties with more rapid declines in PM2.5 and O3 had more rapid declines in mortality than those with slower declines, or where concentrations increased.

Granger Causality and Negative and Positive Controls

A more general approach than studying associations between changes in exposure concentrations and changes in mortality rates over a single time interval is to use time series analysis to test whether past values of exposure help to predict present and future mortality rates more accurately than they can be predicted from past mortality rates alone (Chap. 9). This is the basic idea of the *Granger causality test* (Eichler and Didelez 2010). If the future of a mortality rate time series is conditionally independent of the past and present exposure time series, given the past and present mortality rate series, so that knowing exposure does not improve ability to predict future mortality rates, then exposure is not a Granger-cause of mortality. The Granger causality test produces a p-value for the null hypothesis that one time series does not improve prediction of another compared to using lagged values of the dependent variable itself.

We performed the Granger tests, using the *grangercausalitytests* function in the Python *statsmodels* module, for each county and age category combination described above, with the restriction that the combination must have at least 10 consecutive annual values available for analysis. We tested lags of 1–3 years, as many previous studies suggest that reductions in PM2.5 and other pollutants lead to almost immediate reductions in mortality rates, e.g., within as little as a few days, and certainly well within a year or two (e.g., Friedman et al. 2001; Clancy et al. 2002; Lepeule et al. 2012; Yang et al. 2013). The Python Granger function *grangercausalitytests* provides p values for each of four separate test statistics (two based on the F distribution and two on the chi-square distribution), all of which yield closely similar results. We evaluated the proportion of counties for which these tests produced a p-value of 0.05 or less; random variation alone could explain this occurring in about 5% of counties. Significantly higher levels would be suggestive of a Granger causality effect.

In addition to formal test statistics, we also compared the statistical association between changes in exposures and changes in disease-related mortality rates, on the one hand, to the association between changes in exposures and changes in non-disease-related (external-cause) mortality rates, on the other. The external-cause mortality rates include deaths due to accidents and assaults, changes in which are presumably not caused by changes in pollution levels. Such negative controls test whether hypothesized causal associations are stronger than those presumed to be non-causal (Lipsitch et al. 2010). As discussed further later (c.f. discussion of Table 17.7), we also simulated the effects of a positive causal relation between changes in pollution levels and changes in mortality rates. This simulation-based analysis served as a type of positive control to test whether sample sizes are large enough and whether the statistical methods we applied are powerful enough to detect such genuine causal effects if they are present. Finally, we briefly examined the geographic pattern of results to determine whether findings appeared to hold consistently in different parts of the United States.

Results

Descriptive Statistics

Figure 17.1 shows trends in average pollution levels, population, and mortality rates for all counties from 2000–2010. For each time series, values are normalized by dividing by the value in 2000, so that all time series values in 2000 are defined as 1. PM2.5 and CVD mortality rates declined most steeply over this interval (two lowest curves), while population levels and external-cause mortality rates (e.g., from accidents) increased, perhaps reflecting a longer-lived, aging population.

Figure 17.2 shows how the age-specific mortality curve, plotting annual deaths per capita vs. age, has shifted downward over time. (The horizontal positions for the rates have been spread out to allow easy visualization of trends. Vertical bars indicate 95% confidence intervals for the mean mortality rates but are very narrow due to the large sample sizes.) Clearly, age-specific mortality rates have declined for all age groups, but most for the older age groups.

Figure 17.3 shows analogous curves for age groups 55–64, 65–74, 75–84, and 85 or older, abbreviated 55, 65, 75, and 85, respectively, for different average PM2.5 levels in 2000 (left) and 2010 (right). At all PM2.5 levels, age-specific mortality rates declined conspicuously from 2000 to 2010. In both years, mortality rates in the oldest age categories were higher at PM2.5 levels of 12 μg/m^3 than at 3 μg/m^3, suggesting a possible persistent positive association between PM2.5 concentrations and elderly mortality rates.

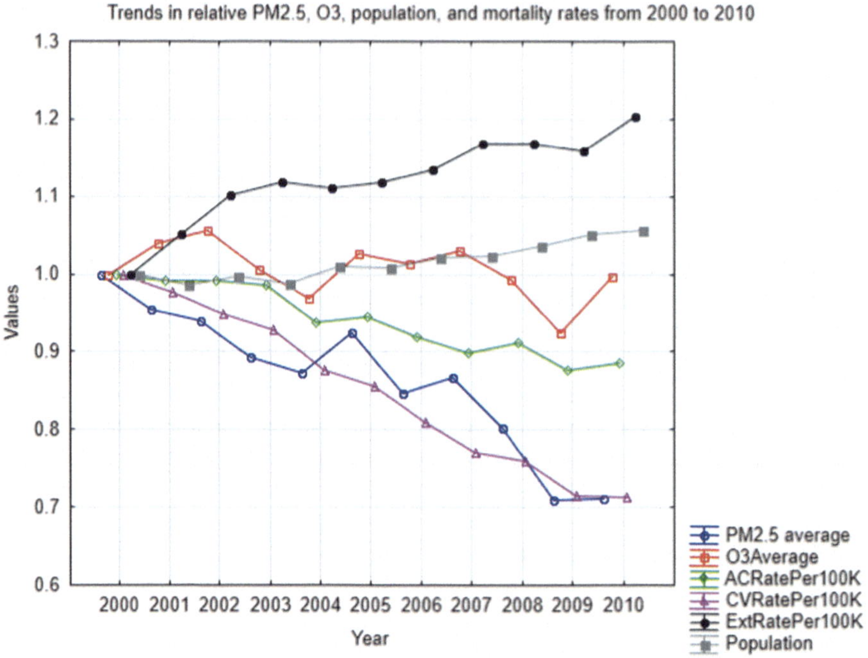

Fig. 17.1 Trends in relative values of pollutants, mortality rates, and population, 2000–2010

There was substantial geographic heterogeneity in both PM2.5 values and CVD mortality rates among the counties in this study, allowing the relation between them to be studied with considerable statistical power despite the smoothing effects of using county-level data (Savitz 2012). PM2.5 average levels ranged from below 2 to above 20 μg/m^3, and cardiovascular deaths per 100,000 people per year ranged from close to zero (for younger age groups) to over 10,000 deaths per 100,000 person-years (for the oldest age group in early years). Even for a single age group (e.g., 75–84 year-olds) and a single year (2010), there is a greater than fivefold variation in CVD mortality rates and a more than eightfold variation in average PM2.5 levels among counties, as shown in Fig. 17.4.

Results on Statistical Associations Between Pollutant Levels and Mortality Rates

Table 17.3 shows the Pearson correlation coefficients between PM2.5 and O3 levels, county population sizes, and all-cause, cardiovascular, and external-cause (non-disease) mortality rates, holding year and age fixed at 2010 and 75–84 years, respectively. Similar correlations hold for other years. All off-diagonal correlation coefficients in Table 17.3 are statistically significant from zero ($p < 0.05$) except for

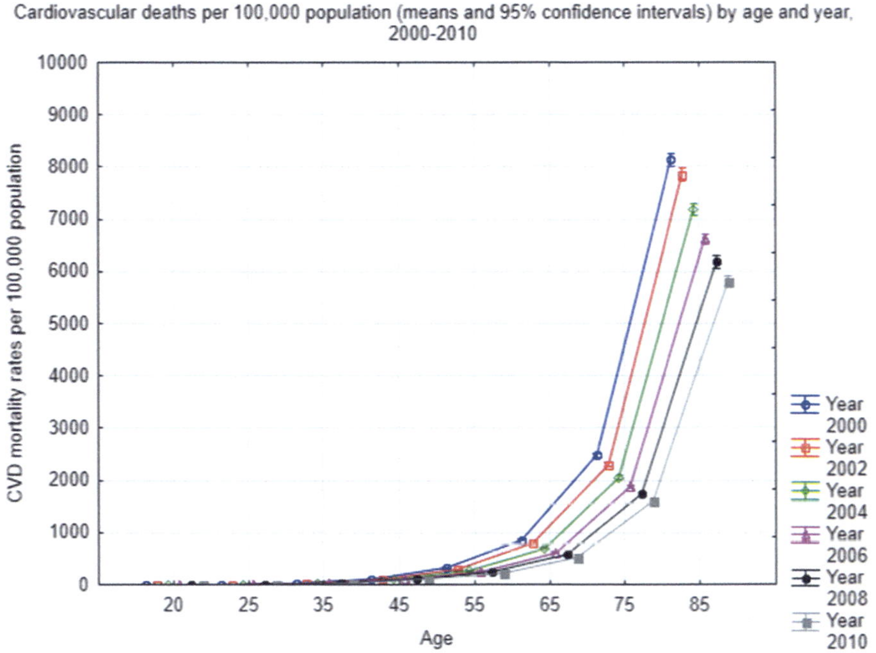

Fig. 17.2 Declines of age-specific cardiovascular disease (CVD) mortality rates over time (top curve is for year 2000, bottom curve is for year 2010)

the −0.09 correlation between PM2.5 levels and non-disease mortality rates (ExtRatePer100k). Specifically, Table 17.3 shows the following significant associations:

- PM2.5 and O3 concentrations are positively associated with each other (correlation r = 0.28)
- Both PM2.5 and O3 concentrations are positively associated with both all-cause and cardiovascular mortality rates.
- O3 is also positively associated with non-disease mortality rates but PM2.5 is not. (All positive correlations in Table 17.3 are significant, but the −0.09 numbers are not.)
- Population size of a county is positively associated with PM2.5 and is negatively associated with O3 and with all mortality rates.
- All mortality rates (disease-related and non-disease-related) are positively associated with each other, but negatively associated with population size.

The associations in Table 17.3 may or may not be causal, but they are not explained by coincident historical trends (since the year is held fixed at 2010) nor by confounding by age category, since the age category is also held fixed at 75–84. Whether confounding by education, income, temperature, or other variables might account for some of these associations—for example, if mortality rates and PM2.5

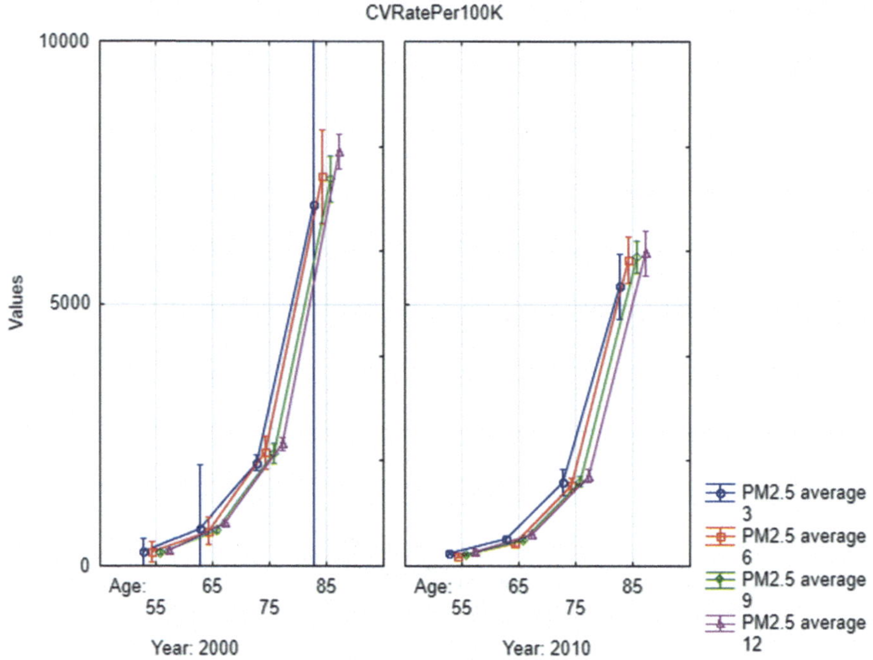

Fig. 17.3 Decline of older age-specific mortality rates over time (left panel is for year 2000, right panel is for year 2010) for counties with different average PM2.5 levels

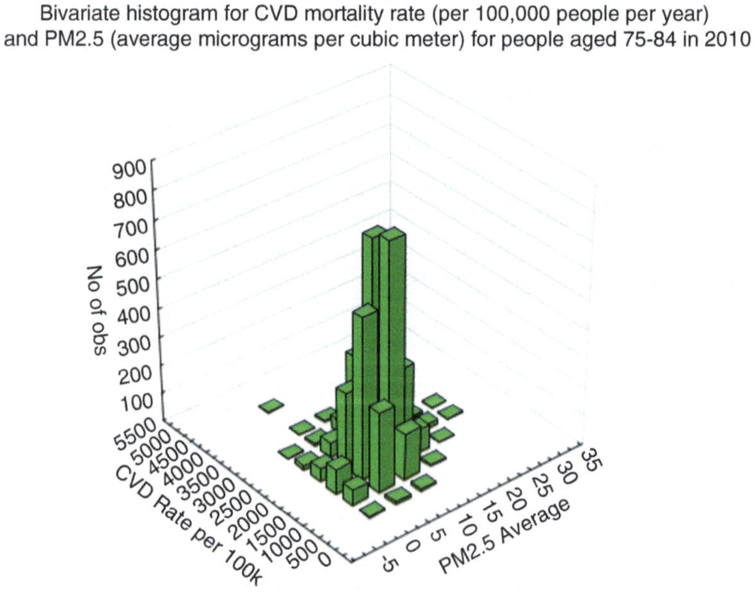

Fig 17.4 There is substantial geographic heterogeneity in PM2.5 levels and CVD mortality rates even within a single age group and year (here, 75–84 year olds in 2010)

Table 17.3 Pearson correlations between pairs of exposure and response variables for elderly (75–84 year-old) people in 2010

	Correlations between county-specific average PM2.5 and O3 concentrations and mortality rates for 75–84 year olds in 2010						
Variable	Means	PM2.5 average	O3Average	Population	ACRatePer100K	CVRatePer100K	ExtRatePer100K
PM2.5 average	9.16	1.00	0.28	0.14	0.17	0.22	−0.09
O3Average	0.04	0.28	1.00	−0.20	0.30	0.14	0.20
Population	15783.16	0.14	−0.20	1.00	−0.33	−0.15	−0.34
ACRatePer100K	4855.06	0.17	0.30	−0.33	1.00	0.72	0.38
CVRatePer100K	1614.13	0.22	0.14	−0.15	0.72	1.00	0.19
ExtRatePer100K	137.28	−0.09	0.20	−0.34	0.38	0.19	1.00

are both elevated on cold days or in colder regions; or if lower-income families tend to live in more polluted areas and also to have higher age-specific mortality rates irrespective of location—cannot be determined from the exposure and mortality rate data alone.

In multiple linear regression modeling of the association between explanatory variables and elderly (75–84 years-old) CVD mortality rate using automated backward stepwise variable selection via F tests, only the regression coefficient between PM2.5 and CVD mortality rate, but not O3 and CVD mortality risk, remains significant. Thus, there is a positive association between PM2.5 levels and CVD mortality rates among the elderly that is not explained by coincident historical trends, nor by confounding by age or population or O3; but the correlations between O3 and CVD mortality rates, and between O3 and all-disease mortality rate, vanish after conditioning (via multiple linear regression) on PM2.5 and population size for all disease-related mortalities. In short, PM2.5, but not O3, passes this conditional independence test for being a potential causal driver of elderly mortality rates. Similarly, for all age categories and years, PM2.5 average levels but not O3 levels help to predict CVD mortality rates.

Table 17.4 shows the results of a multiple linear regression with backward stepwise variable selection; results were also confirmed in multiple disjoint random samples (20% cross-validation samples). The b^* column contains standardized regression coefficients (scaling each variable in terms of standard deviations) and the b column contains the unstandardized regression coefficients. As expected, *Year* is negatively associated with CVD mortality risk, and *Age* is positively associated with CVD mortality risk. *Age* is quantitatively by far the most important predictor of risk. PM2.5 average concentration makes the smallest, but still highly statistically significant ($p < 0.000001$), contribution to predicting CVD values. Population (specific to each county and age group) is also a significant predictor of CVD risk. Results for all-disease-related mortality (AC) risks are similar, with the standardized regression coefficient for PM2.5 increasing to 0.06, with the exception that both ozone (O3) and population size are significantly negatively associated with AC mortality rates (standardized regression coefficients of −0.12 for Population and −0.02 for O3). Interpretively, the coefficient for PM2.5 in Table 17.4 ($b = 33.6$) indicates that CVD mortality risk increases by 33.6 deaths per 100,000 person-years

Table 17.4 County-specific average PM2.5 concentration is significantly positively associated with county-specific CVD mortality rates across all age categories and years

	Regression summary for dependent variable: CVRatePer100K, $R = 0.78$, $R^2 = 0.605$, Adjusted $R^2 = 0.605$					
N = 21,613	b^*	Std. err. of b^*	b	Std. err. of b	t(21,608)	p-value
Intercept			114927.7	7043.87	16.3160	0.000000
Year	−0.08	0.0046	−60.3	3.51	−17.2049	0.000000
Age	0.81	0.0048	120.0	0.70	170.4133	0.000000
pm25Average	0.04	0.0047	33.6	3.54	9.4979	0.000000
Population	0.08	0.0048	0.0	0.00	17.4577	0.000000

for each microgram per cubic meter increase of PM2.5 in air, assuming other variables are held constant. The mean CVD mortality rate averaged over all age categories and years is 1931.6 deaths per 100,000 person-years, so a change in PM2.5 of 10 μg/m^3 corresponds to a change in CVD mortality rate of approximately (10 μg/m^3)*(33.6 deaths per 100,000 person-years per μg/m^3)/(1931.6 deaths per 100,000 person-years) = 336/1931.6 = 17.4%. This slope factor could be described as a 17.4% increase in mortality per 10 μg/m^3 increase in PM2.5 concentration.

Results on Correlations Between Changes in Variables Over Time

Tables 17.5 and 17.6 show correlations between *changes* in all-cause mortality, CVD mortality, and non-disease mortality, respectively (the columns) and different possible predictors (the rows), for all counties included in the study. Table 17.5 presents results for the 75–84 year-old group, and Table 17.6 repeats the analysis for all age groups.

For the 75–85 year old age category, changes in all-cause (AC) and CVD mortality rates are significantly positively correlated with each other, as expected, and with changes in external-cause mortality rates. They are significantly negatively correlated with increases in population. Neither is significantly correlated with changes in PM2.5 or changes in O3. For all age groups, changes in PM2.5 are significantly but weakly positively correlated with changes in external-cause mortality rates. Changes in O3 are significantly positively correlated both with changes in AC mortality rates

Table 17.5 Pearson correlations between changes in variables from 2000 to 2010 for elderly (75–84 year-old) people

	Correlates of changes in mortality rates from 2000 to 2010 for 75–85 year-olds		
Variable	Delta AC mortality	Delta CVD mortality	Delta external rate
Delta PM2.5	−0.07	−0.08	0.04
delta O3	0.03	0.03	0.06
Delta population	−0.59	−0.56	−0.44
Delta AC mortality	1.00	0.99	0.81
Delta CVD mortality	0.99	1.00	0.79
Delta external rate	0.81	0.79	1.00
PM2.5 average	−0.10	−0.12	−0.10
O3 Average	−0.04	−0.04	−0.18
Population 2010	0.07	0.07	0.06
ACRatePer100K	−0.05	−0.07	−0.07
CVRatePer100K	0.06	0.06	−0.02
ExtRatePer100K	−0.16	−0.17	0.14

Table 17.6 Pearson correlations between changes in variables from 2000–2010 for all age groups

	Correlates of changes in mortality rates from 2000 to 2010		
Variable	Delta AC mortality	Delta CVD mortality	Delta external rate
Delta PM2.5	−0.00	−0.01	0.06
Delta O3	0.06	0.08	0.02
Delta population	−0.15	−0.15	−0.17
Delta AC mortality	1.00	0.95	0.69
Delta CVD mortality	0.95	1.00	0.59
Delta external rate	0.69	0.59	1.00
Age 2000	−0.03	−0.14	0.24
PM2.5 average 2000	−0.02	−0.04	−0.04
O3 Average 2000	−0.05	−0.07	−0.05
ACRatePer100K 2000	0.17	−0.01	0.40
CVRatePer100K 2000	0.23	0.02	0.40
ExtRatePer100K 2000	0.21	−0.03	0.56

and with changes in CVD mortality rates. Increases in population are significantly correlated with reductions in all mortality rates.

In multivariate analysis using multiple linear regression, changes in both AC (all-cause) and CVD mortality rates are conditionally independent of changes in both PM2.5 and O3, given changes in population size, changes in external-cause mortality rates, and age in 2010. These three explanatory variables are automatically selected by backward stepwise variable selection, while changes in PM2.5 and O3 are dropped, as they provide no additional information useful for predicting the AC or CVD mortality rates. Thus, by this criterion, changes in PM2.5 and O3 levels do not help to predict or explain changes in CVD or AC mortality rates, undermining a causal interpretation of the positive associations between them in the cross-sectional analysis in Table 17.3.

Other, perhaps unexpected, correlations between changes in variables in Table 17.6 include a strong positive correlation (0.59) between changes in external-cause mortality rates and changes in CVD mortality rates; and positive correlations between baseline levels of mortality rates and changes in their levels. Thus, relatively high-risk areas in 2000 tended to become more risky by 2010. As expected, older age categories saw relatively large reductions in disease mortality rates (but increases in non-disease mortality rates).

Granger Causality Test and Control Results

Granger tests using standard time series regression models with maximum lags of 1, 2, or 3 years show that, for all age categories tested (65–74, 75–84, and 85 or older) and for all mortality outcomes considered (CVD, all-disease, and external-cause mortality rates), both PM2.5 and O3 histories are not useful for predicting mortality rates in most (over 90%) of the counties. PM2.5 and O3 have predictive coefficients for CVD and all-disease mortality rates that are significantly different from zero in only a small minority of counties (7% for AC mortality, 6% for CVD mortality, and 7% for external-cause mortality, which was used as a negative control), roughly consistent with, though slightly higher than, the 5% false-positive error rate that might occur by chance due to the 5% significance level used in the tests. (For 483 counties and a true false-positive rate of 5%, there is about a 26% probability that the sample proportion of false positives would exceed 6% or be less than 4% by chance.) Perhaps more importantly, the negative control (external-cause mortalities) also shows that O3 and PM2.5 histories on time scales of several years are not Granger-causes of CVD or all disease-related deaths any more than they are of external-cause deaths. For example, the age group and lag with the highest fraction of Granger-positive associations between PM2.5 and CVD rate is the 85+ age group with a lag of one year: this fraction is 11%. But the corresponding fraction for Granger-positive associations between PM2.5 and external-cause mortalities is greater, at 14%. Thus, the Granger tests do not support a conclusion of a genuine causal effect, i.e., positive results clearly above what might occur by chance and what is found for the negative controls.

Given the well-known limitations of p-values and significance testing, it may also be useful to consider that, if pollutant levels were detectable causal drivers of increased mortality rates at recent historical levels, then this causal relation should have been visible in a large majority of counties. The fractions in Table 17.7 might all be expected to exceed 50% in the presence of clear Granger-causality, i.e., most counties should have shown evidence of a Granger-positive association between PM2.5 and mortality rates caused by them. Intuitively, as suggested by Fig. 17.1,

Table 17.7 Fractions of counties with positive Granger causality tests for PM2.5 and all-cause (AC), cardiovascular disease (CVD), and external-cause mortality rates, for different age groups and lags (1–3 years)

Age/Lag	AC mortality rate	CVD rate	External rate
65	**0.06**	**0.06**	**0.07**
1	0.09	0.08	0.10
2	0.04	0.05	0.10
3	0.05	0.05	0.02
75	**0.08**	**0.06**	**0.06**
1	0.10	0.08	0.05
2	0.08	0.06	0.08
3	0.04	0.05	0.06
85	**0.08**	**0.06**	**0.08**
1	0.15	0.11	0.14
2	0.06	0.03	0.09
3	0.04	0.03	0.01
Overall	**0.07**	**0.06**	**0.07**

Results averaged over all three lags are shown in bold

although pollutant levels declined substantially in most counties from 2000–2010, declines in CVD and AC mortality rates did not appear to proceed more quickly when PM2.5 declined quickly than when it did not, or than when it increased. The Granger test results confirm this suggestion at the level of individual counties and for time lags of 1–3 years.

Positive Controls: Does Absence of Evidence Constitute Evidence of Absence?

Might the absence of a significant association between county-specific changes in PM2.5 levels and changes in mortality rates between 2000 and 2010, shown in Tables 17.5 and 17.6 and in corresponding multiple linear regression models, be due to limited statistical power to detect changes in the presence of substantial heterogeneity and variability in the data? To check the statistical power of these methods, we modified the observed data by adding a known "signal"—a 2.6% decrease in CVD mortality rate per μg/m^3 decrease in PM2.5 concentration, based on the slope estimate of Lepeule et al. (2012). We then tested whether this known signal is detectable through the noise in the data using the methods we have applied.

Table 17.8 shows the results of multiple linear regression applied to the artificial data set with a simulated known causal impact of exposure. The simulated effect of changes in PM2.5 on changes in CVD mortality rates, based on the 2.6% slope coefficient for change in mortality rate per μg/m^{3}change in PM2.5) was successfully detected. (All predictors remain significant using backward stepwise variable selection.) This suggests that an effect of this size would probably have been detected in the real data if it had been present. This type of positive control gives some reassurance that the substantial variability and heterogeneity in county-level time series data would not hide causal effects of the sizes that have sometimes been estimated from standard associational (regression-based) models by assuming that slope coefficients are causal, if such causal effects were actually present.

Table 17.8 Multiple linear regression detects PM2.5 effects on mortality rates of the sizes predicted from previously published regression slope coefficients (Lepeule et al. 2012)

	Regression for CVD mortality rate with simulated effect of PM2.5, R = 0.76465, R^2 = 0.58469, Adjusted R^2 = 0.5838 F(31,421) = 666.84 p < 0.0000 Std. error of estimate: 1178.0					
N = 1425	b*	Std. err. of b*	b	Std. err. of b	t(1421)	p-value
Intercept			541.5	74.8	7.2	0.000
Delta PM2.5	0.04	0.017	37.9	16.5	2.3	0.022
CVRatePer100K_2000	−0.75	0.017	−0.5	0.0	−43.9	0.000
Delta population	−0.16	0.017	−0.0	0.0	−9.2	0.000

Finally, we briefly examined the geographic distribution of associations. Previous investigators have reported that chronic exposure to PM2.5 is associated with mortality in the eastern and central regions of the United States, but not in the western region (Zeger et al. 2008). In our data set, for the main elderly population (75–84 year-olds) in 2010, PM2.5 was statistically significantly positively correlated with CVD mortality in Florida and overall in pooled data from counties in all states. It was statistically negatively correlated with all-disease (AC) mortality rate in Arizona and statistically positively correlated with AC mortality rate in Florida and overall. Otherwise, state-specific correlations in 2010 were not individually statistically significant at the conventional 0.05, significance level, and were a mix of non-significant positive and negative correlations with no obvious geographic distribution.

Discussion: Caveats for Causal Interpretations of Regression Coefficients

The epidemiological and risk assessment literatures on human health effects of air pollution contain dozens of studies that attribute reductions in mortality risks to reductions in air pollution levels, and that estimate the slope of the concentration-response association between exposures to pollutants and corresponding mortality rates (e.g., Pope 1989; Clancy et al. 2002; Lepeule et al. 2012; Cesaroni et al. 2013; Fann et al. 2012; Dai et al. 2014). The findings in this chapter support previous findings of positive PM2.5-mortality associations, based on PM2.5 (and O3) and age-specific mortality data, using county-level data from the 15 largest U.S. states over the years from 2000 to 2010. Confirming earlier studies such as Lepeule et al. (2012), we find a statistically significant positive association between PM2.5 (and also O3) concentrations and both all-disease related and CVD mortality rates, as well as a significant positive association between O3 and external-cause mortalities, which we used as a negative control (Tables 17.3 and 17.4).

However, such associations between historical levels of exposure and response variables do not necessarily describe *causal* relations. In our examination of historical changes in pollutant levels and mortality rates (Tables 17.5 and 17.6 and multiple regression models and Granger causality tests), actual *changes* in PM2.5 and O3 levels over time did not significantly help to predict or explain corresponding observed *changes* in all-disease or CVD mortality rates over time. This argues against facile causal interpretations of the significant statistical associations between pollution levels and mortality rates. Such causal interpretations of slope coefficients are commonly made in air pollution health effects (and other) epidemiology. For example, the study of Lepeule et al., updating the important Harvard Six Cities Study, offers the important causal interpretation that “These results [i.e., that each 10 μg/m^3 increase in PM2.5 was associated with a 26% increase in cardiovascular mortality risk] suggest that further public policy efforts that reduce fine particulate matter air pollution are likely to have continuing public health benefits." But such policy-relevant causal conclusions are unwarranted if the exposure-response association discussed is not a causal relation, and if the changes referred to are only the hypothetical ones implied by a slope coefficient, rather than actual changes in the levels of exposure and mortality time series.

Study Limitations

The analyses and conclusions presented so far have several limitations. Although our analysis of county-level data does not provide evidence that the roughly 30% reduction in PM2.5 levels from 2000 to 2010 (Fig. 17.1) caused any detectable effect on disease-related mortality rates, it remains possible that such an effect was present that is too small to detect. For example, if each 10 μg/m^3 change in PM2.5 concentration causes only a 1.03% change in CVD mortality rate, as estimated by Dai et al. (2014), then the power of our data set would not be great enough to distinguish this from zero. In addition, like many other studies, this one lacks individual-level exposure data. The basic units of observation are death counts, by cause, within age categories, years, and counties; finer resolution would require a different data set. Age and death are available at the individual level, making this a semi-individual design (Künzli and Tager 1997), rather than a purely ecological design; but other individual covariates are not available. On the other hand, following the same counties over multiple years contributes one of the strengths of a panel study design: the effects of fixed (or slowly changing) possible confounders or effect modifiers, such as differences in income or education or regional climate, cancel out when changes (deltas) in mortality rates are calculated for the same locations in successive years. In addition, the data analyzed here substantially meet several criteria for useful ecological studies (Savitz 2012): marked variation across geographic units (counties); unlikely confounding (due to the longitudinal panel design, in which counties serve as their own controls for purposes of subtracting out fixed effects of confounders when computing changes over time); opportunities to include negative controls (external-cause mortalities); and simulated positive controls (via simulation of postulated causal impacts).

A remaining question is, if the significant associations between PM2.5 and O3 on the one hand and CVD and all-cause mortality on the other are not due to a causal relationship between pollutant exposure and disease, then what does explain them? Our analyses have ruled out coincident trends (since the associations hold even within single years) and chance (since the correlation and regression coefficients reported are statistically significant), as well as fixed confounders (due to the panel design) as plausible explanations. Possible confounders that might co-vary with exposure levels over time, and thus offer explanations, range from co-pollutants to temperature (e.g., if very hot or very cold areas have higher levels of PM2.5, perhaps due in part to coal-fired power plants that power air conditioning or heating, and independently have higher mortality rates). Attaching more variables to the county-specific mortality rate and pollution level data, such as daily temperature (high and low), could potentially help to answer this question. But at present, the answer is unknown.

Finally, by focusing on changes in annual average pollutant levels and mortality rates at the individual county level, we have foregone opportunities to model or "adjust" for effects of seasonality, more granular spatial variations, and measured or latent confounders. As Dominici et al. (2014) suggest, it is not uncommon for different regression models based on different modeling choices and assumptions to produce very different answers. For example, regression coefficients that are significantly positive in one model may be significantly negative in another, depending on which variables and interaction terms are included. By using several different approaches (conditional independence tests, Granger tests, positive and negative controls, automated variable selection) as well as relatively simple measures of association (correlations and linear regression coefficients, fractions of counties with Granger-positive associations) computed using standard, widely available software for all tests), we have sought conclusions that are more robust and objective by minimizing opportunities for manual intervention to shape the results.

Comparisons to Conclusions from Other Individual Studies

The coefficient for PM2.5 in Table 17.4 ($b = 33.6$), corresponding to a 17% increase in mortality per 10 μg/m^3 increase in PM2.5 concentration, is well within the range of other recent association-based estimates based on regression relations. For example, Dai et al. (2014), in a study of 75 U.S. cities between 2000 and 2006, reported a 1.03% (95% CI: 0.65%, 1.41%) increase in CVD mortality with each 10 μg/m^3 increase in PM2.5, averaged over a 2 day period. In their update of the Harvard Six Cities Study, Lepeule et al. (2012) estimated a 26% (95% CI: 14%, 40%) increase in CVD mortality for each 10 μg/m^3 increase in PM2.5, averaged over the three prior years. Thus, our value of 17.4% falls between these two estimates, and is within the 95% CI of the Lepeule study. Some other studies have not detected clearly significant associations between PM2.5 levels and most CVD or all-cause mortality rates (Beelen et al. 2014), or found no association between local trends in mortality and local trends in yearly average PM2.5 after adjusting for national trends and local differences (Greven et al. 2011). For the U.S. county data set we have analyzed, our main conclusions are that (a) There are statistically significant associations between PM2.5 and both all-disease and CVD mortality risks; but (b) There is no clear evidence of a causal relation between PM2.5 and O3 concentration levels and mortality rates. These results differ both from studies that do not find clear associations, and also from some authoritative opinions, including views in an Expert Elicitation Study for the U.S. EPA (2006), that statistically significant exposure-response associations between PM2.5 and CVD mortality are probably causal.

While our results do not support some previous expert judgment-based assessments of causality, this is consistent with studies showing that firmly expressed opinions of key experts about air pollution health effects associations being causal (e.g., Harvard School of Public Health 2002) have later proved to be unwarranted (e.g., Health Effects Institute (HEI) 2013). The practice of applying human judgment using weight-of-evidence considerations to measures of association (such as relative risks, odds ratios, population attributable fractions, burden-of-disease estimates, and regression coefficients) to determine whether an inference of causality is supported has been widespread in epidemiology, even though some methodologists have argued that logically valid causal inferences cannot be derived from such associations in purely observational studies without interventions (Ward 2009). This makes natural experiments, where interventions such as pollution reductions occur differently for different subpopulations, potentially valuable aids to understanding causation.

Comparison to Conclusions from Accountability Studies

Dozens of studies have examined changes in mortality or other adverse health effects associated with interventions that reduce air pollution. As in the Dublin coal-burning ban example, these studies have generally not found compelling evidence of a direct causal impact of reductions in particulate air pollution on reducing mortality rates. Henneman et al. (2017), after reviewing available evidence from such studies, noted that "Two trends are apparent in the accountability assessments above. First, often, one study will assess a regulatory action and determine that the intervention led to a statistically significant change in the response of interest.... Later, using additional data, updated methods, and/or accounting for additional factors, those results are found to be less definitive and potentially invalid .... Second, spatial and temporal scales of accountability studies have expanded, which has increased the number of potential confounders. Both patterns demonstrate the importance of appropriately accounting for confounding factors when developing models and understanding as much as practicable responses in each step of the accountability chain." Similarly, Burns et al. (2020), after reviewing 38 individual interventions from 42 studies, concluded that "Given the heterogeneity across interventions, outcomes, and methods, it was difficult to derive overall conclusions regarding the effectiveness of interventions in terms of improved air quality or health. Some evidence suggests that interventions are associated with improvements in air quality and human health, with very little evidence suggesting interventions were harmful. The evidence base highlights the challenges related to establishing the effectiveness of specific air pollution interventions on outcomes. ... Researchers should also focus on improving the internal validity of future evaluations. Studies assessing interventions aiming to reduce ambient air pollution are, like other epidemiological studies, susceptible to confounding, and the choices surrounding study design and analysis methods are critical. It is particularly important that future evaluations focus on the use of appropriate comparison populations or outcomes (i.e. negative controls) unaffected by the intervention, and account for underlying background trends in outcomes."

Comparison to Conclusions from Pro-regulation Advocates

Findings that reductions in particulate air pollution have had little clear *causal* impact on reducing mortality rates, despite clear *associations* between air pollution levels and mortality rates, have had little impact on news stories and weight-of-evidence (WoE) judgments calling for tighter regulation. In Ireland, the discovery that reducing particulate air pollution by dozens of micrograms per cubic meter had no detectable effect on all-cause mortality (Health Effects Institute (HEI) 2013; Henneman et al. 2017), despite well-publicized earlier claims to the contrary (Harvard School of Public Health 2002), has simply been ignored by many scientists, academics and environmentalists. For example, a 2019 new story (Horgan-Jones 2019) begins as follows: "Irish citizens' lives are being endangered by Government inaction on smoky coal and must be put ahead of 'vested commercial interests', a group of scientists, academics and environmentalists has said. In a letter to Minister for the Environment Richard Bruton, members of the Clean Air Alliance said that four people die in Ireland every day from the effects of air pollution."

Similarly, in the United States, many researchers and advocates for tighter regulation have simply disregarded the mixed empirical evidence on success of pollution-reducing interventions (Henneman et al. 2017; Burns et al. 2020), instead making public claims such as that "The epidemiologic evidence is consistent across studies with diverse designs, populations, pollutant mixtures, locations, and statistical approaches." (Independent Particulate Matter Review Board 2020). Rather than acknowledging the conflicts and ambiguities in real data, such groups often present hypothetical scenarios based on unverified modeling assumptions in computer simulation models as if they were factual data: "Furthermore, the risk assessment conducted by the EPA shows that, in a sample of people 30 years of age or older living in 47 urban study areas, a large number of premature deaths are attributable to PM2.5 exposure under the current standard" (*ibid*). Such dramatic and misleading claims not only do not reflect that the epidemiologic evidence is far from clear and consistent, but they also substitute computer-generated results based on assumptions, having no demonstrated predictive validity, for empirical evidence that tells a very different story.

This chapter, and much of this book, have presented a different perspective that seeks to be more open to, and expressive of, empirical experience. From this perspective, sound science requires the following steps, which do not support the foregoing dramatic claims.

- *Consider all relevant, high-quality studies before forming conclusions.* High-quality studies reporting that large reductions in pollution levels have not detectably reduced all-cause mortality rates and that previously claimed exposure-response associations do not survive more careful scrutiny and control for confounding (Henneman et al. 2017; Burns et al. 2020) should not be ignored. The false narrative of consistency across studies should be replaced by a more realistic recognition of conflicting evidence. Openness to discordant observations is a crucial part of scientific method that helps to overcome the confirmation bias in weight-of-evidence judgments.

- *Provide explicit, independently verifiable derivations of conclusions from data.* Sound scientific conclusions are derived from observations by explicit reasoning that can be checked by others, rather than being asserted as judgments by selected experts (Chap. 18). They are based on predictive generalizations that can be tested independently for predictive validity—and, if need be, corrected—with data not used in deriving them. For example, a conclusion that PM2.5 increases mortality risk could be supported empirically by tests that reject the null hypothesis that risk is conditionally independent of PM2.5, given the values of other variables (such as year and region), in multiple large data sets. Conversely, finding no such evidence (Vitolo et al. 2018) does not support this causal conclusion.
- *Correctly qualify causal conclusions and interpretations.* For example, if the epidemiologic evidence that reducing current ambient levels of PM2.5 would reduce mortality rates consists of significant positive regression coefficients in selected regression models, then tests of other possible explanations for these positive coefficients (e.g., omitted confounders, residual confounding, errors in variables, or model specification errors) should be presented, rather than simply assuming the coefficients represent causal impacts.

From this perspective, the dramatic conclusion that current NAAQS standards cause avoidable "premature deaths" expresses a widely shared and often repeated intuition (Chap. 19), but it does not follow from applying the previous steps to available facts and data.

Many academic researchers have lent support to claims that current levels of air pollution are known to increase mortality rates by publishing papers, such as those discussed in Chaps. 1, 2, 7, 8, and 15, that misinterpret significant positive regression coefficients as if they were manipulative causal relationships. Recently, an increasing proportion of such papers have started to explicitly claim that they are breaking new technical ground by demonstrating causation (e.g., Wei et al. 2020), but most are simply repeating old statistical fallacies and errors, as discussed in preceding chapters. Among the most common flaws in recent papers presenting causal claims about air pollution health effects are the following (e.g., Wei et al. 2020):

- *Omitted predictors and confounders.* Causal conclusions are stated assuming that all confounders have been adequately adjusted for, but the regression models omit important confounders and modifiers, such as percentage of minorities in local populations, daily high and low temperatures over the weeks preceding mortality, and summer and winter average humidity (Armstrong 2006; Assari et al. 2020; Barreca 2012).

- *Uncontrolled residual confounding.* Analyses use spatial and temporal averages of weather and sociodemographic variables without correcting for residual confounding, with authors sometimes arguing that consistent results across different study designs suggest that positive C-R coefficients are not due to residual confounding (Wei et al. 2020). But consistency provides no such assurance: residual confounding can create biases that are consistent without being small.
- *Unmodeled interactions among variables.* Regression model are used that assume no interactions among effects of air pollution exposures and confounders, even though such interactions are well documented (Armstrong 2006).
- *Untested and incorrect modeling assumptions.* Regression models and related models (e.g., generalized propensity score models) are assumed to be correctly specified, often with exposure assumed to be linearly related to covariates and mortality probability assumed to be linearly related to exposure and a generalized propensity score (Wei et al. 2020). These linearity assumptions are usually not valid: nonlinearities matter (Chap. 2 and Armstrong 2006). As shown in Chap. 2, falsely assuming linearity can create significant positive regression coefficients for exposure in the misspecified models, even if mortality risk does not depend on exposure.
- *Unmodeled exposure uncertainties.* It is common in this literature to claim that adverse health effects are found at exposure concentrations below current standards. For example, Wei et al. (2020) state that "the risk difference associated with long-term exposures to PM2.5 and O3 increased substantially at increasingly stringent concentration levels, including those well below the current National Ambient Air Quality Standards…." However, the concentration levels to which people who died were exposed are unknown. Ignoring errors and uncertainties in estimates of exposures and covariates can induce significant positive regression coefficients for exposure even if changing exposure would not affect mortality risk (Chaps. 2 and 16).
- *Unjustified interventional causal interpretation of regression coefficients.* This has been a dominant theme through much of this book, including Chaps. 1, 2, 7, and 8. As a recent example, Wei et al. (2020) state that "The coefficient $\hat{\alpha}_1$ [the C-R coefficient for exposure] is the estimated causal effect of air pollution on mortality, which can be interpreted as the average change in the probability of death if all the observations were to [be exposed to a] 1-unit increase in the exposure level." This misinterprets a statistical effect (a regression coefficient) as an interventional (manipulative) causal effect: the coefficient does not show how an intervention that *increases* exposure by 1 unit would change average probability of death, but rather how *observing* an exposure that is 1 unit higher would change the conditional expected value of the observed mortality variable. These are distinct concepts, as emphasized in Chap. 9.
- *Inappropriate use of technical methods* such as propensity scoring and generalized propensity scoring to make strong causal claims without doing the causal analysis and modeling needed to justify them. For example, propensity scoring

neither identifies the variables that should be conditioned on to estimate causal effects (the adjustment sets discussed in Chap. 9 and subsequently), nor accounts for effects of omitted confounders, measurement errors, and misspecified models; in practice, it often increases, rather than decreases, model-dependence and bias (Pearl 2009; Rodríguez De Gil et al. 2015; King and Nielsen 2019).

In summary, a formidable array of technically flawed papers and unsound statistical analyses have generated a large technical literature supporting the intuitively plausible claim that reducing air pollution further would create additional health benefits. This conclusion may be true, but it is not demonstrated through models and analyses that make the foregoing errors. Natural experiments and intervention studies, such as those considered in this chapter, can provide a more direct way to test the prediction that reductions in ambient air pollution levels should be followed by corresponding improvements in public health. To date the empirical support from intervention studies for this plausible-seeming belief is, at best, mixed and weak (Henneman et al. 2017; Burns et al. 2020), in line with the main results from the analyses in this chapter (Tables 17.5, 17.6, and 17.7).

Conclusions

The calculations in this chapter illustrate that a significant positive association between historical levels of PM2.5 and historical mortality rates does not necessarily provide a sound basis for inferring a positive association between changes in levels of PM2.5 and changes in mortality rates. This methodological point confirms the importance of using quasi-experiments or other appropriate formal methods of causal study design and analysis (Table 17.2) to draw causal conclusions. Free, publicly available data sets such as the EPA and CDC data sets used in this study, and free, publicly available software such as R and Python or the CAT software from Chap. 9, now make it relatively easy to test whether changes in PM2.5 and O3 help to predict changes in disease mortality rates, on time scales from days to over a decade. We hope that this will encourage others to investigate further the relation between longitudinal changes in pollutant levels and changes in mortality rates, and to clarify the crucial distinction between positive statistical associations and evidence of causality in air pollution health effects epidemiology.

References

Angrist JD, Pischke J-S. Mostly harmless econometrics: an empiricist's companion. Princeton: Princeton University Press; 2009.

Armstrong B. Models for the relationship between ambient temperature and daily mortality. Epidemiology. 2006;17(6):624–31. https://doi.org/10.1097/01.ede.0000239732.50999.8f.

Assari S, Chalian H, Bazargan M. Race, ethnicity, socioeconomic status, and chronic lung disease in the US. Res Health Sci. 2020;5(1):48–63. https://doi.org/10.22158/rhs.v5n1p48.

Barreca A. Climate change, humidity, and mortality in the United States. J Environ Econ Manag. 2012;63(1):19–34. https://doi.org/10.1016/j.jeem.2011.07.004.

Beelen R, Stafoggia M, Raaschou-Nielsen O, et al. Long-term exposure to air pollution and cardiovascular mortality: an analysis of 22 European cohorts. Epidemiology. 2014;25(3):368–78.

Buka I, Koranteng S, Osornio-Vargas AR. The effects of air pollution on the health of children. Paediatrics & Child Health. 2006;11(8):513–516.

Burns J, Boogaard H, Polus S, Pfadenhauer LM, Rohwer AC, van Erp AM, Turley R, Rehfuess E. Interventions to reduce ambient air pollution and their effects on health: an abridged Cochrane systematic review. Environ Int. 2020;135:105400. https://doi.org/10.1016/j.envint.2019.105400.

Campbell DT, Stanley JC. Experimental and quasi-experimental designs for research. Chicago: Rand McNally; 1966.

Centers for Disease Control and Prevention (CDC) (2014) Wonder "compressed mortality, 1999–2010" database. http://wonder.cdc.gov/cmf-icd10.html

Cesaroni G, Badaloni C, Gariazzo C, Stafoggia M, Sozzi R, Davoli M, Forastiere F. Long-term exposure to urban air pollution and mortality in a cohort of more than a million adults in Rome. Environ Health Perspect. 2013;121(3):324–31.

Clancy L, Goodman P, Sinclair H, Dockery DW. Effect of air-pollution control on death rates in Dublin, Ireland: an intervention study. Lancet. 2002;360(9341):1210–4.

Cox LA Jr, Popken DA, Berman DW. Causal versus spurious spatial exposure-response associations in health risk analysis. Crit Rev Toxicol. 2013;43(Suppl 1):26–38.

Cox LA Jr. Improving causal inference in risk analysis. Risk Anal. 2013;33(10):1762–71.

Dai L, Zanobetti A, Koutrakis P, Schwartz JD. Associations of fine particulate matter species with mortality in the united states: a multicity time-series analysis. Environ Health Perspect. 2014;122(8):837–42.

Dominici F, Greenstone M, Sunstein CR. Particulate matter matters. Science. 2014;344:257–8.

Eichler M, Didelez V. On Granger causality and the effect of interventions in time series. Lifetime Data Anal. 2010;16(1):3–32.

EPA (2006) Expanded expert judgment assessment of the concentration-response relationship between PM2.5 exposure and mortality. www.epa.gov/ttn/ecas/regdata/Uncertainty/pm_ee_report.pdf

EPA (U.S. Environmental Protection Agency). The benefits and costs of the clean air act from 1990 to 2020. Final report – Rev. A. Washington DC: Office of Air and Radiation; 2011.

Fann N, Lamson AD, Anenberg SC, Wesson K, Risley D. Estimating the national public health burden associated with exposure to ambient PM2.5 and ozone. Risk Anal. 2012;32(1):81–95.

Freedman DA. Graphical models for causation, and the identification problem. Eval Rev. 2004;28(4):267–93.

Friedman N, Goldszmidt M. Learning Bayesian networks with local structure. In: Jordan MI, editor. Learning in graphical models. Cambridge: MIT Press; 1998. p. 421–59.

Friedman MS, Powell KE, Hutwagner L, Graham LM, Teague WG. Impact of changes in transportation and commuting behaviors during the 1996 Summer Olympic Games in Atlanta on air quality and childhood asthma. JAMA. 2001;285(7):897–905.

Gilmour S, Degenhardt L, Hall W, Day C. Using intervention time series analyses to assess the effects of imperfectly identifiable natural events: a general method and example. BMC Med Res Methodol. 2006;6:16.

Greenland S, Brumback B. An overview of relations among causal modelling methods. Int J Epidemiol. 2002;31(5):1030–7.

Greven S, Dominici F, Zeger S. An approach to the estimation of chronic air pollution health effects using spatio-temporal information. J Am Stat Assoc. 2011;106(494):396–406.

Hack CE, Haber LT, Maier A, Shulte P, Fowler B, Lotz WG, Savage RE Jr. A Bayesian network model for biomarker-based dose response. Risk Anal. 2010;30(7):1037–51.

Harris AD, McGregor JC, Perencevich EN, Furuno JP, Zhu J, Peterson DE, Finkelstein J. The use and interpretation of quasi-experimental studies in medical informatics. J Am Med Inform Assoc. 2006;13(1):16–23.

Harvard Law Today (2014). http://today.law.harvard.edu/improving-the-pollution-mortality-link/

Harvard School of Public Health (2002) Press release: ban on coal burning in dublin cleans the air and reduces death rates. www.hsph.harvard.edu/news/press-releases/archives/2002-releases/press10172002.html

Health Effects Institute (HEI) (2010) Impact of improved air quality during the 1996 summer olympic games in Atlanta on multiple cardiovascular and respiratory outcomes. HEI Research Report #148. Health Effects Institute, Boston, MA. http://pubs.healtheffects.org/getfile.php?u=564

Health Effects Institute (HEI) (2013) Did the Irish Coal Bans Improve Air Quality and Health? HEI Update, Summer, 2013. http://pubs.healtheffects.org/getfile.php?u=929. Last Retrieved 1 February 2014

Helfenstein U. The use of transfer function models, intervention analysis and related time series methods in epidemiology. Int J Epidemiol. 1991;20(3):808–15.

Henneman LR, Liu C, Mulholland JA, Russell AG. Evaluating the effectiveness of air quality regulations: a review of accountability studies and frameworks. J Air Waste Manag Assoc. 2017;67(2):144–72. https://doi.org/10.1080/10962247.2016.1242518.

Hernan MA, Robins JM. Causal inference. London: Chapman and Hall/VCFC Press; 2011.

Hill AB. The environment and disease: association or causation? Proc R Soc Med. 1965;58(5): 295–300.

Horgan-Jones J (2019) Failure to ban smoky coal endangering citizens' lives - lobby group Clean Air Alliance urges Government to put public health ahead of 'vested interests' The Irish Times. https://www.irishtimes.com/news/ireland/irish-news/failure-to-ban-smoky-coal-endangering-citizens-lives-lobby-group-1.3921130

Imberger G, Vejlby AD, Hansen SB, Møller AM, Wetterslev J. Statistical multiplicity in systematic reviews of anesthesia interventions: a quantification and comparison between Cochrane and non-Cochrane reviews. PLoS One. 2011;6(12):e28422.

Independent Particulate Matter Review Board. The need for a tighter particulate-matter air-quality standard. NEJM. 2020;11:9.

Ioannidis JPA. Why most published research findings are false. PLoS Med. 2005;2(8):e124. https://doi.org/10.1371/journal.pmed.0020124.

Kelly F, Armstrong B, Atkinson R, Anderson HR, Barratt B, Beevers S, Cook D, Green D, Derwent D, Mudway I, Wilkinson P, HEI Health Review Committee. The London low emission zone baseline study. Res Rep Health Eff Inst. 2011;163:3–79.

King G, Nielsen R. Why propensity scores should not be used for matching. Polit Anal. 2019;27:4.

Künzli N, Tager IB. The semi-individual study in air pollution epidemiology: a valid design as compared to ecologic studies. Environ Health Perspect. 1997;105(10):1078–83.

Lamm SH, Hall TA, Engel E, White LD, Ructer FH. PM 10 particulates: are they the major determinant in pediatric respiratory admissions in Utah County, Utah (1985-1989)? Ann Occup Hyg. 1994;38:969–72.

Lehrer J (2012) Trials and errors: why science is failing us. Wired. http://www.wired.co.uk/magazine/archive/2012/02/features/trials-and-errors?page=all

Lepeule J, Laden F, Dockery D, Schwartz J. Chronic exposure to fine particles and mortality: an extended follow-up of the Harvard six cities study from 1974 to 2009. Environ Health Perspect. 2012;120(7):965–70.

Lipsitch M, Tchetgen Tchetgen E, Cohen T. Negative controls: a tool for detecting confounding and bias in observational studies. Epidemiology. 2010;21(3):383–8.

Maclure M. Taxonomic axes of epidemiologic study designs: a refutationist perspective. J Clin Epidemiol. 1991;44(10):1045–53.

Moore KL, Neugebauer R, van der Laan MJ, Tager IB. Causal inference in epidemiological studies with strong confounding. Stat Med. 2012;31(13):1380–404. https://doi.org/10.1002/sim.4469.

NHS (2012) Air pollution 'kills 13,000 a year' says study. www.nhs.uk/news/2012/04april/Pages/air-pollution-exhaust-death-estimates.aspx
Ottenbacher KJ. Quantitative evaluation of multiplicity in epidemiology and public health research. Am J Epidemiol. 1998;147:615–9.
Pearl J. Letter to the editor: remarks on the method of propensity score. Stat Med. 2009;28:1415–24.
Pelucchi C, Negri E, Gallus S, Boffetta P, Tramacere I, La Vecchia C. Long-term particulate matter exposure and mortality: a review of European epidemiological studies. BMC Public Health. 2009;9:453.
Pope CA. Respiratory disease associated with community air pollution and a steel mill, Utah Valley. Am J Public Health. 1989;79(5):623–8.
Powell H, Lee D, Bowman A. Estimating constrained concentration–response functions between air pollution and health. Environmetrics. 2012;23(3):228–37.
Rodríguez De Gil P, Bellara AP, Lanehart RE, Lee RS, Kim ES, Kromrey JD. How do propensity score methods measure up in the presence of measurement error? A Monte Carlo study. Multivar Behav Res. 2015;50(5):520–32. https://doi.org/10.1080/00273171.2015.1022643.
Rothman KJ, Greenland S. Causation and causal inference in epidemiology. Am J Public Health. 2005;95(Suppl 1):S144–50.
Sarewitz D. Beware the creeping cracks of bias. Nature. 2012;485:149.
Savitz DA. A Niche for ecologic studies in environmental epidemiology. Epidemiology. 2012;23(1):53–4.
Stebbings JH Jr. Panel studies of acute health effects of air pollution. II. A methodologic study of linear regression analysis of asthma panel data. Environ Res. 1978;17(1):10–32.
The Economist (2013). Trouble at the Lab: scientists like to think of science as self-correcting. To an alarming degree, it is not. www.economist.com/news/briefing/21588057-scientists-think-science-self-correcting-alarming-degree-it-not-trouble
Vitolo C, Scutari M, Ghalaieny M, Tucker A, Russell A. Modeling air pollution, climate, and health data using Bayesian Networks: a case study of the English regions. Earth Space Sci. 2018;5:76–88. https://doi.org/10.1002/2017EA000326.
Ward AC. The role of causal criteria in causal inferences: Bradford Hill's aspects of association. Epidemiol Perspect Innov. 2009;6:2.
Wei Y, Wang Y, Wu X, Di Q, Shi L, Koutrakis P, Zanobetti A, Dominici F, Schwartz JD. Causal effects of air pollution on mortality in Massachusetts. Am J Epidemiol. 2020;125(1):23–9. https://doi.org/10.1093/aje/kwaa098.
Wittmaack K. The big ban on bituminous coal sales revisited: serious epidemics and pronounced trends feign excess mortality previously attributed to heavy black-smoke exposure. Inhal Toxicol. 2007;19(4):343–50.
Yang Y, Li R, Li W, Wang M, Cao Y, Wu Z, Xu Q. The association between ambient air pollution and daily mortality in Beijing after the 2008 Olympics: a time series study. PLoS One. 2013;8:10.
Yim SH, Barrett SR. Public health impacts of combustion emissions in the United Kingdom. Environ Sci Technol. 2012;46(8):4291–6.
Zeger SL, Dominici F, McDermott A, Samet JM. Mortality in the Medicare population and chronic exposure to fine particulate air pollution in urban centers (2000-2005). Environ Health Perspect. 2008;116(12):1614–9. https://doi.org/10.1289/ehp.11449.

Chapter 18
Improving Causal Determination

Introduction

Chapter 1 noted that expert judgments about causality are widely used in current regulatory risk assessment and policy making. They are often expressed within a weight-of-evidence (WoE) framework, with causal determination categories being used to summarize huge amounts of complex evidence and to help inform and drive major regulatory decisions. This chapter takes a closer look at causal determination categories and judgments. It identifies some limitations of current practices in regulatory risk assessment, and suggests possibilities for making them more useful to risk analysts and policy makers. In current practice, the causal determination categories used typically cover a relatively narrow range (e.g., from "causal relationship," "likely to be a causal relationship," or "suggestive of a causal relationship" to "inadequate to infer a causal relationship" and "not likely to be a causal relationship"). Other categories, such as "not a causal relationship" or "likely to not be a causal relationship," are omitted entirely. The first part of this chapter notes that a few categories cannot encode most of the wide variations in evidence about risks and causal exposure-response relationships found in both theory and practice (Chap. 17). "Causal relationship" is usually left undefined, and may be interpreted very differently by different people, especially since there are several quite different technical concepts of causation (Chap. 9). Whether it refers to direct, indirect, or total causal effects is seldom specified. Drawing on the technical methods reviewed in Chap. 9 and illustrated in subsequent chapters, propose that causal partial dependence plots (PDPs) of predicted risk against exposure, calculated from conditional probability tables (CPTs) or models that satisfy an empirically testable condition of invariant causal prediction (ICP) across studies, can provide much more useful and clearly defined information to decision-makers. This alternative framework treats causal relationships, and evidence about them, as continuous and quantitative rather than categorical and qualitative. This is not only advantageous for clarity and realism, but it encourages better use of data and scientific method,

L. A. Cox Jr., *Quantitative Risk Analysis of Air Pollution Health Effects*,
International Series in Operations Research & Management Science 299,
https://doi.org/10.1007/978-3-030-57358-4_18

including applying independently verifiable tests to inform conclusions about how and whether changes in exposures would change individual and population health risks.

Scientific Judgment and Causal Determinations

Current regulatory risk assessments in the United States and Europe rely heavily on scientific judgment. For example (all emphases added),

- In 2018, the United States National Institute of Occupational Safety and Health (NIOSH), in explaining how regulatory risk assessments should be done, stated that "*Furthermore, **data from most studies are imperfect and potentially incomplete; therefore, models may require a number of assumptions based on scientific judgment***. ... The choice of modeling approach can markedly influence risk estimates. Moreover, ***limitations in available data often require scientific judgment in order to fill gaps in model specifications***." (NIOSH 2018).
- The United States Environmental Protection Agency (EPA), in its preface to a Draft Integrated Science Assessment (ISA) on Particulate Matter released for public comment in 2018, stated that "*In the evaluation of the evidence **determinations are made about causation, not just association, and are based on judgments** of aspects such as the consistency of evidence within a discipline, coherence of effects across disciplines, and biological plausibility of observed effects as well as related uncertainties.*" (US EPA 2018)
- In Europe, "***A group of experts** convened by WHO Regional Office for Europe reviewed the accumulated primary research evidence together with some commissioned reviews and **recommended concentration–response functions** for air pollutant–health outcome pairs for which there was sufficient evidence for a causal association. ... The project results **provide the scientific basis for formulating policy actions** to improve air quality **and thereby reduce the burden of disease associated with air pollution** in Europe.*" (Héroux et al. 2015)

At present, reliance on expert judgment is especially important in deciding whether and to what extent observed exposure-response associations warrant a causal interpretation and a belief that reducing exposures will reduce burdens of disease.

Table 18.1 shows the weight-of-evidence (WoE) framework used by the United States Environmental Protection Agency (US EPA) as part of a "structured and transparent process for evaluating scientific information and determining the causal nature of relationships between air pollution exposures and health effects" (US EPA 2018). The following sections scrutinize this WoE framework. They argue that its classification criteria are ambiguous, its category boundaries are not well defined, and its conclusions lack testable implications and cannot easily be independently verified or tested empirically. In these respects, the WoE framework departs from the principles for sound science outlined at the beginning of Chap. 1. However, such

Table 18.1 US EPA's Weight of evidence framework for causality determinations

Causal determination category	Health effects evidence
Causal relationship	Evidence is sufficient to conclude that there is a causal relationship with relevant pollutant exposures (e.g., doses or exposures generally within one to two orders of magnitude of recent concentrations). That is, the pollutant has been shown to result in health effects in studies in which chance, confounding, and other biases could be ruled out with reasonable confidence. For example: (1) controlled human exposure studies that demonstrate consistent effects, or (2) observational studies that cannot be explained by plausible alternatives or that are supported by other lines of evidence (e.g., animal studies or mode of action information). Generally, the determination is based on multiple high-quality studies conducted by multiple research groups
Likely to be a causal relationship	Evidence is sufficient to conclude that a causal relationship is likely to exist with relevant pollutant exposures. That is, the pollutant has been shown to result in health effects in studies where results are not explained by chance, confounding, and other biases, but uncertainties remain in the evidence overall. For example: (1) observational studies show an association, but copollutant exposures are difficult to address and/or other lines of evidence (controlled human exposure, animal, or mode of action information) are limited or inconsistent, or (2) animal toxicological evidence from multiple studies from different laboratories demonstrate effects, but limited or no human data are available. Generally, the determination is based on multiple high-quality studies
Suggestive of, but not sufficient to infer, a causal relationship	Evidence is suggestive of a causal relationship with relevant pollutant exposures but is limited, and chance, confounding, and other biases cannot be ruled out. For example: (1) when the body of evidence is relatively small, at least one high-quality epidemiologic study shows an association with a given health outcome and/or at least one high-quality toxicological study shows effects relevant to humans in animal species, or (2) when the body of evidence is relatively large, evidence from studies of varying quality is generally supportive but not entirely consistent, and there may be coherence across lines of evidence (e.g., animal studies or mode of action information) to support the determination
Inadequate to infer a causal relationship	Evidence is inadequate to determine that a causal relationship exists with relevant pollutant exposures. The available studies are of insufficient quantity, quality, consistency, or statistical power to permit a conclusion regarding the presence or absence of an effect
Not likely to be a causal relationship	Evidence indicates there is no causal relationship with relevant pollutant exposures. Several adequate studies, covering the full range of levels of exposure that human beings are known to encounter and considering at-risk populations and lifestages, are mutually consistent in not showing an effect at any level of exposure

Source: US EPA (2015a)

causal determination systems can be improved by recognizing that evidence for causation is quantitative rather than qualitative and by applying the traditional (hypothetico-deductive) scientific method to observations from multiple studies to make testable (and potentially falsifiable) quantitative predictions about how changes in exposure concentrations change risks of health effects. We propose that the recently articulated principle of invariant causal prediction (ICP) (Chap. 9, Peters et al. 2016; Heinze-Deml et al. 2018a) provides a useful framework for assessing evidence of causation across multiple studies and diverse bodies of evidence.

Some Limitations of Causal Determination Categories

Ambiguous Categories

Ideally, a set of causal determination categories could be used to classify available evidence to reveal its implications for whether reducing exposure would reduce health risks. They might also seek to provide information about the sizes of effects, how certain or uncertain they are, the exposure levels at which they occur, and their homogeneity or heterogeneity within and across exposed populations. Table 18.2 summarizes nine factors considered in assigning evidence to the top two categories in Table 18.1. The categories in Table 18.1 (including the top two) are not mutually exclusive. This makes the selection of any particular category somewhat arbitrary if more than one describes available evidence equally well. For example, consider a pollutant-health effect association for which the literature has mixed evidence, with all of the following descriptions being true:

1. Multiple high-quality observational studies show a positive association. However, copollutant exposures are difficult to address, and other lines of evidence are inconsistent or unavailable.
2. Several other adequate studies, covering the full range of levels of exposure that human beings are known to encounter and considering at-risk populations and lifestages, are mutually consistent in not showing an effect at any level of exposure (perhaps after conditioning on other information such as historical trends and potential confounders).
3. Controlled human exposure studies do demonstrate consistent effects. However, they do so only for unrealistically high exposures, and not for realistic exposures.

According to Table 18.1, condition (1) warrants a label of "likely to be a causal relationship" for the observed exposure-response association; condition (2) implies a label of "not likely to be a causal relationship;" and condition (3) exemplifies a "causal relationship." Such mixed evidence is common in practice. If the totality of evidence is compatible with multiple causal determination categories, due to such mixed evidence across different high-quality studies, then selecting

Table 18.2 Factors considered in Table 18.1 in determining that an exposure-response association is causal or likely to be causal

Exposure-response association is observed based on multiple high-quality studies
Conducted by multiple research groups
At relevant pollutant exposures
Chance ruled out with reasonable confidence
Confounding ruled out with reasonable confidence
Other biases ruled out with reasonable confidence
Copollutant exposures adequately addressed
Other lines of evidence (controlled human exposure, animal, or mode of action information) are available and consistent
Human data are available (e.g., clinical, controlled exposure)

any single category to characterize the evidence, as is done in practice (US EPA 2018), does not reveal the full range of categories that are compatible with it.

The five category labels on the left side of Table 18.1 are asymmetric. They allow a determination of "causal relationship" but not one of "not a causal relationship;" of "likely to be a causal relationship" but not "likely to not be a causal relationship" (presumably a stronger conclusion than "not likely to be a causal relationship"); and of "suggestive of, but not sufficient to infer, a causal relationship" but not "suggestive of, but not sufficient to infer, not a causal relationship." The difficulty of proving a negative does not explain this asymmetry: showing that an observed association is *not* explained by unmeasured confounders or other threats to validity of causal inference (Campbell and Stanley 1963) in determining that there is a "causal relationship" Is not necessarily easier than showing that it *is* explained by confounding, biases, or other threats to validity in arriving at a conclusion of "not a causal relationship."

Logical relationships among the subset of categories in Table 18.1 are left to the imaginations of users. Is evidence that is "inadequate to infer a causal relationship" a subset or a superset (or neither) of evidence that is "not likely to be a causal relationship"? To what extent does either overlap with "suggestive of, but not sufficient to infer, a causal relationship"? Such questions have never been answered in over a decade of using the causal determination categories in Table 18.1. Thus, the practical use of these categorical labels does not appear to be to communicate about real world health consequences implied by exposures, but to encode implicit policy judgments by assigning the labels "causal relationship" or "likely to be a causal relationship" to exposure-response relationships for substances that are to be considered for additional regulation.

Missing Definitions: What Does "Causal Relationship" Signify?

The nine factors in Table 18.2 do not address *whether or to what extent available evidence supports the hypothesis that reducing exposures would reduce risks of adverse health effects* (interventional causation) (Chap. 9 and Li et al. 2017), since they do not consider changes in exposure, changes in health effects, or the extent to which the latter depend on—or would not occur without—the former. Table 18.1 equates "chance, confounding, and other biases could be ruled out with reasonable confidence" with "causal relationship," but the two are very different. (The examples in Table 18.1 make clear that the promising-sounding, if somewhat circular, requirement that "the pollutant has been shown to result in health effects" entails nothing more than the conditions in Table 18.2.) For example, suppose (for purposes of discussion only) that air pollution levels and heart attack rates are both declining in a developed country even though neither causes the other; that their decreases have no common causes, i.e., there are no confounders; and that both human and animal data show effects at unrealistically high doses in controlled chamber experiments. Then all nine of the factors in Table 18.2 might hold at the "yes" or "adequate" level, leading Table 18.1 to classify the pollution-heart attack association as a "causal relationship," even though, by hypothesis, it is not.

The considerations on the right side of Table 18.1 do not address standard threats to internal validity (i.e., is the inference sound that exposure causes adverse effects in particular studies?), such as the "history" threat that exposure and effect might both have declining trends even if neither causes the other; nor do they address external validity (i.e., can causal conclusions be generalized beyond the original studies?) (Chap. 9; Campbell and Stanley 1963). Yet, both must be addressed to discover predictively valid general causal conclusions from observational data. Conversely, suppose (unrealistically, but to make a point) that outdoor air pollution levels on high pollution days in a certain city are promptly lethal; knowing this, no one goes outside on such days, and so no excess deaths are seen. Then none of the factors in Table 18.2 might hold, despite the strong direct causal relationship between exposure and mortality risk. The considerations in Tables 18.1 and 18.2 do not determine whether preventing or reducing exposures would prevent or reduce adverse effects; in this sense, they do not determine causation.

Discrete Classification of Continuous Variables Creates Vague and Uninformative Categories

Both exposure and response probability, as well as other common confounders and modifiers of health risk, such as age, income and lagged daily temperature extremes, are continuous variables. The causal determination categories in Table 18.1 are discrete; hence, there is some arbitrariness and loss of information when they are used to classify evidence about underlying continuous variables. For example, consider the following template, in which exposure shortens the life expectancy of some people but not others:

- Exposure increases mortality rates for individuals with a rare gene polymorphism, but has no effect on anyone else.

- Frequency: The polymorphism occurs in a fraction *X* of live births in the exposed population.
- Severity: The effect of exposure on people with this polymorphism is to shorten life expectancy by *Y* years if their income is less than *Z*; otherwise it has no effect.

Should the values of continuous parameters *X*, *Y*, and *Z* matter in determining whether to apply the "causal relationship" designation to exposure and mortality risk? If the answer is no, then the designation would not distinguish between an exposure that shortens life expectancy by arbitrarily short amounts (e.g., one trillionth of a second) for arbitrarily few or even non-existent people (e.g., if $X = 1$ trillionth of live births and $Z < \$100$ per year) and an exposure that shortens life expectancy by decades for most people in a population. Applying the same categorical label to such disparate exposure-response relationships makes the label relatively uninformative about the risk being categorized. On the other hand, if the answer is yes, that the quantitative values of *X*, *Y*, and *Z* should affect the causal determination category used to describe the evidence, then the question arises of where to draw the boundaries—what specific values of *X*, *Y*, and *Z* should lead to one designation rather than another. Similarly, suppose that the conditional probability of an adverse health effect in an exposed individual has the simple form $Risk \approx q*Exposure$ at low levels of exposure, where *Risk* = probability of the adverse health effect developing over an interval of time, *Exposure* = average exposure concentration, and $q > 0$ is a positive slope factor. Suppose that $w\%$ of the value of q reflects a direct causal effect of *Exposure* on *Risk*, and the rest $(100—w)\%$, reflects non-causal sources of association (e.g., confounding, collider stratification, or modeling biases). Thus, if *Exposure* were to decrease by 1, this change would cause *Risk* to decrease by wq. Should the causal determination category depend on w? Again, if the answer is no, then the causal determination category will not distinguish between *de minimis* risk (if $wq \approx 0$) and very material risks (if wq is large) caused by exposure; but if the answer is yes, then the question arises of where the (necessarily somewhat arbitrary) dividing line for w should be drawn between different categories. WoE frameworks such as Table 18.1 offer no guidance on such definitional issues, so users of the frameworks must devise their own (not necessarily mutually consistent) answers, which are seldom or never explicitly stated. Hence, agreement on a causal determination category for a body of evidence need not signify agreement on what it means.

Doing Better: Scientific Method, Causal Discovery, and Hypothesis Testing

What is usually considered to differentiate *scientific* evidence from other types of evidence—expert opinions, confirming anecdotes and observations, consensus views from respected authorities, long-held traditional beliefs, personal convictions and willingness to bet, and so forth—is the possibility of *independently empirically testing* (and potentially falsifying, if they are false) the claims made. Scientific content consists of *predictive generalizations* (regarded as scientific hypotheses that can be tested by comparing their predictions to observations), together with empirical evidence for and against them. This evidence commonly consists of statistical tests of null hypotheses that predicted and observed data values are not significantly different. Table 18.3 summarizes some null hypotheses and statistical tests from Chap. 9 that often used in modern causal discovery and modeling software; see Chap. 9 for further discussion. In general, scientific predictions are formulated using clear operational definitions, so that others can independently verify the test results, given the data. The data needed to test scientific hypotheses about exposure-response relationships can come from many sources and diverse studies, including toxicology, controlled human trials, and natural experiments or quasi-experiments in which exposures are reduced by different amounts for different people, and subsequent changes in risks are observed and compared to predicted changes.

In recent decades, the traditional scientific method just described has been increasingly assisted in certain application areas, such as systems biology, by the use of causal discovery algorithms and software that automate the tasks of identifying and testing scientific hypotheses that are consistent with observed data (Lagani et al. 2016). These methods are discussed in Chap. 9 and illustrated in the Bayesian networks and causal partial dependence plots in other chapters. To illustrate the process, Table 18.4 shows the first 14 of 158 records of a data set on air pollution and mortality. The variables, from left to right, are monthly average fine particulate matter (PM2.5) concentrations (in micrograms per cubic meter); daily average, maximum, and minimum temperatures; dew point; average daily mortality counts for people over 18, people 18 to 74, and people 75 or older; and year and month. (The full data set is bundled with, and can be downloaded from, the Causal Analytics Toolkit (CAT) free in-browser causal discovery software at http://cloudcat.cox-associates.com:8899/; it is the "Boston.csv" data set, discussed in Cox Jr. 2017.)

From these data, and from background knowledge, an unaided human investigator might formulate and test various more-or-less plausible hypotheses, such as that elderly mortality rates (*mortality_75*) depend on month of the year (perhaps being higher during cold winter months) and increase with PM2.5. A mental model consistent with this hypothesis is the following:

$$PM2.5 \rightarrow mortality\_75 \leftarrow month,$$

where any interaction between *month* and *PM2.5* in affecting the conditional probability distribution of *mortality_75* would be captured in a conditional probability table (CPT) for *mortality_75*. To bring empirical data from Table 18.4 into this framework, Fig. 18.1 shows how the average value of *mortality_75* (vertical axis) depends on the value of *PM2.5* (horizontal axis), averaging over the empirical conditional frequency distribution of *months* (given PM2.5in the data set. (The CAT software automatically generates this "partial dependence plot" (PDP), representing the "natural direct effect" of *PM2.5* on *mortality_75*, using a non-parametric model

Table 18.3 Methods for testing consistency of data with a causal dependency of response variable *Y* on explanatory variable (e.g., exposure) *X* (see Chap. 9 for details)

Criterion	Test of null hypothesis to provide evidence on whether exposure X causes response Y	Methods and software
Mutual information	Intuition: causes are informative about their effects Test: Reject null hypothesis that response *Y* is conditionally independent of exposure *X*, given the values of other variables	Tests for conditional independence, e.g., chi-squared tests, tests in statistics packages such as *bnlearn* in R (which connects two variables by an arrow if they are informative about each other)
Directed dependence	Intuition: information flows from causes to effects over time Granger causality test for longitudinal data: Reject null hypothesis that future values of *Y* are conditionally independent of past values of *X* after conditioning on past values of *Y* For cross sectional data: Reject null hypothesis that direction of dependence is undetermined by data	Reject null hypothesis if past/present *X* values point to future *Y* values in a dynamic Bayesian network (DBN) or dynamic information graph (DIG) learned from data (e.g., via the *bnstruct* package). Granger tests can also be used for time series data Reject null hypothesis if arrow directions can be identified from data (e.g., using LiNGAM, BACKSHIFT, or Simon-Iwasaki causal ordering)
Internal consistency	Do not reject null hypothesis that effects of *X* on *Y* (e.g., partial dependence plots of *Y* vs. *X*) estimated from different adjustment sets are significantly different	Reject null hypothesis if confidence bands for effects estimated from different adjustment sets do not overlap
External consistency (ICP)	Do not reject null hypothesis that response CPTs estimated from different studies or data sets are the same	Reject null hypothesis if the study identifier is a parent of response *Y* when data are pooled across studies
Causal coherence and biological plausibility	Reject the null hypothesis that identified biologically plausible pathway(s) directed from *X* to *Y* cannot explain the observed dependence of *Y* on *X*	Reject the null hypothesis if one or more biologically plausible coherent causal explanations (pathways in a causal biological network) are identified that explain the dependence of *Y* on *X*
Causal mediation confirmation	Do not reject the null hypothesis that variations in *Y* caused by variations in *X* are explained by resulting variations in mediating variables	Reject null hypothesis if variations in mediating variables do not explain variations in *Y* for different values of *X*
Refutation of non-causal explanations	Reject the null hypothesis that the observed statistical dependence of *Y* on *X* has a non-causal explanation	Reject the null hypothesis if threats to validity are refuted by data (Campbell and Stanley 1963)

Table 18.4 Excerpt from example data set

PM2_5	Tavg	Tmax	Tmin	Dewpoint	mortality_18	mortality_18to74	mortality_75	Year	Month
16.44	28.10	36.27	19.30	14.35	16.71	6.06	10.65	2000.00	1.00
16.74	34.41	41.41	26.90	22.45	15.17	5.38	9.79	2000.00	2.00
13.84	43.55	51.06	35.55	30.74	14.42	5.97	8.45	2000.00	3.00
12.07	47.57	54.40	40.23	36.63	13.77	5.60	8.17	2000.00	4.00
10.25	57.42	65.90	48.52	47.23	13.13	4.90	8.23	2000.00	5.00
13.42	66.65	74.50	58.23	57.65	13.33	5.70	7.63	2000.00	6.00
12.79	70.19	77.06	62.84	59.32	12.61	5.35	7.26	2000.00	7.00
14.20	69.90	77.10	63.43	62.10	13.10	5.19	7.90	2000.00	8.00
12.17	63.77	71.63	55.37	55.30	13.50	5.57	7.93	2000.00	9.00
13.14	54.16	61.68	46.03	44.42	13.74	5.39	8.35	2000.00	10.00

Source: http://cloudcat.cox-associates.com:8899/ "Boston.csv" data set

Variables are daily average fine particulate matter concentration (PM2_5) in micrograms per cubic meter; average, maximum, and minimum temperatures (Tavg, Tmax, Tmin), dewpoint (Dewpoint), mortality counts for people over 18 (mortality_18), 18 to 74 (mortality_18to74), and 75 or older (mortality_75), as well as the year and month ensemble (random forest) to predict the conditional expected value of *mortality_75* for each value of *PM2.5*, given the empirical frequency distribution of *month* in the data set.)

A Spearman's rank correlation test confirms that the estimated average value of elderly mortality increases significantly with PM2.5 in Fig. 18.1 ($p < 0.0001$).

To complement this hypothesis-driven analysis, Fig. 18.2 shows a result from computer-aided analysis of the same data set using the CAT causal discovery software; as explained in Chap. 9, in such a Bayesian network (BN), and two variables are connected by an arrow if and only if the null hypotheses is rejected that they are conditionally independent, given the values of other variables. (In this analysis, *month* is treated as a categorical variable, and *month* and *year* are specified as input variables for purposes of determining arrow directions.) This diagram indicates that the positive association between PM2.5 and elderly mortality in Fig. 18.1 is not explained by a direct effect of PM2.5 on mortality (there is no arrow between them, violating the mutual information criterion at the top of Table 18.3). Rather, both depend on year. Further investigation shows that both declined significantly between 2000 and 2013, the years in the data set (creating a "history" threat to valid causal inference based on the positive association between them (Campbell and Stanley 1963)). Thus, in this analysis, the data set examined does not provide evidence supporting a hypothesis of a direct causal relationship between PM2.5 and elderly mortality, notwithstanding the clear positive association between them in Fig. 18.1. As mentioned in Chap. 17, similar results using a much larger data set have been

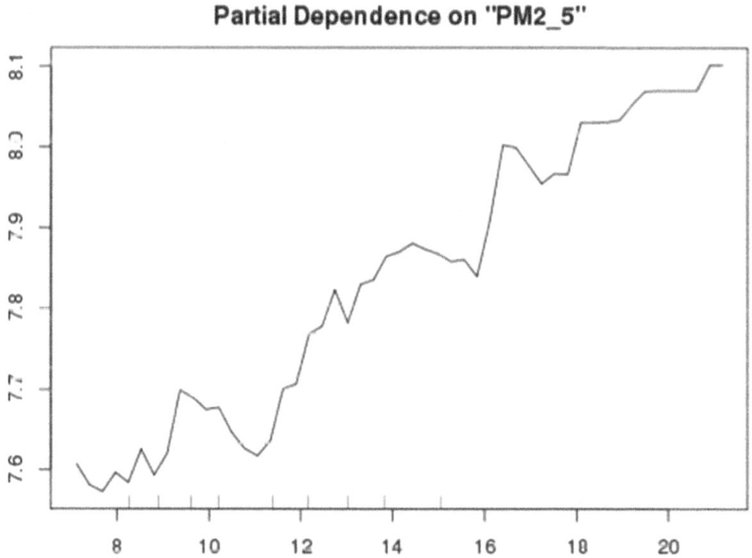

Fig. 18.1 Partial dependence plot (PDP) for dependence of average elderly mortality (*mortality_75*) value on PM2.5 value, averaging over the empirical frequency distribution of *month*

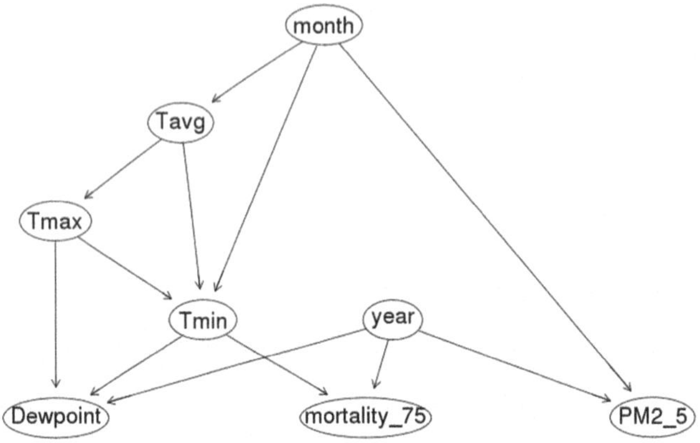

Fig. 18.2 Bayesian network (BN) for the data set in Table 18.4

reported in the United Kingdom: PM2.5 (and ozone) and cardiovascular mortality rates among people over 65 both depend on region and year, but are not found to be directly dependent on each other (Vitolo et al. 2018).

Causal discovery algorithms complement traditional (hypothetico-deductive) scientific method by not requiring an investigator to formulate specific hypotheses to test. Rather, they begin with data, and then seek to identify models of dependencies among variables that are consistent with the data, independent of any favored (or opposed) starting hypotheses. In this sense, causal discovery algorithms avoid the perils of confirmation bias, anchoring, over confidence, p-hacking, and so forth that often distort human scientific judgment (Dingledine 2018). Ideally, causal discovery software can help scientists notice patterns and discover and explain useful predictive generalizations that might otherwise escape detection, while avoiding or reducing biases imposed by preconceptions. Expert judgments about associations such as those in Fig. 18.2 often reach much stronger subjective conclusions than do causal discovery algorithms, e.g., that, in Boston, "the association between PM2.5 and deaths is almost certainly causal," even while noting that such causal interpretations depend on untested assumptions (Schwartz et al. 2015). Computer-aided methods that generate testable hypotheses from models (e.g., Textor et al. 2011) and that automatically test them and report the results can help to reduce such over-confidence and identify conclusions that the data alone warrant, independent of the hopes or untested assumptions and beliefs of investigators.

These modern causal discovery methods have been much less applied than the WoE framework for causal determination. They might therefore be regarded with prudent skepticism unless and until they are further tested and validated in important applications. On the other hand, expert judgment usually compares unfavorably, in multiple important application domains, to more algorithmic methods in terms of performance metrics such as predictive accuracy, in part because of well established heuristics and biases that impair individual and collective expert judgment under uncertainty (Camerer and Johnson 1991; Dingledine 2018; Fisher and

Keil 2018; Sarewitz 2012). Thus, fields such as evidence-based medicine rank even the best consensus judgments of well respected authorities as weaker than even relatively weak data-based evidence (Tenny and Varacallo 2018).

Given these strengths and limitations of algorithmic and judgment-based methods, we propose the following steps to help overcome the limitations of causal determination categories and provide more valuable information to policy makers.

1. *Specify the causal question to be answered.* The questions of greatest policy relevance are usually about effects caused by interventions, e.g., "How would the risk of adverse health effect(s) change if exposure were reduced by different amounts?" To specify this question in adequate detail, it is necessary to specify what is to be assumed about the levels of other variables that affect risk. For example, if a regulated pollutant has both a direct effect on health risk and also an indirect effect mediated by a co-pollutant, is the causal question of interest how risk would change if the pollutant level is reduced by various amounts *holding the level of the co-pollutant fixed* (and, if so, at what level?), or is it how risk would change if the pollutant level is reduced by various amounts, *allowing the level of the co-pollutant to adjust* correspondingly? There is no right or wrong technical answer to this question: it is about what the investigator wants to know. Since the answers to these distinct questions may be entirely different, clarity demands specifying which one is to be answered.
2. *Collect relevant data to answer the question.* Two key technical constructs for answering counterfactual questions about how risk would change if exposure were changed are as follows (Heinze-Deml et al. 2018b):

    a. *Conditional probability tables* (CPTs) (or other conditional probability models, which we refer to generically as CPTs). These give the conditional probability distributions for the response variable(s) of interest as a function of the values of their direct causes (i.e., their parents in a causal network model such as Fig. 18.2).
    b. The *invariant causal prediction* (ICP) property. This is the property that the conditional probability distribution of a variable is determined by the values of its direct causes and is the same (i.e., "invariant") across different settings, experiments, or interventions (Peters et al. 2016). ICP expresses the idea that causal laws are universal and that they can be used to predict the effects of interventions (where "effect" is conceptualized as a change in the probability distribution of observed outcomes).

    To *answer* causal questions for policy makers, it is typically necessary to collect data from which CPTs satisfying ICP can be estimated. Causal discovery software packages (such as *bnlearn* in R; see Chap. 9) automatically estimate CPTs from data consisting of observed joint values of variables in multiple cases, e.g., for many individuals in an exposed population. Study designs can facilitate estimation of CPTs by collecting observations for responses at different levels of causally relevant factors (e.g., the parents of the response variable), and algorithms are available to automatically determine which direct and total effects can be estimated from observations of different sets of variables in a graph model such as Fig. 18.2 (Textor et al. 2011).

3. *Test the invariant causal prediction (ICP) property for the estimated response CPT.* CPTs satisfying ICP provide predictive generalizations of observations from past studies needed to predict exposure-response relationships for new policy interventions under new conditions. The CPT for a health response enables counterfactual *prediction* by the straightforward process of conditioning on assumed values of its parents (direct causes). The ICP property enables *generalization* of CPTs to new situations by identifying CPTs that do not depend on (i.e., are invariant across) the studies and conditions used to collect data. Tests for ICP, such as the invariant residual distribution test (Heinze-Deml et al. 2018a) of the null hypothesis that distributions of residuals for predictions from a CPT model fit to pooled data from multiple studies are not significantly different, are available in free statistical software (e.g., https://rdrr.io/cran/nonlinearICP/). ICP tests address the notorious problems of scientific induction and external validity of causal conclusions (i.e., generalization from past observations to future ones) by testing whether ICP holds, i.e., the null hypothesis is not rejected that the same response CPT holds in different studies, for different conditions (values of the parents) and interventions (changes in the values of some of the parents). If ICP holds for a response CPT across a variety of studies, then the response CPT is tentatively accepted as a predictive generalization, or scientific hypothesis, supported by the data used to test it. It can be used to predict response risks for new situations (joint distributions of causes of the response) and interventions that set the values of one or more of these causes. On the other hand, if the ICP hypothesis is rejected, this failure is also informative: it suggests that additional variables affecting the response probabilities remain to be identified. Techniques are available to estimate and bound the effects of such unmeasured (latent) variables (Albert et al. 2016; Kim et al. 2018; Salway et al. 2010 for time series). Finally, ICP tests constructively address the practical challenge of *model-dependent conclusions* in regulatory risk assessments, in which predictions of how changing exposures would change risks depend on modeling assumptions. The ICP perspective is that models (CPTs) are not used to make counterfactual predictions about how different exposures would affect health risks until they have been validated by ICP tests across multiple studies.
4. *Use data and tested models (response CPTs satisfying ICP) to calculate answers to the causal question of interest and to characterize remaining uncertainties.* The answers to causal questions can be succinctly displayed via partial dependence plot (PDP) curves similar to the one in Fig. 18.1, with exposure values on the horizontal axis and corresponding predicted conditional expected values of the response variable on the vertical axis. More generally, current software estimates *conditional distributions* of the response variable to be estimated and uncertainty bands corresponding to user-specified quantiles to be included in the PDP plot. Unlike a regression curve through a scatterplot, a PDP curve calculated from CPTs satisfying the ICP property has a clear causal

interpretation: that changing the exposure (while holding other variables fixed or allowing them to vary, as specified in the causal question being addressed) causes the predicted conditional distribution of the response to change as shown by the curve.

Discussion and Conclusions

Causal relationships between exposure and response risks are usually quantitative and continuous. We propose that they are therefore better described by continuous curves, as in Fig. 18.1 (perhaps augmented with confidence bands), rather than by discrete causal determination categories as in Table 18.1. This avoids a number of conceptual and definitional challenges, including the previously discussed ambiguity and arbitrariness of category boundaries. Modern computational statistical methods for causal inference quantify such causal partial dependence plot (PDP) curves, provided that sufficiently plentiful and diverse data are available to estimate CPTs satisfying ICP. They can automatically identify effects that can and cannot be estimated from available observations (Shpitser and Pearl 2008) and compute appropriate adjustment sets and instrumental variables for estimating those effects that can be identified, as well as calculating the empirically testable implications of causal models represented as directed acyclic graphs (Textor et al. 2011) They can also quantify direct and total causal exposure-response curves for different user specifications about which other variables are to be held fixed, and at what levels (Chap. 9; Heinze-Deml et al. 2018b). Individual risks are quantified by specifying all of the causally relevant covariates for an individual (e.g., age, sex, co-morbidities) that are parents of the response CPT, and then tracing how conditional response probabilities change with exposure. Population risks are quantified by allowing individual covariate values to be sampled from their empirical joint distribution in a population, as in interventional effect estimation (Vansteelandt and Daniel 2017).

A powerful advantage of causal determination categories is that they provide simple terms for describing complex evidence about causality in exposure-response associations. A disadvantage is that the meanings of the terms are not well defined, and any small number of categories must fail to draw distinctions among very different evidence profiles. The category labels may mean quite different things to different people, and it is often unclear what type of causal effect, if any, they refer to—e.g., direct vs. total, interventional vs. attributive, and so forth. We propose that using causal graph models such as Fig. 18.2, causal exposure-response curves (e.g., partial dependence plots as in Fig. 18.1), and ICP tests to verify whether causal inferences hold consistently across multiple diverse studies, retains most of the desirable simplicity needed to communicate clearly and effectively about causal exposure-response relationships, while promoting greater clarity and shared understanding about what they mean—specifically, the types of effects (e.g., direct or total) being displayed, and the assumptions made about the levels of other variables. Such curves can potentially inform policy deliberations better than any small set of categorical labels by revealing details such as thresholds and nonlinearities and bypassing the need to apply a small number of labels to a potentially infinite variety of different relationships. Finally, we hope that wider use of empirical causal discovery, estimation of CPTs from data, and testing for the ICP property across diverse studies can help to advance the use of data-driven scientific method in regulatory risk assessments and characterizations of causality, thereby contributing to improved understanding and communication about potentially preventable health risks.

References

Albert JM, Geng C, Nelson S. Causal mediation analysis with a latent mediator. Biom J. 2016;58(3):535–48. https://doi.org/10.1002/bimj.201400124.

Camerer CF, Johnson EJ (1991) The process-performance paradox in expert judgment: how can experts know so much and predict so badly? https://authors.library.caltech.edu/22525/1/334945.pdf

Campbell DT, Stanley JC. Experimental and quasi-experimental designs for research. Boston: Houghton Mifflin Company; 1963.

Cox LA Jr. Do causal concentration-response functions exist? A critical review of associational and causal relations between fine particulate matter and mortality. Crit Rev Toxicol. 2017;47(7):603–31. https://doi.org/10.1080/10408444.2017.1311838.

Dingledine R. Why is it so hard to do good science? eNeuro. 2018;5(5):0188. https://doi.org/10.1523/ENEURO.0188-18.2018.

Fisher M, Keil FC. The binary bias: a systematic distortion in the integration of information. Psychol Sci. 2018;29(11):1846–58. https://doi.org/10.1177/0956797618792256.

Heinze-Deml C, Maathuis MH, Meinshausen N. Causal structure learning. Annu Rev Stat Appl. 2018a;5:371–91.

Heinze-Deml C, Peters J, Meinshausen N. Invariant causal prediction for nonlinear models. J Causal Inf. 2018b;6(2):16. https://doi.org/10.1515/jci-2017-0016.

Héroux ME, Anderson HR, Atkinson R, Brunekreef B, Cohen A, Forastiere F, Hurley F, Katsouyanni K, Krewski D, Krzyzanowski M, Künzli N, Mills I, Querol X, Ostro B, Walton H. Quantifying the health impacts of ambient air pollutants: recommendations of a WHO/Europe project. Int J Public Health. 2015;60(5):619–27. https://doi.org/10.1007/s00038-015-0690-y.

Kim C, Daniels M, Li Y, Milbury K, Cohen L. A Bayesian semiparametric latent variable approach to causal mediation. Stat Med. 2018;37(7):1149–61. https://doi.org/10.1002/sim.7572.

Lagani V, Triantafillou S, Ball G, Tegner J, Tsamardinos I. Probabilistic computational causal discovery for systems biology. In: Geris L, Gomez-Cabrero D, editors. Uncertainty in biology: a computational modeling approach. Studies in mechanobiology, tissue engineering and biomaterials, vol. 17. Cham: Springer; 2016. p. 33–73.

Li S, Ernest J, Bühlmann P. Nonparametric causal inference from observational time series through marginal integration. Econometr Stat. 2017;2:81–105.

NIOSH (2018) Current intelligence bulletin: NIOSH practices in occupational risk assessment external review draft. https://www.cdc.gov/niosh/docket/review/docket316/pdfs/Draft-Document-NIOSH-Practices-in-Occupational-Risk-Assessment-6-19-18.pdf

Peters J, Bühlmann P, Meinshausen N. Causal inference by using invariant prediction: identification and confidence intervals. J R Stat Soci Ser B. 2016;78(5):947–1012.

Salway R, Lee D, Shaddick G, Walker S. Bayesian latent variable modelling in studies of air pollution and health. Stat Med. 2010;29(26):2732–42. https://doi.org/10.1002/sim.4039.

Sarewitz D. Beware the creeping cracks of bias. Nature. 2012;485(7397):149. https://doi.org/10.1038/485149a.

Schwartz J, Austin E, Bind MA, Zanobetti A, Koutrakis P. Estimating causal associations of fine particles with daily deaths in Boston. Am J Epidemiol. 2015;182:644–50.

Shpitser I, Pearl J. Complete identification methods for the causal hierarchy. J Mach Learn Res. 2008;9:1941–79.

Tenny S, Varacallo M. Evidence based medicine (EBM). Treasure Island: StatPearls; 2018.

Textor J, Hardt J, Knuppel S. DAGitty: a graphical tool for analyzing causal diagrams. Epidemiology. 2011;22:745.

US EPA (2015a) Preamble to the integrated science assessments U.S. Environmental Protection Agency, Office of Research and Development, National Center for Environmental Assessment, RTP Division Location Research Triangle Park, NC Report Number EPA/600/R-15/067. https://cfpub.epa.gov/ncea/isa/recordisplay.cfm?deid=310244

US EPA (2018) Integrated science assessment for particulate matter (External Review Draft) October 2018 National Center for Environmental Assessment—RTP Division Office of Research and Development U.S. Environmental Protection Agency Research Triangle Park, NC

Vansteelandt S, Daniel RM. Interventional effects for mediation analysis with multiple mediators. Epidemiology. 2017;28(2):258–65. https://doi.org/10.1097/EDE.0000000000000596.

Vitolo C, Scutari M, Ghalaieny M, Tucker A, Russell A. Modeling air pollution, climate, and health data using Bayesian Networks: a case study of the English regions. Earth Space Sci. 2018;5:76–88. https://doi.org/10.1002/2017EA000326.

Chapter 19
Communicating More Clearly About Deaths Caused by Air Pollution

Introduction

How can scientists and risk assessors best communicate with each other, the media, the public, and policy makers what is known, what is guessed, and what is still unknown or uncertain about how changes in air pollution affect human mortality? Current practice includes emphatic pronouncements, striking headlines, and colorful infographics about deaths attributed to air pollution. Important qualifications, uncertainties, and unverified assumptions are often buried deeply in technical papers if they are stated at all. As discussed in Chap. 17, although some sensational claims about air pollution mortalities have proved to be mistaken, errors and corrections receive little or no attention from media, policy makers, or interest groups who helped to spread the original sensational claims. More encouragingly, substantial recent progress has been made in technical methods for quantifying and communicating what is known about how reducing exposure affects health risks and uncertainties. These advances make it possible to better describe and explain both what we now know and what we are still uncertain about for human health effects of interventions that reduce pollution levels. This concluding chapter reviews progress in accurate communication about deaths caused by air pollution. It discusses how advances in quantitative risk assessment and risk communication can provide a more useful basis for informing public opinion and policy deliberations—one that is more accurate and that answers more relevant questions than the infographics and sensational claims that dominate risk communication about air pollution deaths today.

To reduce the chances of being misunderstood and misreported, scientific articles should be careful to draw essential distinctions in reporting their findings about exposure-associated health risks, including distinctions between assumed or modeled risks and empirically observed risks (Chap. 17); between association and causation (Chaps. 1, 2, and 9); and between deaths per year attributed to pollution and

L. A. Cox Jr., *Quantitative Risk Analysis of Air Pollution Health Effects*, International Series in Operations Research & Management Science 299,
https://doi.org/10.1007/978-3-030-57358-4_19

deaths per year preventable by reducing pollution (Chap. 9). Many articles about air pollution heath effects in both the news media and scientific publications omit the essential caveats, detail, and clear definitions needed to inform effective risk management decisions, or even to convey accurately what has been found and what remains unknown about whether and how much reducing air pollution would change risks to human health. Examples are given shortly. Measures of statistical association or apportionment of blame are often misrepresented in both scientific and news articles as if they were with measures of actual causal impact, although the motivation for this conflation—whether sensationalism, ignorance, advocacy, or other reasons—is not always clear and might well differ between scientific and popular news sources. Distinctions between direct causes of observed changes in health effects and factors that are correlated with, but not directly causal of, those changes are often lost; conversely, distinctions between effects of air pollution *per se* and effects of factors that are correlated with air pollution (Chap. 13) are seldom drawn with precision. The true explanations for observed changes in health effects (or the lack of such changes) following changes in air pollution levels may be left obscure for decades while the changes themselves are attributed as modelers see fit and used in risk calculations and predictions that have no know validity. These errors, gaps, and ambiguities in current risk communication about air pollution can be avoided in part by applying advances in epidemiology, data science, and causal analytics developed and tested in other fields over the past half century.

The following sections discuss current gaps and opportunities to improve upon current practice in communicating more accurately about human mortality caused by air pollution. Examples of unclear communication in news stories are presented, followed by advice to authors of scientific articles to help improve the technical clarity of communication about causality. Although this addresses only part of the large challenge of communicating risk information more clearly, improved accuracy and precision at the source—the scientific articles presenting and interpreting findings—may provide some of the preconditions for improving the accuracy of downstream communications in the press and in policy advocacy.

News Stories: Bold Claims Without Essential Caveats

Headlines about the human health effects of air pollution usually emphasize the large number of deaths or other adverse consequences associated with or blamed on (i.e., "attributed" to, in epidemiological parlance) pollutants such as fine particulate matter (PM2.5) and ozone (O3). For example, it is common to claim that millions of deaths per year world-wide and in large countries are caused by, or linked to, air pollution. Examples include the following:

- March 12, 2019: "Air Pollution Causes 8.8 Million Extra Deaths a Year." www.usnews.com/news/national-news/articles/2019-03-12/air-pollution-causes-88-million-extra-deaths-worldwide-each-year-study-says

- November 16, 2018: "Air pollution causes 4 million deaths per year and restricts children's lung growth."www.technology.org/2018/11/16/air-pollution-causes-4-million-deaths-per-year-and-restricts-childrens-lung-growth/
- February 13, 2016: "Polluted air causes 5.5 million deaths a year new research says—BBC." www.bbc.com/news/science-environment-35568249
- April 18, 2018: "Air Pollution Contributed To More Than 6 Million Deaths In 2016" www.forbes.com/.../air-pollution-contributed-to-more-than-6-million-deaths-in
- May 1, 2018 "Air Pollution Kills 7 Million People a Year, WHO Reports—Bloomberg." www.bloomberg.com/news/articles/2018-05-01/air-pollution-kills-7-million-people-a-year-who-reports
- May 7, 2018: "How air pollution contributes to 8 million deaths each year—CBS News." www.cbsnews.com/news/how-air-pollution-makes-people-sick-8-million-early-deaths-each-year/
- October 20, 2017. "Air Pollution Kills 9 Million, Costs $5 Trillion Per Year—EcoWatch."www.ecowatch.com/air-pollution-deaths-2498942499.html

Such claims command attention, and perhaps alarm, but they have critical deficiencies for informing policy making to protect public health. From a scientific point of view, the claim that air pollution causes or contributes to *X* million deaths per year makes no testable, verifiable, or falsifiable predictions. There is no way to determine from facts and data whether it is true. By contrast, more technically precise statements that do make potentially testable predictions are less easy to communicate in short, interesting headlines. For example, a statement of the following form would probably have much less impact on most readers and policy makers than the above headlines, even if it were more accurate and well qualified: "A new study showed that a Poisson regression model that (a) includes PM2.5, age, and daily minimum and maximum temperatures over the past 2 days as predictors of elderly mortality; (b) excludes income and temperatures at longer lags; (c) assumes that estimates of exposure are perfectly accurate (i.e., ignores exposure estimation errors); and (d) assumes that findings from the data analyzed, for a large metropolitan area in the northeastern United States, can be applied nation-wide, predicts that, with 95% statistical confidence, reducing fine particulate air pollution by 10 μg/m^3 would reduce mortality rates for people over 70 years old by at least 1% and increase average life expectancy in the population by 1 year. Using a quasi Poisson model instead, or including income, daily maximum and minimum temperatures over the past 2 weeks, or exposure measurement errors in the model, reduced the estimated effect of fine particulate matter on elderly mortality so that it was no longer significantly different from zero." Such an elaborate statement does not make an effective headline. An abbreviated version, such as "Some models predict that reducing fine particulate air pollution could increase life expectancy by a year or more; others do not" is considerably less newsworthy than the claims made without caveats. But such cautious and qualified statements are often necessary for accuracy.

Unclear Meaning

Second, it is unclear what a statement such as "Air pollution causes *X* million extra deaths a year" actually means. One possible interpretation that might seem plausible is that, if exposure to air pollution causes *X* million extra deaths per year, then there would be *X* million deaths fewer per year in the absence of exposure. This interpretation is incorrect, however. Since each person born dies only once, if exposure causes more members of a birth cohort to die in any given year, then necessarily fewer of them die in later years. Increased deaths in some years must be exactly offset by corresponding reductions in deaths in other years. This makes it impossible for air pollution (or anything else that does not increase birth rates) to permanently increase the average number of deaths per year. Doing so would require increasing the *total* number of deaths, which is impossible for a given number of births. Harmful exposures might shorten life lengths and thus redistribute deaths among years, but they cannot increase the average *number* of deaths per year, as the above headlines suggest. (Inserting "premature" before deaths, as is sometimes done, does not help: the iron laws of arithmetic still decree that more premature deaths in one year implies fewer survivors to experience premature deaths in subsequent years.) Thus, what the headlines actually mean is not clear.

Association Presented as Causation

The explanation for this apparent violation of arithmetic in numerous headlines is that deaths-per-year attributed to a cause such as air pollution are usually estimated from statistical measures of *associations* between pollution levels and mortality rates. For example, they may be estimated from the slope of a regression line through a scatter plot of observed annual deaths vs. pollution levels. Such statistical associations are not constrained by the physical conservation laws governing numbers of deaths per year. To illustrate using a simple hypothetical example, suppose that the mortality rate for individuals in a population is given by

$$DEATHRATE = 0.01 + 0.1 * AGE * POVERTY * POLLUTION$$

where $DEATHRATE$ = expected deaths per person per year, AGE is coded as 1 for people over 70 and 0 otherwise; $POVERTY = 1$ for people with annual incomes below some level and 0 otherwise; and $POLLUTION = 1$ for people living in areas with air pollution above some level and 0 otherwise. Suppose for simplicity that there are only two types of people: elderly people who live in impoverished and polluted areas, for whom $AGE = POVERTY = POLLUTION = 1$; and people with none of these risk factors, for whom $AGE = POVERTY = POLLUTION = 0$.Then the mortality incidence rate is 0.11 deaths per person-year in the former (high-risk) subpopulation and 0.01 in the latter (low-risk) subpopulation. A standard univariate epidemiological calculation of the mortality risk attributable to air pollution yields:

$$\begin{aligned}&\textit{Attributable risk}(AR)\textit{ for mortalities due to air pollution among exposed people}\\&=\frac{(\textit{incidence rate for exposed})-(\textit{incidence rate for unexposed})}{\textit{incidence rate for exposed}}\\&=(0.11-0.01)/0.11=91\%.\end{aligned}$$

That is, 91% of deaths in the high-risk group are attributed to air pollution. (Symmetrically, 91% of the mortality risk in the high-risk subpopulation might also be attributed to *POVERTY* using analogous calculations.) Such calculations shed no light on how much, or whether, risk would be reduced by reducing the factors to which risk is attributed.

Epidemiologists understand that attributable risk calculations and similar association-based calculations for population attributable risk (PAR), population attributable fraction (PAF), probability of causation (PC), burden of disease (BOD), relative risk (RR) ratios, and so forth can readily yield attributable risks that sum to far more than 100% of deaths among risk factors that occur together. But this important nuance is usually lost in communicating to the public and policy makers messages such as that some percent (e.g., 91% in this simplified example) of mortalities are attributable to air pollution. The misconception that if *X*% of deaths are *attributable* to air pollution, then *X*% of deaths would be *prevented* if air pollution were removed results from the fact that epidemiological measures are based on association rather than causation (Chap. 9). This misconception gives exaggerated estimates of the harm that would be prevented by reducing one factor when other factors that are correlated with it contribute to the risk of harm.

Ambiguity and Conflation of Causal Concepts and Language

As discussed in Chap. 9, specialists in modern causal analysis and inference distinguish among several distinct concepts of causation. These include probabilistic (Suppes 1970), attributive (Murray and Lopez 2013; Prüss-Üstün et al. 2003; Lo et al. 2016), counterfactual/potential outcomes (Höfler 2005; Glass et al. 2013; Lok 2017; Li et al. 2017; Galles and Pearl 1998), predictive (Wiener 1956; Granger 1969; Kleinberg and Hripcsak 2011; Barnett and Seth 2014; Papana et al. 2017), structural (Simon 1953; Simon and Iwasaki 1988; Hoover 2012), manipulative (Voortman et al. 2010; Spirtes 2010; Hoover 2012; Simon and Iwasaki 1988), mechanistic (Keele et al. 2015; Imai et al. 2011; Menzies 2012; Simon and Iwasaki 1988), and but-for causation (Cox Jr. 2018). These are essentially different concepts. For example, nicotine-stained fingers might be a predictive cause of lung cancer (assuming that people acquire nicotine-stained fingers by smoking), but not a manipulative cause—unless the only way to keep fingers unstained is not to smoke. In that case, nicotine stained fingers would also be a manipulative cause, but not a mechanistic cause or a but-for cause, of lung cancer. Smoking is a but-for cause of lung cancer if lung cancer would not occur but for smoking. These and other distinctions among the different concepts of causation are discussed more thoroughly in Chap. 9 and its references.

Likewise, epidemiologists distinguish among different concepts of "effect" when multiple causal factors interact to increase risk of an adverse outcome. Returning to the simple hypothetical model *DEATHRATE* = 0.01 + 0.1**AGE***POVERTY***POLLUTION*, the effect of decreasing *POLLUTION* from 1 to 0 depends on the levels of *AGE* and *POVERTY*: it reduces the risk of mortality per person per year from 0.11 to 0.01 for someone with *AGE* = 1 and *POVERTY* = 1 but it has no effect if either of them is 0. Thus a claim about the "effect" on mortality of reducing air pollution depends on the levels of other factors. This is also true in the real world, although the interrelationships among age, income, other covariates (such as education, occupation, and residential location) and health risks are more complex than in this simple example. Recognizing that the effect on risk of changing one risk factor depends on what is assumed about levels of other factors, epidemiologists distinguish among concepts such as "controlled direct effect" (e.g., the change in death rate if air pollution is changed from 0 to 1but mediators are held constant), the "natural direct effect" (mediators are held fixed at their zero-exposure levels for each individual while air pollution is increased from 0 to 1), "natural indirect effect" (mediators are set at what they would be when exposure is 1, as exposure change from 0 to 1), "total effect" (the sum of natural direct and natural indirect effects), and others (Robins and Greenland 1992; Pearl 2001; Petersen et al. 2006; Vander Weele 2011; Tchetgen and Phiri 2014). It is easy to explain these effect concepts in layman's language, but such precision about exactly what type of "effect" is intended is not often used in current scientific and media communications about health risks.

Even without delving further into these concepts of cause and effect, each of which has a large technical literature, (Cox Jr. 2018), it is clear that "Air pollution kills *X* million people per year" is a highly ambiguous statement. It might be (and typically is) true for some kinds of causation but not for others, and it may also be true for some kinds of effects (e.g., total effects) but not others (e.g., natural or controlled direct effects). Unfortunately, neither media accounts nor underlying scientific articles are usually specific about the kind(s) of causation that have been investigated and the types of causal effects that have been quantified. However, what policy makers and the public most need to know, and often mistakenly assume they are being informed about, is *manipulative or interventional causation* (Chap. 9), i.e., what fraction of adverse health effects (per person-year for individuals, or per year for populations) would be prevented if air pollution were reduced by a stated amount. Current practice in much regulatory risk assessment is to make a judgment (a "causal determination" judgment, as discussed in Chapter 18) about whether confounding and biases are believed to be small enough to warrant calling an empirically estimated concentration-response (C-R) association "causal" and, if so, to use it to predict public health benefits of the reduction in exposure. (As discussed in Chaps. 2, 15, and 17, although regression models are sometimes used to adjust for known and measured confounders, some common confounders are often simply ignored in air pollution health effects modeling, and regression-based modeling delivers unreliable results in many cases because they depend on modeling assumptions and choices (Dominici et al. 2014)). Predictions in scientific articles and news stories based on C-R models usually do not warn that the effects of exposure-reducing interventions are model-dependent and not actually known, as the model specification is usually under-determined by available (empirical association) concentration-response data.

Both scientific articles and media accounts often make causal claims, such as "Implementing the US air quality standard for PM2.5 worldwide can prevent millions of premature deaths per year" (Giannadaki et al. 2016) or "Ban on coal burning in Dublin cleans the air and reduces death rates" (Harvard School of Public Health 2002), even when the underlying research only addresses associations and does not permit valid inferences about manipulative causation—specifically, about whether or by how much reducing air pollution has reduced or would reduce death rates. Although such conflation of association with manipulative causation has been well recognized and warned against countless times by methodologists in the epidemiology community (e.g., Petitti 1991), some of whom have also stressed the importance of well defined interventions or manipulations for answering causal questions of interest to policy makers (Hernán and Vanderweele 2011), common current practice continues to communicate association-based results as if they addressed manipulative causation. This creates large and needless gaps between what policy makers and the public are routinely told that science has found and what it actually has found about how changes in exposures affect human health.

As an example of this gap, in Ireland in late 2017 and 2018, a planned rollout of a nation-wide policy to ban selling or burning smoky coal was announced with public risk communication messages such as the following: "The original ban in Dublin is cited widely as a successful policy measure both nationally and internationally as best practice within the international clean air community. It is estimated that approximately 8000 premature deaths have been averted in Dublin since the introduction of the smoky coal ban in 1990. Further health, environmental and economic benefits (estimated at €53million/year) are expected, when the ban is extended nationwide." (Ireland EPA 2018) Essentially the same deceptive message—a clear manipulative causal interpretation of observed declines in particulate air pollution and cardiovascular mortality rates in Dublin and other counties in Ireland—has been repeated for decades. As discussed in previous chapters, as early as 2002, researchers announced that a "Ban on coal burning in Dublin cleans the air and reduces death rates" (Harvard School of Public Health 2002). This message was reinforced to policy makers and the public many times over the next two decades through enthusiastic headlines such as "How the coal ban dealt with Dublin's burning issue. The prohibition of 'smoky' coal in 1990 resulted in 350 fewer annual deaths in city" (Kelly 2016). However, all of these claims are based on associations and attribution: the original study data consisted simply of observations that both pollution levels and all-cause and cardiovascular mortality rates were lower after the ban than before it (Clancy et al. 2002).

The story for manipulative causation is simpler: as discussed in Chap. 17, the bans had no detectable effect on mortality risks, for either all-cause mortality or cardiovascular mortality (Wittmaack 2007; Zigler and Dominici 2014). Instead, "Long-term trends in cardiovascular health spanning implementation of the ban—not the coal ban itself—contributed to apparent effects on cardiovascular mortality" (Zigler and Dominici 2014). This was made clear in an accountability study by the Health Effects Institute (Health Effects Institute (HEI) 2013) that found that mortality rates (both all-cause and cardiovascular, although not respiratory) fell just as much in areas not affected by the ban, where particulate air pollution had not been dramatically reduced, as in areas that were affected by the ban, where particulate air pollution was reduced by some 40–70%. Although both pollution levels and mortality rates were indeed significantly less in later years than in earlier years (association), this reflected trends in the developed world toward lower cardiovascular mortality rates due to better prevention, diagnosis, and treatment; the reduction in mortality rates was not detectably affected by the bans (manipulative causation).

In this example, reported estimates of numbers of deaths averted by reducing air pollution attracted widespread media coverage and became widely cited and influential in shaping risk perceptions and communications used in justifying environmental policies. By contrast, follow-up research—including some by the original authors—reporting that previously claimed health benefits of the bans did not exist after all (Health Effects Institute (HEI) 2013, Zigler and Dominici 2014) was comparatively neglected: it played no role in reshaping Ireland's new national coal ban policy or in communications to the public about mortality attributed to air pollution. The original research is still cited to support nation-wide bans (Ireland EPA 2018). This pattern of selective and incomplete reporting is very common (Dumas-Mallet et al. 2017). It undermines collective ability to perceive the effectiveness of interventions accurately, and hence to learn from experience.

Over-Generalized, Model-Dependent, and Assumption-Dependent Conclusions

The Irish coal ban risk communication experience illustrates the perils of using unverified models and modeling assumptions in interpreting data to tell a useful-seeming, policy-relevant causal story. The data collected were uncontroversial: air pollution levels and all-cause mortality rates were smaller after the bans than before them. Misleading conclusions arose only when the data were interpreted using a counterfactual causal model that incorrectly assumed that mortality rates would have remained constant in the absence of the bans. The subsequent accountability study used mortality data from relevant comparison groups (populations in other counties) to discover that this assumption was not true. Use of comparison groups is a key principle for valid causal inference from quasi-experimental studies—those in which different treatments (e.g., ban vs. no ban) are not randomly assigned (Campbell and Stanley 1963). Had it been used in the original study, the erroneous causal conclusions and headlines could have been avoided.

Designing quasi-experimental studies and analyses to facilitate testing of modeling assumptions and to avoid erroneous conclusions about causality is well developed for threats to *internal validity*, e.g., potential non-causal explanations for observed exposure-response associations in a given data set. However, threats to *external validity*—valid generalization of causal conclusions from the specific conditions for which data were collected to different conditions elsewhere—must also be addressed (Campbell and Stanley 1963). Technical methods for doing so have been developed (Chap. 9; Schwartz et al. 2011; Bareinboim and Pearl 2013; Hernán and Vanderweele 2011; Lee and Honavar 2013), but have seldom been applied in air pollution health effects research. Instead, extreme generalizations are often made from narrowly focused data. For example, the uncontroversial observation that particulate air pollution and mortality rates are higher on the north side of a river in China than on the south side has recently been interpreted with the help of a traditional quasi-experimental design (regression discontinuity, which models discontinuities in exposure and response on two sides of a boundary or threshold (Thistlethwaite and Campbell 1960)) to develop an "Air Quality Life Index" (AQLI) that is interpreted causally to imply sensational policy-relevant conclusions. These include statements that "Particulate matter (PM) air pollution is the greatest current threat to human health globally" and that "Particulate matter air pollution cuts global life expectancy by nearly 2 years," or about 1 year of life lost per person per 10 $\mu g/m^3$ of particulate air pollution (AQLI 2018). These conclusions attracted worldwide headlines. The supporting web site and scientific articles present colorful maps and interactive graphics showing excess deaths attributed to particulate matter in countries around the globe.

To communicate such exciting-looking scientific results accurately, scientists (and scrupulous reporters) should ideally address the extent to which their validity has been established. In the ALQI example, two key unverified assumptions underlying the analysis are as follows:

1. *Internal validity:* The estimated difference between mortality rates on the two sides of the river was *caused* by differences in particulate pollution.
2. *External validity:* The reduction in risk per unit reduction in PM is the same everywhere else as it is estimated to be on the two sides of the river.

In documenting their methods, the authors explain that "Following the epidemiology literature, the AQLI assumes a linear relationship between long-term exposure to $PM_{2.5}$ [fine particulate matter] and life expectancy throughout the observed $PM_{2.5}$ distribution. Though it is possible that the pollution-life expectancy relationship is nonlinear over certain ranges of $PM_{2.5}$ concentrations and/or that there is a threshold below which $PM_{2.5}$ has no effect, we are unaware of credible empirical evidence that would cause a rejection of the linearity assumption" (AQLI 2018, methodology section). In other words, the key assumption that life expectancy increases in direct proportion to reductions in particulate air pollution, which drives all of the reported conclusions about life-extending benefits of reducing air pollution, rests on the authors being unaware of credible evidence to the contrary. That the conclusions depend on this linearity assumption should be made clear when they are communicated, along with available information about the extent to which the assumption has been empirically tested and validated. In this example, contrary evidence presumably could include the findings from Ireland that large reductions in particulate air pollution by about 40–70% and several dozen μg/m^3 had no detectable effect on life expectancy. It might also include toxicological evidence that particulate health effects are mediated by inflammation processes with clear thresholds (Cevallos et al. 2017; Du et al. 2018; Xin et al. 2018; Xu et al. 2018; Zheng et al. 2018); or evidence from natural experiments involving severe smog episodes that caused no resulting increases in mortality rates in urban areas (Zhou et al. 2015; Zu et al. 2016).

In summary, in this example, the modeling assumption that reducing PM causes proportional increases in life expectancy, with the same constant of proportionality everywhere, drives the dramatic conclusions and worldwide headlines that communicated these predictions to the public as if they were scientific facts. Lost in this process was the crucial caveat that the conclusions rest entirely on unverified modeling assumptions that are demonstrably false for locations where accountability studies have been done, such as Dublin. More generally, claims such as "Reducing exposure to air pollution increases life expectancy by *Y* years per unit concentration of pollution reduction" usually depend on a mix of data and assumptions. It is not clear from the claim alone how sensitive it is to assumptions and how plausible they are. Responsible reporting requires addressing the extent to which key assumptions have been verified before the results are reported as if they were facts.

Doing Better

Table 19.1 summarizes important challenges to accurate risk communication about air pollution health effects in its left column and suggests possible guidance for more accurate communication in the right column. The main focus is on suggestions for how to report findings in the scientific literature to avoid advocacy and to include appropriate caveats. Most of the suggestions are simple but important: state the caveats and assumptions needed for a claim to be valid; be specific about the types of cause and the types of effects that are being claimed; focus wherever possible on manipulative causality, i.e., on the fraction of adverse effects that have been shown empirically to be *prevented* by reducing air pollution (as opposed to the fraction *blamed* on or *associated* with air pollution), or else state that it was not addressed; emphasize the actual observation on which claims are based; and exercise due diligence and critical thinking in reporting whether the internal and external validity of claims have been tested. For accurate scientific and technical communication, the following more specialized technical concepts from modern epidemiology and causal analytics are useful (Chap. 9):

Table 19.1 Challenges and suggestions for communicating more accurately about health effects of air pollution

Challenge	Possible solutions
Bold claims about health effects of exposure made without essential caveats	• State essential caveats, especially any untested (or false) assumptions to which the stated claims are known to be sensitive. (Example: "These findings depend on assuming that, worldwide, loss of life expectancy is proportional to exposure concentration. This assumption is currently not supported by data") • If a claim no longer appears newsworthy when essential caveats are added, forego the claim rather than the caveats
Unclear meaning of claimed causal effects of exposure	• Specify the type of causal effect for which evidence is provided (e.g., controlled direct, natural direct, total, mediated etc.) (Cox Jr. 2018; Robins and Greenland 1992; Pearl 2001; Petersen et al. 2006; Vander Weele 2011; Tchetgen and Phiri 2014) • If possible, address manipulative causation: by how much would reducing exposures change effects? Under what assumptions about other contributing causes? How do we know and how sure can we be? • If a study does not address manipulative causation, say so in plain language (e.g., "These risk predictions are made by assuming that reducing air pollution would proportionally reduce mortality rates. No data were provided on actual changes in mortality rates observed following actual changes in air pollution levels (accountability study data) to validate this assumption")
Association presented as manipulative causation	• Don't present association-based results as if they had manipulative causal implications • Report results of tests of threats to internal validity (Campbell and Stanley 1963). If threats to internal validity have not been addressed, say so (e.g., "No data were collected from comparison groups to test the assumption that mortality rates would not have decreased if air pollution levels had not been reduced")
Ambiguity and conflation of causal concepts and language	• Distinguish clearly between manipulative causation and other types of causation (e.g., attributive) in reporting findings • Focus specifically on manipulative causation if possible, as this is what most policy audience believe that causal statements address
Over-generalized, model-dependent, and assumption-dependent conclusions	• Report *observations* (e.g., "Air pollution levels and mortality rates were both significantly higher just north of the river than just south of it") rather than untested generalizations or extrapolations (e.g., "Each 10 μg/m^3 reduction in ambient particulate air pollution would increase life expectancy by 1 year worldwide") • Do not assume external validity unless it has been demonstrated • For technical and scientific communication, report results of tests of the invariant causal prediction (ICP) property (Heinze-Deml et al. 2017; Peters et al. 2016). If ICP has not been tested, do not assume it • For technical and scientific communication, report whether appropriate transportability formulas have been applied to support generalizations of study findings to other settings

1. Several different types of causal effects are well defined in epidemiology (e.g., controlled direct, natural direct, and total effects of changes in air pollution levels on probability or frequency of adverse health effects in a population). These should be distinguished, explained, and used in communicating about human health effects of changes in exposure. They are distinguished by the different assumptions they make about other contributing causes—e.g., whether they are assumed to be fixed at their historical levels, or at specified hypothetical levels, or at new levels that are more consistent with the postulated changes in exposures.
2. Threats to internal and external validity of causal conclusions in observational studies should be understood and clearly addressed in reporting claims about causal effects of exposures on health effects. These threats are important and are easy to understand and explain in plain language (Campbell and Stanley 1963; Behi and Nolan 1996; Bärnighausen et al. 2017). They include "history," i.e., coincident historical trends in causes and effects; "maturation," i.e., changes in exposure-response patterns as a population ages; and "regression to the mean," i.e., exceptional observations that trigger an intervention are expected to be followed by less exceptional observations whether or not the intervention works. Such threats should be addressed when causal claims are reported. For example, the researchers and reporters communicating claims from the original Dublin study could and should have inquired whether relevant comparison groups had been used to protect against the "history" threat, and reported that the answer was no, and hence that the claims being made were based on untested and unverified assumptions—specifically, that observed declines in mortality rates would not have happened in the absence of reductions in air pollution. Those reporting research claims, as well as those making them, should take responsibility for verifying and reporting whether common threats to internal and external validity of causal claims have been addressed in the study design and analysis before making or accepting causal interpretations of observations.
3. It is often relatively easy to estimate the ratio of (a) differences between health response incidence rates to (b) differences in exposure concentrations estimated in a study—e.g., on two sides of a river in the recent AQLI work, or before and after coal burning bans in the Irish studies. The hypothesis that this "concentration-response" (C-R) ratio can be applied elsewhere to predict how changing exposure concentrations would change response incidence rates is closely related to a technical property of causal models called *invariant causal prediction* (ICP) (Heinze-Deml et al. 2017; Peters et al. 2016). The essence of the ICP property is that a valid causal model or law predicts the same probabilities for effects from the same values of their direct causes, no matter where they occur. When this holds, modern epidemiological calculations ("transportability formulas") can be used to adjust predictions from the original study of the changes in health effects caused by changing exposures to make corresponding predictions under new conditions or in new areas where other contributing causes (e.g., sociodemographic, weather, and co-morbidity variables) have different joint distributions

in the exposed population (Bareinboim and Pearl 2013; Hernán and Vanderweele 2011; Lee and Honavar 2013; Schwartz et al. 2011). This is a very useful technique for generalizing beyond study-specific conclusions. It can even help to deal with unobserved confounders and errors in estimated or measured values of variables. Although the technical (but straightforward) details of ICP tests and transportability formula calculations are not needed to assure accurate reporting of results, it is important to understand whether they have been applied—and, if so, what the results were.

Conclusions

Communicating accurately about the effects on human health of reductions in air pollution is essential for informing rational and effective policy making about how much reduction is needed to protect human health with an adequate margin of safety. Present practice does not emphasize accurate communication or verification and appropriate qualification of causal claims. It tends instead to the sensational, in both scientific and media accounts, emphasizing dramatic claims about millions of lives per year being lost to air pollution. Taken at face value, such claims are often nonsensical (requiring violations of arithmetic or an assumption that air pollution retroactively increases birth rates), mistaken (as in the case of claims about lives saved by the Ireland coal burning bans), or wildly speculative (as in the case of recent AQLI clams about worldwide changes in life expectancy attributed to air pollution based on comparisons of pollution levels and life expectancies on two sides of one river in China.)

It is possible, and important, to do much better in communicating more accurately about changes in health effects caused by changes in ambient air pollution levels. Perhaps the single easiest and most important improvement is to shift the focus of communication away from *blame*—how many deaths or illnesses per year do authorities or investigators choose to attribute to air pollution?—and toward potential *change*: what fraction of these adverse effects would be prevented by reducing air pollution? Clear, accurate, useful communication to inform rational policy-making begins by being careful not to conflate these distinct concepts. Traditional epidemiological measures based on statistical associations, such as attributable risks and population attributable fractions, do not answer the question about change, although they are frequently misinterpreted as doing so; indeed, although they are still widely used, such associational methods are now regarded by specialists as being often unreliable, yielding assumption-dependent conclusions (Dominici et al. 2014). Applying more modern causal analytics methods and concepts developed and applied in epidemiology and other fields mainly since the 1960s (Chap. 9) can greatly clarify the causal conclusions that can be validly inferred from well-designed and well-conducted accountability studies, natural experiments, and other studies about how changes in exposures and other causally relevant variables would cause health effects that depend on them to change.

Being aware of these methods and of guiding principles for risk communication based on them (Table 19.1) can help scientists, reporters, and policy makers to distinguish among importantly different concepts of causation and effects and to communicate more responsibly and accurately about what is known, what is guessed or assumed, and what is still unknown about how human health effects change in response to changes in air pollution. Such clear communication provides an essential foundation for assessing the effectiveness of regulations and for designing regulations that will be more causally effective in truly protecting human health.

References

AQLI (2018). https://aqli.epic.uchicago.edu/wp-content/uploads/2018/11/AQLI-Report.111918-2.pd; https://aqli.epic.uchicago.edu/

Bareinboim E, Pearl J (2013) Causal transportability with limited experiments. In: Proceedings of the 27th AAAI Conference on Artificial Intelligence, pp 95–101

Barnett L, Seth AK. The MVGC Multivariate granger causality toolbox: a new approach to Granger-causal inference. J Neurosci Methods. 2014;223:50–68.

Bärnighausen T, Tugwell P, Røttingen JA, Shemilt I, Rockers P, Geldsetzer P, Lavis J, Grimshaw J, Daniels K, Brown A, Bor J, Tanner J, Rashidian A, Barreto M, Vollmer S, Atun R. Quasi-experimental study designs series-paper 4: uses and value. J Clin Epidemiol. 2017;89:21–9. https://doi.org/10.1016/j.jclinepi.2017.03.012.

Behi R, Nolan M. Causality and control: threats to internal validity. Br J Nurs. 1996;5(6):374–7.

Campbell DT, Stanley JC. Experimental and quasi-experimental designs for research. Boston: Houghton Mifflin Company; 1963.

Cevallos VM, Díaz V, Sirois CM. Particulate matter air pollution from the city of Quito, Ecuador, activates inflammatory signaling pathways in vitro. Innate Immun. 2017;23(4):392–400. https://doi.org/10.1177/1753425917699864.

Clancy L, Goodman P, Sinclair H, Dockery DW. Effect of air-pollution control on death rates in Dublin, Ireland: an intervention study. Lancet. 2002;360(9341):1210–4.

Cox LA Jr. Modernizing the Bradford Hill criteria for assessing causal relationships in observational data. Crit Rev Toxicol. 2018;15:1–31. https://doi.org/10.1080/10408444.2018.1518404.

Dominici F, Greenstone M, Sunstein CR. Science and regulation. Particulate matter matters. Science. 2014;344(6181):257–9. https://doi.org/10.1126/science.1247348.

Du X, Jiang S, Zeng X, Zhang J, Pan K, Zhou J, Xie Y, Kan H, Song W, Sun Q, Zhao J. Air pollution is associated with the development of atherosclerosis via the cooperation of CD36 and NLRP3 inflammasome in $ApoE_{-/-}$ mice. Toxicol Lett. 2018;290:123–32. https://doi.org/10.1016/j.toxlet.2018.03.022.

Dumas-Mallet E, Smith A, Boraud T, Gonon F. Poor replication validity of biomedical association studies reported by newspapers. PLoS One. 2017;12(2):e0172650. https://doi.org/10.1371/journal.pone.0172650.

Galles D, Pearl J. An axiomatic characterization of causal counterfactuals. Found Sci. 1998;3:151–82.

Giannadaki D, Lelieveld J, Pozzer A. Implementing the US air quality standard for PM2.5 worldwide can prevent millions of premature deaths per year. Environ Health. 2016;15(1):88.

Glass TA, Goodman SN, Hernán MA, Samet JM. Causal inference in public health. Annu Rev Public Health. 2013;34:61–75.

Granger CWJ. Investigating causal relations by econometric models and cross-spectral methods. Econometrica. 1969;37(3):424–38.

Harvard School of Public Health (2002) Press release: ban on coal burning in dublin cleans the air and reduces death rates. www.hsph.harvard.edu/news/press-releases/archives/2002-releases/press10172002.html
Health Effects Institute (HEI) (2013) Did the Irish Coal Bans Improve Air Quality and Health? HEI Update, Summer, 2013. http://pubs.healtheffects.org/getfile.php?u=929. Accessed 1 February 2014
Heinze-Deml C, Peters J, Meinshausen N (2017) Invariant causal prediction for nonlinear models. https://arxiv.org/pdf/1706.08576.pdf
Hernán MA, Vanderweele T. On compound treatments and transportability of causal inference. Epidemiology. 2011;22:368.
Hoover KD (2012) Causal structure and hierarchies of models. Stud Hist Philos Biol Biomed Sci. https://doi.org/10.1016/j.shpsc.2012.05.007.
Höfler M. The Bradford Hill considerations on causality: a counterfactual perspective. Emerg Themes Epidemiol. 2005;2:11.
Imai K, Keele L, Tingley D, Yamamoto T. Unpacking the black box of causality: learning about causal mechanisms from experimental and observational studies. Am Polit Sci Rev. 2011;105:4.
Ireland EPA (2018). www.askaboutireland.ie/enfo/irelands-environment/health-and-wellbeing/air/nationwide-smoky-coal-ban/
Keele L, Tingley D, Yamamoto T. Identifying mechanisms behind policy interventions via causal mediation analysis. J Policy Anal Manage. 2015;34(4):937–63.
Kelly O (2016) How the coal ban dealt with Dublin's burning issue. The prohibition of 'smoky' coal in 1990 resulted in 350 fewer annual deaths in city. The Irish Times. Saturday, September 26, 2015. www.irishtimes.com/news/environment/how-the-coal-ban-dealt-with-dublin-s-burning-issue-1.2367021
Kleinberg S, Hripcsak G. A review of causal inference for biomedical informatics. J Biomed Inform. 2011;44(6):1102–12.
Lee S, Honavar V (2013) m-Transportability: transportability of a causal effect from multiple environments. In: Proceedings of the twenty-seventh AAAI conference on artificial intelligence. www.aaai.org/ocs/index.php/AAAI/AAAI13/paper/viewFile/6303/7210
Li J, Ma S, Le T, Liu L, Liu J. Causal decision trees. IEEE Trans Knowl Data Eng. 2017;29(2):257–71.
Lo WC, Shie RH, Chan CC, Lin HH. Burden of disease attributable to ambient fine particulate matter exposure in Taiwan. J Formos Med Assoc. 2016;116(1):32–40.
Lok JJ. Mimicking counterfactual outcomes to estimate causal effects. Ann Stat. 2017;45(2);461–99. https://doi.org/10.1214/15-AOS1433.
Menzies P. The causal structure of mechanisms. Stud Hist Phil Biol Biomed Sci. 2012;43(4):796–805. https://doi.org/10.1016/j.shpsc.2012.05.00.
Murray CJ, Lopez AD. Measuring the global burden of disease. N Engl J Med. 2013;369(5):448–57. https://doi.org/10.1056/NEJMra1201534.
Papana A, Kyrtsou C, Kugiumtzis D, Diks C. Assessment of resampling methods for causality testing: a note on the US inflation behavior. PLoS One. 2017;12(7):e0180852. https://doi.org/10.1371/journal.pone.0180852.
Pearl J. Direct and indirect effects. In: Proceedings of the seventeenth conference on uncertainty in artificial intelligence. San Francisco: Morgan Kaufmann; 2001. p. 411–20.
Peters J, Bühlmann P, Meinshausen N. Causal inference by using invariant prediction: identification and confidence intervals. J R Stat Soc Ser B. 2016;78(5):947–1012.
Petersen ML, Sinisi SE, van der Laan MJ. Estimation of direct causal effects. Epidemiology. 2006;17(3):276–84.
Petitti DB. Associations are not effects. Am J Epidemiol. 1991;133(2):101–2.
Prüss-Üstün A, Mathers C, Corvalán C, Woodward A. Introduction and methods: Assessing the environmental burden of disease at national and local levels. Environmental burden of disease series No. 1. Geneva: World Health Organization (WHO); 2003.
Robins JM, Greenland S. Identifiability and exchangeability for direct and indirect effects. Epidemiology. 1992;3:143–55.

Schwartz S, Gatto NM, Campbell UB. Transportabilty and causal generalization. Epidemiology. 2011;22(5):745–6.
Simon HA. Causal ordering and identifiability. In: Hood WC, Koopmans TC, editors. Studies in econometric method, Cowles commission for research in economics monograph no. 14. New York: Wiley; 1953. p. 49–74.
Simon HA, Iwasaki Y. Causal ordering, comparative statics, and near decomposability. J Econ. 1988;39:149–73.
Spirtes P. Introduction to causal inference. J Mach Learn Res. 2010;11:1643–62.
Suppes P. A probabilistic theory of causality. Amsterdam: North-Holland Publishing Company; 1970.
Tchetgen EJ, Phiri K. Bounds for pure direct effect. Epidemiology. 2014;25(5):775–6. https://doi.org/10.1097/EDE.0000000000000154.
Thistlethwaite D, Campbell D. Regression-discontinuity analysis: an alternative to the ex post facto experiment. J Educ Psychol. 1960;51(6):309–17. https://doi.org/10.1037/h0044319.
Vander Weele TJ. Controlled direct and mediated effects: definition, identification and bounds. Scand Stat Theory Appl. 2011;38(3):551–63.
Voortman M, Dash D, Druzdzel MJ. Learning causal models that make correct manipulation predictions with time series data. Proc Machine Learn Res. 2010;6:257–66.
Wiener N. The theory of prediction. In: Beckenbach EF, editor. Modern mathematics for engineers, vol. 1. New York: McGraw-Hill; 1956.
Wittmaack K. The big ban on bituminous coal sales revisited: serious epidemics and pronounced trends feign excess mortality previously attributed to heavy black-smoke exposure. Inhal Toxicol. 2007;19(4):343–50.
Xin L, Che B, Zhai B, Luo Q, Zhang C, Wang J, Wang S, Fan G, Liu Z, Feng J, Zhang Z. 1,25-dihydroxy vitamin D_3 attenuates the oxidative stress-mediated inflammation induced by $PM_{2.5}$ via the p38/NF-κB/NLRP3 pathway. Inflammation. 2018;42(2):702–13. https://doi.org/10.1007/s10753-018-0928-y.
Xu F, Qiu X, Hu X, Shang Y, Pardo M, Fang Y, Wang J, Rudich Y, Zhu T. Effects on IL-1β signaling activation induced by water and organic extracts of fine particulate matter ($PM_{2.5}$) in vitro. Environ Pollut. 2018;237:592–600. https://doi.org/10.1016/j.envpol.2018.02.086.
Zheng R, Tao L, Jian H, Chang Y, Cheng Y, Feng Y, Zhang H. NLRP3 inflammasome activation and lung fibrosis caused by airborne fine particulate matter. Ecotoxicol Environ Saf. 2018;163:612–9. https://doi.org/10.1016/j.ecoenv.2018.07.076.
Zhou M, He G, Fan M, Wang Z, Liu Y, Ma J, Ma Z, Liu J, Liu Y, Wang L, Liu Y. Smog episodes, fine particulate pollution and mortality in China. Environ Res. 2015;136:396–404. https://doi.org/10.1016/j.envres.2014.09.038.
Zigler CM, Dominici F. Point: clarifying policy evidence with potential-outcomes thinking--beyond exposure-response estimation in air pollution epidemiology. Am J Epidemiol. 2014; 180(12):1133–40.
Zu K, Tao G, Long C, Goodman J, Valberg P. Long-range fine particulate matter from the 2002 Quebec forest fires and daily mortality in Greater Boston and New York City. Air Qual Atmos Health. 2016;9:213–21.

Index

L. A. Cox Jr., *Quantitative Risk Analysis of Air Pollution Health Effects*, International Series in Operations Research & Management Science 299,
https://doi.org/10.1007/978-3-030-57358-4

The manufacturer's authorised representative in the EU is Springer Nature Customer Service Centre GmbH, Europaplatz 3, 69115 Heidelberg, Germany. If you have any concerns regarding our products, please contact ProductSafety@springernature.com

Printed and bound by CPI Group (UK) Ltd, Croydon, CR0 4YY
30/09/2026
02225452-0020